AF684978

PLANNING AND ARCHITECTURAL DESIGN OF INTEGRATED SERVICES DIGITAL NETWORKS
Civil and Military Applications

THE KLUWER INTERNATIONAL SERIES IN ENGINEERING AND COMPUTER SCIENCE

COMMUNICATIONS AND INFORMATION THEORY

Consulting Editor

Robert Gallager

Other books in the series:

WIRELESS INFRARED COMMUNICATIONS, John R. Barry
ISBN: 0-7923-9476-3

COMMUNICATIONS AND CRYPTOGRAPHY: ***Two sides of One Tapestry*****,** Richard E. Blahut, Daniel J. Costello, Jr., Ueli Maurer and Thomas Mittelholzer
ISBN: 0-7923-9469-0

WIRELESS AND MOBILE COMMUNICATIONS, Jack M. Holtzman and David J. Goodman
ISBN: 0-7923-9464-X

INTRODUCTION TO CONVOLUTIONAL CODES WITH APPLICATIONS, Ajay Dholakia
ISBN: 0-7923-9467-4

CODED-MODULATION TECHNIQUES FOR FADING CHANNELS, S. Hamidreza Jamali, and Tho Le-Ngoc
ISBN: 0-7923-9421-6

ELLIPTIC CURVE PUBLIC KEY CRYPTOSYSTEMS, Alfred Menezes
ISBN: 0-7923-9368-6

SATELLITE COMMUNICATIONS: Mobile and Fixed Services, Michael Miller, Branka Vucetic and Les Berry
ISBN: 0-7923-9333-3

WIRELESS COMMUNICATIONS: Future Directions, Jack M. Holtzman and David J. Goodman
ISBN: 0-7923-9316-3

DISCRETE-TIME MODELS FOR COMMUNICATION SYSTEMS INCLUDING ATM, Herwig Bruneel and Byung G. Kim
ISBN: 0-7923-9292-2

APPLICATIONS OF FINITE FIELDS, Alfred J. Menezes, Ian F. Blake, XuHong Gao, Ronald C. Mullin, Scott A. Vanstone, Tomik Yaghoobian
ISBN: 0-7923-9282-5

WIRELESS PERSONAL COMMUNICATIONS, Martin J. Feuerstein, Theodore S. Rappaport
ISBN: 0-7923-9280-9

SEQUENCE DETECTION FOR HIGH-DENSITY STORAGE CHANNEL, Jaekyun Moon, L. Richard Carley
ISBN: 0-7923-9264-7

DIGITAL SATELLITE COMMUNICATIONS SYSTEMS AND TECHNOLOGIES: Military and Civil Applications, A. Nejat Ince
ISBN: 0-7923-9254-X

IMAGE AND TEXT COMPRESSION, James A. Storer
ISBN: 0-7923-9243-4

VECTOR QUANTIZATION AND SIGNAL COMPRESSION, Allen Gersho, Robert M. Gray
ISBN: 0-7923-9181-0

THIRD GENERATION WIRELESS INFORMATION NETWORKS, Sanjiv Nanda, David J. Goodman
ISBN: 0-7923-9128-3

SOURCE AND CHANNEL CODING: An Algorithmic Approach, John B. Anderson, Seshadri Mohan
ISBN: 0-7923-9210-8

ADVANCES IN SPEECH CODING, Bishnu Atal, Vladimir Cuperman, Allen Gersho
ISBN: 0-7923-9091-1

SWITCHING AND TRAFFIC THEORY FOR INTEGRATED BROADBAND NETWORKS, Joseph Y. Hui
ISBN: 0-7923-9061-X

21.382
Ince
c.1

PLANNING AND ARCHITECTURAL DESIGN OF INTEGRATED SERVICES DIGITAL NETWORKS

Civil and Military Applications

by

A. Nejat Ince
Istanbul Technical University Foundation
Centre for Defence Studies

Dag Wilhelmsen
ALCATEL Telecom Norway

Bülent Sankur
Boğaziçi University
Istanbul-Turkey

KLUWER ACADEMIC PUBLISHERS
Boston / Dordrecht / London

11-99

Distributors for North America:
Kluwer Academic Publishers
101 Philip Drive
Assinippi Park
Norwell, Massachusetts 02061 USA

Distributors for all other countries:
Kluwer Academic Publishers Group
Distribution Centre
Post Office Box 322
3300 AH Dordrecht, THE NETHERLANDS

Library of Congress Cataloging-in-Publication Data

İnce, A. Nejat.
Planning and architectural design of integrated services digital networks : civil and military applications / by A. Nejat İnce, Dag Wilhelmsen, Bülent Sankur.
p. cm. -- (The Kluwer international series in engineering and computer science ; SECS 308)
Includes bibliographical references and index.
ISBN 0-7923-9554-9
1. Integrated services digital networks--Planning. 2. Computer network architectures. I. Wilhelmsen, Dag. II. Sankur, Bülent. III. Title. IV. Series.
TK5103.75.I55 1995
621.382--dc20 94-48207
CIP

Copyright © 1995 by Kluwer Academic Publishers

All rights reserved. No part of this publication may be reproduced, stored in a retrieval system or transmitted in any form or by any means, mechanical, photo-copying, recording, or otherwise, without the prior written permission of the publisher, Kluwer Academic Publishers, 101 Philip Drive, Assinippi Park, Norwell, Massachusetts 02061

Printed on acid-free paper.

Printed in the United States of America

TABLE OF CONTENTS

PREFACE

The main subject of this book is Integrated Services Digital Networks (ISDN); how to plan and design them architecturally and how to implement them so that they meet certain given user requirements ranging from a variety of service demands to transmission performance, security, reliability/availability, capability for growth, interoperability with other ISDN and non-ISDN networks and, of course, cost.

The network application that are used as the model in this book is that of a defence network serving the strategic communications requirement of a government. The main purpose is, however, to describe a system design approach that can be applied more generally given that the network required uses the general principles of ISDN with a defined user community (e.g. a major industrial corporation, a government or an international organization). The considerations concerning the commercial operation of an ISDN network (for example as experienced by a public network operator, PTT etc.) have not been the intention of this book, even if the technical concepts described are still valid in this case.

In the last two decades some 6000 papers and over 40 books have been published on ISDN and related topics covering the aspects of standards, concepts, protocols, architecture and services. We shall assume in this book that the reader is either familiar with the ISDN concept and standards or can have access to the rich literature that exists on the subject. The book concentrates, instead, on the application of ISDN concepts and standards to the planning and design of real costed networks to meet certain specified user requirements. This includes where there are multiple options, considerations and rationale on the choice of network aspects and standards. The unique feature of the book, compared with existing books on ISDN, is that it expounds an original methodology which starts from an assumed or given set of complete user requirements and proceeds to designing a complete network taking into account the technology and standards of ISDN as well as some constraints including cost which may be imposed on the network.

The book describes computer-aided design tools employed for dimensioning the network for various traffic loads and for assessing its traffic carrying performance for assessing different precedence categories and network configurations, transmission conditions and routing algorithms which may be static-deterministic or dynamic-adaptive. Aspects such as surveillance and control, security, survivability and EMP protection are also addressed in this book.

The content of the book is based on the results of studies which were carried out in the period 1987-1989, for the architectural design and master plan of a national strategic communications network called TAFICS (Turkish Armed Forces Integrated Communications System) which benefitted from earlier works done for an international network called NATO integrated Communications Network (NICS), and for a national infrastructure network called Norwegian Defence Digital Network (NDDN). The strategic networks carry military and government traffic and are often designed and implemented, as it were, from scratch. They allow therefore the application of appropriate and logical design implementation procedures, without undue compromises which may otherwise be imposed on them by existing plant and organisation. These networks are required to provide a wider choice of services (supplementary services such as precedence and pre-emption, security and dedicated networks) and more stringent features (such as redundant grid network structure, multi-homing, adaptive routing, hardening etc.) than those demanded from commercial public networks.

Some of these "military aspects" are gradually becoming requirements also for civil public and private networks. Consequently it can be said that the only significant difference between the *defence* strategic and commercial public networks is the higher traffic carrying capacity of the latter and the requirement for operation under stringent and chaotic conditions for the former.

Many public network operators around the world are implementing the Integrated Services Digital Networks for the reason that internationally agreed CCITT standards exist today.

Rather than adopting some other standards such as the military tactical EUROCOM standards which are based on CVSD modulation and that were available in the seventies, it was decided in the mid-eighties that it would be better (massive international standardisation effort in CCITT) and more economic (economies of scale of the civil market) to follow the CCITT IDN-ISDN standards and recommendations for the design of military-government strategic networks adopting also the International Standards Organisation's (ISO) Open System Interconnection Reference Model (OSI/RM).

The ISDN that will be treated in this book has the full features of strategic military systems. These networks which are being planned and implemented today, will therefore involve aspects and technologies which are wider in scope and more onerous to obtain than those required by the civil systems. This approach will allow us to discuss almost all conceivable features, architectures, techniques and technologies as well as network evolution aspects and system design methodology in relation to a real system rather than in the abstract.

The type of ISDN that is in question here is the so-called narrow band ISDN (NB-ISDN) which is intended to provide circuit-switched and packet-switched connections at the rate of 64 kbit/s which is the fundamental building block of ISDN. Even though the word ISDN was coined in 1971 at the CCITT study group XI (Switching and Signalling), focused and well organised study efforts started at CCITT in the study period 1981-1984 which produced the so-called Red Book ISDN Recommendation of 1984. They presented the basic framework of the ISDN concept, network architecture, user-network interface protocols, and common channel signalling protocols, and most important, showed what an ISDN was, and its tremendous potential and feasibility, albeit on paper only. In the following Study Period of 1985-1988, the ISDN studies at CCITT were further advanced over a wider area, including for example, supplementary services and telematic services, and the so-called Blue Book Recommendations of 1988 were produced. The definition of ISDN was reaffirmed in the 1988 document as:

> "An ISDN is a network, in general evolving from a telephony IDN, that provides end-to-end digital connectivity to support a wide range of services, including voice and non-voice services, to which users have access by a limited set of standard multi-purpose user-network interfaces".

It was during the above mentioned Study Periods that NATO member countries and many PTT Administrations, without waiting for the promulgation of the 1988 Recommendations by CCITT, decided to adopt the NB-ISDN standards wherever they exist, for the planning, design and implementation of their future networks making use also of the results of the standardisation of other international standards bodies such as ETSI (European Telecommunications Standards Institute) and ISO (International Standardisation Organisation) on matters such as OSI protocols.

Several countries are now moving from the planning stage to prototypes and actual implementations. The present book deals fully with how a system of this kind which is to be implemented is planned, designed and specified. It is believed that NB-ISDN's, once implemented, will have adequate capability to take us well into the next century before a system capable of switching signals with rates greater than 2 Mbit/s becomes commercially and operationally a necessity.

Since 1988, much of the planning and design effort by CCITT has become directed toward a network concept that will be, it is claimed, far more revolutionary than ISDN itself. This new concept is referred to as broad-band ISDN (B-ISDN) which is defined as a "service requiring transmission channels capable of supporting rates greater than 2 Mbit/s". This subject will receive only a passing reference in this book.

ACKNOWLEDGMENTS

Most of the material used in this book has come or derived from studies carried out under Prof. Ince's direction in NATO for the NATO Integrated Communications System (NICS) and in Turkey for the Turkish Armed Forces Integrated Communications System (TAFICS) and benefitted from the work done in Norway for the Norwegian Defence Digital Network (NDDN).

The authors are most grateful to many engineers and scientists who worked in the projects above and therefore contributed indirectly to the book. It would be invidious to single out names but we would like to mention those who played important roles in the design of WIS and to whom we owe thanks and appreciation: Prof. Dr. Nadir Yücel, Mr. Cahit Büçkün and Mr. Okay Çakıroğlu. Our thanks and gratitude go also to the staff of Kluwer Academic Publishers as well as to Mr. Z. Ener and Ms. K. Gül for the help given in the production of this book.

CHAPTER 1

INTRODUCTION

1.1 OBJECTIVE

In the available literature and university courses sufficient coverage is given to the detailed design and structure of all functions building up to complete ISDN. This book will therefore assume that the reader is well acquainted with these subjects. It is believed that a problem experienced by implementors of ISDN today is the lack of system engineering experience and methodologies.

Compared with existing books on ISDN, this book is unique in that it expounds an original approach of system engineering to guide a system developer from a given set of complete user requirements and constraints to design and implement a complete network, taking into account the technology and standards of ISDN.

The primary objective of the book is to give practical support to academic and industrial R&D engineers as well as to civil and military system planners and designers involved with planning, design and implementation of "non-commercial" ISDN-type networks with requirements and constraints similar to those of a defence strategic communications network.

1.2 TERMINOLOGY

The book is about designing a network with certain attributes which conforms to the CCITT ISDN Recommendations and methodologies and meets some constraints (financial, time etc.) which may be imposed on it.

There may be several "authorities" involved in the process of acquiring this network:

a) the network owner,
b) the network operator,
c) the design authority,
d) the network user.

It is often the case that some of these may be combined into one authority; e.g. the network owner may be the network user and the operator.

In this book we shall use the term "Authority" to mean the authority ((a) and/or (d) combined) who is empowered to state the operational/user requirements and network attributes as well as financial and other constraints for the network. The "Design "Authority" is identified with the reader (the writers of this book were in fact designers), who takes the requirements etc. given by the "Authority" and designs a network to meet the requirements with the constraints imposed.

1.3 CONTENT OF THE BOOK

The subjects that are treated in the book are presented in a sequence which is usually followed by a system designer. This starts in Chapter 2, from some specific or generic requirements (users, traffic, availability, security etc.) and taking into account the ISDN concept and the relevant CCITT Recommendations as summarised in Chapter 3. From this point under given or assumed constraints (financial, time, industrial and political) it proceeds to obtain, in Chapter 4, the minimum-cost network from competing dimensioned and costed networks with differing number of nodal switches based on the use of a specified or assumed routing system.

In Chapter 5, circuit, packet and message switching methodologies are presented in terms of functions, features, interfaces, synchronisation, O&M, services, integrity, security, and grade of service. Signalling systems are discussed and compared and rationale is given for the preference given to a signalling system which has to cope with the requirements imposed on it by an adaptive routing system and Network Management Subsystem. Different numbering methodologies are described and their design to suit specific needs is discussed.

Different timing and synchronisation methods are analysed and compared in Chapter 6 to meet certain performance criteria in term of bit count integrity for a hypothetical reference connection, considering availability, timing instabilities and inaccuracies, interoperability with other networks, echo control, radio path-length variation, the influence of cryptographic equipment and, of course, cost.

Both hierarchical-static and dynamic-adaptive routing schemes and their hybrids are defined and compared in Chapter 7 in terms of grade of service, network damage and multi-level precedence and pre-emption, taking into account itinerant and mobile users and zonal operations e.g. autonomous sub-networks. It is shown that dimensioning and cost of networks for a given traffic (load and precedence) and grade of service are very dependent on the adaptiveness of the routing systems used and that the network dimensioning, performance and cost are not significantly different for a well designed, intelligent and adaptive routing system.

One of the most important subsystems, transmission, is discussed in Chapter 8 under the main headings of digital transmission modelling (HRX, HRDL, HRDS), performance objectives in terms of quality and availability, voice encoding, multiplexing and timing and synchronisation.

Chapter 9 to 13 deal with the subsystems that affect approximately equally the cost of all the networks competing to the minimum-cost network.

In Chapter 9, the user requirements are discussed in all their dimensions, from voice/data and message handling to interoperability and availability, followed by considerations for the terminal area architectural design embracing the subjects of integrated terminals, automated message processing systems based on CCITT X.400 data processing facilities, local area network and security issues.

The important question of interoperability, particularly in the transition period from analogue to end-to-end digital working is treated in Chapter 10, between similar ISDN networks and with analogue switched networks and with digital switched networks using standards totally different from those of the ISDN. Inter-working functions in the various networks and gateway concepts including functions and features are also treated in this chapter.

The subject of Chapter 11 is the Network Surveillance and Control (NSC) which is required to ensure communications availability to the users at all times, to maintain the network at a stipulated level of communications performance, to provide and restore services to users in their appropriate priorities under various operational conditions and to support centralised and decentralised modes of operation. The NSC subsystem that is discussed and defined in this chapter meets the above requirement by supporting in real and near real-time several control functions related to the status and performance monitoring of all network elements, fault detection, location and isolation, maintenance of data on users/resources/assets and for long-term planing, effective planning and direction of logistics, electronic distribution of encryption keys etc.

Chapter 12 addresses issues related to identification of threats to civil and military communications security, describing general principles of COMSEC/COMPUSEC policy, system and software security, physical and electronic security for the network installation (TEMPEST), bulk and end-to-end encryption for protection of information, authentication, key management and COMSEC related issues. Security in military OSI networks, software security of switches, and transmission performance in relation to the security device performance are also treated in this chapter.

One of and probably the most important requirement for a strategic network is the survivability of the whole system against a range of physical and electromagnetic threat including electromagnetic pulse (EMP), jamming and physical attack. The subject of survivability including protection against EMP is treated in Chapter 13 which shows how to produce a balanced distribution of strength in all components of the system so that a uniform level of protection is available in the entire network with no unnecessarily weak unjustifiably over-designed parts. It may be thought that the requirement for rugged construction and survivability treated in the book may not now be considered very realistic either for civil or even for military strategic networks in the present world political climate and difficult to justify for implementation. On the other hand, considering the uncertainty and a lack of stability together with the high number of local conflicts that exist in the world today, one may think differently. In any case it was decided to include in the book the material on survivability with all its different aspects as they were developed for the original study in the belief that there would be readers wishing to see ramifications of survivability either because they would like to implement it or to have an appreciation of what cost they are saving by not having to implement survivability of a certain degree.

1.4 CURRENT NATO COMMUNICATIONS SYSTEM AND PLANNING

This book presents the system design process of ISDN networks implemented for non-commercial network applications. A number of such networks are being planned and implemented for the use of national governments and their defence forces as strategic defence networks. Inside the NATO alliance a number of member nations have built such networks, that for the reasons of interoperability will be interconnected to serve the alliance's interests. NATO as an organisation has therefore for a number of years studied the possibility to implement a multi-national overlay network to provide this inter-connectivity.

The planning for NATO communications goes back to the start of the alliance, and in the first decades it was believed that the best solution was to implement a NATO-owned and operated communications infrastructure to serve the multi-national command structure of NATO forces as well as supporting the need for political consultations between the governments.

In this phase, NATO implemented a number of communications systems, like the NATO ACE-High Troposcatter System, various Line-of-Sight Radio Relay Systems [1.1] and the NATO SATCOM system containing NATO owned and operated satellites and a number of Satellite Ground Terminals throughout the alliance [1.2].

While the traffic in the first stages were mainly manually switched dedicated services, a later development led to the establishment of switched voice, data and telegraphy services. The services were expanded to cover narrow-band secure voice services, message services as well as limited special purpose data services.

It was the intention in the late 70's and early 80's to evolve these transmission and switching facilities into modern digital operation, as the early systems started to become outdated and no longer cost-effective to operate and maintain. Significant studies were undertaken, and a complete network design called NICS Stage 2 (NATO Integrated Communications System) was presented to the nations.

The changes that have come about since then have changed the basis on which the planning in NATO today is being performed. These changes can be summarised as follows:

- The political environment that have emerged after the CFE treaty and the collapse of the Soviet Union and the Warsaw pact has led to requirement for a significant reduction in defence spending.
- The commercial markets have developed the commercial networks to a point where they today represent a significant capability available to all users throughout the alliance, at highly competitive prices.
- The technology used in commercial networks have given a number of nations a cost-effective way of developing and implementing national defence networks that, if they were offered to NATO, should give a significantly lower cost than any NATO-owned alternative.

On this basis, the current concept for NATO communications embraces three complementary elements:

1) A NATO owned and operated SATCOM system that will provide a digital, though thin, backbone transmission infrastructure, also useful in supporting the currently envisaged NATO out of area peace-keeping missions, by allowing the interconnect of deployed headquarters to the fixed political and military infrastructure.
2) The rationalised use of National Defence Networks (NDN) and switched services provided by these in providing interconnections between NATO command elements within the alliance.
3) The commercial use of the international ISDN services of the public network operators throughout the alliance.

The integrating element between these complementary systems will be a small network of NATO owned and operated ISDN switches, making up a NATO core network. This network is currently being planned, and will when implemented constitute a cost-effective solution to the NATO communications requirements in an environment that has led to fundamentally different solutions.

1.5 REFERENCES

[1.1] Ince, A. N. "EW and NATO Communications", Signal, March 1978.

[1.2] Ince, A. N. et. al., "Digital Satellite Communications Systems and Technologies", Kluwer Academic Publishers, 1992.

CHAPTER 2

USER REQUIREMENTS AND SERVICES FOR ISDN

2.1 CURRENT NATO COMMUNICATIONS SYSTEMS AND PLANNING

An outline of the current NATO communucations network is given below as a model of a modern strategic system incorporataing both terrestrial and space segments with some military features. The objective is to show what exists today and what the present thinking is for the development in the next decade or so of both the terrestrial and space segments. It is to be noted however that the infiormation provided in this chapter may undergo change as a result of the political changes in Europe and elsewhere (Appendix 2A).

Over the next two decades communications within NATO are to undergo a fundamental change. At present the NICS (NATO Integrated Communications System) [2.1] is implemented using NATO-owned bearer and switching facilities and rented PTT circuits, but is planned to phase these out in favour of a scheme whereby NATO will rely principally on national military communication networks to meet its needs. This will be facilitated by the widespread introduction of digital switching technology based on CCITT standards. In the near term, NATO will provide gateways and cross-border links to form a robust system of inter-connected NATO and national networks. NATO subscribers will be progressively transfered to national digital networks and NATO networks such as IVSN (Initial Voice Switched Network) and TARE (Telegraph Automatic Relay Equipment) will eventually cease to exist as separate entities. Provision of a common user interface based on the Integrated Services Digital Network (ISDN) concept will allow all existing and foreseen NATO subscribers' voice and data requirements to be met. To the greatest extent possible, NATO users will be served by this common-user system. However, it is recognised that independent "emergency" communications systems will probably still be required to provide highly survivable circuits to vital users in tension and war. Such systems must be capable of interfacing with the common-user network. The totality of NATO communications systems, common-user and emergency, will form the "NATO C3 Architecture".

Two developments have prompted this change in philosophy. Firstly, the ACE-High system of microwave and tropo-scatter links which has supported many of the NICS trunks since the 1960s has reached the end of its useful life and has come up against serious frequency-coordination problems. It is being replaced by the NATO Terrestrial

Transmission System (NTTS) consisting of all-digital cross-border links. Secondly, the widespread introduction of digital switching and high-capacity fibre-optic bearers promises to give national networks adequate connectivity and capacity to meet NATO's communication needs in additiion to national needs.

It is intended to introduce ISDN standards at an early stage in the evolutionary process. According to current thinking [2.2], the internetwork gateways will form switching nodes within the core of a "NATO ISDN", the concept of which is illustrated in Fig. 2.1. The inter-gateway links will be supported by a mix of NTTS, NATO SATCOM, and digital circuits established through nationally-owned sub-networks. Initially, the majority of NATO users will access the NATO ISDN via "Access Networks", each of which could be a national military network or an exiting NATO network such as IVSN. "Access Interfaces" will be provided to implement the protocol and signalling conversion needed between the Access Network and the NATO ISDN. A-to-D and D-to-A conversion will also be performed here if necessary, as will data rate conversion. In the long term, most national military networks are expected to adopt ISDN standards, making the full range of ISDN services available to NATO users. This will also facilitate network interconnection, though it is likely that the NATO ISDN will be preserved so as to police and optimally route inter-network calls.

The ISDN concept has evolved to meet commercial needs and therefore does not provide all the facilities required of a military communications system. In particular, it does not provide encryption, Traffic Flow Security, precedence or pre-emption. It is envisaged that as far as possible these additional functions will be provided at the boundary of the NATO ISDN, for example by "Virtual Terminals" implemented within the Access Interfaces or inter-network gateways.

2.2 IMPLICATIONS OF THE C3 ARCHITECTURE FOR NATO SATCOM

The implications of the transition to an ISDN-based architecture for NATO SATCOM are still being determined. However it seems possible that there will be an increase in the demand for SATCOM in the early stages, since SATCOM will be required to provide some of the bearers in the NATO ISDN while still supporting the IVSN and TARE as "Access Networks". This will be offset by the progressive removal of selected IVSN and TARE SATCOM links as the subscribers they serve are transferred to national networks.

ISDN offers the user both digital voice and data facilities. There are avilable simultaneously if required. Speech is encoded using 64 kb/s PCM. The standard "2B+D" interface provided to subscribers comprises two 64 kb/s channels and one 16 kb/s channel for signalling and packet data. Within switched network capacity is normally provided in multiples of 64 kb/s. End-to-end transparency is a basic principle of ISDN.

It seems likely, then, that the adaption of ISDN standards will imply a phasing-out of 32 kb/s CVSD SATCOM circuits as IVSN subscribers transfer to national networks. At the same time, there will be an increasing demand for 64 kb/s SATCOM bearers for the NATO ISDN. The number and connectivity of such bearers is as yet unknown, but is unlikely that the total number of point-to-point SATCOM circuits provided via the NATO inter-static network (consisting of several 12 - and 7 - meter terminals) can exceeed that currently supported even if all SGTs (Satellite Ground Terminal) are broad-banded, since the increase in bandwidth of the NATO IV satellite compared with that of NATO III will be taken up by the transition from 32 kb/s to 64 kb/s links[2.3]. In the post-NATO IV era, it may be possible to accomodate additional common-user circuits at SHF through frequency re-use, but this will have to be shown to be cost-effective vis-a-vis use of alternative communications media.

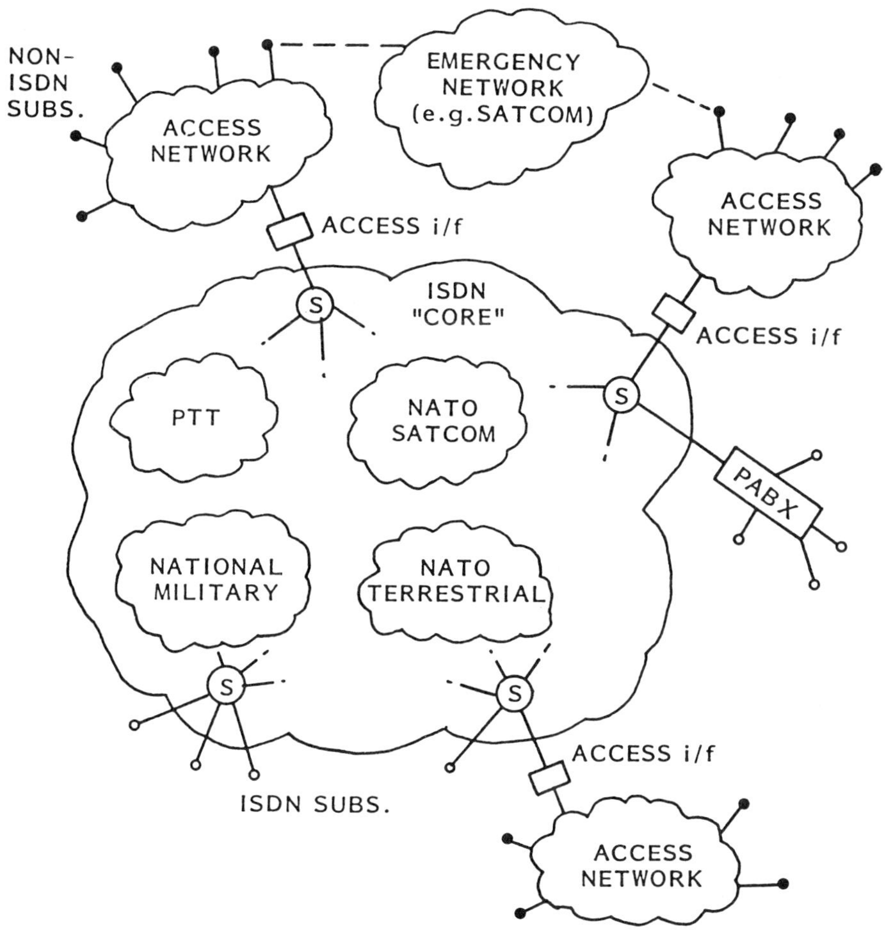

Figure 2.1 A NATO ISDN Concept

When jamming is present, NATO SATCOM will fall back to an ECMM mode of operation using spread-spectrum modems to support the more essential users. Data rates must be minimised in this mode so that each terminal can provide an adequate signal-to-jammer ratio for as many users as possible. Thus, for example, 2.4 kb/s vocoded speech would be used in place of 64 kb/s PCM. It may be possible for links within an ISDN to revert to such an ECCM mode without violating basic ISDN principles, on the basis that "information transparency" rather than "data transparency" is the requirement for voice circuits. However, this is arguable and a subject for further study. The alternative approach is to treat ECCM SATCOM as a separate non-ISDN system which can be accessed from the NATO ISDN through suitable interfaces. ECCM data circuits may more easily be accomodated within the ISDN concept since reversion to a low rate is possible without losing bit integrity, provided adequate buffering and flow control is provided within the network.

A number of technical problems will arise when SATCOM bearers are used in a NATO ISDN. Several of these are due to the propagation delay. This means that the signalling protocol (CCITT No.7 see chapter 5) used on terrestrial links will not work on links via geosynchronous satellites. A suitable protocol for such link has been specified, but protocol conversion between the terrestrial and satellite standards will be needed at each SGT. Although the CCITT recommendations for ISDN allow for single-hop satellite delays, multiple hops are barred and the system control will have to prevent these. Another difficult area is network timing: SATCOM modems will have to synchronise bit timing with that of the ISDN, and provide buffering to absorb variations in propagation delay while maintaining bit integrity.

In the long term, interconnected national military networks sharing common standards will provide a highly connected common-user system for NATO, at least in peacetime and tension. Use of fibre-optic trunks will ensure adequate capacity, excellent circuit quality and immunity from interception. As the common-user system evolves towards this goal, it may become increasingly difficult to justify continued use of a "permanent" SATCOM overlay: The high cost of maintaining and manning the terminals of the NATO inter-static SATCOM network may well make it no longer cost-effective to continue using this network as a complementary transmision medium to the terrestrial bearer system. Where SATCOM will have a part to play is in war when the common-user network has become fragmented. Transportable SATCOM terminals can be rapidly deployed to establish survivable ECCM links, albeit of low capacity, between ISDN "islands" in an effort to restore full connectivity. And of course, SATCOM will continue to provide access links into the common-user network from mobile users such as ships.

The reader should refer to [2.3] for a fuller account of NICS and the NATO SATCOM.

2.3 REQUIREMENTS AFFECTING NETWORK DESIGN

2.3.1 Introduction

This section develops specific operational and systems requirements, relating them to the systems architecture for WIS. Flexibility with the capacity to incorporate new as yet unforeseen or unvalidated requirements is a major consideration.

This section details the following requirements:

(a) Integration
(b) Survivability

(c) Communications Security
(d) Readiness
(e) Interoperability and Interconnection
(f) Standardization
(g) Reliability
(h) Availability, Flexibility and Responsiveness
(i) Maintenance Support
(j) WIS Users
(k) WIS Services
(l) Sub-Systems Requirements
(m) Network Surveillance and Control

The requirements are those that arise from the functions of the individual users (e.g. OPS, INT, LOG, MET etc. in the likely wartime conditions) in the expectation that these are certain to be the most exacting. That is to say, the overriding deficiencies in existing systems to be corrected in the system being designed (WIS) are usually survivability and security. This section also addresses additional requirements that are specific to peacetime, to a period of crisis, and to an adequate state of readiness.

In this section reference is made to such terms as "TARE" "DWS" and others, recognizing that they may change or disappear in time.

2.3.2 Assumptions

The requirements are written on the assumption that:

a) WIS which is assumed to be implemented in phases, is complete, fully operational and meeting all the original design requirements.

b) A SATCOM system will be established as foreseen or justified by survivability and other considerations.

c) Any command control and communications (C3) system (or any electronic office system) that is to utilize WIS will be designed to take account of its impact on that latter system, i.e. to be within the capacity of the WIS as outlined in this requirements analysis, taking into account the degree of flexibility and expansion as stated in sections 2.3.1 and 2.3.10.

2.3.3 Integration

WIS is to be developed in phases into a fully integrated grid nodal system. It is by definition a basic requirement that WIS forms one continuous network with common technical, operational and procedural standards, and with a unified set of rules for such aspects as precedence levels, security, control and reporting, restoration and maintenance. In order to provide multiplicity of circuit routes and to extend the network to minor and remote users, it is required that this basic network makes full use of the military transmission media, PTT and other existing systems.

2.3.4 Survivability

2.3.4.1 General

This section should be read in conjunction with chapter 13 "Survivability Design".

Survivability is the overriding requirement. WIS is to be a highly survivable system in all conditions of crisis and war, certainly no less than the user community (e.g. command and control elements) that it supports. The analysis of this degree of survivability is made in chapter 13 where the required degree of survivability of WIS system is assessed against the mission and degree of survivability of the related user facilities.

When subject to disruption, WIS is to degrade progressively and be as insensitive to damage as practical; practical in the sense that the degree of survivability will inevitably have to be a compromise between absolute protection and a reasonable allocation of the total resources available to WIS as a whole. In any event WIS is to provide under all, but the most extreme conditions a minimum of communications for the vital traffic of the primary users (see Definitions at Appendix).

Protection, in the fullest sense of the word, is to be provided to mitigate against the effects of:

- Physical damage by natural causes.
- Physical damage by enemy attack or sabotage.
- Disruption through civil unrest.
- Electromagnetic interference (ECM), in particular accidental or deliberate jamming.
- Electromagnetic pulse (EMP).
- Enemy generated "dummy traffic".

Protection is to be provided for WIS as a whole, although it is recognised that initially it may not be cost-effective to upgrade fully all the old facilities/systems.

2.3.4.2 Protective Measures

Such protection is in general to be provided by a mix of related measures at each WIS site and in the WIS system functions and concept; the mix depending upon the threat and local conditions, and the specific role of that site in terms of both local users and the WIS as a whole. However, the commitment of resources to the achievement of survivability is generally to be in the following order of preference:

The development of a grid nodal network with an effective control and restoration system so that the complete disruption of a few WIS installations will not sectionalize the network, and the destruction of any one installation will not isolate any major user location from the WIS. This, for example, is primarily to be provided by:

(a) The progressive development of the transmission media into a grid network with independent paths. using protected and secure facilities, together with a flexible control organization capable of rapid detection of damage and restoration of communications.

(b) The optimum mix of switches and inter-switch trunks, with a minimum per node in general i.e. CS, PS, MS or combinations of three independent terrestrial inter-switch routes, or two independent terrestrial routes and one independent interconnection to a satellite ground terminal if this exists.

(c) The mutual interconnection of various communications systems to the maximum extent possible.

(d) Multiple access for the primary users over at least two different transmission routes, from the access switch to two different nodes.

Protection against types of jamming that would affect simultaneously a number of links or stations, such as jamming of the satellites or the access links of primary users.

Local duplication e.g. provision of on-site spare components and repair facilities, stand-by power and redundant equipment.

Establishment of a survivable support organization for repair and resupply.

Protection against fire and where necessary extreme climatic conditions.

Local physical protection and/or dispersion, such as:

(a) Control of access by physical restrictions, guards and alarms.

(b) Siting away from likely target areas, or within areas already well protected.

(c) Hardening but only where this is cost effective, e.g. where the protection provided by diversity or dispersion is inadequate; where the site would otherwise be less protected than the local facility that it serves; where the operational value of the facility merits it; but not where the value of such hardening is negated by external components, such as antennas, that are not open to protection.

Protection against EMP and associated effects caused by nuclear bursts which may simultaneously affect all electronic installations over a wide area.

Provision of back-up transportable facilities.

2.3.4.3 Dispersion

The location of most of the WIS installations is to be predetermined by:

(a) The user locations to be served and their geographical concentration.

(b) The availability and length of access links.

(c) The intersection points of the grid network.

(d) The terrain.

(e) Goverment consent and real-estate availability.

(f) Existing facilities.

However, these considerations apart, WIS facilities are in principle to be sited away from likely target areas. A nodal switch is not to be sited within the likely damage area of a primary target. Concentration of communications facilities which will make such a location a likely target area are to be avoided.

2.3.4.4 Physical Site Protection and Hardening

a) Bearing in mind the inherent survivability of a grid nodal network, priority in the allotment of resources to physical site protection and hardening is to go to the weakest elements of that network, i.e. WIS facilities at the primary user locations and the access links from them into the network. The required degree of protection for these is to be provided, where practical, by:

Installing the WIS equipments of a single user inside the locak hardened facility of that user or by providing protection to the same level as that facility.

Multiple access to at least two WIS nodal switches, for all primary users.

b) As to other WIS sites, the order of priority in resource allocation to protection is to be:

By the nature of their place in the network, those WIS nodal switches, MS/PS and transmission media sub-system control centres, the disruption of which would cause inordinate damage to the network as a whole.

The remaining WIS nodal switches and MS's/PS's.

The major WIS transmission stations, e.g. satellite ground stations and transmission nodes.

2.3.4.5 Protection Against Electromagnetic Interference and Deliberate Enemy Electronic Counter Measures (ECM)

a) Adequate protection for the WIS against accidental electromagnetic interference will be attained by:

- Equipment designed with minimum vulnerability to interference and jamming.
- Full coordination with the Authorities for frequency management.
- Site-by-site design that will be compatible with local conditions and national requirement.

b) However, the threat of enemy ECM (notably jamming) will still be high. This is to be overcome as far as is practical by, for example:

Diversity of routing.
The provision of jammer locating equipment at transmission nodal sites and SATCOM, SGT's.

The provision of electromagnetic counter-counter-measures (ECCM) for those links which are particularly vulnerable.

2.3.4.6 Protection Against EMP

a) All major WIS installations are to be protected against the EMP threat according to the following criteria:

For nodes and MS's/PS's, no interruption of service nor corruption of system software.

For transmission media, short interruption of service in the order of five minutes but no damage beyond that which can be restored within minutes by on-site facilities/spares.

For access switches and other WIS installations in user locations, the protection will be commensurate with that of the individual user location, taking the requirement in the paragraph above as the minimum.

All new communications buildings are to be provided with an EMP shielding to a level as specified in chapter 14, where the requirements above do not entail higher levels of protection.

Satellites used in WIS are as a minimum to be protected against high energy radiation particles from nuclear bursts at altitudes of about 1,000 km.

2.3.5 Communications Security

a) The ultimate requirement is the total security of WIS between all users; that is to say security of the communications sub-sytems, the traffic on them, the signalling information and the communications processors and their programming functions. This requirement is to be achieved in manageable phases with priority going to:

- The most vulnerable elements, i.e. those which are most open to long-range interception.
- Expansion and improvement to the secure voice service.

b) To this end:

A system of bulk/link encryption along with "RED" switches and "approved" circuits is to be realized as early as possible.

All those manning "RED" areas are to be security cleared to the highest security level.

There will be a continuing requirement for both narrow and wide band voice encryption, these two systems to be interoperable.

End-to-end and off-line encryption is to be provided for special and exclusive users (see Definitions at Appendix). The former is required for up to about 10% of the DWS at each user location.

- End-to-end encryption will also be required for a few percent of all users located outside a "RED" area. This encryption is to include signalling information.
- Conference calls (see section 2.3.13) are required to be secure.
- Access to and from the civilian PTT systems into the WIS is to be strictly controlled. A method of warning is to be provided to alert subscribers that they are no longer using a secure system.
- The impact of the hard copy keying material resulting from the increased number of crypto equipments is to be kept to the minimum through the use of electronic key distribution. Additional security is to be ensured through key compartmentation and the development of authentication procedures suited to an automated, integrated system.
- Access points required for mobile and itinerant users are to be physically secured. Furthermore, booking-in authentication and crypto procedures are to be designed to prevent unauthorized use of an access point.
- The above considerations apart, the personnel involvement of communications staff is to be reduced to the minimum by the progressive reduction in the number and size of RED areas.
- Access to all kinds of communications processors, including the stored data and the programs, is to be strictly controlled.

c) WIS Security and Cypto Plan

All the above requires the development of an overall WIS security and crypto plan, taking into account:

- The different forms of encryption.
- The mix of analogue and digital sub-systems.
- Increasing automation.
- The continuing reliance on PTT (Public networks) and other insecure systems.
- The need for maximum interconnection and interoperability with other systems.
- The plan is to include a common crypto logic for all the systems that are required to interoperate. Furthermore, the keying system adopted is to be as flexible as possible.

2.3.6 Readiness

It is accepted that for reasons of economy and manpower, certain elements of the WIS may in peacetime be at a reserve status, others at a minimum manning level and others unmanned. However, it is a requirement that the WIS be capable of being brought to full operational status and to a crisis level of manning within a few days. This is to include where necessary the deployment of transportable elements and where the security plan requires, the manning of unmanned stations for both technical and security purposes.

2.3.7 Interoperability and Interconnection

2.3.7.1 General

The ultimate requirement in order to optimise both flexibility and survivability, is that the WIS should be capable of interconnections in a secure, automatic mode at a large number of widespread gateways i.e. on a multi-channel basis, with civil and military strategic and national tactical systems; albeit that the civil systems may not necessarily be in the secure mode. However, there is a number of major constraining factors that will make the achievement of this ultimate requirement a long-term goal.

Interconnection is therefore to be viewed as an evolutionary process recognizing that in the short and medium term the number of gateways and the degree of automatic and secure interconnection will be limited.

2.3.7.2 The Types of Interconnection

a) With Static WIS Users: The interconnection of the organic communications elements with the WIS is to be either direct to an access or nodal switch, or through a local PABX which itself will be connected directly to either an access or nodal switch.

b) With Mobile Land WIS Users: These interconnections are to be achieved with assets organic to the mobile users, or by direct physical connection to WIS pick-up points.

c) With National Tactical Communications Systems: The ultimate goal is the interconnection of the WIS with the national tactical networks and systems (land, air and maritime) in an automated and secure mode. This is required to link these national networks to reach one another through the WIS.

d) For ships at sea equipped with satellite terminals.

e) Interconnections with the PTT: These interconnections, which may be on a leased circuit/group basis, are required:

To reach users who for economic and other reasons cannot be otherwise served.

To enhance the survivability and the services provided by the WIS.

To make use of the extensive availability and resources of the PTT and their special services.

f) Access to PTT Switched Network: WIS users, IWS's and DWS's will have the capability to make a connection via the WIS to a gateway to a PTT network through their local PABX. This capability is to be available both in peace and crisis and is to be under local control.

2.3.8 Standardization

The requirement of an integrated system, of economy and of ease of management and operation, call for the maximum degree of standardization possible; standardization of rules, procedures, transmission parameters, processes and techniques as much as of equipment. Standardization of equipment is to be as total as the timeliness and phasing of the implementation of the various sub-systems and equipments will allow. WIS standards will basically comply with appropriate CCITT Recommendations NATO/Regional Standards, appropriate TEMPEST Regulations and survivability requirements.

2.3.9 Reliability

A very high level of reliability will be a design feature of the WIS, supplemented by automated re-routing and rapid restoration of service in cases of failure. Specifically this is to entail:

- Duplication of important equipments or subassemblies so that major outages are not caused by failure at a single point.
- Availability of adequate spare parts, spare circuit boards, equipment assemblies, tools and test equipment on site (see section 2.3.11)
- The establishment of an efficient and responsive maintenance support system.
- Provision of standby power and no-break power systems.
- Provision at both manned and unmannned stations of fault alarm systems, easy fault location methods and where cost-effective, automatic switch-over to standby equipment.
- Protection against extreme climatic conditions, lightning and high voltages.
- A high level of training and operational and technical competence, including pre-training of crisis and wartime augmentation manpower.

2.3.10 Availability, Flexibility and Responsiveness

The WIS is required to have a very high level of availability for each specific user-to-user service.

The WIS is also required to be a highly flexible and responsive system in terms of operational changes, traffic capacity and patterns, and redeployment of resources. All this will generally be attained by modular design and as a consequence of the achievement of a high degree of survivability (section 2.3.4.1), reliability (section 2.3.9) standardization (paragraph 2.3.8) and required user service (paragraph 2.3.13). In designing the separate elements and subsystems of the WIS, the requirements for interoperability, flexibility, responsiveness and general ability to absorb new requirements, traffic sources, systems performance, etc. are to be kept to the fore. Furthermore, it is a requirement that the WIS users' dependence on inflexible dedicated systems is progressively reduced as the common user, integrated, standardized WIS evolves. However, the design of this common user system is to ensure a service that is responsive to the separate "dedicated function" requirements of the individual users.

2.3.11 Maintenance Support

All WIS equipments and installations are to be designed for ease of maintenance, i.e.:

2.3.11.1 Hardware Maintenance. It is a requirement that:

Equipment and installations are designed to give each site the maximum possible in-house self-sufficiency commensurate with cost-effectiveness.

This in-house maintenance is within the capacity of moderately skilled personnel trained to "unit" and "field" levels of maintenance.

The level of operational performance and the identification and location of faults, are easily and quickly determinable.

Test points in the equipments and suitable in-station test jack fields are readily accessible.

Faulty components and sub-assemblies are readily accessible and easily replaced.

Spares, coded and easily identifiable, are available in adequate quantities on site and at the maintenance and repair depots.

Comprehensive technical documentation including equipment manuals, on-the-job training packages and "as-built documentation" is readily available on site and at the maintenance and repair depots.

Adaquate numbers of tools, test equipment and instruments for measurements are provided on site and at the maintenance and repair depots.

2.3.11.2 Software Maintenance. It is a requirement that:

The Authority has full proprietary rights over all the software of the WIS operational systems.

WIS personnel is trained and competent to manage these programs without contractor assistance.

The contractors provide the Authority with the complete documentation of WIS software.

All WIS systems software is centrally documented, controlled and maintained at a WIS Software Maintenance Development Centre (SMDC).

Detailed procedures are established for local changes to data bases, and for local testing and fault finding.

2.3.12 WIS Users

2.3.12.1 General

a) The legitimate WIS users will be defined by The Authority in terms of geographical locations.

b) The User List indicates the number of:

- Static WIS users (assumed to be about 200 in this book) dispersed at about 250 user locations.

  Land mobile and some mobile afloat WIS users (assumed here to be about 40).

- Some 25,000 subscribers; of these approximately; 3,000 DWS's and 22,000 IWS's.

It is possible that the differentiation in WIS between DWS and IWS will eventually disappear as WIS progresses.

c) The broad categories of WIS users are to be:

i) Primary Operational WIS users which consist of important, high level military and civil users. These users will depend primarily on the WIS for all external communications for all functions (i.e. operational, logistic and administrative communications) supplemented by their organic, not-WIS communications. The civil users will depend on the WIS only for political and military consultation, for communications with the military users above.

ii) Subsidiary Operational WIS Users. This category embraces generally all eligible military users and agencies not included in the paragraph above, but who have important operational functions, e.g. meteorological centres; communications management organisations; NBS control and warning centres; Regional Commands, Naval Commands, NCSOs, Coastal Radars, naval bases and control of shipping authorities.

iii) National Logistic, Administrative and Special WIS Users. This category comprises special national units and agencies; technical and scientific centres or agencies, and schools and training installations. They have, however, not the same needs as the operational users for survivability and to a greater extent, rely on the civil PTT systems.

iv) Civil Wartime Agencies (CWA). These users, particularly in a crisis, will need the same WIS services as the primary operational WIS users, in terms of survivability and responsiveness. However, by their nature, a substantial amount of their communications will be with civil national authorities or industry who are not WIS users. They will therefore have to rely on the PTT systems to a large extent.

2.3.12.2 Mobile and Itinerant Users

a) The WIS is primarily a static system serving a static user community. There will, however, be two exceptions to this:

- Land-mobile WIS user and,
- Afloat WIS users.

b) Land-Mobile WIS Users are to be served through appropriately located WIS pick-up points with adequate access capacity. These users are to be provided with appropriate access links to be interconnected with the pick-up points by transmission means organic to their organization. The planning requirement for an organic radio relay interconnect from the mobile user to the access point is to be no more than 50 km.

c) Transportable WIS Facilities. There is a limited but essential requirement for a reserve of transportable facilities (satellite ground terminals, nodal switches, line-of-sight stations) primarily to replace fixed facilities that are out of operation, or to strengthen or extend the WIS services in contingency areas. It is a requirement that these facilities be easily and economically transportable by land, sea or air, in terms of both size and speed. Their readiness is to be no less than that specified in section 2.3.6.

d) Afloat WIS Users. These are to be served by SATCOM through compatible shipborne satellite terminals or by HF.

2.3.13 WIS Services

2.3.13.1 General

a) WIS services are in principle to be provided on a common user, precedence controlled, automated basis. However, certain circuits routed through the WIS may still be assigned on a dedicated basis where this is either operationally or technically required, or where it is the most cost-effective solution. All the same, dedicated functions are mainly to be served by "hot-line" and "off-hook" over the common user system.

b) Primary operational WIS users and the CWA's are to be provided with the full range of WIS services to both their static and mobile locations. Subsidiary operational WIS users, national logistics and administrative users, and special users are in the longer-term to have the same range of services, although it is recognized that in the short term, these may, for reasons of economy and availability, have to be limited. WIS services on a restricted scale are also to be extended to meet certain national requirements of the users as agreed on a case-by-case basis in accordance with The Authority.

2.3.13.2 Precedence

a) The following precedence levels, assigned to each call or message by the originator, are to be provided for voice, telegraph and data:

- FLASH (P1)
- IMMEDIATE (P2)
- PRIORITY (P3)
- ROUTINE (P4)

b) For telephone subscribers, precedences above ROUTINE are only to be afforded to DWS. If others require higher precedence, this is to be provided by operator assistance and controlled by local procedures. Higher precedence voice is to have, when necessary, rights of pre-emption over lower precedence. Data and telegraph traffic are to be handled in order of precedence and only FLASH is to have a right of pre-emption. Hot-line and off-hook calls are to be automatically assigned a fixed, predetermined precedence (FLASH or IMMEDIATE for hot-lines and PRIORITY or ROUTINE for off-hook calls) and be routed to predetermined distant subscribers.

2.3.13.3 Types of Services

a) The following services, which comprise the minimum military requirements, are to be provided in WIS.

Multiple Access (Homing): This is the provision of two or more independently routed access links utilizing different transmission paths. For telephone circuits, these links are to connect the access switch of the user location with two or more nodal switches. Where a WIS switch is collocated within a user location, this user location is to obtain its multiple access through this switch which itself will be connected to at least two nodal points.

b) Atomatic Switching of WIS Telephone Circuits: This is to include:

- Automatic selection of trunk and access routes.
- Automatic access with direct subscriber dialling.
- Direct in-dialing to subscribers, as aminimum to all DWS's, but with operator assistance when requested.
- Automatic pre-emption on trunk and access circuits, and pre-emtion of wanted subscriber lines for DWS.
- Automatic hot-line and off-hook services to replace the needs for dedicated voice circuits and to supplement or replace the voice circuits dedicated for data or message transmission. Selection by the originating subscriber between a small number of predetermined destinations may be included.
- Short code dialling and/or abbreviated dialling for selected DWS's.
- Facilities for transfer of incoming calls to another local subscriber and, for selected DWS's, automatic call transfer to a predetermined alternative number or the operator, if the primary number does not answer.
- Indication and warning signals to subscribers. In addition to the dialling, ringing and busy tones or signals, indication or warning signals are to be automatically provided to subscribers to indicate called number not available call pre-empted, call not secure, and other conditions.
- Facilities for recorded voice announcements to calling subscribers to provide information not covered by the standard tones and signals.

c) Automatic and Rapid Forwarding of Telegraph and Data Message. This is to include:

- Rapid and reliable delivery to user terminals of single and multi-address messages inserted into MS's in ACP 127 NATO Supplement 3 format, including dual precedence handling of multi-address messages.
- Full accountability of all format messages, such that the probability be very low (10 to minus 6) that a message be either lost, wholly or in part, or misdirected.

- Interfaces with a wide range of user facilities ranging from single access, low speed telegraph circuits and manually operated Tape Relay Centers (TRCs), to automated user facilities (e.g. Message Handling Terminals).

d) Secure Voice Communication. It is to provide with preference to DWS's:

- User-to-user (see Definitions at Appendix 2A) cryptographic protection.
- Secure extentions to an increased number of subscribers within the users' locations, accessible by automated telephone calls, preferably without duplication of subscriber circuits and telephones.
- The capability for secure conference calls.
- The capability for secure facsimile.
- Authentication methods and procedures for calling and called subscribers.
- Warning signals to subscribers if the complete subscriber-to-subscriber connection is not secure.
- Cryptographic key compartmentation so that general compromise of keys is avoided and groups of users can communicate on an exclusive basis.

e) Non-Secure Voice Communication: There will remain a requirement for unclassified telephone communication with non-WIS users (i.e. PTT subscribers), and until the long term goal of a totally secure WIS is achived for voice communications with WIS subscribers who have not yet been provided with secure voice facilities.

f) Data Transmission: The requirements for the (Command Control Information System) CCIS/ADP will be supported by WIS. The requirement for handling ADP systems traffic will have an impact on the WIS, both on the transmission network and on the switching techniques. The characteristics of the ADP communications traffic will govern the cost and complexity of communications arrangements necessary to handle this traffic. These characteristics will depend largely on how the ADP system is organized; in particular, on the manner in which the data files and processing loads are distributed. In order to optimise the overall ADP/communications costs, it is essential that the ADP system is organized so that communications costs are kept low, commensurate with the operational requirements. This implies that, in determining the preferred organization for the ADP system, due account must be taken of the impact of the various ADP options on the comunications requirements. Present planning (NATO/National) calls for up to 64 kbit/s transmissions between host computers and rates according to CCITT Rec.X.1 between host computers and local terminals.

g) Special Services

- Closed Networks: These are required for determined functional groups with controlled access for non members of the group. The WIS system design is to facilitate changes to the numbers and compositions of the groups.
- Conferencing: There is a requirement for the automatic and manual establishment of conference calls on both a predetermined and ad-hoc basis.
- Switching of ADP Traffic: It is recognized that this is required for reasons of economy, reliability and survivability. However, in the initial phases of WIS, and despite the cost, it is recognized that reliance may have to be placed on dedicated circuits.

h) Manual Interfaces and Operator Assistance

- Manual interfaces are to be provided with the PTT systems and with other systems, wherever required, for legal, security or technical reasons. Manual interfaces may also have to be provided for IWS's who can not be provided with automatic interfaces.

- Telephone operator assistance is to be provided for:

  Directory iniformation service, in particular, information to incoming calls on local subcriber numbers.

  Assistance to incoming precedence calls to obtain wanted subscribers, in particular, IWS's who might be busy.

  Assignment, within given regulations, to a subscriber of a higher precedence than his normal entitlement for individual calls justifying such precedence.

  Setting up of conference calls.

- Wherever possible, the manual interface functions and the operator assistance functions are to be combined and associated with existing operator functions.

i) Directories and Instructions for Use

The users and subscribers are not to be required to know the geographical location of the other users, in particular not those of mobile or itinerant users, or their accesses into the WIS in order to reach them. All users are to be provided with:

Instructions for the use of the automated WIS services, including descriptions of the types of services, the numbering system, extracts of procedures and the meanings of the different tones and signals.

Appropriate up-to-date number directories necessary to make good use of the system, including the circuit switched telegraphy and data services.

Immediate updating of important directory changes and additions by messages on a need-to-know basis.

j) Other As Yet Unvalidated Services:

Section 2.3.10 states requirement for the design of WIS to be sufficiently flexible to be able to absorb additional as yet unvalidated requirements. Such requirements might include, for example, fast high-resolution facsimile, slow-scan TV, and communications to serve the dedicated functions of new weapons systems.

2.3.13.4 Quality of Service

a) The system is to provide:

- For voice communications, a good intelligibility and speaker recognition in the clear and secure mode between any two subscribers of the WIS.

- A voice circuit quality in the internodal network corresponding to CCITT recommendations for international circuits and, hence, capable of carrying data traffic at up to 64 kbit/s.

b) For telegraph and data communications:

- A bit error rate better than 10 to the minus 5 between any two fixed terminal equipments in the WIS.
- A bit error rate better than 10 to the minus 4 between any two terminal equipments when one of these serves a mobile user.
- A routing and accountability system giving a very low probability (10 to the minus 6) that a telegraph or data message be either lost, wholly or in part, or misdirected.

2.3.13.5 Speed of Service

a) The system is to provide rapid services to its users:

i) Circuit Switched Connections:

- 90% of all calls attempted are not to encounter a dial tone delay of more than 1 second.
- The maximum connection set-up time for circuit switched connections in the undamaged system is to be:

1) < 5 sec. for 95% of all routine DWS calls
2) <10 sec. for 95% of all IWS calls
3) < 1 sec. for 95% of Hot Line and Flash calls
4) < 3 sec. for 95% of priority calls

The remaining 5% of the calls should be completed within twice the delay times stated above for each category.

ii) Telegraph Traffic: Messages passed through the store-and-forward network are to be delivered rapidly to all addressees, with reporting to the originator in all cases of non-delivery to user terminals within the maximum times specified below; the time counting ifrom completion of the message transmission into the first MS until the start of delivery to the destination terminal:

- FLASH	(ZZ)	5	minutes
- IMMEDIATE	(OO)	20	minutes
- PRIORITY	(PP)	60	minutes
- ROUTINE	(RR)	300	minutes

iii) Data Traffic:

- Data speeds up to 2.4 kbit/s are to be easily accommodated within the WIS.
- Speeds above this up to 64 kbit/s will be required once overall digitalization is achieved. This initial restraining factor is unlikely to cause problems except for CCIS inter-host computers transmissions where, for a time, it may be necessary to resort to rental PTT circuits.

2.3.13.6 Capacity and Grade of Service

a) General:

i) The Transmission Network: The main considerations are to be extra capacity expressed not only in additional circuits bu additional routes.

ii) Dimensioning: The WIS subsystems are to be designed to meet the estimated wartime peak load as assessed on major exercises with a 20% extra allowance for contingencies and the unforeseen. In any dimensioning, the load occasioned by reserve circuits and reserve or non-exercised facilities, is to be taken into account. For planning purposes, the dimensioning of the network will be based on the required grade of service and the traffic loads defined in the following paragraphs, assuming an undamaged network.

iii) Degredation: When damage occurs in the network, the service is to degrade progressively in such a manner that it is maintainde as far as possible for the increasingly higher (vital) precedences of the primary users, affecting the lower priority traffic and subsidiary users first.

b) Grade of Service:

In an undamaged network, the grades of service are to be as follows or better:

(a) FLASH OVERRIDE	0 % (i.e.non-blocking)
(b) FLASH	0.01 %
(c) IMMEDIATE	0.1 %
(d) PRIORITY	1.0 %
(e) ROUTINE	4% for DWS's and 10% for IWS's.

c) Estimated Traffic Loads for Planning Purposes:

i) Voice Traffic: A total of about 1000 Erlangs. This corresponds to an average telephone use in the busy hour of 8 minutes per DWS and 4 minutes per IWS.

ii) Telegraph Traffic: A total of about 300 Erlangs based on an avarage transmission speed of 64 kbit/s.

iii) Data Traffic: A total of about 500 Erlangs.

iv) Total Traffic: Based on the above, the total busy hour traffic load is, for planning purposes, assumed to be about 2000 Erlangs.

2.3.13.7 Scale of Service

The minimum military requirement is in general:

a) Voice

The number of DWS's shall in general not exceed 15% of the number of officers on the establishment of a Peace Headquarters (PHQs) or 30% of the officers on one shift in a War Headquarters (WHQs). Similarly, the number of IWS's shall not exceed the total number of officers on the establishment of a PHQs or the number of officers on one shift in a WHQs.

b) Telegraph and Data

All WIS users are to have access into the WIS network with formal messages in current ACP 127 NATO Supplement 3 format, normally directly but where unavoidable, indirectly via a tape relay centre/MS. Transmission speeds will vary from single low speed circuits (50-75 Baud) to several circuits at medium speed (600 bit/s) and up to higher speeds of 2.4 kbit/s or more. Automatic or semi-automatic accesses and interfaces for circuit switched and/or packet switched telegraphy or data on voice circuits is to be provided into access switches or modes for:

i) The message centres, Computer Aided Message Processig System and medium high speed message terminals of all primary operational WIS users at medium and high sepeeds.

ii) The CCIS terminals of all primary operational WIS users at speeds up to 64 kbit/s.

iii) These communications for medium and high speed users will form closed networks.

c) Precedence Scale of Entitlement. The maximum entitlement is:

i) FLASH : 15% of the DWS entitlement.

ii) IMMEDIATE : 25% of DWS entitlement.

iii) PRIORITY : 30% of the DWS entitlement

iv) ROUTINE : All other subscribers, noting that these will have the possibility of justifying higher precedences through operator assistance.

d) Scale of Special Services

i) Closed Networks. The number is to be restricted to no more than 20 networks, each of no more than 100 subscribers.

ii) Conference Facilities. Each switch is to be designed to permit up to two secure conference calls at any one time, each with up to 20 distant subscribers

e) Pick-Up. The minimum requirement (number of points, locations and capacities) for planning purposes for WIS pick-up points (see Definitions at Appendix) for mobile users is given by the Authority as about 40 locations.

2.3.14 Subsystem Requirements

2.3.14.1 General

The requirements stated in other sections of this chapter have largerly been of the user. There are however a number of requirements that WIS will have to meet that are subsystems as opposed to user associated.

2.3.14.2 SATCOM

a) WIS will incorporate SATCOM links to be used for node skipping and access functions.

b) The WIS SATCOM subsystem is to meet the following requirements.

   i) At least one alternate SATCOM path through another satellite. This satellite to be stationed at a distance sufficient to ensure that it and the primary sattelite can not be destroyed together by the same enemy action.

   ii) Several large transportable and small mobile terminals capable of transportation by air and lifting by helicopter. These terminals are to have a capability of providing a maximum of 3 voice and 3 telegraph channels being acceptable.

   iii) It is a requirements that the SATCOM be extended to the maximum number of National and NATO assigned ships possible.

   iv) Continuance of the cooperative programmes for the sharing of facilities with NATO and the nations, to the maximum possible degree of standardization and exchange of circuits.

2.3.14.3 Radio Relay

a) Radio relay systems, existing or to be established, for WIS will form the backbone of WIS transmission susbsystem and will also provide links between forward user locations and the major WIS transmission media.

b) These radio relay systems are to conform to the requirements already stated in this book in terms of reliability, survivability services and interoperability. In particular, these radio relay systems are to have the capability at each location to interface with the organic radio relay systems of mobile and itinerant users.

2.3.14.4 PTT

a) The PTT (Public Networks) will in all probability continue to provide a proportion of the total WIS capacity by way of leased circuits and groups. As such, these systems will form essential element of the WIS. It is therefore a requirement that the WIS is fully interoperable with the public circuit and packet-switched networks.

b) It is also a requirement that in the design of the WIS full recognation is taken of PTT procedures and regulations. Also the progress of PTT, individually and collectively, towards progressive digitalization is to be closely monitored.

2.3.14.5 Switches

The essential functions of the WIS switches, in terms of survivability and responsiveness of the network, require the development of processor controlled adaptive routing methods, computer memory storage of up-to-the-minute network information, and automatic testing.

2.3.15 Network Control

WIS is to include a Network Surveillance and Control Subystem based on the principles and requirements as stated by the Authority.

2.4 REFERENCES

[2.1] Ince, A. N. "EW and NATO Communications", Signal, March 1978.

[2.2] "The C3CS Goal Architecture", STC Technical Memorandum TM-867 (Draft), 1989.

[2.3] Ince, A. N. "Digital Satellite Communications Systems and Technologies", Kluwer Academic Publishers, Boston, 1992.

APPENDIX 2A

DEFINITIONS

1.1 WIS User: Any authority (i.e. headquarters, organization, unit base or station) who is entitled use the communications services of the WIS (See paragraph 2.3.12.1 (c) Broad Categories).

1.2 Itinerant WIS User: A user who may operate from alternative static locations from different access points into the WIS.

1.3 Mobile WIS User: A user who has the facility to change his geographical location frequently and freely.

1.4 Primary WIS User: Short term for Primary Operational WIS User detailed at Paragraph 2.3.12.1 (c).

1.5 NATO WIS User: A user oparating as a NATO body in peace and/or war with NATO international manning wholly or in part.

1.6 National (NICS) User: A user operating as a national body in support of NATO with national manning.

1.7 Remote User: A WIS user who is located at a significant distance from the area of other WIS access facilities.

1.8 Subscriber: An individual terminal instrument within a WIS user establishment (e.g. a telephone, data terminal or teleprinter) or the person using it.

1.8.1 Direct WIS Subscriber (DWS): A WIS subscriber who is authorized precedence ROUTINE and above, WIS hot lines, circuit switched telegraph or data services or other special WIS services and who is therefore required to be directly connected to an access switch.

1.8.2 Indirect WIS Subscriber (IWS): A WIS subscriber who is not authorized the service of DWS and therefore can be connected indirectly (e.g. through a PABX) to an access switch.

1.9 User Location: A geographically defined area where subscribers of one or more users are located sufficiently near each other that they can make use of the same local communications facilities. In some cases one WIS user may need WIS services at several user locations, such as peace, war and alternative HQs.

1.10 WIS Pick-up Point: A fixed circuit termination point where access circuits into the WIS network are avaliable for use by itinerant or mobile WIS users.

1.11 Transmision Node: A point where different transmission paths meet and where some demultiplexing may take place.

1.12 Nodal Switch: A nodal point with a switch where a significant amount of demodulation takes place.

1.13 Access Switch: A switch through which any one user location obtains access to the nodal network.

1.14 Packet Switch (PS): A switch performing termination, routing and through-connect of packet formatted data communication.

1.15 Message Switch (MS): A store-and-forward message switch which has the capability of receiving and distributing messages via the circuit switched telephone network as well as via the packet switched telephone network.

1.16 Access Link: A number of circuits connecting a user or user location with WIS nodal switch or MS/PS via one particular route.

1.17 User Multiple Access: The provision of two or more independent access links between a user or user location and different WIS nodes or MS/PS.

1.18 Internodal Link (trunk group): A number of circuits between two nodes.

1.19 Grade of Service: The expected percentage of calls failing to obtain a succesful connection in the busy hour (NOTE: A connection is considered succesful, when the calling subscriber obtains a ringing tone or a busy tone from the subscriber's local extension.).

1.20 Busy Hour: A period of one hour during which the peak traffic load corresponds to "alert" circumstances.

1.21 Connection Set-up Time: The time elapsing between reception of the last digit of the delivery number transmitted from the calling terminal by the parent switch of that terminal and the start of transmission of the ringing tone from the parent switch of the called terminal.

1.22 Off-Hook Service: A connection to a predetermined distant subscriber obtained automatically without dialling.

1.23 Hot-Line: An Off-hook service with precedence FLASH or IMMEDIATE.

1.24 Vital Traffic: Communication traffic with FLASH or FLASH OVERRIDE precedence originated by a primary user.

1.25 User-to-User Encryption: Cryptographic protection of a connection between the access switches or the secure voice switchboards of two different user locations without decryption at any intermediate point.

1.26 End-to-End Encryption: Cryptographic protection of a connection between the two subscriber instruments without decryption at any intermediate point.

1.27 Primary Targets: Those key command and control locations that the enemy would consider it cost-effective to destroy.

APPENDIX 2B

POST-CFE NATO CIS ARCHITECTURE

At the time of going to print, NATO was in the process of developing a new concept for its overall Communications and Information Systems (CIS) architecture which reflects NATO's changed role following the arms reduction agreements and profound political changes of recent years. Uppermost in the minds of planners is the need to serve a more flexible and mobile force at a time when defence budgets are being drastically reduced. This has led particularly to a critical re-apraisal of the need for all military-style system features including ECCM, nuclear hardening and communications security (COMSEC).

The new concept is still at an early stage of formulation and all except the highest-level aspects are still under discussion. It is generally agreed however that NATO CIS can be divided into two areas: the User Area consisting of end-instruments, local area networks and local exchanges and the Transport Area providing the means of interconnection between geographically separate locations. The Transport Area is divided into two segments, called the General Purpose Segment (GPS) and the Special Purpose Segment (SPS). The GPS will support the bulk of NATO communications in peacetime and low-level conflict situations. It will have few, if any, military features apart from COMSEC and will use commercial equipments and commercial standards wherever possible. A measure of survivability will be provided through richness of connectivity, and a variety of media will be used, including national military communication networks, PTT networks, NATO SATCOM and commercial SATCOM. The SPS, on the other hand, is seen as an austere and highly survivable overlay able to support low-capacity emergency communications in war when access to the GPS cannot be guaranteed. The benefits of a multi-media approach in this environment is recognised, and ECCM SATCOM is seen as an essential element.

This new view of the future NATO CIS architecture does not really invalidate the earlier "Goal Architecture" approach, but recognises that universal provision of ISDN services throughout NATO is probably unaffordable in the medium term. At the same time it accepts that practical emergency communications systems are likely to offer a very limited range of user services that are incompatible with ISDN standards, and that therefore it is sensible to treat them as more or less independent facilities.

CHAPTER 3

ISDN CONCEPTS AND STANDARDS

3.1 TRENDS IN TELECOMMUNICATIONS AND ISDN

The trends of telecommunications indicate an evolution toward "information management and movement technology". In this process one can witness the following facts:

- Increase in information productivity for users necessitated by the facts that the number of calls to be answered, databases to be consulted, information flow to be dealt with, media conversions executed, data storage and retrievals to be carried out etc. are all on the rise.
- Tele-industries, such as telemarketing, tele-education, telerecreation, teleconsultation, teleworking etc. are all information processing activities that are implemented by means of computer mediated communication forms between humans and/or via various man-machine dialogues.
- A more cost-effective and user friendly approach to the control and assignment of communication resources whereby the users' demands have a definite role in the negotiation of service attributes.
- Ubiquitous and universal delivery of communication services.

The role of ISDN in these evolutionary trends will be:

- In faster rendering of new services, since ISDN offers generic structures over which various services can be easily constructed,
- In offering the universal multiservice plugs over which data, voice, image, video, graphics services can be reached,
- In out-of-band signalling channel capacity which allows a very rich repertoire of call and service control messages, through which the customer can exercise control on the network resources,
- In migrating the network intelligence toward the user terminal, resulting in an easier network management and administration,
- In incorporating OSI principles down to the end user's access protocols, which makes possible evolutionary changes in terminals.

3.2 PRINCIPLES OF ISDN

CCITT Recommendation I.120 states the following general principles of ISDN:

- ISDN will support a wide range of voice and non-voice services in the same network. A key element of service integration in ISDN will be the provision of a range of services using a limited set of connection types and multipurpose user-network interface arrangements,
- ISDN will support a variety of applications, both switched and non-switched. Switched connections in ISDN include both circuit-switched and packet-switched connections and their tandeming.
- New services introduced in the ISDN must be compatible, as far as practicable, with 64 kbit/s switched digital connections,
- An ISDN will contain intelligence for the purposes of providing service features, maintenance and network management functions. Part of this intelligence may reside within terminals, especially for new services where network intelligence alone may not suffice,
- A layered protocol structure should be used for the specification of the access to an ISDN. Access from a user to ISDN resources may vary depending upon the service required and upon the status of implementation.
- A variety of ISDNs is possible depending upon the state of technology, and to the needs and existing equipment of the customer base.

On the user premises ISDN has the following implications:

- The basic access for a user provides two 64 kbit/s basic channels (B channels) and one 16 kbit/s signalling channel (D channel) in each direction; the connections established over the two 64 kbit/s channels can be to different destinations. The primary rate access comprises twenty four or thirty 64 kbit/s channels and one 64 kbit/s signalling channel. Basic and primary rate access can be provided on the copper wire pairs of existing subscriber lines,
- Subscribers are assigned a single directory number over which all the individual voice, text, data and image communication services can be reached,
- The universal user-network interface defined for the ISDN allows different terminals to be connected by a standard "communication socket", hence standard user procedures for call set-up and clear-down are used,
- The network not only establishes connections between the user stations, but also between the compatible terminals in the user's premises. These user terminals can be connected in a star or bus configuration,
- Subscribers in existing networks, as in the analog switched telephone network or in a telex network can be reached via ISDN by means of interworking units.

ISDN benefits accruing on the side of the customer are:

- Cost savings, since the user does not have to buy multiple services to meet multiple needs, and furthermore the same services will be offered at a lower cost due to the economies of scale,

- Flexibility, whereby the user tailors the network resources according to his actual needs,
- Product diversity and better and wider availability of services due to a competitive multivendor environment,
- Reduced risk of obsolescence due to open-ended evolutionary nature of products.

ISDN benefits accruing on the side of the network provider are:

- Larger potential markets since standards lead to universality,
- Smooth and compatible technical evolution and innovation without service disruptions,
- More flexible and extensive management and control of network resources and customer premises equipment,
- Enhanced services providers such as videotext, information retrieval, or transactions services providers can be expected to thrive on this network infrastructure and on the simplified network access.

Finally the benefits accruing to the manufacturers are:

- the assurance of broad potential markets,
- possibility to exploit specialized market niches,
- economies of scale.

EVOLUTION OF ISDN

Since all telecommunication infrastructures are very capital intensive, the backward compatibility of new technologies becomes an imperative. In this context ISDN also will evolve in a backward compatible manner, from IDN (Integrated Digital Network). The IDN was initially conceived to improve the existing telephone network, hence the "I" in IDN is limited to the integration of digital switching and transmission. However the "I" in ISDN has a much broader scope and it involves the integration of customer digital access and voice and data services. An important aspect of the transition from IDN to ISDN is that the end-to-end digital connectivity is extended to user-to-user.

The following remarks are relevant in assessing the evolution of ISDN:

- The IDN technology will form the foundation for the services provided by ISDN.
- The evolution of ISDN will last a few decades, within which there will be a migration toward new technical bases, and new applications. However existing networks will coexist for a long to come with ISDNs, which in turn implicates the deployment of interworking units. The coexistence of ISDN and non-ISDN services on an ISDN switch is illustrated in Fig.3.1.
- Existing networks (e.g., packet switching) and user-network arrangements (modems over analog networks) will continue to be used and reached at via a set of standard interfaces. As new technologies evolve, then new interfaces may have to be defined.
- The 64 kbit/s connection may loose its preponderance in time, as lower or much higher rate connections may be commonplace.

ACCESS of ISDN SUBSCRIBERS TO NETWORKS

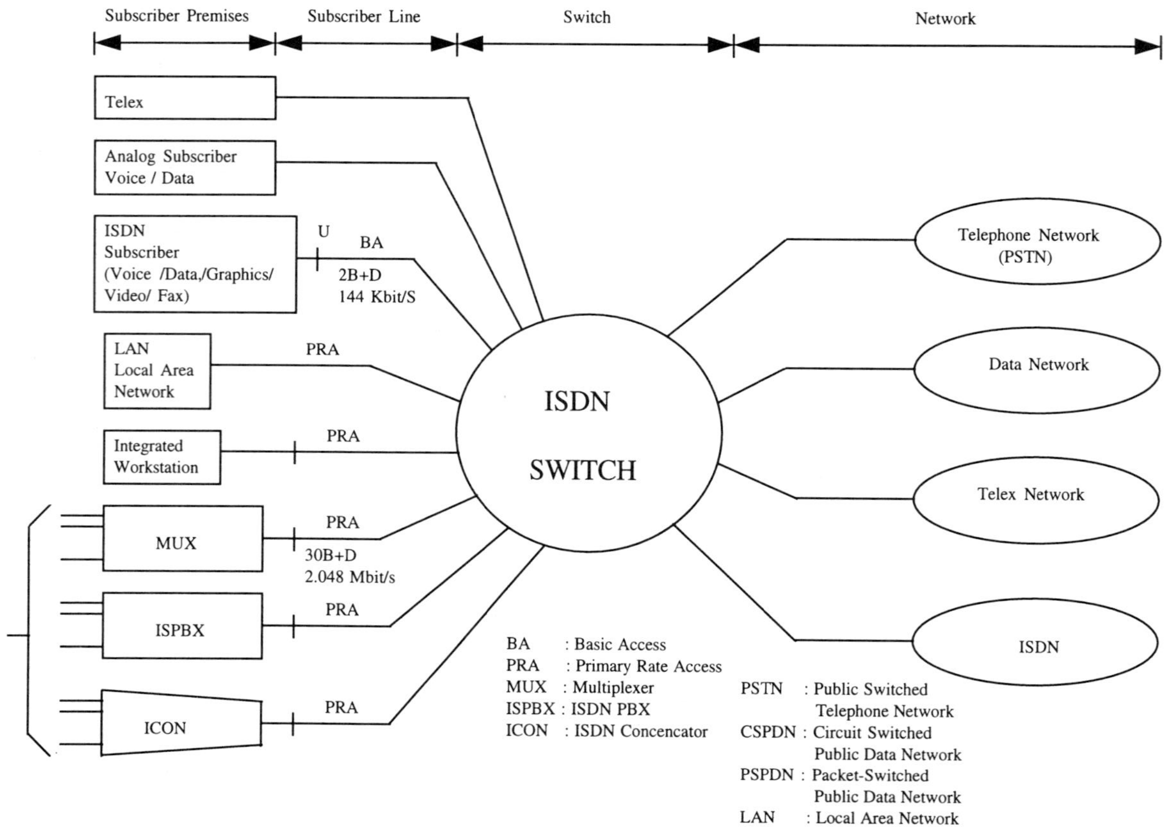

Figure 3.1 User Access to Networks Via an ISDN Switch

3.3 ISDN STANDARDIZATION

International standards are of fundamental importance in the development and realization of telecommunication infrastructures, and for wide-scale deployment and acceptance of the services they offer, both for the professional customer with specialized needs and the many small users with routine requirements. Standardization in the context of ISDN will imply that ISDN-standard equipment can be moved from one location to another location and yet be plugged into the network anywhere. Furthemore customers will be able to chose functionally equivalent equipment in a competitive multivendor environment ISDN standardization is primarily concerned with international interfaces, i.e., interexchange signalling, user-network interfaces, services and supplementary attributes. CCITT, I-series recommendations detail the various aspects of ISDN. Noteworthy among them are:

- I.100 General Structure and Terminology, ISDN Network Capabilities and Characteristics of ISDN Communication Services,
- I.200 Service Capabilities (Bearer Services, Teleservice, Supplementary Services),
- I.300 Overall Network Aspects and Functions, Protocol Reference Models, Terminal Selection, Connection Types,
- I.400 User-Network Interface Principles, Support of Existing X- and V-Series Devices,
  - I.43x User-Network Interface: Physical Layer,
  - I.44x User-Network Interface: Data Link Layer,
  - I.45x User-Network Interface: Network Layer,
  - I.46x Rate Adaptation and Multiplexing,
- I.500 Internetwork Interfaces, Between ISDNs and Between ISDN and Telephone/Dedicated Networks,
- I.600 Maintenance Principles.

Other relevant recommendations from Q-, V-, X-, T-, G-, and E-series are listed below:

Q.500 Digital Exchanges,
Q.700 Interexchange Signalling (Switching System No. 7),
Q.920-Q.930 User-Network Interface (the same as I.440 and I.450 above),
Q.71-Q.99, Q.932 Supplementary Services.

X.30-X.31 Support of X-Series Data Terminals by an ISDN,
V.110 Support of V-Series Data Terminals by an ISDN,
E.164 Numbering Plan for the ISDN Era,
T.90 ISDN Telematic Terminals,
G.700, G.800, G.900 Digital Transmission.

3.3.1 Service Capabilities

Services are the means and resources that the telecommunication network provides to the users. The services are subdivided into two categories, namely, bearer services and teleservices.

3.3.1.1 Bearer Services

The bearer services describe the transportation of information between locations, in other words provision of a transparent transmission and switching medium to enable communication. The technical specifications of these services cover the transmission functions of OSI reference model layers 1 to 3. In the offering of bearer services, such higher layer OSI functions as terminal compatibility checking remains in the responsibility of the user. Examples of bearer services are:.

a) Circuit mode, 64 kbit/s, 8 kHz structured, unrestricted,
b) Circuit mode, 2 Mbit/s, 8 kHz, structured, unrestricted and clock transparent,
c) Bidirectional, circuit-mode 64 kbit/s, 8 kHz structured channel usable for speech. This is a nontransparent channel and where typical of the signal processing operations are μ-law/A-law conversion, speech transmission over analog sections and DSI (Digital Speech Interpolation) type speech compression whenever satellite links are involved.
d) Bidirectional, circuit-mode, 64 kbit/s, 8 kHz structured channel usable for 3.1 kHz audio information transfer. This connection is needed for modem signals transmitted in the voice frequency band as the modem outputs should avoid devices such as speech compressors and echo suppressors intended for speech signals.
e) Circuit mode, 384 kbit/s, 8 kHz, structured, unrestricted and clock transparent,
f) Circuit-mode 2x64 kbit/s circuit mode unrestricted,
g) D channel packet mode, virtual circuit and permanent virtual circuit,
h) B channel packet mode, virtual circuit and permanent virtual circuit,
i) Connectionless packet mode over a B or D channel.

The two bearer services (cases c and d) can be expected to be phased out eventually, as,inall probability, analog internodal links will not exist necessitating speech conversion operations. Also A/μ-law conversion will take place at the gateways, and with end-to-end digital connectivity speech digitization and compression will take place at the terminal area.

3.3.1.2 Teleservices

Teleservices are services for user-to-user and user-to-host communication, including the specification of the communication functions of the terminals. The communication functions comprise not only all the transmission functions and communication protocols of OSI layers 1-3, but also, if necessary, higher layer functions such as communication oriented editing, presentation and reproduction of information. In other words teleservices ensure that terminals dedicated to a particular service are compatible, e.g., as far as structure of the user information or the character set employed is concerned. The compatibility checking may focus on the higher layer or lower layer functionality of the terminal. The lower layer characteristics could be user rate, rate adaptation mechanism used, parity information, asynchronous or synchronous operation; the higher layer functionality could be telephony, Group 3 or 4 fax, teletex, videophone etc. Examples of teleservices are:

- Telephony, provides the ability for two-way real-time speech conversation,
- Videophone, provides the ability for two-way real-time near synchronous speech and video communication,
- Teletex, provides the ability of exchanging information in the form of documents containing Teletex coded information,
- Videotex, provides the ability to access a videotex center for retrieval and mailbox functions for text and graphic information,
- Telex, provides the user with the ability for interactive text communication,
- Telefax 4, provides the ability of exchanging information in the form of documents containing facsimile coded information,
- Mixed mode, provides the ability of exchanging combined text and facsimile information,

- Electronic directory, provides a distributed database application about persons, organizations, systems and applications, e.g., for addresses and telephone numbers,
- Transactions, provides the ability to read and/or write some information, that is data collection, on a remote device, such as alarms, telemetering,
- Electronic data interchange, provides the ability of exchanging, filling in formatted and computer generated documents.

3.3.1.3 Service Attributes and Supplementary Services

The service attributes of a communication service can be dealt with in two classes:

1) User attributes,
2) Carrier attributes.

The user attributes can be further classified as general attributes, basic attributes and supplementary services.

General Attributes describe the ISDN features relating to the subscriber side, and usable for all sevices, such as:

- Characteristics of user connection device,
- Type of connection (switched or permanent),
- Establishment of communication, i.e., on demand, by reservation or on a permanent basis.

Basic Service Attributes describe the service configurations, such as:

- Type of switching, i.e., circuit or packet,
- Bit rate for circuit switching and throughput for packet switching,
- Information transfer capability, e.g., A/μ-law speech, 3.1 kHz audio, video, unrestricted digital information etc.,
- Communication configuration, i.e., point-to-point, multipoint, broadcast,
- Symmetry, that is rate and direction between two or more points, which could be unidirectional, bidirectional symmetric, bidirectional asymmetric,
- Service quality parameters, e.g., service reliability and service availability,
- Performance such as slip rate, bit error probability.

Supplementary Services are defined as those services that modify and augment the basic attributes and characteristics of a service. They are not offered on a stand-alone basis, but they accompany bearer and tele-services. The supplementary services can be grouped as follows:

- Number Identification Supplementary Services, e.g., Direct-dialing-in, Subaddressing, Calling Line Identification,
- Call Offering Supplementary Services, e.g., Call Transfer, Call Forwarding, Call Deflection, Line Hunting,
- Call Completion Supplementary Services, e.g., Call waiting, Call Hold, Completion of Calls to Busy Subscribers,
- Multiparty Supplementary Services, Conference Calling, Three Party Service,
- Community of Interest Supplementary Services, e.g., Closed User Group, Private Numbering Plan,
- Information Transfer Supplementary Service, e.g., User-to-User Signaling.

More detailed descriptions of some of these services are as follows:

- "Closed User Groups" (CUG), that is the possibility for a group of users to intercommunicate only amongst themselves with either outgoing calls barred and/or incoming calls barred. For example, subscribers with special NBSV (Narrow Band Secure Voice) equipment can be accommodated in WIS through networkwide CUG classes.
- "Direct Dialing In" (DDI) is a facility to enable calling directly to a subscriber in a PABX without operator intervention. The access switch to which the PABX is connected must store and forward the extension number for further automatic dialling to set-up the connection. For a closed numbering scheme the transit network needs to know, and has its numbering scheme integrated with all PABXs in the network.
- "Abbreviated Dialing" is the possibility to make a call by dialing a short code instead of the full subscriber number.
- "Hot-Line" and "Delayed Hot Line" where the call set-up process for a predefined subscriber number is initiated automatically either immediately or after a fixed delay.
- "Conference Facilities" for both secure and non-secure calls. The conference set-up procedures can be of two types, i.e.,"add-on conference" through progressive dialing and "preprogrammed conference".
- "Operator Positions" to handle operator assisted calls, to provide trunk offering, to set-up special connections, to intercept a call with notification to the called PABX subscriber, to complete the call when all in-dialing lines are busy.
- "Indication and Warning Signals" to indicate to the subscribers such conditions as network congestion, network not accessible for technical or operational reasons. Also "Announcements" to calling subscribers to provide information not covered by the tones and cadences. These announcements should comply with CCITT Rec. E.183.
- "Call Waiting" which enables a user to wait (camp) on a busy subscriber until he is free. The called and busy subscriber will be notified by an audible "call waiting" tone indicating a subscriber is waiting on him. The user is free to continue or to terminate his ongoing call.
- "Group Number" where one number may be allocated to more than one subscriber line. Various methods for hunting and diversion will be defined for such groups.
- "Calling Line Identification" enabling a called user to be informed of the address of the calling party for incoming calls.
- "Called Line Identification" is a supplementary service for circuit switched connections, which enables the calling user to be informed for outgoing calls of the actual identity of the called user to which he has been connected. Note that these last two supplementary services necessitate subscriber loops and terminal equipment with ISDN or similar features.
- "Call Transfer" to transfer a call to another local subscriber and "Transfer on no Answer "that make possible automatic call transfer to a predetermined alternative number or to the operator if the normal number does not answer. This service may be accompanied by such features as "Call Diversion to Announcements".

- "Stockbroker's Call" where the calling subscriber can alternate between two independently called subscribers without disconnecting any of them during suspension intervals.

3.3.1.4 Interactive and Distribution Services

Another CCITT classification covering both the narrowband and broadband services distinguishes two categories:

1) Interactive services,
2) Distribution services.

The interactive services are further subdivided into:

a) Conversational services which provide the means for bidirectional dialog communication with bidirectional, real-time end-to-end information transfer user-to-user or user-to-host. Examples are telephony, audio conference, teletex, telefax, data transmission. For these services ISDN has to provide bidirectional and unrestricted bearer capability, without any store-and-forward capability.

b) Messaging services provide user-to-user communication by means of store-and-forward capability, mailbox facilities, and message handling functions including editing, processing and conversion options.

c) Retrieval services provide access to information stored in information centers. These services can be offered both by the network provider or as a value added service by third party providers. A typical example is videotex where documents including text, data, graphics, audio information, image and video can be retrieved.

Distribution services are intended to distribute a stream of information from a central source to an unlimited number of authorized users in the network. The rendition of distribution services can be:

a) Without user-individual presentation control, i.e., broadcast services such as television and sound programs. The user can access to the flow of information without being able to control the start and the termination of it.

b) With user-individual presentation control, i.e., broadcast videography, teletext, video-on-demand. The user can individually select the distributed information and the beginning of the stream as well.

Note that while most of the interactive services are encompassed by the ISDN (also referred to as narrowband ISDN), distribution services, especially the video variety, necessitate broadband ISDN (B-ISDN).

3.3.2 Network Aspects and Functions

3.3.2.1 Reference Models

A reference configuration has been defined for the user-network interface as shown in Fig. 3.3.1. In this model the following reference points and functional units have been defined:

- NT1: Unit provides the network termination functions necessary for the network operations. These functions pertain to the proper physical and electromagnetic termination of the network connection, and tasks such as performance monitoring, timing, power transfer, physical layer multiplexing, and transmission line termination.

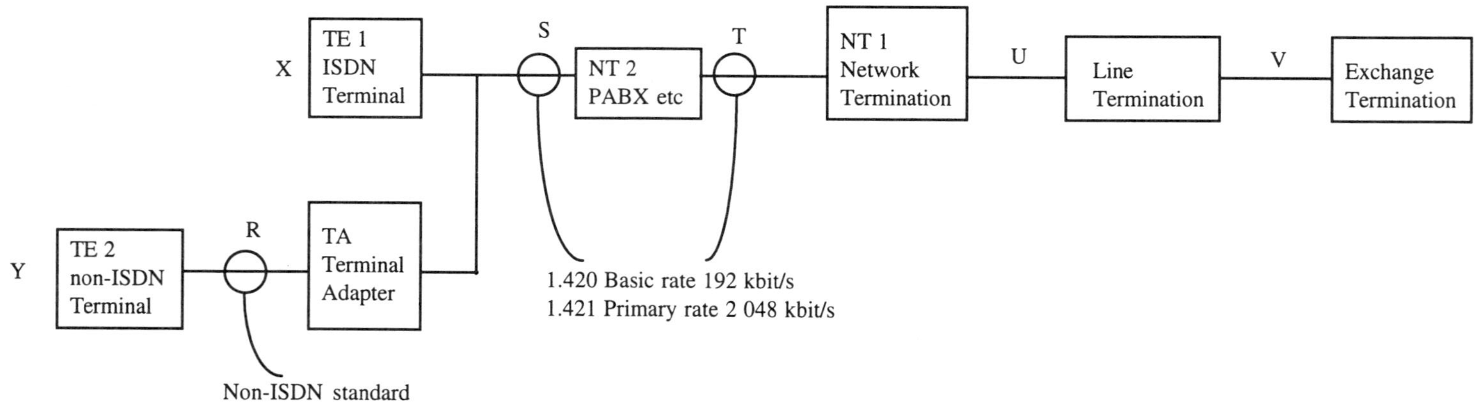

Notes :

(1) Bearer services are provided at S and/or T
(2) Teleservices are provided at X and/or Y
(3) Other CCITT services are provided at R

Figure 3.2 **ISDN Reference Model. Note That Bearer Services are Provided at S/T Reference Point, Teleservices are Provided at X/Y Point.**

- NT2: Network Termination-2 contains layer 2 and 3 protocol handling, switching, concentration, and maintenance functions. NT2 is typically a more sophisticated device as compared to NT1, such as a PABX, a terminal controller, or a LAN. The NT2 device will typically include all the functions of the NT1 as well.
- TE1: The TE1 box represents terminal equipment which connects to the network at the S reference point which is an ISDN interface. It provides the functions of access protocol handling, maintenance functions and interface functions.
- TE2: This functional unit represents a non-ISDN terminal such as an analog dial pulse telephone, and it performs the normal TE functions by means of a terminal adapter unit. TE2 connects to the TA at the R reference point.
- TA represents the Terminal Adapter unit and provides the user-network access functions for the TE2 unit.
- S-interface is the normal reference point at which the ISDN user connects to the network, while if an ISDN PABX or LANdevice is involved, then they connect to the network via the T reference point.
- U reference point is the interface between the NT1 equipment and the ISDN exchange line termination.

3.3.2.2 Reference Configurations

The reference model illustrated in Fig. 3.3 can assume different configurations in the implementations of ISDN. For example the NT2 device, which is actually a PBX or LAN will usually incorporate the NT1 functions as well, hence the "NT1 + NT2" configuration (Fig. 3.3a). Similarly the NT2 device will also possess a pool of ISDN terminal adaptors (TAs) to cater for the non-ISDN devices that terminate with R interface. It is more pragmatic to have an ISDN PBX (ISPBX) that has the terminal adaptation functions rather than trying to retrofit all the individual non-ISDN devices with adaptors (Fig. 3.3b). For ISDN at home or in small businesses there will not be an NT2 device, but the ISDN terminal will be connected to the NT1 device (Fig. 3.3c). Finally an example of a private network acting as an NT2 is shown in (Fig. 3.3d).

3.3.2.3 Channel Types and Structures

The channel types available in ISDN are as follows:

- B-Channel: The B-channel operates at 64 kbit/s and carries user information during the active phase of a call. The B-channel itself does not impose any restrictions on the type and format of information, which is solely determined by the nature of user equipment and the type of call.
- D-Channel: The D-channel operates at 16 kbit/s for basic rate interfaces and 64 kbit/s for primary rate interfaces. This channel carries signalling information for all calls, and in addition may carry packetized data or telemetry information.
- H-Channel: This is a family of channels consisting of certain integer bundles of B-channels, e.g., H0 formed with 6B channels with a total bandwidth of 384 kbit/s, H12 (24 B-channels or 1536 kbit/s as in USA and Japan) and H12 (30 B-channels or 1920 kbit/s as in Europe).

The channel structures are as follows:

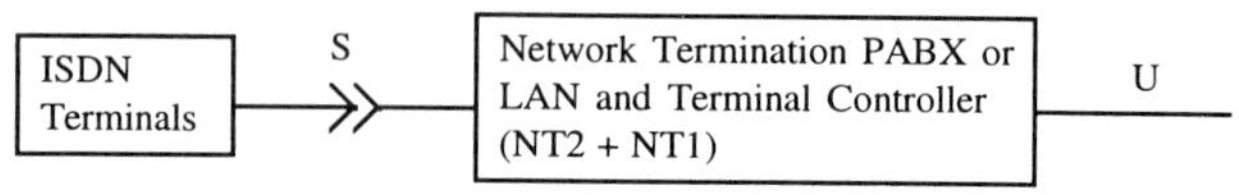

3a : Configuration where NT1 and NT2 functions are performed in one device

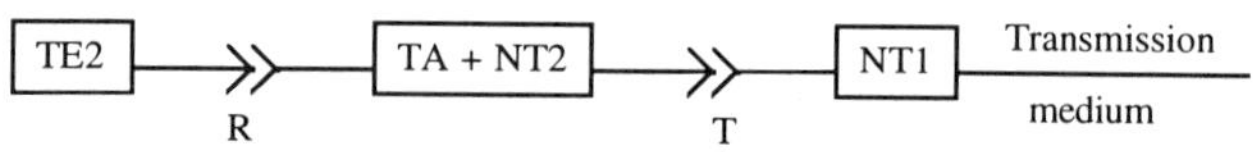

3b : Configuration where NT2 incorporates bank of terminal adaptors

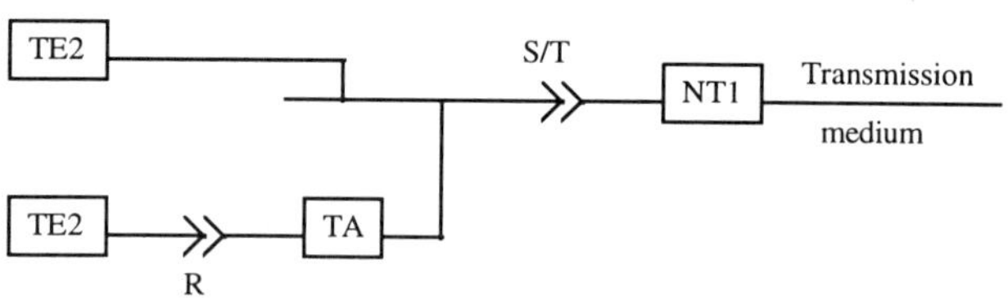

3c : Customerpremises equipment directly connected to an NT1

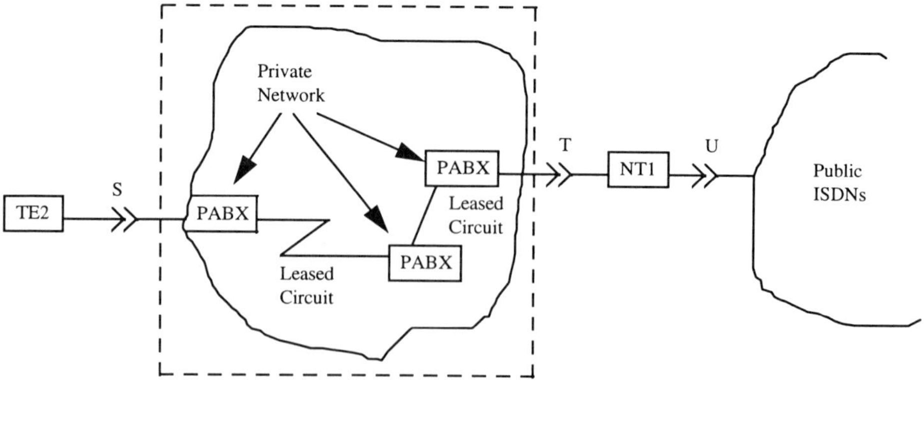

3d : Private network as an NT2

Figure 3.3 Examples of Reference Configurations

- Basic Rate Channel Structure: The basic rate interface has 144 kbit/s information carrying capacity, which is split in a 2B+D structure. On a particular interface, however, one or both of B-channels may remain idle. Whether the B channels are active or not, on the user premise to ISDN switch connection, the U-interface continues to operate at about 160 kbit/s (CCITT G.961).
- Primary Rate Channel Structure: Because of existing differences in digital hierarchies there exist several variants of primary rate structure. The two commonest ones are 23B + D structure at 1544 kbit/s and 30B + D running at 1920 kbit/s. In some cases more than one primary rate interface connects between equipment. For example three primary rate interfaces in the 31B + 31B + 30B + D manner may connect a PABX to the access switch for a total of 92 channels. Similarly, the user may opt to use nx64 kbit/s ($2 \leq n \leq 23$ or 30) channel capacity, e.g., for such applications as videoconferencing, multimedia and imaging. This option is called multi-rate ISDN. The bandwidth of a call at increments of 64 kbit/s over the Primary Rate Interface is determined by the originating end user at the call setup.

3.3.3 User-Network Interfaces (UNI)

3.3.3.1 UNI at the Physical Layer

The ISDN user-network interface physical layer protocol takes place on the S/T reference point and is implemented on a balanced, metallic, bidirectional transmission medium supporting 192 kbit/s rates. This rate is made up of the 2B + D channel structure occupying 144 kbit/s and an additional 48 kbit/s for control purposes such as synchronization and maintenance. The four-wire S/T bus (two wires in the receive and two wires in the send direction) is designed to operate on the twisted copper pair as may be provided for traditional analog sources.

The connection types in the S/T interface can be point-to-point or point-to-multipoint. In the point-to-point mode the TE can be as far as 1 km from the NT. In the point-to-multipoint mode up to 8 different terminals can be connected, though the bus is then limited to 200 m due to differential timing constraints. The terminals are high impedance devices (2500 ohms) so that they do not load the bus.

Power is provided to the devices by the power supply available within the NT, which is itself fed by the mains. There is also a provision for power feeding from the network for at least one telephone device in case of local mains failure.

Layer 1 is responsible for transferring information between terminals and the NT1, and hence must provide the following functions:

- Must support for each direction of transmission two independent 64 kbit/s channels,
- Must support for each direction of transmission a 16 kbit/s signalling channel,
- Must ensure in the point-to-multipoint configuration that the access contention on the D channel is resolved,
- Activate the terminal devices, that is their functioning in the normal power mode, and deactivation, whereby the NT and TE equipment are placed in the low power mode, both to save power and to reduce radiation and crosstalk effects.

Binary organization of Layer 1 frame: The structure of Layer 1 frame is shown in Fig. 3.4. A frame is 48 bits long and lasts 250 us, so as to make up 192 kbit/s total rate. The

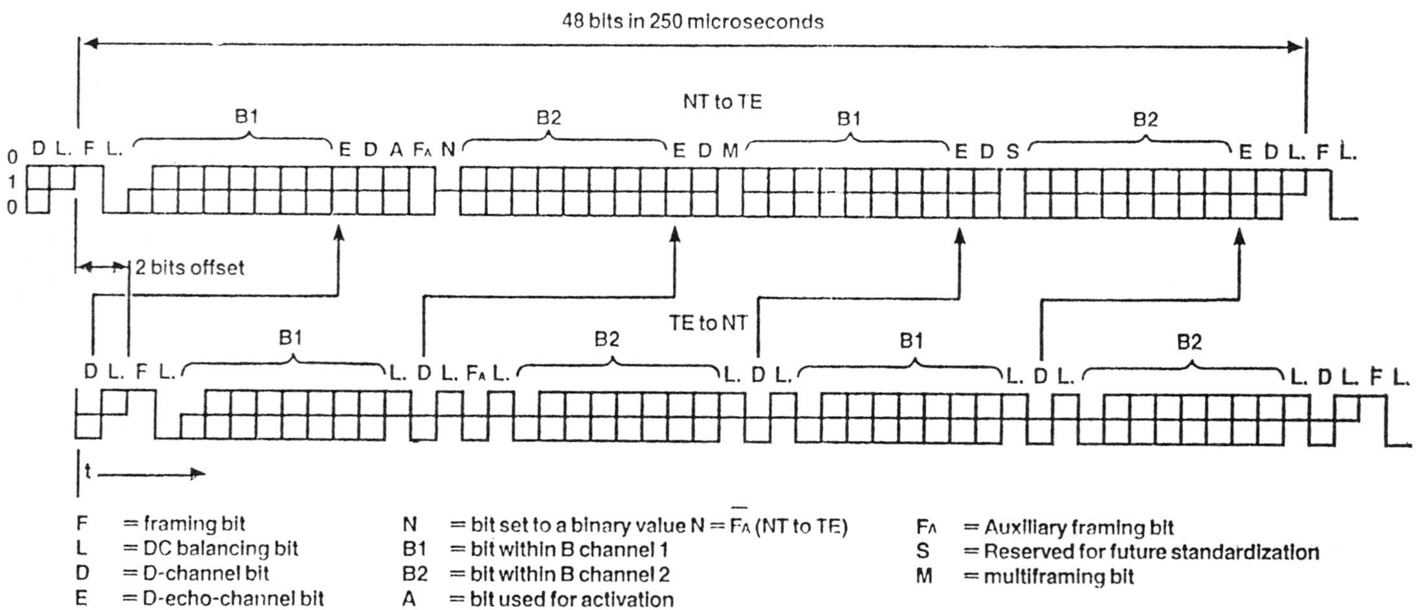

Figure 3.4 **Layer 1 Frame Structure. Note That the Nominal Two Bit Offset is as Seen From the TE. The Corresponding Offset at the NT May be Greater Due to Configuration or Delay in the Interface Cable.**

transmission employs AMI (Alternate Mark Inversion) line code. Beside the B- and D-channel bits, each frame has several specialized bit groups such as:

- L bits for balancing the frame so that no d.c. builds up on the line,
- F/L pair used for frame alignment procedure, which is based on AMI violations,
- A bit used in the activation procedure to indicate to the terminals that the system is in synchronization.

The contention between different terminals for access on the D-channel is based on the fact that the information to be transmitted consist of Layer 2 frames delimited by the 01111110 flags. The interframe time fill consists of binary 1s which are represented by zero voltage as generated by terminals going to high impedance. To access the D-channel a terminal looks for a predetermined number of consecutive 1s. This number can be incremented or decremented to set the priority level of the terminals.

3.3.3.2 UNI at the Data Link Layer

CCITT I.440 Layer 2 Recommendation describes the high level data link procedures applicable in ISDN, called Link Access Protocol on the D-channel (LAPD). The objective of LAPD is to provide a secure, error-free connection between two end-points so as to reliably transport Layer 3 messages. LAPD detects and handles lost or duplicated frames also. LAPD itself is a derivative of the LAPB protocol commonly used in the packet network, the main distinction being that LAPD allows multiplexing of Layer 2 connections on the same physical connection. In fact it is this multiplexing that allows the multipoint terminal configurations on the customer's S-bus.

LAPD protocol operations are based on the exchange of Layer 2 frames, which are groups of bytes delimited by flags, and possessing an address field, a control field, a frame check sequence and an information field as shown in Fig. 3.5. The Layer 2 address has significance only between the two end points of the LAPD connection, in other words it is not part of the network addressing. The LAPD address consists of two parts, namely the Service Access Point Identifier (SAPI) and TEI (Terminal Identifier).

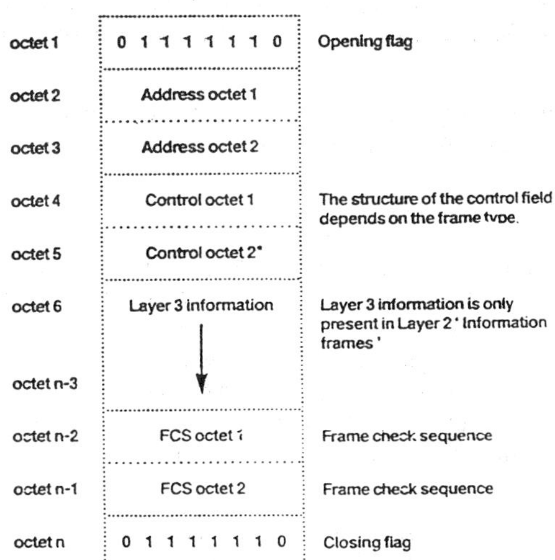

Figure 3.5 Layer 2 Frame Structure

*** Note That the Second Octet of the Control Field May Not Always be Present**

The SAPI is used to identify the service that the signalling frame is intended for, e.g., packet switching. The values of SAPI are fixed for given services. The advantage of using a service identifier is that the network can handle the signalling associated with the service in the same module. The TEI takes values associated with the customer's terminals on the S-bus. The TEI is either assigned by the TEI management entity (automatically) or by the user; in either case no two TEIs can be the same. Thus a unique TEI combined with a service specific SAPI results in a large number of terminal possibilities using various services. The TEI manager has the ability to allocate, remove, check and verify the TEIs that are in use at the customer's premises. The manager, in order to communicate with terminals which have not yet have a TEI assigned, transmits all management frames on a broadcast TEI.

The function of Layer 2 has been defined as the task of delivering Layer 3 frames error-free and in sequence, across Layer 1 interface. The Layer 2 operation starts with a Layer 3 request to establish a call, which prompts Layer 2 to demand Layer 1 to initiate the start-up procedure. Afterwards Layer 2 initializes itself through the SABME (Set Asynchronous Balanced Mode Extended) procedure. The SABME is conducted between two peer processes, i.e., one at the terminal site and the other at the network end, by means of unnumbered frames. Initialization guarantees that the correct sequence numbers are used in processes on both ends.

Layer 2 protocol is designed to maintain the correct sequence of information in the face of many different types of error, such as:

- Frames corrupted by errors due to noise are controlled via the Frame Check Sequence (FCS) values and consequently discarded. The FCS used is able to detect all odd errors and a very large percentage of even errors.
- Lost frames in the transmit direction are detected by a timer mechanism and preset values of timeout intervals. Since the lost frame can be either a response or a command frame, the distinction is made by soliciting an acknowledgment that reveals the state variables at the receiver.
- Frames received out of sequence denote a lost frame in the receive direction. The sender rejects the incoming frame, meanwhile announcing the correct state variables.
- Frames lost repetitively, i.e., more than three consecutive times, cause the re-establishnent of the LAP under the assumption that the connection has failed.

3.3.3.3 UNI at the Network Layer

ISDN Layer 3 protocol has the duty of establishing, modifying and terminating network connections as defined in CCITT Rec. I.450. In implementing these functions, Layer 3 employs the services of the Data Link Layer for reliable transport of messages. Layer 3 performs the functions of exchanging messages between the called and calling subscriber /switches, and extract information from the data base about the called subscriber. Furthermore it has to multiplex network connections, to relay information subscriber-to-subscriber or network-to subscriber, to check the compatibility between user demands and network services.

Layer 3 protocol is effected by means of structured messages and message sequences (Fig. 3.6). It is worthwhile to note that in these messages:

- The first octet contains a protocol discriminator, which gives the capability to simultaneously support several communication protocols,

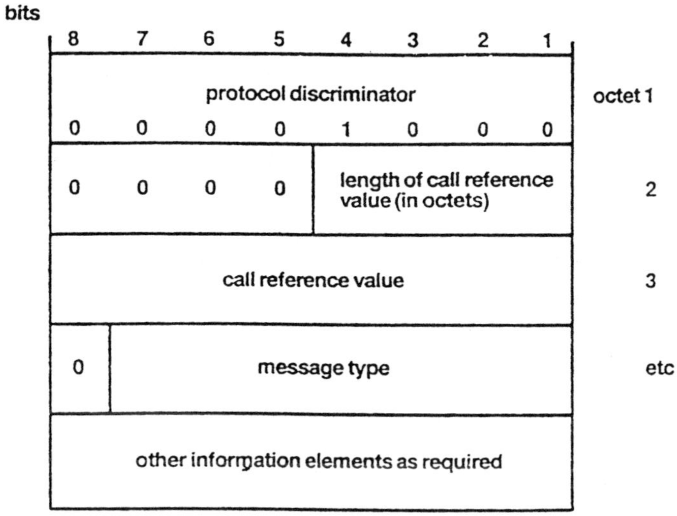

Figure 3.6 ISDN Signalling Message Structure

- The call reference value identifies a call independently of the communications channel on which it is supported,
- The message type code in the fourth octet defines the scope of the message, such as call establishment messages, call progress messages, call clearing messages, and miscellaneous other messages. For example, typical of call establishment messages are alerting, call proceeding, connect, connect acknowledgment, progress, setup and setup acknowledgement.

3.3.3.4 ISDN Primary Rate Interface

Interconnectivity of PBXs, LANs and multimedia terminals involving larger bandwidths will be effected by means of the Primary Rate Interface operating at 1,544 or 2,048 kbit/s rates. The D signalling channel is 64 kbit/s and occupies slot 24 and 16, respectively, in the 1,544 and 2,048 kbit/s schemes. Sometimes one D channel carries signalling information for more than one PRI, in which case some PRIs can use that extra B channel. Primary Rate Interface resembles mostly the Basic Rate Interface, but with the following differences:

- The PRI connection is simply between the network and a single device like PBX. In other words there is no provision for multipoint connections at the customer's premises. The user can, of course, route the individual PRI channels to whatever destination is required within its premises.
- The network does not activate or deactivate the link in order to economize power.
- The PRI supports both switched and non-switched connections.

3.4 TERMINAL ADAPTATION

ISDN will suport the well established non-ISDN terminals, such as in the X- and V-series Recommendations. The support given to the X-terminals, V-terminals and packet mode terminals are discussed in the sequel. The terminal adaptation details are covered in CCITT Rec. I.460 series.

3.4.1 Support of X-Series Terminals

A terminal adaptor designed in compliance with Rec. I.461 encompass three sets of functions:

- Mapping of X.21 call control procedure to the D-channel protocol,
- Bit rate adaptation from the X.1 rates to ISDN basic access rate,
- Ready for data alignment in the 64 kbit/s channel.

Call Establishment and Clearing Procedures: The mapping of the X.21 signalling events to the D-channel messaging events are detailed in the I.461 Recs. The following remarks are to be made:

- On call setup the TA accumulates all X.21 messages, before sending them en bloc to the network,
- On call clearing, with the clear request from the X.21 the TA initites the release procedures both on the D and B channels.

Rate Adaptation: The terminals running at 8 kbit/s, 16 kbit/s and 32 kbit/s are directly rate adapted into 64 kbit/s streams by bit stuffing. For example the bits of a 8 kbit/s source take place in the first position of the bytes of a 64 kbit stream, while the rest of the bits of these bytes are filled with stuffing 1s. The terminals running at rates other than the above, e.g., 4.8, 9.6 etc., are adapted in two stages. In the first stage the source is adapted to the rate just higher , i.e., 8, 16 or 32 by repetitionof the information bits and bit stuffing; in the second stage the 8, 16 or 32 kbit/s stream is converted to a 64 kbit/s as described above. Thus, for example, terminals at the X.1 rates of 0.6, 2.4, or 4.8 are rate adapted first to 8 kbit/s, and then to 64 kbit/s. For rates higher than 32 kbit/s, i.e., 48 or 56 kbit/s, the source is rate adapted to 64 kbit/s on one step directly.

Terminal Multiplexing: The 8, 16, and 32 kbit/s sources are multiplexed to the B channel rate by bit interleaving. Any other bit positions are stuffed with 1s. For example the bits of a 16 kbit/s source are placed in the bit positions either (1,2), or (3,4), or (5,6) or (7,8) of the bytes. For sources other than 8, 16, or 32 kbit/s, the source is rate adapted to one of the standard rates, and then multiplexed.

For packet-mode services the stuffing is done not with 1s, but by repeating the HDLC flag sequence. On the other hand, the multiplexing function is automatically provided by the Layer 3 virtual circuit mechanisms (multiple virtual circuits).

Ready For Data Alignment: Procedures between TAs and X.21 terminals are based on the exchange of status information provided at the X.21 interface for call control.

3.4.2 Support of V-Series Data Terminals

Non-voice terminals with interfaces conforming to the CCITT V Series Recommendations for use in analog telephone networks can be adapted to the S/T interface of the ISDN user access via two alternate techniques:

1) Conversion to the tip/ring interface of the telephone network: In this case the decadic pulses or the touch-tone signals are converted to the D-channel protocol. The terminal adaptor performs the A/D conversion of the modem signals. In this mode the network interworking capabilities of the conventional voice telephone can also be used for data traffic.

2) V-Series Interface: The terminal adaptor (TA-V), as in Rec. V.110, performs a direct conversion of the V.24 interface signals to the S-interface using purely digital signals, thus bypassing the need for modems. There exist also mapping functions to convert automatic calling/answering procedures of CCITT Recs. V.25 and V.25b to the ISDN D-channel protocol. In summary the TA-V adaptor performs the functions of:

   a) Conversion of the electrical, mechanical, procedural and functional characteristics of the V-series interface to those of the S/T interface of the ISDN,
   b) Bit rate adaption of the V-series user data to the 64 kbit/s rate;
   c) End-to-end alignment of entry to and exit from the data transfer phase.

3.4.3 Support of Packet Mode Terminals

The accommodation of packet switching terminals into the ISDN will be as follows:

1) ISDN will incorporate internal X.25 switching functions, called packet handlers (PH). The packet handlers can be colocated within the ISDN exchange, or can be located remotely from the ISDN circuit-switching exchanges. In this case the packet handlers are connected by 64 kbit/s ISDN connections using packet exchange X.75 protocol. In essence this is like an overlay network for packet switching. ISDN procedures for packet switching services are given in Rec. X.31, which apply both to the future ISDN terminals to be directly attached to S/R bus and present terminals TE2 adapted via TA.

2) ISDN will give a circuit switched access to a PSPDN (Packet Switched Public Data Network) according to X.32 Rec., whereby all packet switching services and supplementary services will be performed by this "external network".

With the ISDN virtual circuit bearer method, a packet terminal at the ISDN 2B+D access can either use the B-channel or the D-channel for end-to-end exchange of X.25 control information and exchange of data packets with a PH. The choice of the B or D channel is either preset as part of the customer access profile or is made by the network individually for each incoming call.

3.5 ISDN INTERWORKING

Since there will probably never exist a single homogeneous ISDN network throughout the world, there will always be a definite need for interworking functions. The main interworking functions between the ISDN and another network, e.g., analog telephone network, or between ISDN varieties, e.g., two national implementations, are as follows:

- Interworking of numbering plans as discussed in Section 3.6
- Checking if the non-ISDN network can adequately meet the service demands of the calling party,
- Ensuring service and connection compatibility, that is both low-level and high-level compatibility checking,

- Signaling interworking whereby signal messages such as service identification, call status, channel identification between calling ISDN network and called network are mapped,
- Maintaining synchronization (error control, flow control) on different networks,
- Coordinating operation and maintenance procedures.

Interworking functions will take place in interworking units (IWU) typically but necessarily hosted by the ISDN network. The following interworking interfaces, as illustrated in Fig. 3.7, are defined:

- N Interface between two ISDN networks to ensure service compatibility,
- K Interface with a non-ISDN network, such as a telephone network, where the interworking functions are performed by the ISDN network itself,
- L Interface with a non-ISDN network where the interworking functions are performed by the non-ISDN network,
- M Interface with a specialized network such as teletex or message handling system where the adaptation is performed by the said network,
- S/T Interfaces occurs between ISDN terminals and ISDN network.

The present candidate networks that will support/implement the above interfaces are:

- Another ISDN,
- Public Switched Telephone Network (PSTN),
- Circuit Switched Public Data Network (CSPDN),
- Packet Switched Public Data Network (PSPDN),
- Telex.

The following ISDN interworking cases can be discussed:

ISDN-ISDN Interworking: Even two ISDN implementations cannot be expected to possess identical attributes for the bearer services and teleservices. A service agreement is said to be reached if the maximum common service that can be provided across the network equals or exceeds the minimum service that the user is willing to accept. Such an interworking function would take place in two steps, namely:

- Control phase where the interworking functions on the two networks determine if the requested service of the calling user is supported by the network of the called party, and if compatibility is not achieved a negotiation starts with the calling party to change or abandon the service request.

- User phase starts after the connection between the two TEs has been established; in this phase the low-level and high-level compatibilities are examined on an end-to-end basis.

ISDN-PSTN Interworking: Since ISDN is born out of digital telephone network, and is using its network components, interworking between PSTN and ISDN is of paramount importance. The interworking between ISDN and PSTN is reasonably straightforward Table 3.1 lists ISDN and PSTN characteristics involved in interworking.

ISDN-CSPDN Interworking: There has been been significant progress for interworking between ISDN and circuit switched public data networks. The main interworking functions will focus on the numbering interworking, matching the data rates (e.g., between 64 kbit/s and 2.4 kbit/s) and mapping between call control protocol of X.21 and ISDN.

SDN-PSPDN Interworking: A packet switching public data network can present two interworking cases, namely:

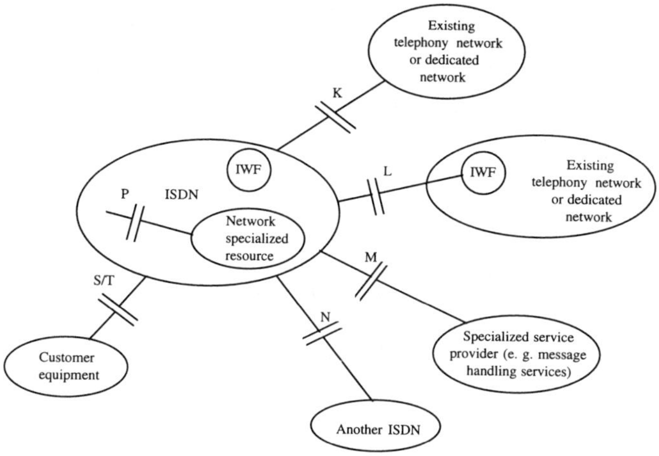

Figure 3.7 Reference Points for Interworking Functions Between an ISDN and Another ISDN or a Non-ISDN Network

Table 3.1 ISDN and PSTN Features in The Interworking Functions

Subscriber Interface	**Digital**	**Analog**	**ISDN-PSTN Conversion**
User-network signaling	Out-of-band	In-band (DTMF)	Mapping PSTN subscriber access signaling to I.451 interexchange messages, and mapping I.441 and I.451 to PSTN interexchange signaling
User terminals	NT,TE1,TE2, TA	Analog telephone modem DTEs etc.	Interworking between modem equipped DTEs and ISDN terminals
Interexchange signaling	SS No. 7 ISDN user part	In band (R1,R2, No.4 or 5)	Mapping between ISUP and PSTN signaling system, and ISDN access signaling and PSTN interexchange signaling
Transmission facilities	Digital	Analog	D/A and A/D conversion
Information transfer mode	Circuit/Packet	Circuit	-
Information transfer capability	Speech, 3.1 KHz audio, video, digital unrestricted...	3.1 KHz audio	-

- A circuit-mode bearer service is used in ISDN to connect the packet user with the packet handler in the ISDN network (minimum integration scenario),
- A packet-mode bearer service is used in ISDN (maximum integration scenario) whereby the packet user employs normal X.25 access protocol and the networks employ X.75 protocol for splicing together the rest of the connection between ISDN and PSPDN.

ISDN-Private Network Interworking: Private networks will be based on PBXs and LANs, which can meet the many varied communication requirements of establishments, like companies, businesses, military headquarters. The service requirements of office communication can be text, video, image, graphics, data, voice, facsimile. ISDN PBXs can cater for these services on a circuit switching base, while LANs offer these services on a packet switching base. LANs provide the high speed interconnectivity between workstations, high performance terminals, computer systems. Gateways provide the interoperation capability between LANs and PBXs by performing the signalling conversion and rate adaptation functions. Finally PBXs are connected to the ISDN network via basic or primary access schemes.

3.6 NUMBERING AND ADDRESSING

The ISDN numbers and addresses are loaded onto call setup messages (LAYER 3) and transported on the common channel signalling. The format of an ISDN address is shown in Fig. 5.18, where one can denote the following variable length elements:

- Country code which specifies the destination country,
- National destination code, which specifies an area (trunk) code, a region, or a particular ISDN network (if the country is served by more than one ISDN administration) within a country,
- ISDN subscriber number, which is used to reach a subscriber in the same local area or network,
- ISDN subaddress provides additional addressing information, though this part does not take a role in the network routing.

The national code plus the subscriber number form a unique national address number within a country.

ISDN Address vs ISDN Number: An ISDN number relates to the ISDN network and numbering plan, and contains sufficient information to route a call. The ISDN number denotes typically the T reference point, that is the subscriber's attachment point to the network. The ISDN address, on the other hand, comprises both the ISDN address and additional information to distinguish a terminal beyond the T-reference point, that is on the S-bus. The subaddress part is not handled by the network, but on the subscriber's premises, e.g., by the PBX.

There are two methods to reach to a subscriber's terminal, as illustrated in Fig.3.8, i.e.,

1) Single ISDN number determines the T interface, and subaddress selects one of the terminals on the S-bus (the latter part is transparent to the network),
2) An ISDN number is assigned to each terminal so that terminals can be reached by Direct In Dialing. The terminal selection part of this network address is broadcast to all the terminals on the S-bus, and only the terminal for which the call handling process has been programmed to the matching number responds. This scheme is

tantamount to the assignment of multiple numbers to the same network termination point. Note that the DDI practice must be used sparingly so that the network is not exhausted of available numbers.

There will exist in the transition period several numbering schemes, such as telex numbering, telephone numbering, packet data numbering, ISO's international numbering scheme, which are not all entirely compatible. ISDN numbering interworking implies that ISDN subscribers are able to setup calls in the non-ISDN networks. The numbering interworking can be effected in two ways:

1) Single-stage approach whereby the address contains sufficient information both to route a call to the destination non-ISDN network and the non-ISDN network to route the call to the called party. The destination network would constitute in this case the prefix of the address.

2) Two-stage approach whereby the subscriber first selects an interworking unit (IWU) by using an ISDN number; after a delimiter during which the connection is established the calling party proceeds to dial the specific number in the non-ISDN network.

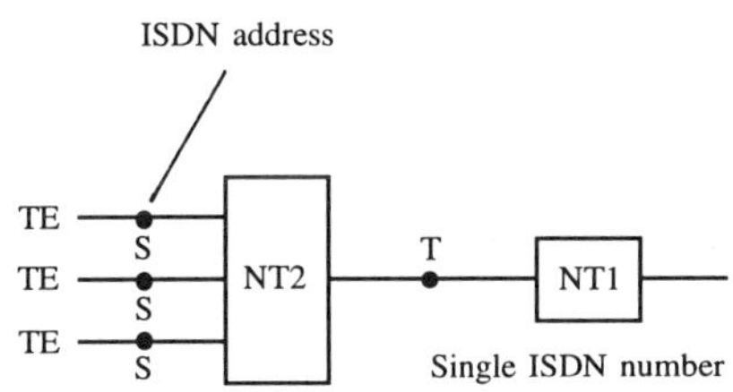

(a) Single ISDN number at T interface

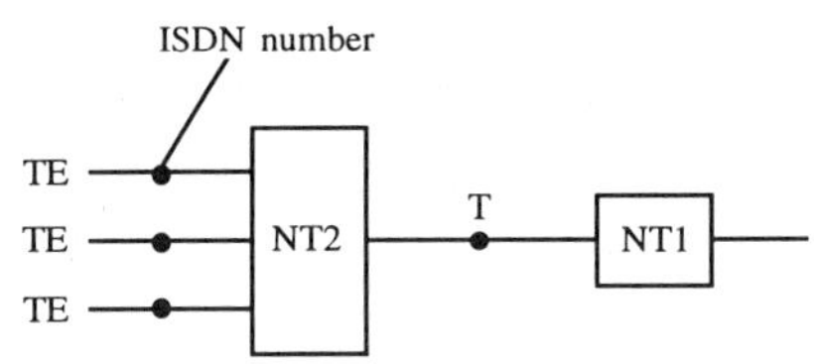

(b) Direct dialing-in

Figure 3.8 Determination of an ISDN Terminal on the S-bus

3.7 BROADBAND ISDN (B-ISDN)

The 64 kbit/s basic rate offered by ISDN (sometime referred to as narrowband ISDN) may represent a substantial augmentation of the channel capacity, e.g., as compared to 2.4 kbit/s modem rates of a decade ago. Yet the narrowband ISDN remains inadequate for some the emerging and/or contemplated services, such as videoconferencing, color facsimile at the photocopier speed, music and hi-fi sound, interconnection of LANs, high speed data transfer, group communication with virtual private networks and video distribution. Among the many applications driving the demand evolution for B-ISDN two are the leading factors:

- LAN Interconnectivity: The dramatic growth of local area networking in large corporations and the development of distributed applications, client-server environment, access to supercomputing facilities or remote servers, and the need for cooperative working environment provided by shared screens and LAN interconnectivity.
- Multimedia: Support for multimedia communications involving both data, video, advanced graphics, and video distribution.

The bit rate requirements of services vary from a few bit/s to several hundreds of megabits, as illustrated on a logarithmic scale in Fig. 3.9 and a possible implementation scenario for B-ISDN is illustrated in Fig. 3.10. It can be argued at this stage that the bandwidth requirements of a significant number of the above services can be met by nx64 kbit/s schemes. In fact CCITT H.221 Recommendation provides the control and allocation of bandwidth for services carried on such assemblies of 64 kbit/s channels. Some of the difficulties encountered in this solution are:

- The multitude of basic access channels may not all follow the same route and trunk groups, hence being subject to various amounts of delay, some as much as one frame long, with obvious handicaps for, e.g., video or multimedia services that require time synchronization. The handicap of differential delays can be bypassed by ensuring that all channels are kept within the same time division multiplex and hence follow a common route. In this case, however, the grade of service may deteriorate significantly due to much higher blocking probabilities. For example to obtain a bundle of six 64 kbit/s channels, e.g., channels 6 to 12 in the group for videophony within a 30 channel group the blocking probability increases significantly.
- Some of the services, such as video distribution will require much higher bandwidth, and in high speed LAN interconnectivity for shared screens with advanced graphics the bursty nature of the traffic and the access times may render nx64 kbit/s ISDN solution impractical.

Enabling Technologies: A number of technical solutions have been considered for the implementation of high speed networks, among which one can mention Synchronous Digital Hierarchy (SDH), Frame Relay (FR) and Asynchronous Transfer Mode (ATM), MAN and the IEEE 802.6 Distributed Queue Dual Bus (DQDB) protocol. A brief review of some of these technical solutions is given below.

3.7.1 Synchronous Digital Hierarchy

Current PDH (Plesiochronous Digital Hierarchy) transmission network multiplexes channels into higher bit rate structures on a stage-by-stage basis, with each stage using its own multiplexing and framing methods. To access individual channels, converting between different channel rates, removing/adding channels are complicated in that total demultiplexing of the high rate signal is needed. In contrast SDH provides the following advantages:

- One worldwide standard for multiplexing and interworking,
- Direct access to lower rate signals without having to demultiplex the entire signal, thus permitting add/drop multiplexing,
- Simplified evolution to higher bit rates,
- Comprehensive provision for network management,
- Interconnection between independent networks without introducing frame slips.

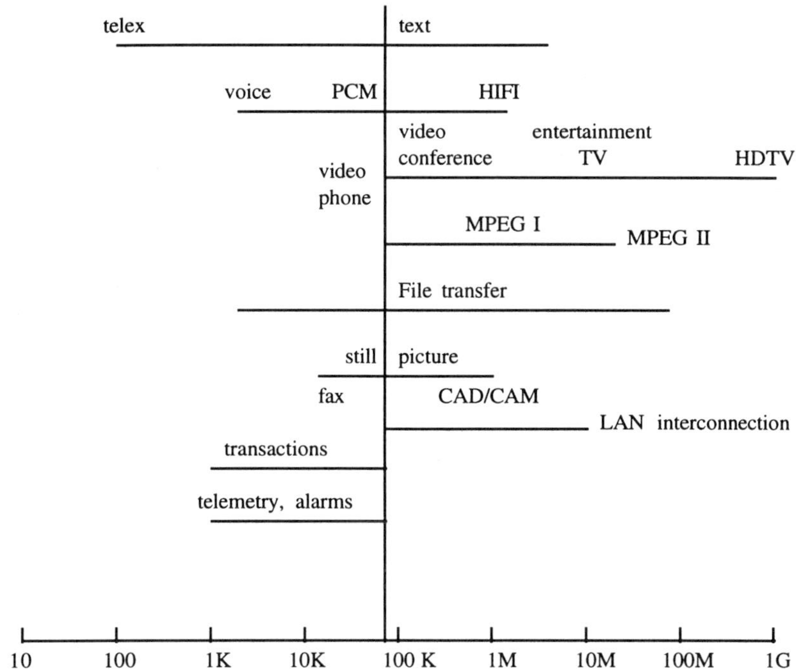

Figure 3.9 Bandwidth Requirements of Services

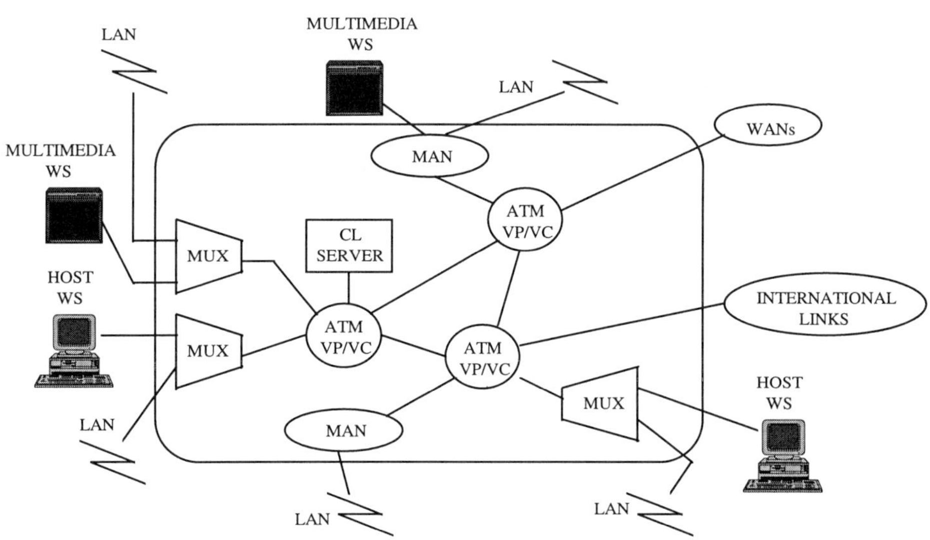

Figure 3.10 Implementation Scenario for the B-ISDN

SDH has evolved in order to solve for the shortcomings of the PDH, and is presently standardized in the CCITT Recs. G.707, G.708, G.709. The SDH protocol's path layer provides for the end-to-end transport of information. In the path layer the information is encapsulated inside structures called Virtual Containers, which consist of two parts:

- The container that holds the data to be transported,
- The Path Overhead, which provides maintenance channels and control information associated with the path across the network.

Different classes of virtual containers have been defined to support the service rates existing in the PDH, namely, 1.544, 2.048, 6.312, 34.368, 139.264 kbit/s rates. The lower bit rate virtual containers, however, can also carry channels that are synchronized to the SDH data stream, which, in turn, enables direct access to individual 64 kbit/s channels.

3.7.2 Asynchronous Transfer Mode

ATM is an alternative message-based multiplexing strategy, that is believed to satisfy the transfer requirements for:

- Handling both narrowband and broadband rates,
- Handling both continuous and bursty traffic,
- Being sensitive to both delay and/or loss types of quality aspects,
- Accommodating unforeseen new demands.

ATM uses short, fixed length cells (hence "cell switching") with minimal headers, which are routed at high-speed by hardware-based routing tables at each switch. The international agreement is that the cells consist of 53 bytes, with a payload of 48 bits and 5 bits of header information. Note that 48 cells correspond to about 6 ms of speech or 200 us of MPEG-I video. The header contains two elements of routing information, namely:

- Virtual Path Identifier denotes a node to node route established at the beginning of each call via signalling messages. The path does not possess a fixed capacity.
- Virtual Call Identifier denote the calls set up over the virtual paths indicated by the respective VPI tag.

The virtue of ATM is that it can handle flexibly all types of traffic, such as voice at 64 kbit/s, hi-fi sound at 705 kbit/s or alternately at variable rate, videoconference at 384 kbit/s, video streams according to MPEG-I or MPEG-II at respectively 1.5 and 10 Mbit/s, telemetry and signalling at rates of a few kilobits, but prioritized to avoid congestions etc. The B-ISDN protocol reference model is shown in Fig. 3.11. The main features are as follows:

Physical Layer: Two interface rates have been adopted, namely 155.52 Mbit/s and 622.08 Mbit/s in accordance with SDH. These high rates should satify all the user bit rate requirements, even for a number of high-definition digital television channels. Among the differen physical media available for local access two emerged as viable solutions, namely:

- Coaxial cable for the 155 Mbit/s and within reaches of 100 meters and
- Optical fiber for larger reaches and/or higher bandwidths.

The physical layer also includes two alternative framing structures, that is

- Cell-based structure consisting of 53 octets,
- Cells mapped into the payload of the STM-1 frame of the SDH.

ATM Layer: ATM Layer is concerned with the cell structure and coding. This Layer is common to all services and provides cell transfer capabilities.

ATM Adaptation Layer (AAL): ATM Adaptation Layer has the functions of adapting and mapping the higher layers supporting various services onto the resources of the ATM layer. In other words this layer is service dependent. These services can all have different characteristics and requirements, such as:

- Connection mode: Connection-oriented or connectionless,
- Bit rate: Constant bit rate or variable bit rate,
- Timing relation between source and destination: required or not-required.

In principle each application may require a different ATM Adaptation Layer solution in order to optimally handle the traffic and service requirements. Examples of services provided by AAL include handling of quantization effect due to cell information field size, transmission errors, lost and misdelivered calls, flow control, and timing control.

Higher Layers: The higher layers of each service or application in the user plane have different specifications so that new services and applications that can be supported by ATM can be defined.

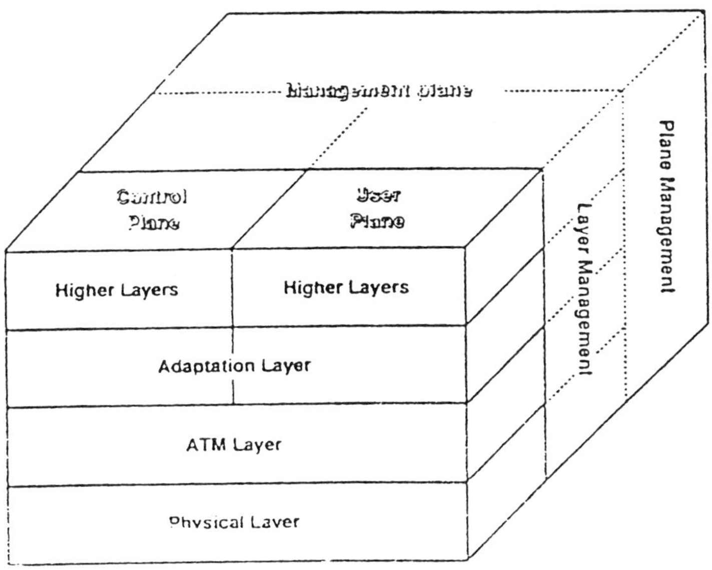

Figure 3.11 ATM Protocol Reference Model

B-ISDN Standardization: Standardization efforts proceed in the direction of specifying the UNI (User-Network Interface) and NNI (Network Node Interface). The NNI standard is based on the SDH. The SDH specifies 155.52 Mbit/s as the worldwide unique interface bit rate. This bit rate is the result of mutual concessions of the parties presently employing different rates in the PDH, and is high enough to carry a single High Definition TV. He lower bit rate channels such as 1.5 Mbit/s and 6.3 Mbit/s are packed into the multiplexing structure of Virtual Containers.

3.8 G-SERIES RECOMMENDATIONS FOR DIGITAL TRANSMISSION

With regard to digital access section CCITT has drawn up two recommendations which describe the functional behaviour and the aspects involving maintenance of the base and primary access of ISDN. The following recommendations are concerned with the section between the V-interface on the switch side and the S/T reference on the user side.

G.960 Basic Rate Access Digital Section: Generic description of maintenance and of the activation and deactivation procedures of the basic ISDN access.

G.96Y Primary Rate Access Digital Section: Description of the maintenance and management of the alarms of the ISDN primary access. It refers to maintenance and operating procedures according to Rec. I.604, which envisages processing of the CRC in the NT also.

G.96X Section for Broadband Access (to be completed by 1996):

G.961 Digital Transmission System: Is concerned with problems of digital transmission on symmetric copper pairs? This recommendation is especially non-binding to allow for free competition among suppliers and manufacturers, and for alternate national solutions. Among several annexes two are significant, in that they describe the 2B1Q and 4B3T line codes, adopted, respectively, in USA and Germany.

3.9 T-SERIES RECOMMENDATION FOR TELEMATIC SERVICES

Major telecommunication standards for telematics include teletex, fax and videotex.

Teletex: The Teletex service provides for document interchange between special terminals which cancreate, manipulate and print document text and graphics. The teletex service is described in CCITT Rec, F.200 (service) and T.60 (terminal functions).

Facsimile: This service provides for raster information transmission. The relevant standards are CCITT F.160 (service), CCITT T.4 (G3 encoding), and CCITT T.6 (G4 encoding). While Group 3 fax does not use any OSI protocol, Group 4 fax uses the OSI stack DTAM (Document Transfer and Manipulation). Group 4 fax is organized in three classes, namely:

- Class 1: Only raster information is transmitted,
- Class 2: Can transmit raster images and receive mixed mode, text plus graphics, information,
- Class 3: Transmits and receives mixed mode information.

Videotex: Videotex is an interactive service which enabls users equipped with special terminals to access a database through the telecommunications network. The interchange process involves such information elements as characters, graphic, geometric and photographic elements. Videotex service is defined in CCITT F.300 and encoding aspects in CCITT T.50.

3.10 REFERENCES

[3.1] W. Stallings, "ISDN and Broadband ISDN", Macmillan, 1992.

[3.2] J.Griffiths, "ISDN Explained", Newyork, Wiley, 1990.

[3.3] P.Bocker, "ISDN The Integrated Services Digital Network, Concepts, Methods, Systems", Springer-Verlag, 1992.

[3.4] M. de Prycker, "Asynchronous Transfer Mode", Ellis Horwood, 1991.

[3.5] S. Kano, K. Kitami, M. Kawarasaki, "ISDN Standardization", Proc. IEEE, 79, 118-123, February 1991.

[3.6] K.D. Kovarik, P. Maveddat, "Multi-Rate ISDN", IEEE Comm. Magazine, 48-54, April 1994.

[3.7] N. Dağdeviren, J.A. Newell, L.A. Spindel, M.J. Stefanick, "Global Networking with ISDN", IEEE Comm. Magazine, 26-32, June 1994.

[3.8] H.A. Scott, "Teleaction Services: An Overview", IEEE Comm. Magazine, 50-53, June 1994.

[3.9] L. Kleinrock, "ISDN -The Path to Broadband Networks", Proc. IEEE, 79, February 1991.

[3.10] CSELT Technical Reports, Vol. 23, No. 2, Special isue on "Status and Trends of Telecommunications Standardiation", June 1993.

[3.11] R. J. Horrocks, R. W. Scarr, "Future Trends in Telecommunications", Wiley, 1992.

CHAPTER 4

NETWORK DESIGN AND DIMENSIONING

4.1 DESIGN METHODOLOGY

4.1.1 System Acquisition Process

The system acquisition process for our system (that we have called WIS), or indeed any similar system is divided into four phases:

i) Establishment of the concept: Taking into account requirements, analysis, assumptions, state of technology and standards, objectives, design rules leading to network characteristics.

ii) Development of the architecture: Definition of the Performance requirements of the subsystems and their integration in a manner which satisfies the end to end performance required. This involves the study of each aspect of the concept in sufficient detail to produce costed proposals for budgetary cost estimates and to identify the major parameters and subsystem characteristics. This will then allow the project to pass into Phase iii), which will include preparation of detailed cost estimates, technical specifications and implementation planning

iii) Development of System Engineering Plans and Preparation of Specifications and Implementation Plan: When approval is given from the Authority to the designer's proposals and recommendations developed in phase ii above, it will be possible to enter phase iii) activities of specification writing and implementation planning (e.g. site surveys and plans). In this phase, the design authority will be required to produce a detailed cost estimate.

iv) Procurement, Installation and Commissioning: The last phase of system acquisition when the equipment and services as specified in phase ii) will be procured and installed according to the time phased plan and consistent with the operational requirements and funding constraints.

4.1.2 Design Approach

For any design authority that is tasked to design an ISDN type system for a specific application, like in our case a military network, we believe that the approach would be similar to the approach reported in this book. The results and conclusions may vary because of different user requirements, constraints or available technologies, but the described design approach should be applicable in most cases. Some general comments should be made at this stage about our system, we call it WIS, and its characteristics. The system was defined as a communications system to be owned and operated by its main user. The network was therefore primarily a single user system, although it was planned that this user was going to include capacity to be rented to or co-owned with at least two other user communities.

It was decided early that the design and dimensioning for these other users were going to be pursued in parallel. The method of rationalisation was decided to be to share the transmission resources, without combining the traffic in the switched part of the network. For the explanation of the design approach for the rest of this chapter, these other traffic sources, and how parallel networks using the same transmission carriers were designed and dimensioned will not be discussed further. It is today likely that most network designs of this type would be solved with any secondary user communities catered for as VPN's (Virtual Private Networks) inside the same switched network architecture. The described approach will in this case also be applicable by adding the traffic from all sources and user locations and design the network as a multi user common resource for all user communities.

The network architecture for our system was influenced by several key operational user requirements, principally the need for enhanced security and survivability, for the users to be served and the traffic to be carried. The detailed operational requirements must be developed as an iterative process involving both the user/owner of the system as well as the design authority. This is to ensure that the requirements can be stated in terms which would lead to a feasible, cost-effective system capable of meeting the operational requirements. In order to develop the architecture, methodologies have been devised which enable the cost-effectiveness of the proposals in relation to the requirements to be assessed. In figure 4.1 the principle of this methodology as well as sequencing of the tasks are shown. Computer aided design tools should be employed to dimension the network for various traffic loads and to assess its traffic carrying performance (transient and steady state) for different network configurations, transmission conditions, routing algorithms and under different scenarios of hostile actions from a potential opponent. The result of the studies is a set of costed options for implementing the system over a time scale consistent with the users requirements and within funding and other constraints.

The approach taken to develop our system consists of establishing a "minimum-cost-network" capable of carrying the traffic specified between the users of the system for a given grade-of-service and then it should be assessed against a set of damage scenarios. If the minimum-cost network does not satisfy the desired survivability (which is fairly likely), then additional nodes and links can be added and the resulting network is then re dimensioned and costed. The cost difference that is obtained in this way can be regarded as the cost of survivability.

4.1.3 System Concept

As can be seen from figure 4.1, the main driver for the whole iterative process of designing our system is the so-called "WIS concept". This is based on the existing knowledge of technology and trends, civil and military standards and the operational requirements as stated by the user of the system. The concept was developed by the design authority and reviewed and approved by the operational user early in the design process. To explain how such a concept can be defined and promoted the main ingredients of the developed concept are described below:

i) The architecture will be based on a common user circuit switched area-grid network whereby users have access to the network via "access switches" and "access links" to the system through two or more switching centres installed at the nodal points which will give the survivability required in an economic manner.

ii) Automatic switching will be employed to meet the requirements for responsiveness, survivability, flexibility, including progressive transition from analogue to digital working.

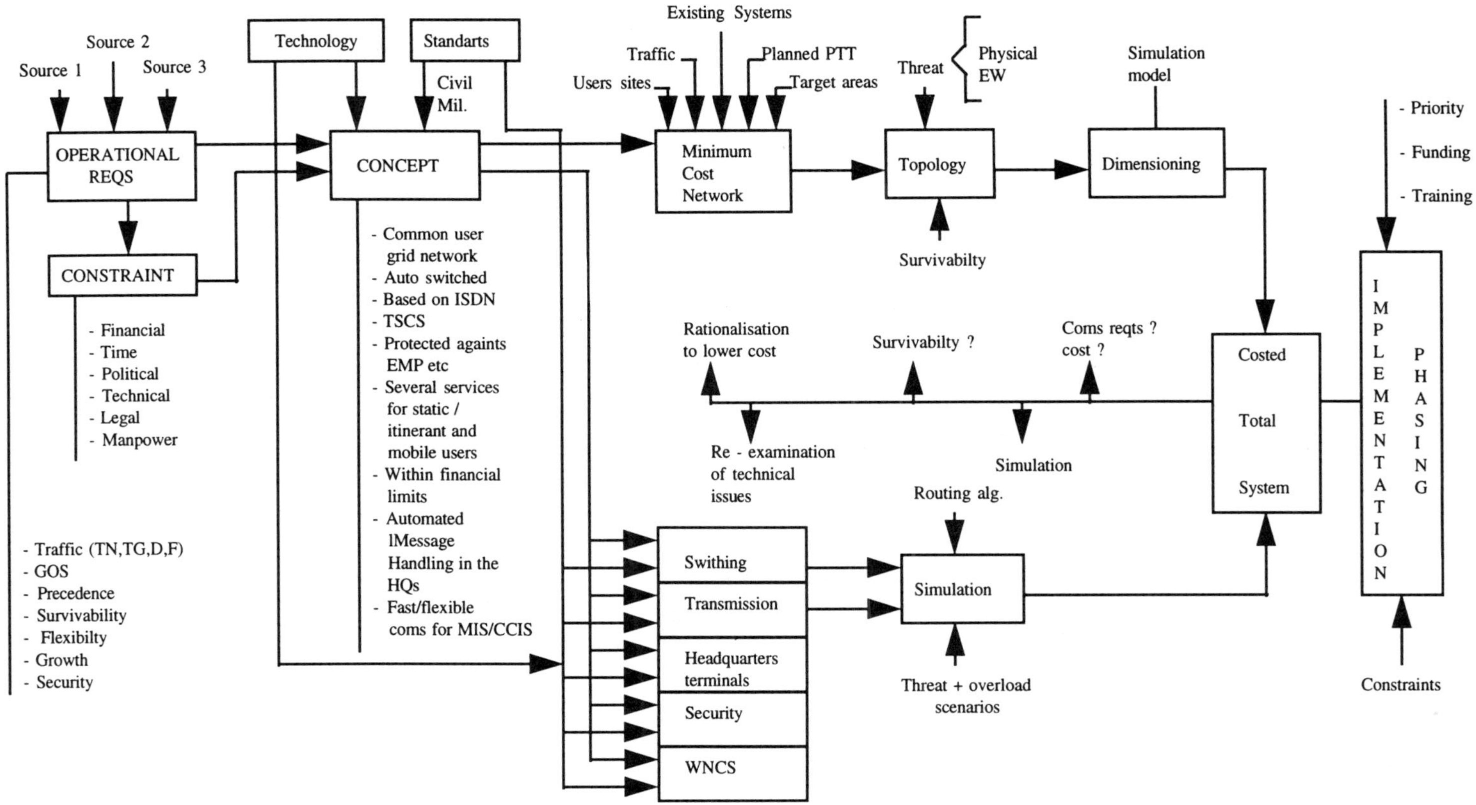

Figure 4.1 WIS Design Methodology

iii) The system will cater for the following services which will be provided for static, itinerant and mobile users:

Bearer and Teleservices:

- Plain Voice,
- Secure Voice,
- Data at various rates,
- Telegraphy,
- Conference,
- Hot lines.

Supplementary Services:

- Precedence,
- Pre-emption,
- Abbreviated Dialling,
- Dedicated Networks.

iv) The architecture will be inherently flexible to allow continuous adaptation in a cost-effective manner to meet evolving user requirements in capacity and service.

v) The subsystems will, as far as possible, be capable of unattended operation so as to achieve minimum requirements for manning

vi) The services described in (iii) above, will be provided using circuit, packet and message switching techniques.

vii) Within legal, financial economic and technical constraints, planning will give due regard with respect to previously installed resources, to the maximum cost-effective use and integration of these capacities and functionalities in each successive implementation stage.

viii) The system must be capable of being implemented in phases over a period of 15 to 20 years. Each phase in this plan must involve a noticeable increase in capacity and functionality over the previous phase.

ix) The system will be digital in its basic concept, but must be capable of operating in a mixed analogue/digital environment.

x) The cost of the system must be within the financial limits as defined by the user/owner.

xi) The network will have a number of static nodal switches, transportable switching centres, access switches to provide concentration and other facilities at user locations.

xii) The system will offer simple interfacing facilities with other networks such as the public PTT network and other parallel system to which the user has a traffic interest.

xiii) Transmission media to be used will be, as appropriate, LOS microwave radio, cable systems (copper or optical fibre) and satellite links.

xiv) CCITT-ISDN recommendations and methodologies will be employed to the greatest extent possible in order to minimise the cost of acquisition, risk of delays in implementation and cost overruns, and the problems of interoperability with PTT networks.

4.2 NETWORK DESIGN

4.2.1 Traffic Requirements

The network planning for our system was done mainly based on the specific communications requirements defined by the user of the system. The network defined was a switched service integrated network of potentially about 30.000 user located at about 250 locations over a geographical area of several hundred thousand square kilometres. The users were to be served with a variety of traffic types, traffic profiles as well as traffic volumes, and all these requirements would be changing over time and during different phases of operation:

- Normal daytime operation in peace,
- Exercise situation and
- During periods of tension and war.

In the latter phases the system would also be exposed to significant stress, both intentional (such as damage and interference) and non-intentional (such as overload and mis-management).

To perform a network design in such a case it is obvious that the initial data on traffic interest and traffic volume must be as correct as possible. It is also necessary at this early stage to make room for a certain level of flexibility and growth potential. The future requirements may be very difficult to assess and the traffic variations, especially during periods of stress, may be difficult to assess and quantify. The planning of such a network is also in most cases complicated by the fact that the existing communication facilities, and hence, the user's experience is restricted to the use of dedicated circuits, manual switchboards and manual message handling procedures.

To assemble data on user traffic and interest must be done early on in the project and should be used as an opportunity to acquaint the future user with the concept of his new communication solution. Depending on the level of user experience and predictability of traffic interest, the choice of method is here between a number of different options:

i) If the number of users are very high and the user community is reasonably evenly distributed, the simplest approach is to use conventional PTT techniques of traffic estimation. In this case each traffic source, defined as a subscriber is characterised as one of a limited number of subscriber profiles described in terms of traffic intensity (e.g. in calls per hour or average busy hour traffic in Erlang) and traffic interest (e.g. as percentage numbers for local, short-haul, long-haul and network external traffic interest). It would typically require very little user participation quickly to establish a traffic estimate with this method, but sources of errors can become significant if there is typically large variations of users both in mission types, locations and traffic interests.

ii) If the users are already operating with a similar communication arrangement and if their future use of the new system will be comparably similar to their current operations, then it would be feasible to arrange a monitored exercise to map traffic interests, volumes and profiles. This would be an exact method, could give good user participation, and would with careful post-processing give good data for the network planners to dimension the network. It is often the case, however, that the users are not currently in the same situation as they would be when the network is installed, and the introduction of a new system is often linked with changes in their operational environment which would complicate the task of creating a realistic exercise scenario.

iii) The case where the users have little ability to foresee how they would operate in the future system, and where the new technology have little practical exposure at the time of the network planning effort, then a scenario should be established to describe a current communication environment, map the requirements based on the current technology, and then convert the assembled data to the new environment using general conversion estimates. This will for example mean to ask the users to define how many dedicated circuits they would need to which other locations in order to perform their current mission. Having received this as a map of the user spread and logical traffic interest, this can then be modified with estimates of traffic density and traffic type distribution (e.g. data, message, fax and voice).

The traffic interest in our case was given as the third option as traffic interest in terms of number of point to point circuits that would have been required if the system were a non-switched network. The data was presented as single sided traffic relations i.e. if A has the need to talk to B, and B has a need to talk to A, these were reported separately, and the total traffic interest was assumed as the sum of all inputs, giving the double sided traffic interest between all user locations in the system.

Since our network was going to be a switched network, the dimensioning rules were then applied assuming that the circuits only would be through-connected during the time of active interchange of data between the parties. The network is a common user network, where all resources except the subscriber's access line to the system will be time-shared. With the given data it was therefore very important to establish a representative user behaviour pattern. This will for dimensioning purposes only, require an assumption of the percentage of time that the above mentioned "point-to-point" circuits would have been in active use. The traffic profile was therefore defined as an Erlang value per circuit. This value would be a number between 1 and 0, with 0,5 E for example meaning that the defined circuit would be occupied for 50% of the time during the "network's busy hour".

For dimensioning of the call handling capacity of the switches it would also be necessary to assume a certain call rate or call holding time. This value need not be addressed in the static state dimensioning phase, but it has been assumed that a nodal switch call rate of 10 calls per second would be an ample capacity for this type of network.

4.2.2 Network Topology

The design of the WIS network topology was initially performed from map studies, taking the following factors into account:

1) The identified and plotted locations of all user sites.

2) The positions of all existing military systems, with emphasis to those that represent the networks interfaces to other networks and facilities.

3) The locations of existing and planned PTT LOS systems, especially those that are digital, giving the starting-point to define the possibility for WIS internodal links using PTT sites and link profiles as collocated WIS transmission elements.

From this initial set of maps and data, a potential internodal network was established using the following criteria:

1) A network node should have some proximity to major clusters of user locations, to make the shortest possible access links.

2) A network node should be located such that it is close to a topologically natural node in the PTT collocated WIS LOS network from above.

3) A network node should have a natural connectivity of at least 3 physically separate links to nodes.

4) Potential new link routes (i.e. without PTT collocation) should be identified when the above criteria cannot be fulfilled, while the topological importance dictates that the connectivity should be established.

Based on these rules and using the information available on adjacent networks from the Authority, a potential set of links and nodes were listed. In the first phase some 60 potential nodal locations and some 91 possible internodal link routes were listed as the initial resource potential.

There may be previously defined networks available as starting points for the design. In the case of WIS two such networks were available, and these were studied initially. One was a 17 node network with 31 internodal links. The other network was a 25 node network with 50 internodal links.

Using the rules and principles discussed above, these networks were used to form a basic and "maximum number of links and nodes" network of 33 nodal switches and 68 internodal links. This was then the nodal network used as a starting point for the minimum cost network design to be described in the next section.

The following topological principles were at this stage established for the access network:

1) All identified headquarters were given dual access to the network. The dual access was planned as independent if the headquarters was isolated from other headquarters.

2) Each of the two access links for all headquarters were brought independently to the nearest two separate internodal sites.

3) In two densely populated regions this was deemed inefficient, and a separate map study was made to establish co-ordinated network for those regions, utilising as appropriate existing transmission media in these regions.

The further topological design of the access network and the internodal network is presented together with the appropriate dimensioning activities in Section 4.5.

4.2.3 Minimum-Cost Network Design

The WIS network consists of access circuits (between the user locations and the nodal switches) and internodal circuits (between nodal switches). By increasing the number of nodal switches, the number of internodal circuits will also increase, and so will the cost. On the other hand, increasing the number of nodes will decrease the average length of the access circuits. In as much as the transmission cost increases with the total circuit-length of the network, this reduction indicates a potential for the cost savings.

However, as the number of switches increases, so do the cost of switching. The possibility of a trade-off between switching and transmission cost indicates that a "minimum-cost-network", in which the number of nodes is at an optimum, can be formulated. Hence it was decided that the appropriate starting point for the study was an examination of the interplay between the number of switches and the transmission cost vis-à-vis the network configuration which would carry the envisaged traffic load at minimum cost.

The costing basis adopted for the study was the initial capital outlay requirement, bearing in mind that the final WIS will be owned and operated by the Authority. The cost elements that were analysed were therefore the cost of switches, together with the cost of transmission systems for both access links and internodal switches, together with the cost of transmission systems for both access links and internodal links.

The WIS network design was initiated while all data on user locations, traffic interests and volumes were not yet available from the Authority. Even if this may seem like a "WIS-specific" and not general issue, it is used in the description here because it may often be the case that the Authority maybe slow in providing input data to the network designer.

Three network configurations (reference networks) were designed, dimensioned and costed with respect to the same unit cost values and the same traffic requirements. The traffic volume was, at this stage, estimated to be about 2300 E, generated from about 130 identified headquarters. This traffic estimate was calculated using 0.8 Erlang per circuit defined by the Authority, giving what at this stage was assumed to be an ample margin for the information on headquarters that was not yet at that stage available from the Authority. When compared with the total estimate of about 1800 E used as basis in the final dimensioning later, from about 250 headquarters, the assumption made seems to be fully justified.

The cost basis was derived from available sources, and included all cost elements related to transmission and nodal switching. The nodal switches were defined in different size ranges, to take account for the fact that when number of nodes were low, the sizes of the switches, and hence their cost, would have to be higher. The following are specific data about the different reference networks that were planned and analysed.

(a) 18 Node Network

This network was designed as close as possible to the first initial 17 node network. The only significant change was the addition of a second node in the most densely populated region, because of the obviously central role of this part of the network, and to reduce the length of the second access link for the high number of headquarters that are located there. Figure 4.2 shows the configuration and the internodal link capacity of this network.

The nodal switch register shown in table 4.1 presents the characteristics of the 18 nodal switches in terms of access and internodal links capacity and estimated cost. The average switch size is about 1200 channels, and the total number of internodal links is 36, or an average of 4 link terminations per switch. The internodal network seems to be topologically reasonably well designed, apart from the obviously too critical node number 9.

The survivability performance of the internodal network was analysed and judged to be reasonably good. The access network performance will, however, show some shortcomings, in particular in regions with very low node density.

(b) 26 node network

This network was derived from the second initial network, with some modifications to cater for the shortcomings of this network in the two most densely populated regions. The network topology and the final dimensioning is shown in figure 4.3. The network has 52 internodal links, giving the same 4 internodal link terminations per node as for the 18 node network. Table 4.2 shows the data and cost estimate on the 26 nodal switches, with an average capacity of 940 circuits per switch.

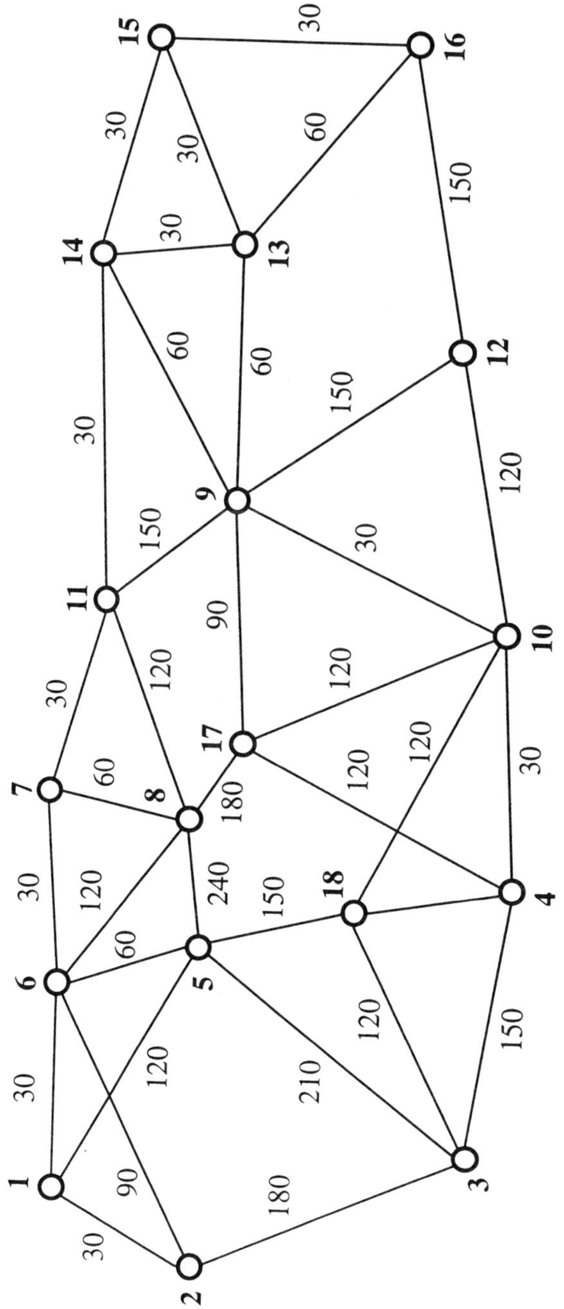

Figure 4.2 18 Node Network Configuration and Link Dimensioning

Table 4.1 18 Node Network Nodal Switch Data

Nodenr	Acclink	Acclcap	Inlink	Inlcap	Totcap	Size	Cost
1	23	30	3	3	33	1200	1587
2	31	41	3	10	51	1800	2672
3	30	40	4	22	62	2100	3395
4	22	30	4	12	42	1500	2090
5	14	38	5	26	64	2100	3395
6	25	31	5	11	42	1500	2090
7	8	13	3	4	17	600	778
8	23	45	5	24	69	2100	3395
9	13	25	6	18	43	1500	2090
10	23	28	5	14	42	1500	2090
11	9	14	4	11	25	900	1123
12	25	37	3	14	51	1800	2672
13	6	9	4	6	15	600	778
14	10	14	4	5	19	600	778
15	4	5	3	3	8	300	467
16	11	20	3	8	28	900	1123
17	23	44	4	17	61	2100	3395
18	8	28	4	15	43	1500	2090
	154	492	36	223	715		36008

NODENR : Node Number
ACCLINK : Access Link
ACCLCAP : Access Link Capacity
INLINK : Internodal Link
INLCAP : Internodal Capacity
TOTALCAP : Total Capacity

Table 4.2 26-Node Network Switch Data

Nodern	Acclink	Acclcap	Inlink	Inlcap	Totcap	Size	Cost
1	21	28	4	14	42	1500	2090
2	15	21	3	17	38	1200	1587
3	3	3	3	7	10	300	467
4	11	17	5	16	33	1200	1587
5	16	21	3	7	28	900	1123
6	24	29	5	17	46	1500	2090
7	17	22	3	7	29	900	1123
8	11	34	5	34	68	2100	3395
9	7	10	4	10	20	600	778
10	10	27	4	18	45	1500	2090
11	13	21	5	11	32	1200	1587
12	20	44	5	26	70	2100	3395
13	7	9	4	11	20	600	778
14	8	11	3	4	15	600	778
15	10	17	4	8	25	900	1123
16	5	12	6	23	35	1200	1587
17	15	24	5	18	42	1500	2090
18	20	28	4	10	38	1200	1587
19	6	7	5	6	13	600	778
20	10	14	4	7	21	900	1123
21	12	20	3	9	29	900	1123
22	11	11	3	7	18	600	778
23	6	9	3	4	13	600	778
24	3	4	3	3	7	300	467
25	9	11	2	5	16	600	778
26	18	39	4	22	61	2100	3395
	154	493	51	321	814		38475

The network topology may be judged as reasonably well balanced, with the only critical comment that the area of nodes 17, 21 and 22 may be reduced to two nodes, and the area with nodes 12 and 26 should be increased with one node. The survivability analysis showed very good results despite this, and all major user locations seemed to have a reasonable access link situation.

(c) 33 node network

This network was formed as an expansion of the 26 node network, the aim of the expansion being to give easier dual access for user locations in the central regions, and take the maximum benefit from the possible transmission network redundancy. The network configuration and the internodal link capacities are shown in figure 4.4. The network has as much as 68 internodal links, as shown in the nodal switch register in table 4.3, and this gives an average of 4,1 link terminations per node.

This is therefore the richest of the networks, and the survivability performance analysed was clearly showing this network to be the most robust. The nodal switches have an average size of 800 circuits, while the total cost of switching here clearly was highest of all the estimated networks.

The three networks were all costed in detail, with the most complex part to estimate properly being the access transmission cost. For the purpose of assessing this, three different access network databases were prepared, with the access information prepared for all the headquarters that were identified at this stage. The number of known locations were about 150. Compared to the final 250 locations, the cost of the access links would be underestimated for all networks, but more importantly the sensitivity for this cost element can be assumed underestimated in the trade off calculations.

The results were compiled and are presented in figure 4.5. It can be seen from this figure that a minimum cost configuration is clearly identified at the 26 node network level. This was used as a conclusion at this stage and the detailed analysis performed later with better traffic and user input data has basically confirmed this conclusion.

The differences in survivability that was observed at the early stages, where the 33 node network appeared as the most survivable may still be correct, but the approach of balanced survivability specifies that the network survivability performance should be neither better nor worse than that of the users it is supporting. With this approach the minimum cost network of 26 nodes was adjusted, cost reduced, improved slightly and analysed to verify that it fulfilled the full survivability requirements for the WIS.

The improvements and adjustments later made to the 26 node network were also directed at making some cost reductions in internodal link cost. The final and proposed network with dimensioned link capacities is shown in figure 4.6.

4.3.4 Tools and Support Systems

The network design, dimensioning and analysis reported in this chapter, were performed primarily using the same basic ideas as was used in the architectural design effort for the NICS Stage 2 network. The design methodology applied to that study provided a very good basis for the methods that were to be developed for WIS. A certain accessibility was also achieved to the suite of simulation and dimensioning programs used at STC, allowing the verification of the in-house operated computer tools that were used.

The additional tools and support programs that now represent the total WIS tools package will be briefly presented and discussed below. The aim of this section mainly being to

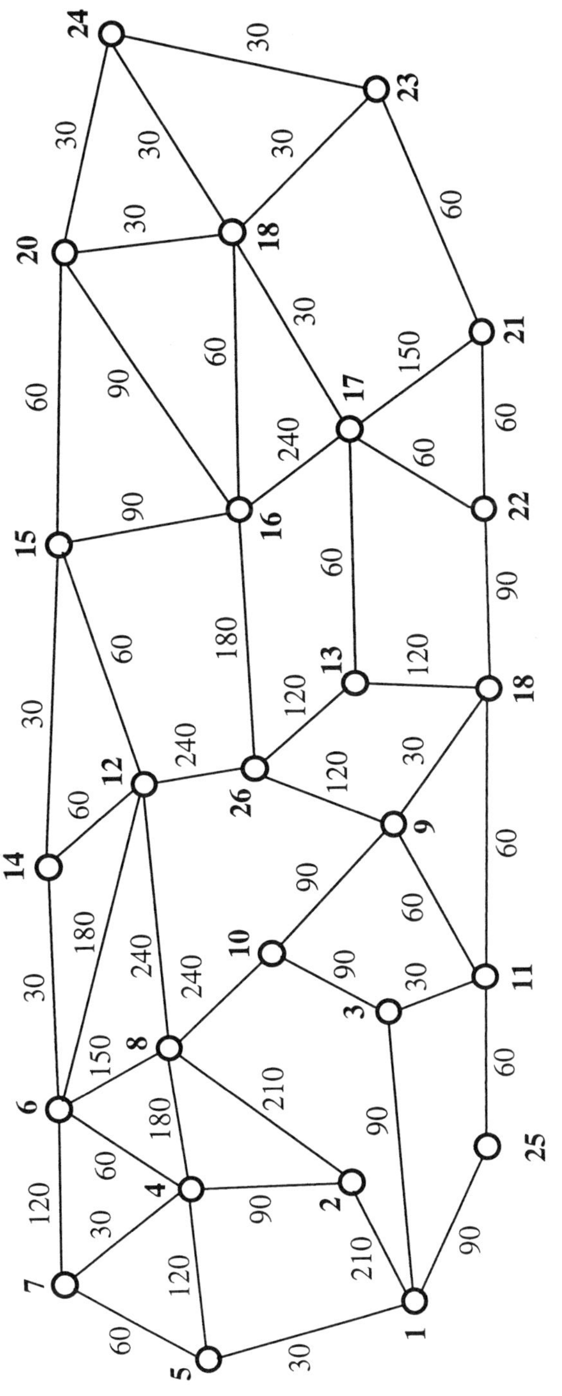

Figure 4.3 26-Node Network Configuration and Link Dimensioning

Table 4.3 33-Node Network Nodal Switch Data

Nodem	Acclink	Acclcap	Inlink	Inlcap	Totcap	Type	Size	Cost
1	7	8	3	5	13	2	600	778
2	15	19	3	7	26	3	900	1123
3	20	27	4	10	37	4	1200	1587
4	9	11	4	12	23	3	900	1123
5	7	11	3	7	18	2	600	778
6	8	13	4	7	20	2	600	778
7	10	14	5	9	23	3	900	1123
8	3	4	3	3	7	1	300	467
9	11	12	3	6	18	2	600	778
10	13	20	6	23	43	5	1500	2090
11	15	36	3	15	51	6	1800	2672
12	4	4	4	12	16	2	600	778
13	5	12	6	19	31	4	1200	1587
14	2	4	4	4	8	1	300	467
15	6	7	4	6	13	2	600	778
16	6	27	6	38	65	7	2100	3395
17	12	31	6	37	68	7	2100	3395
18	7	10	4	13	23	3	900	1123
19	10	10	3	4	14	2	600	778
20	16	26	5	17	43	5	1500	2090
21	10	30	5	18	48	5	1500	2090
22	7	9	4	5	14	2	600	778
23	16	25	6	19	44	5	1500	2090
24	6	9	3	6	15	2	600	778
25	11	13	3	6	19	2	600	778
26	3	3	4	4	7	1	300	467
27	5	7	6	20	27	3	900	1123
28	12	20	3	10	30	3	900	1123
29	9	11	3	7	18	2	600	778
30	10	14	5	10	24	3	900	1123
31	7	10	3	4	14	2	600	778
32	18	25	5	15	40	4	1200	1587
33	10	10	3	5	15	2	600	778
	155	492	68	383	875			41959

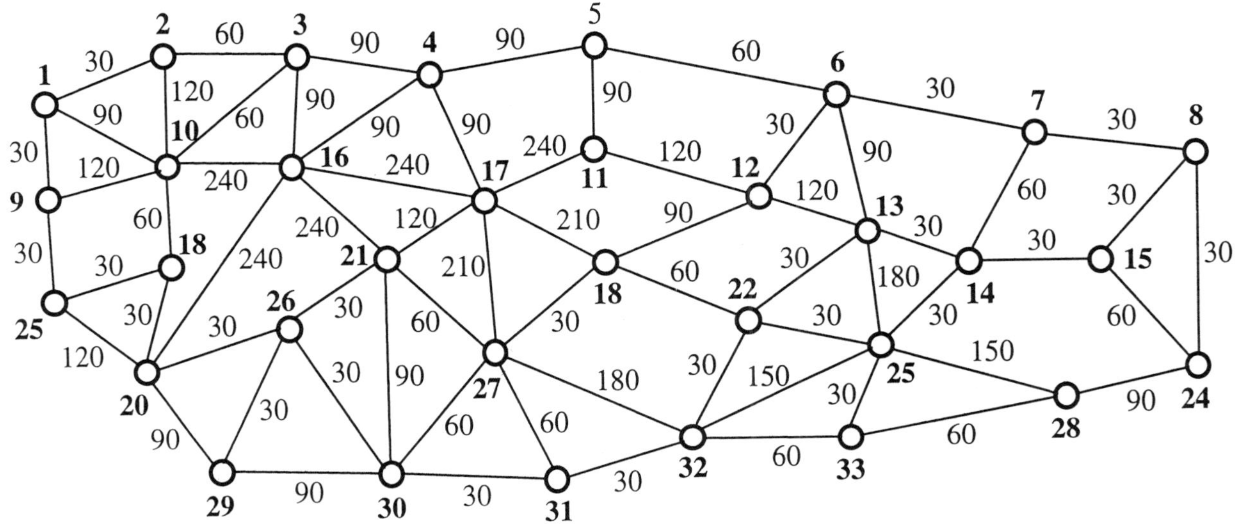

Figure 4.4 33-Node Network Configuration and Link Dimensioning

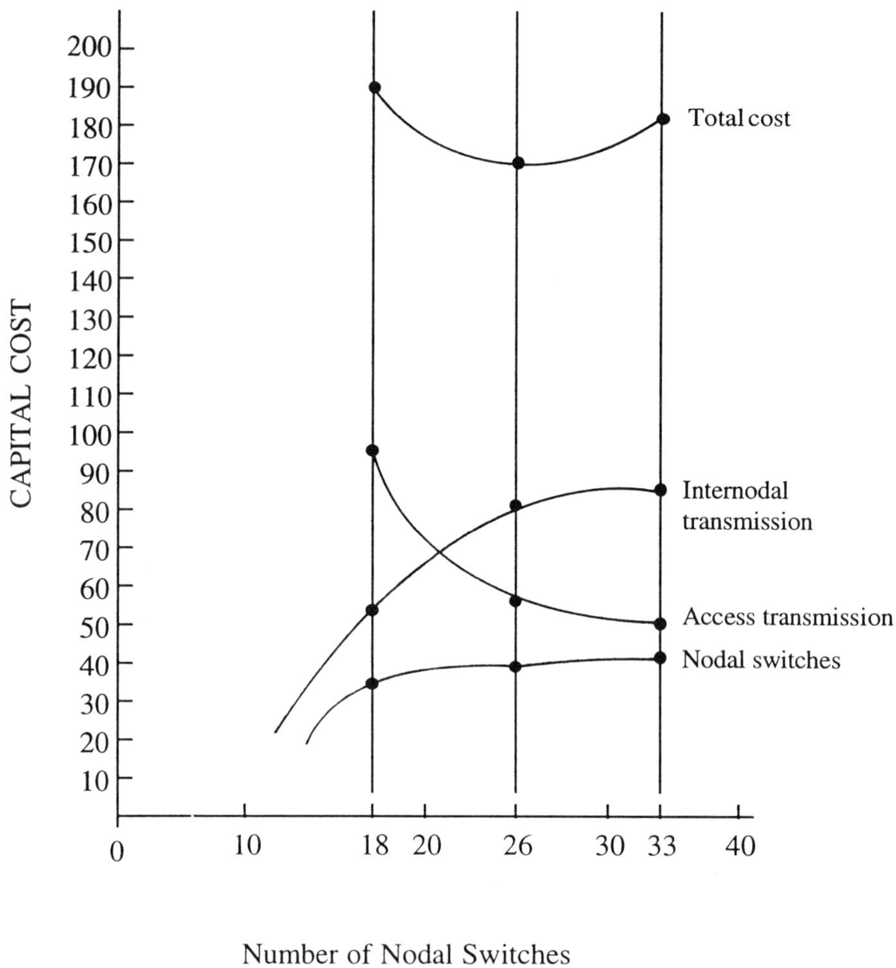

Figure 4.5 Minimum-Cost-Network Comparison

register the result of the "on-the fly" process of providing the tools, and to propose useful extensions and improvements that may be relevant to other network designers and planners..

The network dimensioning, optimisation and analysis tool that was most thoroughly applied to the study, was a tool developed by Dr. R. Bottheim and improved by others. This program was based on an analytical model of traffic flow in a non-hierarchical mesh-grid network. It has today the capability to dimension and analyse a network of up to 40 nodal switches and include features like:

- Optimum allocation of capacity on internodal links for minimum blocking probability under given cost constraints,
- Analysis of resulting blocking probability for up to four levels of precedence in a defined network
- Capability of analysing effects on end-to-end blocking probability following network damage as well as changed traffic volume or distribution of precedence.

In addition to these externally provided tools, a wide range of programs and routines were developed within the WIS development team to support the WIS network design activity. These tools and procedures were developed continuously during the design phase, and can broadly be classed as follows:

1) Tools related to the in-house operated "off the shelf" relational database system. This was extensively used for the management of all data developed and stored in the processes of headquarters information, traffic values and network design parameters.
2) Tools related to the high number of different cost estimate exercises made both in the minimum network design and in the preparation of the various cost estimates in the design process of the WIS. The tools used here were commercial off the shelf spreadsheets and database management programs.

The main methodology and sequence of dimensioning operation, from input of traffic data, network structure and cost, to a dimensioned and costed network is described in sections 4.5 and 4.6. The process of survivability assessment is presented in chapter 13, WIS Survivability Design, and will not be repeated here. The developed sequence of operations in this process is believed to be useful also in the general context of network planning and implementation, with the following ideas for improvements:

1) The handling of the full network with all its 250 user locations could not be effected directly by the program used, because of its limitation of 40 nodes in the network. In both directions (dimensioning and analysis) it was therefore necessary to perform pre and post-processing of traffic and blocking data. It should be studied if the extension of the network to about 300 nodes will create any fundamental problem for a selected tool, since this would improve the quality and ease of its use with large networks like WIS. The requirement for computing power may in this case be the limiting factor, and may demand the transition from the used PC platform to a workstation environment.
2) The algorithms for optimisation, with real cost functions for the links had some shortcomings in the applied dimensioning program. If the planning is to include also commercial PTT rental of internodal groups and satellite links for node-skipping, then it would be necessary to find a way to include these operations.

3) The analytical model does not provide call by call event-driven simulation of the network. This would have been a useful expansion of the analysis tools for WIS, especially since this would also allow the study of various adaptive routing techniques, and would give a better possibility to verify the dimensioning program's analytical model.

4.3 NETWORK DIMENSIONING

4.3.1 Access Network

From the user to user traffic matrix described in section 4.2.1, the access links from each user location can be dimensioned. This is done first by summing up all traffic interests from/to each user location, using the traffic matrix. This gives the total number of Erlangs generated and terminated at each user location.

Headquarters with access switches (i.e. with more than 30 users), will be connected with access links to two nodes. The assumption is that the total traffic will be shared equally between the two links. To find the total traffic volume for each link, the headquarters traffic is divided by two, and then both figures are multiplied by 1,5, to achieve 50% overload margin. If one access link is cut, this will allow the second link to carry about 75% of the total estimated traffic.

Using the Erlang loss formula tables, the number of required servers (i.e. circuits) was found. For the dimensioning of the network, blocking probability was set to be 2%. Having the number of channels required on each access link, the number was finally rounded up to the nearest number of 30 channel groups. The resulting number of groups was then entered into the access network data-base, to be used for link dimensioning and cost estimation..

The headquarters served with remote access units are all dimensioned with 30 channels or one TDM group and with single access.

Having the access capacity in circuits/groups required for each of the user locations, the physical structure of the access network was defined. The design was based on map studies, information from available systems, and the following principles were established, with the final result entered into the access network data-base (each user location having its own data stored as one record):

1) Exact user location was identified.

2) Type of access arrangement was determined. This is either remote access unit with a single 30 channel access link, or an access switch of varying size, with two physically separate access links of size dimensioned as described above.

3) All headquarters in the two most densely populated regions were studied separately and their access network was designed as a combined effort. This is described later in this report. For all other user sites the following rules were applied.

4) All links from remote access units were routed to the nearest other user site or access point to the internodal network. A remote access unit can be connected to either an access switch, a transmission node or a nodal switch. If another user site with an access switch was located closer than any of the backbone network access points, the access link was defined to go to this.

5) For user sites with access switch, the two access links were established to the two nearest access points in WIS, being either access repeater, transmission nodes or nodal switches.

6) If the distance for the access link (from an access switch or a remote access unit) was less than 20 kilometres, the link was defined as fibre optical cable connection. If the distance was more than 20 kilometres but less than 40 km, one LOS hop of microwave link was prescribed. For user locations located more than 40 kilometres away from its WIS access points, LOS microwave systems with through repeaters were prescribed, calculating with a design hop-length of 40 kilometres.

Special combined access networks were designed and documented for the user clusters in the two most densely populated regions. The basic input for the design in these areas were the already existing transmission structure in these areas, together with the planned WIS entry points available in these regions. In the cases where no existing or planned projects could be used directly, the predominant choice for the connection of a cluster of user sites were to design optical fibre ring networks with two WIS nodal switches in each ring.

For the larger access switches the rules applied above resulted in the predominant use of two different physical transmission media for the access links, one being fibre optical (the shortest), and the longest being an LOS radio system. This is also clearly prescribed as an important result of the WIS survivability study reported in chapter 13.

When in this way the total number of defined users were connected to the network, the complete information was kept in a well-organised access network data-base. This data-base was the used for the calculation of the cost of the access circuits.

A special case should finally be discussed as there was a number of defined mobile users. For each of these a number of pick-up points were identified where these users were expected to be connected to the network. Since these users in their operational mobile role would be more likely primarily subscribers of the tactical networks already planned for the same region, it was decided to convert these pick-up points to general gateway points between the tactical network and the WIS. In this way the mobile users are given a switched network access to WIS, using their tactical network terminals, and indirectly allowing more of the tactical users this access as required.

In the study reported here, this question was therefore simplified by assuming that a WIS access switch will be established at each of these defined access points. The cost and complexity of this will be comparable to the final anticipated solution of gateway arrangements, such that for the initial planning purposes this simplification should be acceptable.

4.3.2 Internodal Network

Having the user to user traffic matrix and the allocation of user sites to nodes from the access database, the internodal network was dimensioned by first converting the traffic matrix to become a node to node traffic matrix. The main rules to be applied to this process are the "traffic distribution rules".

It must be noted here that it was assumed at this stage that the WIS access switches would not participate in the routing process, and will not have any information on the destination of outgoing calls they are handling. Since they have two outlets to the transit network, we will assume that they will route the outgoing calls equally distributed over its two access links, regardless of the destination address.

The "traffic distribution rules" from A to B can therefore be summarised as follows:

1) If A and B are connected to the same two access switches, say M and N, half of the A to B traffic will be switched locally in M and N respectively. None of the traffic will therefore result in internodal traffic.

2) If A and B have one nodal switch in common (say M) and one each that is not common (say X and Y), half of their traffic will be local in M and the remaining half will be split into 3 equal parts (1/6 X to Y, 1/6 X to M and 1/6 Y to M).

3) If A and B are connected to four different nodal switches (say M, N, X and Y) the traffic will be split into four parts. 1/4 will be between X and M, 1/4 between X and N, 1/4 between Y and M, and 1/between Y and N.

Using these rules of distribution, a traffic matrix was constructed to represent the undamaged networks traffic load. For the later purposes of network analysis the capability of this process was made such that the matrix can be recalculated, to take account for the effects of damage to the access links or user sites.

The resulting traffic matrix (node to node) was produced in two formats. One format is defined by the planning program (strict ASCII format). The other format is shown in Table 4.4 and is a suitable way to present the internodal traffic interest pattern.

Three input files are then required for the dimensioning phase. The format of these files is defined by the planning program. The first is the traffic matrix as described above. The matrix used for dimensioning is naturally the undamaged network version.

The nest input file is the network structure file, which describes the designed logical internodal network. This file is the output of the topological network design phase described above in Section 4.3.1.

The third input file that the dimensioning program needs is the cost file. The use of this file would allow the dimensioning to take into account cost variations between the different links. This could be useful if the cost structure of the transmission links were such that real cost savings could be obtained by finding optimal loading of each link. This would be the case for example if some of the links were leased from the PTT, or if the cost per channel for the transmission links were for example linear up to a certain capacity and then had a significant increase for expansion.

It would also be a useful feature if the topological network design included certain expensive links added for survivability reasons. The dimensioning process could then analyse if the real cost of these links could be justified also on the account of their traffic value instead of being allocated completely as survivability cost.

The currently performed and reported dimensioning of WIS did not take these issues into consideration. One reason for this is the fact that no PTT leased groups were planned to be used in the final network. Further it was clear that digital LOS links or fibre cable systems have a cost profile where the circuit number cost sensitivity is extremely low. The cost for one channel and 480 channels would be practically the same; the only difference will be in the sub equipping of higher order multiplexers that represent a very small part of the total cost compared with buildings, roads, power and radios. The same fact applies directly to fibre optical cable systems, where the capacity cost sensitivity up to very high circuit numbers is practically non-existing.

It was instead decided that the network initially should be equipped with a certain level of connectivity, and that the resulting survivability should be analysed and compared with the stated requirements. This was done and is reported in chapter 12, and after a certain

Table 4.4 WIS Internodal Traffic Matrix

NODES	1	2	3	4	5	6	7	8	9	10	11	12	13	14	15	16	17	18	19	20	21	22	23	24	25	26	NOWSUM
1	16.6	3.5	1.1	8.5	1.4	0.6	0.2	0.0	9.6	0.9	0.5	0.8	1.1	0.0	3.0	0.0	0.0	0.1	0.0	0.0	0.1	0.2	0.3	0.1	0.3	0.0	49.53
2	3.5	40.4	11.9	2.5	0.5	2.4	0.0	0.0	6.5	0.5	0.1	0.5	3.2	0.0	1.9	0.1	0.0	0.3	0.0	0.0	0.3	0.8	0.0	0.1	0.0	0.0	76.33
3	1.1	11.9	42.2	2.1	0.7	1.2	0.0	0.0	8.0	0.6	0.0	2.5	4.5	0.0	2.3	0.0	0.0	0.1	0.0	0.0	0.0	0.3	0.4	0.2	0.0	0.0	78.80
4	8.5	2.5	2.1	49.4	3.9	0.1	0.0	0.0	10.7	0.1	0.0	1.1	1.2	0.0	0.0	0.0	0.0	0.0	0.0	0.0	0.0	0.0	0.0	0.0	0.0	0.0	80.06
5	1.4	0.5	0.7	3.9	36.0	12.2	2.8	0.1	10.4	0.5	0.4	2.4	3.1	0.0	1.8	0.1	0.0	0.8	3.2	0.0	0.0	0.1	0.0	0.0	0.0	0.0	81.13
6	0.6	2.4	1.2	0.1	12.2	64.4	2.1	0.0	3.9	0.0	1.0	0.3	5.6	0.0	2.6	0.6	0.0	1.6	0.0	0.4	0.8	2.3	0.0	0.0	0.0	0.6	103.00
7	0.2	0.0	0.0	0.0	2.8	2.1	4.2	0.9	14.2	0.1	5.2	0.0	0.4	0.0	1.4	0.1	0.0	0.8	0.0	0.0	0.0	0.0	0.0	0.0	0.0	0.0	33.60
8	0.0	0.0	0.0	0.0	0.1	0.0	0.9	5.0	1.4	0.3	2.4	0.1	0.3	0.0	0.0	0.8	0.0	0.0	0.0	0.0	0.0	0.0	0.0	0.0	0.0	0.0	11.60
9	9.6	6.5	8.0	10.7	10.4	3.9	14.2	1.4	37.2	9.0	15.1	2.4	6.2	0.0	3.2	2.2	0.0	1.1	0.8	0.1	0.7	0.8	0.2	0.1	0.0	0.0	144.40
10	0.9	0.5	0.6	0.1	0.5	0.0	0.1	0.3	9.0	52.0	0.9	0.6	3.6	0.0	0.0	0.0	0.0	0.4	0.0	0.0	0.0	0.7	0.2	0.0	0.0	0.0	70.86
11	0.5	0.1	0.0	0.0	0.4	1.0	5.2	2.4	15.1	0.9	34.2	0.1	2.7	0.0	0.2	3.1	0.2	5.1	0.0	0.2	0.0	0.8	0.0	0.0	0.0	0.0	72.73
12	0.8	0.5	2.5	1.1	2.4	0.3	0.8	0.1	2.4	0.6	0.1	8.4	2.9	0.0	2.4	0.0	0.2	0.0	5.4	1.2	0.0	0.0	0.0	0.0	0.0	0.0	32.60
13	1.1	3.2	4.5	1.2	3.1	5.6	0.4	0.3	6.2	3.6	2.7	2.9	111.8	5.1	14.5	1.1	1.9	2.0	0.8	2.1	5.1	4.8	0.5	1.2	0.5	0.5	187.26
14	0.0	0.0	0.0	0.0	0.0	0.0	0.0	0.0	0.0	0.0	0.0	0.0	5.1	4.4	4.3	0.0	0.1	0.0	0.0	0.0	0.1	0.0	0.0	0.0	0.0	0.0	14.06
15	3.0	1.9	2.3	0.0	1.8	2.6	1.4	0.0	3.2	0.0	0.2	2.4	14.5	4.3	198.8	2.0	0.9	1.6	3.0	0.3	3.1	4.7	1.2	1.4	1.6	0.9	257.66
16	0.0	0.1	0.0	0.0	0.1	0.6	0.1	0.8	2.2	0.0	3.1	0.0	1.1	0.0	2.0	18.8	0.0	2.7	0.0	0.1	0.1	3.1	0.0	0.0	0.0	0.0	35.33
17	0.0	0.0	0.0	0.0	0.0	0.0	0.0	0.0	0.0	0.0	0.2	0.2	1.9	0.1	0.9	0.0	11.4	0.2	0.2	1.9	3.4	1.2	0.0	0.2	0.4	0.0	22.60
18	0.1	0.3	0.1	0.0	0.8	1.6	0.8	0.0	1.1	0.4	5.1	0.0	2.0	0.0	1.6	2.7	0.2	3.2	0.4	0.2	1.6	3.1	0.0	0.0	0.0	0.0	26.06
19	0.0	0.0	0.0	0.0	3.2	0.0	0.0	0.0	0.8	0.0	0.0	5.4	0.8	0.0	3.0	0.0	0.2	0.4	15.6	4.5	0.0	1.6	0.9	0.4	0.2	0.0	37.46
20	0.1	0.0	0.0	0.0	0.0	0.4	0.0	0.0	0.1	0.0	0.2	1.2	2.1	0.0	0.3	0.1	1.9	0.2	4.5	22.0	0.9	6.4	0.3	2.3	0.4	0.1	43.93
21	0.1	0.3	0.0	0.0	0.0	0.8	0.0	0.0	0.7	0.0	0.0	0.0	5.1	0.1	3.1	0.1	3.4	1.6	0.0	0.9	83.2	9.6	0.0	1.3	0.0	0.0	111.00
22	0.2	0.8	0.3	0.0	0.1	2.3	0.0	0.0	0.8	0.7	0.8	0.0	4.8	0.0	4.7	3.1	1.2	3.1	1.6	6.4	9.6	95.0	2.0	2.9	0.5	2.2	143.73
23	0.3	0.0	0.4	0.0	0.0	0.0	0.0	0.0	0.2	0.2	0.0	0.0	0.5	0.0	1.2	0.0	0.0	0.0	0.9	0.3	0.0	2.0	25.8	3.4	1.5	0.2	37.40
24	0.1	0.1	0.2	0.0	0.0	0.0	0.0	0.0	0.1	0.0	0.0	0.0	1.2	0.0	1.4	0.0	0.2	0.0	0.4	2.3	1.3	2.9	3.4	18.6	0.2	1.3	34.26
25	0.3	0.0	0.0	0.0	0.0	0.0	0.0	0.0	0.0	0.0	0.0	0.0	0.5	0.0	1.6	0.0	0.4	0.0	0.2	0.4	0.0	0.5	1.5	0.2	9.4	0.0	15.33
26	0.0	0.0	0.0	0.0	0.0	0.0	0.0	0.0	0.0	0.0	0.0	0.0	0.5	0.0	0.9	0.0	0.0	0.0	0.0	0.1	0.0	2.2	0.2	1.3	0.0	14.8	20.80

number of iterations, the minimum cost network was established also as having the required survivability. It is recommended in the follow on in general network design efforts to study if these added features can be used to enhance the detailed study results.

The maximum capacity on each link is 480 circuits. Since this will represent a resource maximum, and for survivability reasons, it is preferred to align the sizes of the internodal links. The cost function used for dimensioning of the internodal network was therefore defined as described below:

- All links will be assumed with the same cost function.
- Each circuit will cost the same (value 1) between one and 240 channels
- Channel 241 is given a high step value (value more than 1000) to avoid links to be loaded to more than 240 circuits

With the three input files and no damage specified, the internodal links were then dimensioned. The dimensioning started with the initial network topology and with all links equipped with 30 channels each. The program uses an iterative process and allocates capacity according to the defined traffic interest until the defined maximum overall cost figure has been spent. The program will do this in a number of iterations, seeking to find the optimum allocation, by testing the blocking probability achieved by each allocation. It will continue until it has found the capacity allocation under the defined cost constraint that gives the lowest average blocking probability.

At the end of each optimisation a network will be produced by the program. To find the resulting blocking probability, the network is then analysed. The analysis calculates the actual blocking probability for all traffic relations in the traffic matrix, and gives the average value for all traffic. The design blocking probability used in the dimensioning phase was 1%, giving an average end to end blocking probability of 5% (each access link contributes with 2%). If the calculated blocking was found to be below or above 1%, a new optimisation was done with lower or higher cost ceiling respectively. This process continued until the average blocking reached 1%.

The network dimensioned to average 1% blocking in this way is called "the non-rounded dimensioned network". Before further use this network was modified and re analysed by adding circuits to each link until a full number of TDM groups (30 channels) is allocated to each link. This internodal network will give an average blocking probability that will be less than 1% due to the addition of circuits to make up full groups. The network with its final allocated capacity is shown in figure 4.6. This network was then the end result of the study at this stage and was used for the final costing exercise and the subsequent survivability analysis reported in chapter 13.

4.3.3 Dimensioned Transmission Network

The complete WIS backbone transmission network dimensioned to handle the total access and internodal switched traffic is shown in figure 4.7, and has the following main characteristics.

- 61 internodal transmission links made up of 34 Mbit/s (480 channels) digital LOS radio systems, with a total of 239 LOS hops.
- 40 nodes terminating the internodal links, containing in total about 120 LOS terminals.
- The internodal network contains about 180 repeaters, of which about 25 are located in new sites, while the remaining are planned to be collocated with existing PTT installations.

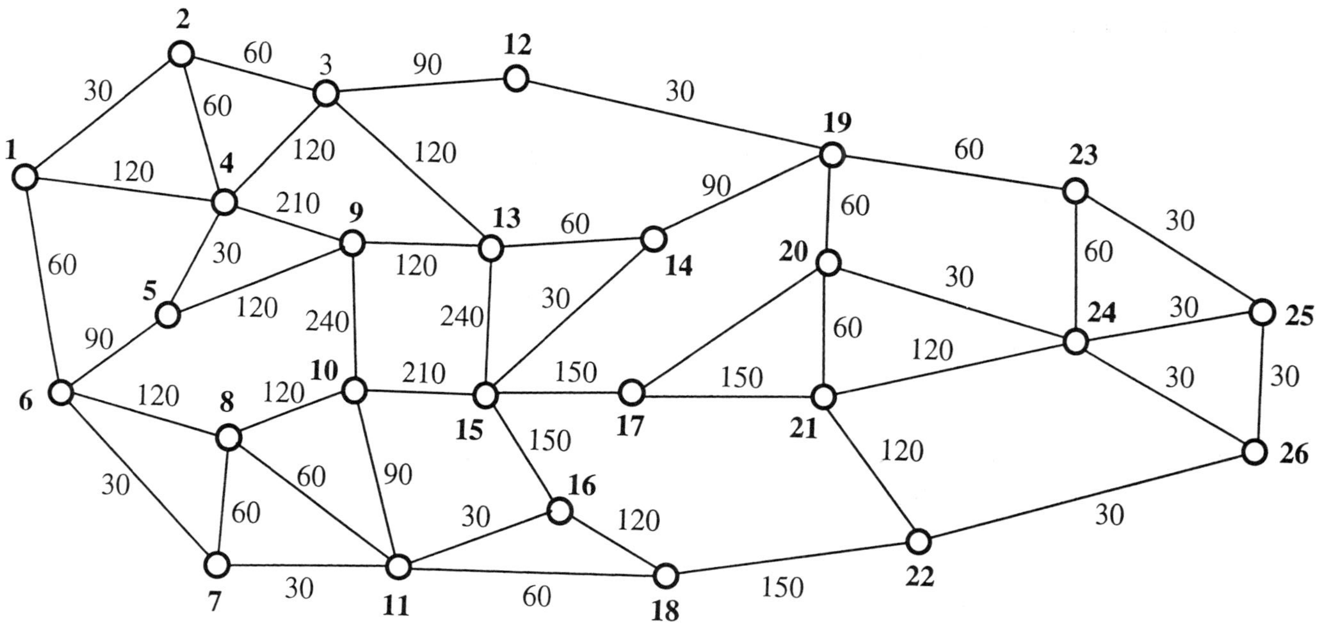

Figure 4.6 Final 26-Node Network

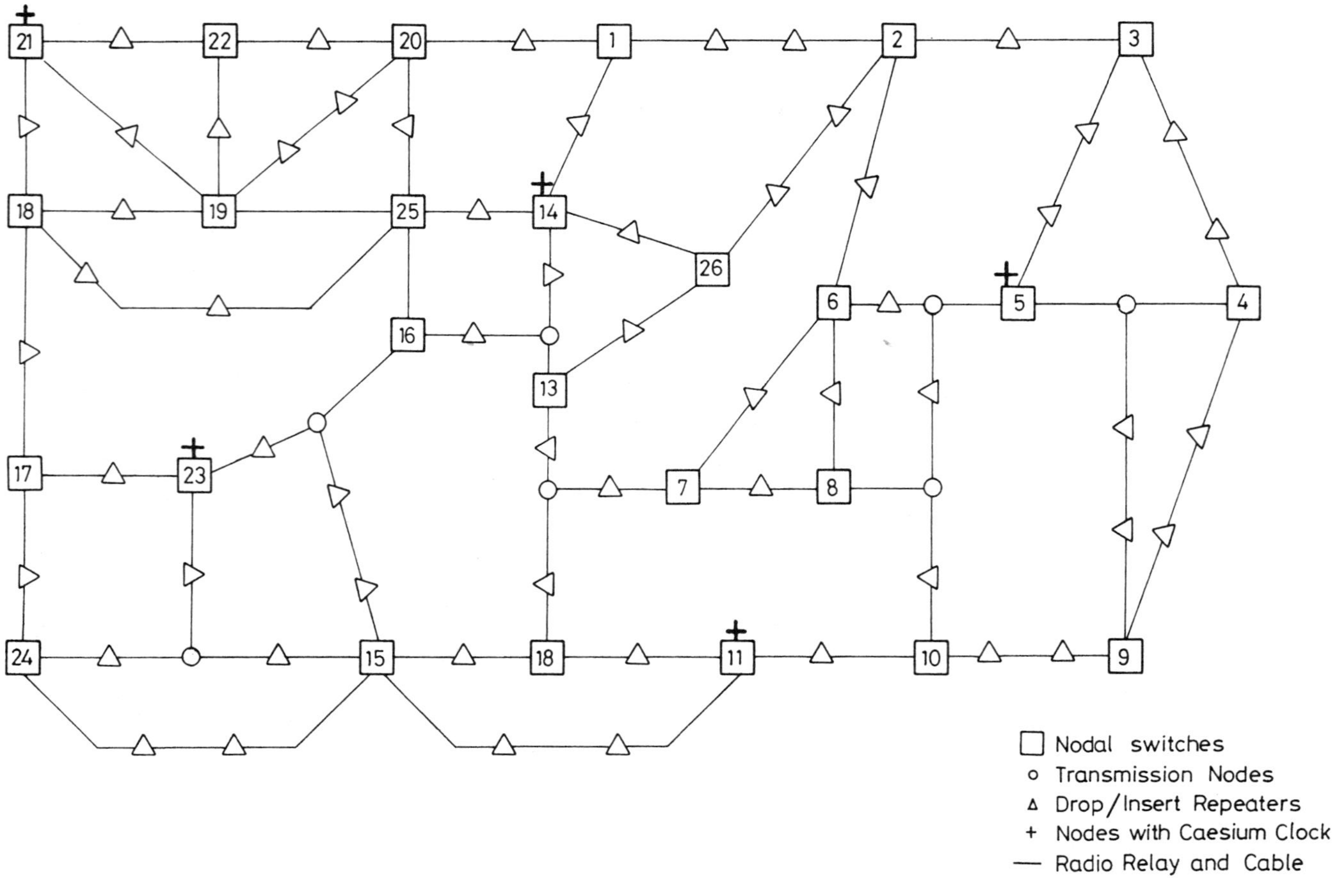

Figure 4.7 Dimensioned Transmission Network

About 90 of these repeaters are pure through-repeaters, while about the same number are also used for group drop and insert to terminate access links from nearby user locations.

In addition to these internodal links, the WIS transmission system contains an access transmission network planned to consist of a combination of fibre optical cable systems and LOS digital radio systems at 8 or 34 Mbit/s. The access network contains the following elements:

- About 280 fibre optical links amounting to a total of about 1800 km of cable
- About 150 LOS radio systems, containing about 300 LOS terminals and 50 LOS through-repeaters.

The total transmission network for access and internodal traffic has been dimensioned as described in this chapter, and was consequently costed using the methodology described below.

4.4 COSTING

The preparation and use of various types of cost estimates has been an important element of the WIS network design. Various types of cost estimates have been prepared to provide the basis for a number of the design decisions that have been made. The cost estimation process was a highly iterative one, starting from a fairly rudimentary estimate prepared just one month after the study started, to the final estimate, which for practical purposes can be viewed as a cost estimate suitable for detailed budgeting purposes.

The main principle followed in the costing of WIS was to take a generalised approach to the different installation types that were to be used. A certain number of standard configurations were identified. Each of these standard configurations were costed to a reasonably high level of detail, and the configuration of these installations were co-ordinated closely with the network design as that evolved.

A summary of the final list of standardised installations are shown in Table 4.5. The installations shown are of two different types. The first type contains all those that can be defined around a certain physical site, and these identical for each type of site. These elements are the various types of repeaters, transmission nodes and nodal switch sites, as well as all the different types of user access sites such as remote access units and access switches.

The other type contains all elements that are optional in the above installations, or are not requiring any specific infrastructure. These elements are for example user terminals, message and packet switches, intra-headquarters networks and the various stand alone elements of the network surveillance and control system.

The overall cost estimate for WIS was assembled by using the configurations listed in Table 4.5, with the unit cost that was calculated, and then multiplied with the number of each type as it was identified during the network design.

Some unit cost estimates like the cost of EMP protection, was subject to separate studies, while other parts of the estimates were based on experience figures collected from various sources, including vendor provided budgetary figures.

The main difficulty encountered in the costing process was to find a way to properly capture the fact that the implementation process would extend over a number of years. This may lead to a number of problems in evaluating the need and cost for interim solutions, and the added cost often experienced when a certain installation can not be completed i one period, but must be revisited a number of times with different subsystems and expansion packages.

Table 4.5 WIS INSTALLATIONS SUMMARY

Installation Type	Nr Of Sites	Total Area	Cap-acity	Power Req	Pers Req	Nbc Prot	Los Ants	Cable Entries
NATO ACCESS LOS	17	20	30	3	2	N/A	1	0
NATO ACCESS F/O	43	20	30	3	2	N/A	0	1
SUBTOTAL	60				120			
TU REM ACC LOS	20	20	30	3	2	N/A	1	0
SH REM ACC LOS	8	20	60	3	2	N/A	1	0
TU REM ACC F/O	79	20	30	3	2	N/A	0	1
SH REM ACC F/0	16	20	60	3	2	N/A	0	1
SUBTOTAL	123				246			
TU ACCESS SW SMALL	87	40	200	10	2	N/A	1	1
TU ACCESS SW MEDIUM	27	60	400	15	2	N/A	1	1
TU ACCESS SW LARGE	10	80	800	20	2	N/A	1	1
SUBTOTAL	124				248			
TERMINALS	10000	0	N/A	0	0	N/A	0	0
MSG PROC SINGLE USER	201	0	N/A	0	0	N/A	0	0
MSG PROC MULTI USER	45	0	N/A	0	0	N/A	0	0
MESSAGE SWITCH	26	0	N/A	0	0	N/A	0	0
PACKET ACCESS UNITS	124	0	N/A	0	0	N/A	0	0
PACKET SWITCH	26	0	N/A	0	0	N/A	0	0
INTRA HQ NETWORK	247	0	N/A	0	0	N/A	0	0
SYNCHRONIZATION UNITS	5	0	N/A	0	0	N/A	0	0
SUBTOTAL	10674				0			
NODAL SWITCH SMALL	8	70	900	35	5	YES	6	2
NODAL SWITCH LARGE	18	140	900	65	5	YES	6	2
SUBTOTAL	26				130			
TRANSMISSION NODE	14	0	480	15	1	YES	5	2
SUBTOTAL	14				14			
TU ACC REPEATER EX	7	20	N/A	10	0	NO	3	2
SH ACC REPEATER EX	76	20	N/A	10	0	NO	3	2
TU ACC REPEATER NEW	1	20	N/A	10	0	NO	3	2
SH ACC REPEATER NEW	3	20	N/A	10	0	NO	3	2
SUBTOTAL	87				0			
TU THR REPEATER EX	17	20	N/A	7	0	NO	2	0
SH THR REPEATER EX	57	20	N/A	7	0	NO	2	0
SH THR REPEATER NEW	17	20	N/A	7	0	NO	2	0
ACC/THR REPEATER NEW	54	20	N/A	7	0	NO	2	0
SUBTOTAL	145				0			
OPERATING UNIT TRANSM	14	0	N/A	0	1	YES	0	0
REGIONAL CONTROL CENT	4	20	N/A	5	12	YES	0	1
MAIN CONTROL CENTRE	2	20	N/A	7	14	YES	0	1
TAFICS CENTRAL OP ADM	1	700	N/A	0	133	N/A	0	0
SUBTOTAL	21				223			
GRAND TOTAL	11274				981			

This problem was attempted resolved by ensuring that the implementation plan to the least possible extent requires interim solutions, and by planing to complete all common activities in an installation, like power, buildings etc. in the same stage of its implementation. This will require that some elements will have to be planned from the beginning with their full capacity, although this capacity may not be usable or required until some years later. This is judged anyway to be the most cost-effective solution, provided that the estimate for the final configuration can be made reasonably accurate from the beginning.

The use of a limited number of standard installations will also give cost saving potentials during the installation period, since it will allow a standardisation to take place in design and specification practices. No savings of this type has been assumed, and the full overhead assumptions as described below have been applied to all types of purchases. It is quite clear that this is a necessary safety margin to include at this stage, since some adjustments may prove necessary when the actual site inspection results and the engineering details are available.

For all equipment cost elements a overhead value is added, using the assumed equipment cost as the basis. This overhead amounts to about 60% of the equipment cost and includes the following elements:

	SUM
A) Equipment total: 100	100
B) Installation: 22% of A	122
C) Special tools and test equipment: 7%of A	129
D) Initial spare parts: 10% of (A + B)	140
E) Codification: 1% of (A + C)	141
F) Documentation: 6% of A	147
G) Transport and insurance: 3% of (A + C + D + F)	150
H) Training: 4% of (A + C)	155
I) Depot level maintenance and test software: 5% of (A + C)	160

CHAPTER 5

SWITCHING AND SIGNALLING

5.1 REQUIREMENTS

At the outset of the WIS System Design study it become apparent that the WIS switching subsystem must satisfy communications requirements arising from the functions of its individual users in war and peacetime conditions as well as in periods of crises. The switching system concept was therefore developed by taking into consideration the operational and performance requirements of the WIS network and services. The WIS network is conceived to be based on a common user, basically circuit switched, area grid network, providing situation oriented, low delay, voice and non-voice services, supporting packet and message switches, and capable of accommodating evolutionary trends in services.

The switching system which will be referred as WIS Switching Subsystem (WSS) is expected to cater for various teleservices required in WIS such as voice, data, facsimile, videophone and message communication in both plain and secure modes. These teleservices should be enhanced by the introduction of various CCITT defined supplementary services, such as closed user groups, call forwarding etc. In addition non-CCITT defined services such as secure-nonsecure warning, precedence, pre-emption and multihoming can be offered.

Such a switching subsystem may also be required to provide wide-area network functions for other subordinate networks. Typical examples are national tactical area communications networks or networks of allied forces. In the case of tactical networks, WSS would provide voice and data, switched and semi-switched interconnections amongst tactical units as well as between them and other national/international networks. For instance, these services could be offered to provide access circuits for remote users of other dedicated networks, their interswitch trunk connections, or point-to-point (voice, data, telegraph) circuits. In the sequel, the subscribers of the WIS network will be called DWS (Direct WIS Subscriber), while users from the other networks will be denoted as IWS (Indirect WIS Subscriber). The context of rationalized resource sharing and service provision for other companion networks imply that WSS should be designed to ease future interoperability requirements.

5.1.1 Service Requirements

WIS Switching Subsystem is to provide voice and non-voice services at stipulated levels of Quality of Service (QOS). WIS is conceived to provide digital switching, accessing and transmission facilities which can be used to support a complete range of services. The minimum service requirements that are to be supported by WSS list as follows:

5.1.1.1 Telephone Service

This service capability is to provide automatic switching of WIS analogue and digital telephone circuits for voice communication purposes and will also include:

- Automatic selection of trunk and access routes,
- Automatic access with direct subscriber dialing, and direct dialling to all Direct WIS Subscribers (DWS) inside a network-wide close numbering scheme, in addition to access to operator assistance when requested,
- Generation of network-wide closed user groups
- Automatic hot-line and delayed hot-line services to replace the needs for dedicated voice circuits and to supplement or replace the dedicated voice circuits used for data or message transmission.
- Short code dialing and/or abbreviated dialing for selected DWS's,
- Facilities for transfer of incoming calls to another DWS and automatic call transfer to a predetermined alternative number or the operator if the primary number does not answer,
- Indication and warning signals to subscribers. In addition to the dialing, ringing and busy tones or signals, indication or warning signal are to be automatically provided to subscribers to indicate called number not available, network congestion, network not accessible due to technical or operational reasons, and other relevant conditions,
- Facilities for recorded voice announcements to calling subscribers to provide information not covered by the standard tones and signals,
- The capability for conference arrangements, or for establishment of semipermanent and permanent lines
- Automatic pre-emption on trunk and access circuits

5.1.1.2 Secure Communication

This capability will include, with preference for DWS users:

- User to user cryptographic protection, i.e., end-to-end encryption capability,
- The capability for secure conference calls,
- The capability for secure facsimile and data communication,
- Authentication methods and procedures for calling and called subscribers,
- Warning signals to subscribers if the complete subscriber-to-subscriber connection is not secure,
- Cryptographic key compartmentation so that general compromise of keys is avoided.
- Secure extensions to IWSs within the users locations.

Unclassified telephone communication with non-WIS users will be effected via a PABX in a nonsecure mode.

5.1.1.3 Telegraph and Data Message

This service capability is to provide automatic and rapid forwarding of telegraph and data messages, and will also include:

- Rapid and reliable delivery to user terminals of single and multi-address, formatted and character-oriented messages. The multi-addressing may involve a separate end-to-end or via store-forward transfer for each copy of the message.
- Rapid and reliable delivery of bit-oriented messages used for a variety of applications such as surveillance data, management/control/status reporting, sensor information etc.,
- Full accountability of all formal messages, such that the probability will be very low (e.g. 10E-6) that a message is either lost, wholly or in part, or misdirected,
- Interfaces with a wide range of user facilities ranging from single access, low speed telegraph circuits, to automated user facilities, e.g., Message Handling terminals,
- Enhancement of Message Handling Services through various supplementary services as related to the Message Transfer Services (MTS) and Interpersonal Messaging (IPMS) Services (see Chapter 9).

5.1.1.4 Data Communication

This capability will include:

- Automatic access with direct subscriber dialing,
- Catering for a large spectrum of required data rates, e.g., as specified in CCITT X.1 Rec.,
- Facilities for different modes of data communication, i.e., automatic hot line or dialed circuit switched, datagram, permanent virtual circuit etc.,

Formatted data communication using both bit-oriented and character-oriented messages.

- Indication and warning signals provided automatically to inform the subscribers about such conditions as network congestion, number not available, incompatible terminals etc. In the case of text or image transmission an appropriate text or graphical indication is to be given.

Another important aspect of data communication will be concerned with the transfer of operational data in WIS, e.g., related to operational situations, battlefield scenarios. These must be speedily transferred. Most of the ensuing file transfer traffic will be due to the transportation of computer files for back-up purposes or for the creation of data bases.

5.1.1.5 Videotelephony/Videoconferencing

Sophisticated video coding algorithms have made it possible to achieve videotelephone and videoconference services at 128 and 384 kbit/s, respectively, with adequate quality. Videophone has been proven to increase the effectiveness of personal communication substantially, and to be a viable tool for teleinstruction, teleeducation, telemonitoring, etc. This low bit rate, high quality service will include:

- Dialed access for synchronous transmission of voice and color video images
- Optional provision of high quality still frames, e.g., as in M-JPEG, that is, intraframe coding of video
- Semipermanent (user programmable) or user dialed videoconference arrangements.

5.1.1.6 Bulk Data Transfer

Bulk data transfer involves usually the transfer of files between one computer and another. The data volumes could be typically of the order of several kilobytes or megabytes. File transfers for the creation of data bases, and data base updates will constitute an important source for this type of traffic. The data flow during one session is basically in one direction, with only protocol messages like acknowledgements flowing in the other direction. The increasing interest in image data file transfer would also be covered under the heading of bulk data transfer. File transfers should preferably be handled in off-peak periods.

5.1.1.7 Messaging Services

The main types of massaging services in WIS can be itemized as follows:

- Electronic Mail or E-mail
- Facsimile or Fax
- Electronic Data Interchange (EDI)
- Transaction Processing
- Telex/Teletex
- Paging

Electronic Mail: E-mail provides an efficient means of communicating textual information between end terminals. E-mail is also claimed to be more user friendly as compared to telex, and more flexible in the body of the message. The main thrust for e-mail usage between user locations will come from the adoption of the X.400 services. There are two options for e-mail, namely mail box delivery, like paste restate, where the user has to interrogate his mail box to reach his mail, or direct delivery. The major advantage of electronic mail is that in terminal areas PAS and word processors are ubiquitous. Furthermore e-mail bypasses message preparation formalism on a separate terminal, time consuming dispatching and monitoring chores.

Facsimile: The advantages of fax is that the original form of the document can be preserved and diagrams can be handled as conveniently as just text. In addition it has a wide range of optional features, and switched telephone network presently provides ample means of intercommunication. The only disadvantage is that fax documents are not word processor compatible.

There is expected a significant growth in the use of facsimile devices, especially, Group 4 devices that have the ability to transmit an A-size document in less than 1 second, and with improved gray scale resolution. On the other hand, since the investment in Group 3 machines is very extensive, one should not expect it to be supplanted by the all-digital fax in the short run. To this effect ISBN interfaces and interworking with conventional Group 3 machines is also being conceived.

Electronic Data Interchange: While the term "e-mail" applies to services that convey information between humans in a free format, a large number of business transactions involve orders, bills, invoices, payments and receipts, which have standard formats, and often are computer generated. These transactions are overall called Electronic Data Interchange (EDI). The CCITT recommendations that cover the EDI operations under X.400 are F.435 and X.435. The potential for EDI in civilian networks is great, if one considers that by the end of 1980s there were close to a billion postal business-to-business transactions. The functionality of EDI in a military network would be limited as opposed to that in a civilian network.

Document Interchange and Data Base Access: Office Document Architecture (ODA) specifies standard electronic document types, that combine text, graphics, scanned images, and audio messages within one framework. ODA holds in principle great potential for conveying many textual documents that presently is being carried by the post, telex, or facsimile.

Another important source of messaging traffic will be constituted from data base accesses, inquiries and updates.

Telex/Teletex: In civilian public networks, telex is the oldest and the most widely established messaging service in the beginning of 1990s. Despite the emergence of several competing services, telex is expected to persist in popularity, and in fact with an increasing use of e-mail services to access telex services. Most telex services throughout the world uses circuit switching to connect terminals. Teletex is a superior telex service with an extended character set, higher data rate and improved layout features. In WIS the function of telex/teletex will mostly be performed by Computerized Message Preparation services.

Transaction Processing: In Transaction Processing the connection is between a human being operating a keyboard and a machine, where the answers are expected to arrive in a few seconds. Typical applications are reservation systems for airlines and electronic point-of-sales activities. It is estimated that in not too distant a future the number of transactions will soon reach the number of telephone calls, although with much shorter holding times, so that the resulting traffic in Erlangs will be smaller. In WIS the relevance of transaction processing will be limited.

Communication for Dedicated Tasks: WIS networks will have to support communication services for dedicated tasks. Typical examples are the support to be given to tactical air operations such as air mission control, airspace management, sensor/air picture data, and air traffic control, offensive/defensive support missions, and coordination with other services, such as command and control resource management, force management, and surveillance.

Paging Services: This is a personal service with the intention to advert the called party (the pagee) of an attending message or of a request to call back a specific number. Paging network could also carry short textual time-stamped messages.

In conclusion, the basic switching and communication functions for telephone/videophone, message, and data supplementary services can be regrouped and assessed as follows:

a) The establishment of connections if called subscribers are busy (e.g., call back, mail box facility, call waiting tone, as well as intrusion warning or subscriber line preemption),

b) The establishment of connections for non-answering subscribers (e.g., call transfer, line group hunting, mail box),

c) The establishment of connections if trunks are blocked (e.g., operator support, trunk pre-emption),

d) Measures for fast establishment of connections (e.g., hot line , short code dialing, simplified packet protocols),

e) The establishment of network-wide functional user communities (e.g., user class marking, closed user groups with different combinations of restrictions),

f) Conferencing and narrowcasting, to be established on both predetermined and ad hoc basis.

5.1.2 Quality Of Service

5.1.2.1 User Friendliness

Services provided by WSS should be easily used and should require minimum user knowledge about the system itself. WSS will also support operator services and facilities to assist WIS users, in particular, in damaged and/or congested network conditions. Operator to operator calls should also be possible for demand services.

Telephone operator assistance will be typically provided for:

a) Directory information service, in particular, information to incoming calls on local subscriber numbers,

b) Setting up of conference calls,

c) Assistance to incoming precedence calls to obtain wanted subscribers,

d) Assignment to a subscriber, within given regulations, of a higher precedence level than his normal entitlement.

5.1.2.2 Directories and Instructions for Use

The users and subscribers are not to be required to know the geographical location of the other users, in particular not those of mobile or itinerant users or their access points into the WIS in order to reach them. All users will be provided with:

a) Instructions for the use of the automated WIS services, including descriptions of the types of services, the numbering system, extracts of procedures and the meaning of the different tones and signals,

b) Appropriate up-to-date number directories necessary to make good use of the system including the switched telegraphy and data services. Directory services should be available as an automatic network service, wherever appropriate, and also through the operator assistance,

c) Immediate up-dating of important directory changes and additions by messages on a need-to-know basis.

5.1.2.3 Service Accessibility

The WIS communication and information processing services should be offered network-wide, that is, the same type and quality of services should be made available everywhere independent of the location and of the user spread.

5.1.2.4 Grade of Service

In all telecommunications networks the resources provided for carrying the offered traffic is necessarily limited, within operational constraints, for economic reasons. This limitation may affect the quality of service to the user of circuit switched services in two different ways:

a) Call processing delays,

b) Blocking, and to the users of packet switched services as delay - throughput relationship.

These impairments are consequences of the finite traffic handling capacity of the network and they constitute the "grade of service". The grade of service together with the factors of malfunctions, loss of service, and transmission performance constitute the "quality of service".

The grade of service parameters for voice, data, and message services in WIS are given below. These figures (valid for normal busy hour period) should be taken as design objectives in network planning and dimensioning.

For voice services it suffices to dimension the WIS network with 5% blocking for end-to-end connections. The blocking probability is partitioned as follows:

- Nodal link blocking 1.0%
- Nodal switch internal blocking 0.002%
- Access link blocking (on each side) 2.0%
- Access switch internal blocking 0.002%.

It is recommended then that "blocking figures of 0.002 % under normal load and 0.01 % under high load" as in CCITT Rec. E.543 has been adopted for the WIS switches. Notice that the 5% blocking figure is an average one for network users, but that prioritized user classes will experience varying degrees of blocking, as detailed in Chapter 2.

Call processing delays, based on the CCITT delay grade of service standards, for WIS switches are as follows:

a) 90% of all calls attempted are not to encounter a dial tone delay of more than 300 - 500 milliseconds

b) The maximum connection set-up time for circuit switched connections in the undamaged system can be calculated assuming the involvement of 5 switches (two access and three nodal switches). The delay consists of the sum of the " dial tone" delay, "exchange call set up" delays encountered at the two access switches and of the "through connection" delays occurring at the nodal switches. Both the through connection delay and the exchange set up delay in the case of common channel signalling, are specified in CCITT Rec. E.543 to be less than 0.5 second under normal load and 1 second under high load conditions. Thus the maximum call set up time can be calculated to be 3 seconds under normal load and 5.5 seconds (worst case) under heavy load. However user and terminal dependent dialling time is not included in the computation of the maximum call set up time. Consequently one can state that the call set up time will be less than 5 seconds for 95% of all routine DWS calls, while the remaining 5% of the calls should be completed within twice the delay stated.

c) The maximum disengagement delay (call clear down time) should be less than 3 seconds for 99% of all calls,

d) The probability of user isolation, i.e., the inability of the subscriber to receive and/or originate calls should be less than 0.02% under no damage to the network,

e) The probability of incorrect call processing should be less than 0.1%. An incorrect call processing occurs whenever, for example, the subscriber is connected to the wrong number.

For telegraph traffic: Messages passed through the store-and-forward network are to be delivered rapidly to all addresses, with reporting to the originator in all cases of non-delivery to user terminals within the maximum times specified below. The time counting is from connection of the message transmission into the first message switch until the start of delivery to the destination terminal. Delivery speeds of telegraph messages are given in Chapter 2. Finally there should be a routing and accountability system giving a very low probability (10 E-6) that telegraph or data messages are wholly or in part lost or misdirected.

For data traffic: The grade of service in circuit switched data services can be considered in two basic areas, namely,

a) call processing delay,
b) failures due to congestion (blocking).

CCITT X.130 specifies in detail data call connection delays. However the recent trend in the ISDN user part is to allow circuit switched data calls to be set up as non-restricted 64 kbit/s channels using the same call processing as voice calls. Thus the data connection delays amount to the delay in ISDN voice calls plus any delay to be encountered in the terminal adapter.

Grade of service for packet switched data services are follows (95% figures):

- Network transit delay: for call request packets is 1600 ms, while for call accepted /connected packets is 1900 ms.
- Data packet network transfer delay is 1600 ms.
- Network clear indication delay is 1900 ms.
- Bounds for other delays not specified here (e.g., queuing delay, DTE processing time) should be as in Rec. X.135.
- Blocking probability in the establishment of virtual circuits should be as specified in Rec. X.136.

5.1.2.5 Service Integrity

Service integrity is characterized by the access, switching and transmission subsystems performances. WSS should meet the following quality of service parameters.

Quality of Service for Voice: For voice communications, a good intelligibility and speaker recognition should be attained both in the clear and secure modes between any two WIS subscribers. Guidelines for the quality objectives both for the clear and secure mode voice between any two WIS subscribers are given below [1]:

- Loudness Loss at 8 dB
- Idle Channel Noise less than 20 dBrnC
- Talker Echo Path Loss greater 60 dB if satellite connections are involved
- Sidetone at 15 dB
- Bandwidth: 3100 Hz (300-3400Hz)
- Total Signal Quantization Noise Ratio greater than 30 dB
- Appropriate measures to be taken if delay exceeds 50 ms for terrestrial connections
- Bit Error Probability as specified in CCITT Rec.G.821 and for satellite links as specified in CCIR Doc.522.

Quality of Service for DataTelegraph: Quality objective guidelines for data are as follows:

- User information error probability, i.e., percent of error-free seconds (EFS): For a 64 kbit/s ISDN connection the value for percent EFS should be as given in CCITT Rec. G.821. A bit error rate better than 10E-5 should be attained between any two fixed terminal equipments in WIS. A bit error rate better than 10E-4 between any two terminal equipments should be attained when one of these serves a mobile user.

- User information transfer denial probability should be 0.15% (or 99.85% availability) based on a similar figure for call cut-off rate.
- User information transfer delay is the value of elapsed time between the start of transfer and successful transfer of a specified user information unit. This parameter should be specified depending upon the type of data service, for example, as in the case of interactive or real-time applications.

5.1.3 Operational Requirements

5.1.3.1 Survivability

Survivability concept has been introduced in Chapter 2 and the issue is addressed in detail in Chapter 13. WSS is to be a highly survivable system in conditions of crisis and war or natural catastrophes, and certainly not less than various infrastructures that it supports. Therefore when subjected to disruption and damage, WIS switching system is to degrade gracefully, and be as insensitive to damage as practical; practical to the degree afforded by a reasonable allocation of resources. In any event WSS is to provide under all but the most extreme conditions, a minimum of the communications for the vital traffic, locally, regionally, on a national scale, and possibly with international links. WIS is required to be fully protected against the effects of:

- Sabotage
- Direct conventional attacks on communications
- ECM
- Electromagnetic pulse
- Collateral effects due to nuclear attacks on other targets.

The consequences of survivability considerations on the WIS Switching Subsystem (WSS) are as follows:

Network Architecture: A non-hierarchical grid network architecture will be adopted at the nodal level, to avoid traffic choke points and to provide maximum routing flexibility through redundancy and adaptive routing techniques. The development of a grid nodal network with an effective control and restoration system is essential to guarantee that the complete disruption of a few WIS switches will not sectionalise the network, and the destruction of any one installation will not isolate any major user location from the WIS network. This is achieved primarily by the following measures:

- Each nodal switch will be connected to at least three others via independent interswitch routes.
  Each access switch will be connected to at least two nodal switches via two different transmission routes.
- Even if the network is disrupted and sectionalized, for groups of users isolated from their access switch, they should be able to continue to communicate in pockets of neighborhoods.

Switches:

- The switch design must be such that it can support an effective network surveillance management actions,
- Local duplication, e.g., provision of on-site spare components and repair facilities, standby power and redundant equipment,
- Adaptive routing: survivability of the network requires that the network topology be reconfigured, in response to changed requirements and, more specifically to

inflicted damage. The efficient use of the surviving reconfigured assets depends on the adaptive routing scheme supported by WSS.

- The signaling facilities should be robust.

5.1.3.2 Communications Security

There is a general requirement for WIS to be a "trusted" system, handling in the clear all the sensitive communication originated from its users while being able to maintain integrity and security up to a stipulated level. One can expect that WIS be able to incorporate secure virtual networks", as can be typically expected for strategic goals or tactical forces. Within these secure networks, one can expect that the majority of the traffic will have classification levels of SECRET and below. (see Chapter 2 for an overview and also Chapter 12) This leads to specific measures taken for:

a) communication services;
b) signalling and network management information;
c) system programs.

In general terms it is required that:

- Designated links connecting an access switch with a nodal switch, and designated internodal links should be protected via bulk encryption. In addition subscriber loops and loop groups traversing non-secure areas to remote sites shall be protected by the use of first order multiplexors incorporating bulk encryption devices.
- Access and nodal switches treat "RED" traffic. The switch design must be such that the probability of passing classified information to unauthorized users or over insecure links is negligible,
- End-to-end encryption to be provided for a certain percentage of DWS's
- WSS common channel signalling system may be required to support electronic key distribution,
- Access to all kinds of communications processors, their stored data and programs is to be strictly controlled through authentication procedures [18, 19],
- A method of warning is to be provided to alert subscribers that they are no longer using a secure system. The warning signals can be in the form of distinctive tones and cadences stored announcements.
- WIS switch software should comply with the requirements of the "secure network" authority.

5.1.3.3 Availability, Flexibility, and Responsiveness

The WIS network is required to have a very high level of availability for each specific user to user service. The WIS network is also required to be highly flexible and responsive system in terms of operational changes, traffic capacity and patterns, and redeployment of resources. All this will generally be attained by modular design and also as a consequence of the achievement of a high degree of reliability and standardization. In designing the WSS the requirements for interoperability, flexibility, responsiveness and a general ability to absorb new requirements, additional traffic sources, varying system performance levels etc. were carefully considered.

5.1.3.4 Interoperability

Chapter 10 addresses the general requirements for interoperability. WSS is to be interoperable with other national networks, and a number of specific networks, such as tactical networks, networks of allied countries etc. The interoperability requirements can be itemized as follows:

i) For reasons of minimum complexity, maximum maintainability, flexibility and robustness WIS should be internally as "clean" and homogeneous as possible. This necessitates gateway solutions and the widescale deployment of gateway units implemented separately from WIS switches. In other words as much of the interworking functions as possible should be relegated to the gateway units.
ii) WIS nodal switches should be capable of interconnections in a secure, automatic mode at gateways on a multichannel basis.
iii) WIS should be able to interchange transmission capacity in the clock transparent mode with the interoperating networks, e.g., the PTT network.

5.1.3.5 Maintainability

Surveillance and Control of WIS Network will be developed in Chapter 11. WIS Switching Subsystem shall be designed so as o provide all information necessary for identification of trouble/fault conditions, the network connectivity, repair and restoration activities and for the traffic management. In short WSS equipment and installations are to be designed for ease of operation and maintenance.

The switches shall be capable of transmitting and receiving maintenance information and responding to commands from on-site and if appropriate, from remote control center(s) on systems over the recommended interfaces and protocols as specified for the WIS Network Surveillance and Control Subsystem. The switches shall use CCITT Z.300 based Man-Machine Language at their input/output terminals. The switches will provide typically the following information:

1) Equipment/system status,
2) Critical load levels,
3) Trouble/fault conditions,
4) Network management controls in effect,
5) Screening of various grade of service.

The hardware and software maintenance features of WSS equipment and installations shall satisfy the following :

Hardware Maintenance:

a) Equipment and installations are to give each site the maximum possible in-house self-sufficiency commensurate with cost-effectiveness.

b) This in-house maintenance should be within the capacity of moderately skilled personnel trained to "unit" and "field" levels of maintenance.

c) The level of operational performance and the identification and location of faults, are easily and quickly determinable; thus FUTs (functional unit tests) and FETs (functional equipment tests) should be available.

d) Test points in the equipment and suitable in-station test jack fields are readily accessible.
e) Faulty components and subassemblies are readily accessible and easily replaced.

f) Spares, coded and easily identifiable, are available in adequate quantities on site and at the maintenance and repair depots.

g) Comprehensive technical documentation including equipment manuals, on-the-job training packages and "as-built documentation" is readily available on site and at the maintenance and repair depots.

h) Adequate numbers of tools, test equipment and instruments for measurement are provided at the maintenance and repair depots.

Software Maintenance:

a) The Authority should have full rights for modifying all the application and system software of the switches.

b) The contractor shall provide the Authority with the complete documentation of WIS software, from technical requirements specifications to individual listings and test specifications.

c) All WIS systems software is centrally documented, controlled and maintained by an agency named, for example, Software Maintenance Development Center (SMDC).

d) SMDC should have adequate software development tools to modify, patch and enhance the available switching software, i.e., a proper software configuration management system.

e) WIS personnel should be trained both at the manufacturer's plant and also at a training school to be set up by the network authority so that software maintenance can proceed without contractor assistance.

f) Detailed procedures should be established for local changes to data bases, and for local testing and fault finding.

g) Proper software replacement procedures should be defined; this will include on-line extension facilities of the switches.

h) A centralized error/problem reporting system shall be established.

5.2 COMMUNICATION SERVICES AND FACILITIES

Communication services and facilities in WIS will be implemented so as to meet the operational and strategic communication requirements as stated in Chapter 2 and Section 5.1. The service types supported by WIS can be classified as:

1) bearer services,
2) teleservices,
3) supplementary services.

Bearer Services: The bearer services describe the transportation of information between locations. They will be available in all access and nodal switches. The bearer services for WIS will be as described in Section 1.3.1.1.

The "circuit-mode, 64 kbit/s, 8 kHz- structured, unrestricted" bearer service is a channel where any bit pattern is allowed and the octet integrity is preserved. This is the basic bearer capability and it can be used to transport such signals as digitized voice, circuit or packet switched data and multiplexed traffic. The "circuit-mode, 2048 kbit/s, 8 kHz structured, unrestricted" bearer capability is required to transport transparently traffic on dedicated trunk groups or trunk groups leased to other networks. If a bearer service is to employ bulk encryption, then clock transparency must be guaranteed. Finally the packet mode bearer service (case c) will in general include the PAD (Packet Assembly Disassembly) facilities to accommodate non-X.25 type terminals.

Teleservices : Teleservices combine the transportation function of communication messages with other information processing functions. The minimum set of teleservices to be implemented in WIS is as follows:

- Digital 3.1 kHz voice telephone
- Facsimile (Group 3 and Group 4)
- Message handling services
- Telemetry and teleaction services (e.g., alarms, telecontrol, weapons control, surveillance, measurement, operational control, and monitoring)
- Videophone/Videoconference.

Supplementary Services: Supplementary services in telecommunications increase the potential and efficiency of teleservices. The supplementary services for voice applications are detailed in Section 5.2.2.2 and the supplementary services for data applications are listed in Section 5.2.3.5.

5.2.1 Implication of Services on Switching Subsystem

The following requirements must be satisfied by WSS in the provision of user services.

Service Deploymen: WIS communication and information processing services should be offered networkwide, that is, the same type and quality of services should be made available everywhere, whenever there is a demand for them, independent of the user location and spread. However, a basic core of teleservices should always be provided to a selected subset of WIS subscribers under all but the most extreme damage conditions (Chapter 2).

Evolutionary: WSS should be open to the maximum range of reasonable options for the evolution of services provided to the users, so as to be able to meet changing requirements in both user traffic capacity, service quality and service types as well as to be able to upgrade services in a compatible manner with new technologies.

Service Integration: The WIS services should be planned and implemented with the spirit of Integrated Services Digital Network (ISDN). This entails pooling of resources, migration toward integrated user terminals, (i.e., voice, data, graphic and video) and integration of functions in the switching and transmission subsystems.

Efficient and Adaptable: The users should be able to efficiently use the full range of potential services and also be able to adapt readily to new services and features. This may be accomplished through various combinations of:

- User friendly operation of the integrated terminals.
- The ability of users to subscribe, to modify and cancel various supplementary service through their own initiatives, e.g., by simply entering certain codes.
- Diagnostic and informative feedbacks to explain the shortcomings and failures in the services. Typical of such feedbacks or call progress signals could be "unavailable bearer service, unknown closed user group, unregistered facility, invalid service parameter" etc.

Flexible and Scenario Independent: The WIS services should be conceived independent of any contemplated scenario and type of operation supported. Therefore these services should be as generic as possible and be provided by a common-user, integrated-resource, integrated capability system.

Intelligence : The processing and information handling intelligence related to the higher layer OSI functions (i.e., presentation and application layers) as a rule will not reside in the switches, except of course, for signalling related services. Message switches, on the other hand, will possess higher OSI layer functions. For example, directory services and other data base systems (application layer) will typically reside in the data processing centers, which will be viewed by the WSS as "just another" terminal element (LAN, PABX, work station etc.). On the other hand data compression, end-to-end encryption services are considered to be terminal specific.

5.2.2 Voice Services

In WIS most of the subscribers will be communicating in the voice mode. The subscriber equipment presently are predominantly of analog type with dial pulse or DTMF signalling in most of the networks. Given the goal architecture of ISDN, it would be best to either upgrade the existing terminals in WIS with TA (Terminal Adaptor) units and/or to plan replacements as well as all new acquisitions and procurements for integrated voice and data terminals.

5.2.2.1 Speech Digitization

The speech digitization scheme will be PCM encoding/decoding rule with A-law companding according to CCITT Rec. G.711 as discussed in Chapter 8. There exist several alternative speech digitization techniques, that can be broadly classified into narrowband codecs rendering synthetic to communications quality speech such as LPC-10 vocoder, mediumband codecs rendering communications to telephone quality speech such as ADPCM or CVSD, and wideband coders rendering broadcast quality speech. Some of these speech codecs may find applications in WIS as follows:

* Wideband speech coding, i.e., 7 kHz speech with ADPCM coding at 64 kbit/s according to CCITT Rec. G.722 for such specific applications as broadcast announcements and special teleconference sessions.
* Mediumband speech coding, i.e., ADPCM at 32 kbit/s according to CCITT Rec. G.721 for such applications as stored announcement messages and for applications where the transmission cost is high, e.g. satellite links, and/or where more channels are to be extracted from an existing transmission capacity.
* Narrowband speech coding, i.e., LPC-10 at 2.4 kbit/s which together with end-to-end encryption is to provide secure speech communication in transmission media which can not support wider bandwidth.

The implications of speech digitization on the WSS are as follows:

- WIS will not be employing sub-64 kbit/s switching. Therefore the narrowband and mediumband speech coders must be either multiplexed (as shown in Fig. 5.1) or rate adapted into 64 kbit/s streams. In either case the problem will be handled in the WIS Terminal Area Subsystem.
- Digitization of speech in the telephone set itself (as opposed to digitization in the subscriber line interface card in the switches) is desirable because this will provide a direct digital interface with the network at the subscriber premises. Changing all analog telephone sets to ISDN versions can not, however, be realized in the short term because of the large investment factor. The implication is that the WIS access switches and PABXs will support both analog and digital (i.e., ISDN) interfaces

on the subscriber side in varying proportions. It is expected that the existing analog outside plant can be used to a large extent (e.g., 90% utilization for subscriber loops up to 4-5 km in extent) in supporting digital transmission at 144 kbit/s for the 2B+D ISDN basic access [8].

5.2.2.2 Supplementary Services for Voice

It will be assumed that certain sets of subscriber categories will be defined with certain access levels to services and with given priority and dialling privileges (i.e., user profiles). Table 5.1 indicates the involvement of WIS nodal switches, access switches and PABXs in the handling of supplementary services. These services are described in detail in Chapter 3.

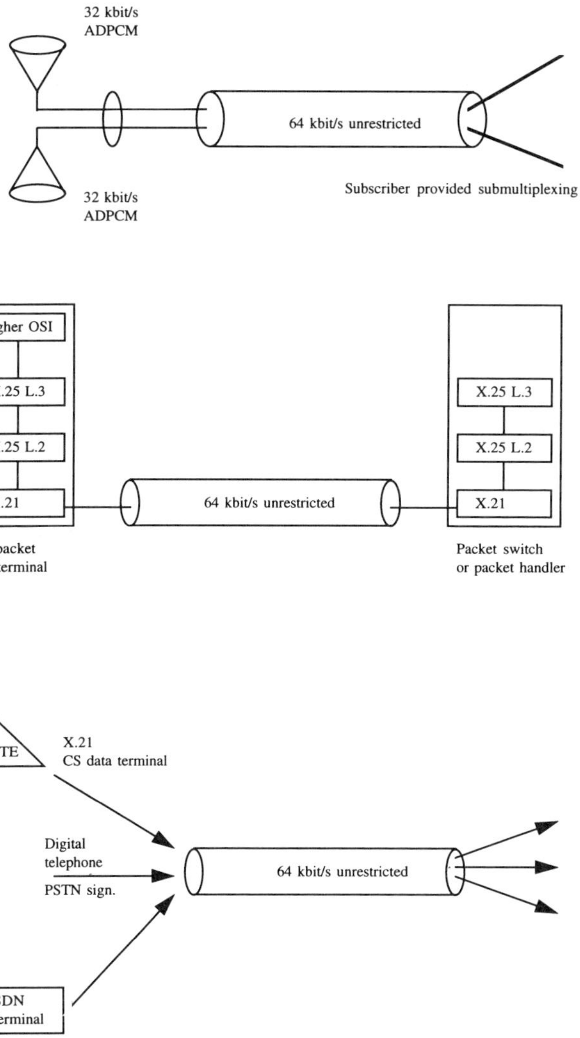

Figure 5.1 **Scenarios For the Use of the 64 kbit/s Unrestricted Circuit Mode Channel**

For WIS, two essential supplementary services are precedence and preemption.

"Precedence" is defined as a subscriber class which avails users with priority to achieve a higher probability of success for call completions. The precedence levels, as detailed in Chapter 2, may be either assigned through fixed class-marking or on a per call basis. In the case of "per call basis" precedence, the subscriber will claim his precedence level by pushing an extra button on his telephone set; otherwise his calls will be treated with routine (default value) precedence. The per call basis precedence makes it possible for the network, through subscriber self-discipline, to treat those calls as routine when they are indeed routine calls. "Preemption" is executed by the routing algorithm (see Chapter 7). Higher precedence preempts lower precedence. The routing algorithm can find out the precedence of the channel(s) to be preempted from the "Call Status Register". Finally the "Closed User Groups" (CUG), can be used network-wide by subscribers with special affiliation to each respective CUG.

Table 5.1 Deployment of Supplementary Services in the PABXs, Access Switches and Nodal Switches

Feature	Access Switch	PABX	Nodal Switch
Closed User Group	Yes	Yes(1)	Yes
Calling Line Id.	Yes	Yes	No
Called Line Id.	Yes	Yes	No
Direct Dialing In	Yes	Yes	No
Call Waiting	Yes	Yes	No
Call Transfer	Yes	Yes(2)	No
Call Forwarding	Yes	No	No
Call Interception	Yes	Yes	Yes
Stockbroker's Call	Yes	Yes	No
Hot Line/Delayed Hot Line	Yes	Yes	No
Group Number	No	Yes	No
Abbreviated Dialing	Yes	Yes	No
Conference Facilities	Yes	Yes	No
Operator Position	Yes	Yes	No
Precedence	Yes	No	Yes
Preemption	Yes	Yes(2)	Yes
Announcements	Yes	Yes	No
Secure/Non-Secure	Yes	No	Yes

(1) Local Only (2) For Incoming Calls Only

5.2.3 Data Communication Services

The traditional data communication needs are met predominantly by leased circuits and/or dial-up modems operating in the voice frequency band at rates from 600 bit/s to 9.6 kbit/s. With the convergence of computer and telecommunication applications, it is expected that data communication services and facilities will be more widespread and the growth in the data traffic volume will be significantly higher in the near future. A major surge of data trafic of all varieties, e.g., resulting from computer mediated work environments, to image and video transport, to file transfers etc. is foreseen.

5.2.3.1 Data Service Types

In WIS the data communication services are envisaged to transport all types of messaging services, such as electronic mail, facsimile, database accesses etc., and teleaction services, such as alarms, telemeasurement (sensor data), telecontrol.

5.2.3.2 Message Services

The requirements for a computerized message preparation and distribution system reflect on the message switches as follows:

- Message switches should be able to handle both single-address and multi-address messages and er them rapidly and reliably.
- Message switches should be able to use efficiently both the circuit-switched and packet- switched media. This is the task of the Transport Layer in the message switching system.
- Message switches will support the range of Message Transfer Services and Interpersonal Messaging Services foreseen by the WIS CMPDS [9,15,16]. Typical of these supplementary services for messages are "Deferred Delivery", "Multidestination Delivery" "Type Conversion rohibition", and "Secure Path".
- The message switches should provide full accountability of all formal messages, such that the probability will be very low (e.g. 10E-6) that a message is lost or misdirected.
- It should be able to transport any formatted and unformatted messages,
- Multimode messages should be allowed, i.e,. messages including text, graphics, picture.

5.2.3.3 Data Services

Communication between computers, their application processes and data processing equipment will constitute an important service category in WIS. The data rates supported will comply to CCITT Rec. X.1, i.e.

a) Circuit switched and semi-switched connections with data terminal equipment operating in start-stop mode, using X.20/X.20bis interface operating at rates of 50 to 200 bit/s and 300 bit/s.

b) Circuit switched and semi-switched connections with data terminal equipment operating synchronously, using X.21/X.21bis interface operating at rates of 600 bit/s, 1200 bit/s, 2400 bit/s, 4800 bit/s, and 9600 bit/s.

c) Packet switched data transmission for data terminal equipment operating in start-stop mode, using X.28 (PAD) interface 50-300 bit/s and 75/1200 bit/s.

d) Packet switched data transmission for data terminal equipment operatingin synchronous mode, using X.25 interface at 2400 bit/s, 4800 bit/s and 9600 bit/s.

e) Data signalling rate supported by ISDN interface at 64 kbit/s and at rates feasible with D channel signalization.

Video services in the form of slow-scan, frame-freeze pictures can also be envisioned. Compressed video signals are being transmitted with the present technology at and below 64 kbit/s rates for such applications as videophone, videoconferencing, and still picture transmission.

5.2.3.4 Teleaction Services

WIS Network Surveillance and Control (WNSC) will either make use of WIS data services or of the WIS signalling infrastructure in order to carry out network surveillance operations, e.g., monitoring network element failures in near real time, detecting degradations in the network (event, performance and status data) and network control operations, e.g., network reconfiguration and restoration, routing and flow control information, as detailed in Chapter 11.

5.2.3.5 Supplementary Services for Data

For circuit and/or packet switched data calls, the operational requirements for supplementary data services are:

* Closed User Group with incoming/outgoing access barring
* Precedence
* Secure/Non-secure Status Indication
* Redirection of Calls
* Multiaddress Calling.

In addition, the following CCITT defined services can be rendered whenever there is an explicit demand for them: Calling Line Indication, Called Line Indication, Date and Time Indication, Connect when Free (Camp on Busy), and for packet switched calls, one can have Extended Packet Sequence Numbering, Nonstandard Default Window Sizes.

5.3 WIS SWITCHING CONCEPT

WIS Switching Subsystem (WSS) will be designed to provide WIS bearer services, teleservices and supplementary services in order to meet operational and strategic communications requirements. WSS system design will be based on ISO's OSI/RM (Open Systems Interconnection Reference Model) and will conform to CCITT ISDN recommendations (see Chapter 3). The military requirements which are not included in the current CCITT recommendations have to be incorporated into an augmented set of standards.

5.3.1 Network Architecture

5.3.1.1 Network Architectural Goals

The following architectural goals will be operationally significant for WIS:

- A layered protocol architecture to provide flexibility and to facilitate interoperability and future evolution,

- Multiple redundant intra-and internetwork links will be used to provide a high degree of resilience and robustness, against failures accidental and deliberate,
- Automated intra- and internetwork management for network surveillance, control as well as selection and maintenance of routes,

Technically, it should provide for minimum proliferation of technical standards, and it should be as application-independent and as scenario-independent as possible. To this effect, a common carrier, common user approach to network design is adopted to achieve cost effective and survivable communications.

WIS will be a two-level network as shown in Fig.5.2. Level 1, called the nodal network, is the area grid network consisting of nodal switches at the nodes in the network, and of the internodal links connecting these switches. Level 2, called the access network, uses the nodal network for its basic interconnections, and comprises the access equipment through which the users are connected to the network. Each nodal switch will be connected to at least three other nodal switches by physically separate routes so as to form a grid network.

5.3.1.2 The Nodal Network

The nodal network will be a switched grid network with ISDN switches in every node. These switches will incorporate packet switching within, initially minimum integration scenario, and eventually within maximum integration framework. Preferably all or the majority of nodal switches will be located within hardened locations and will be EMP protected.

Since survivability is the paramount consideration for this network design, a non-hierarchical dynamic/adaptive routing algorithm is recommended for the WSS. Otherwise from routing and traffic carrying capacity point of view, a hierarchical routing would be more efficient. A common channel signalling system (CCS) is recommended for WSS, which is capable of supporting various voice and non-voice services, and it must be sufficiently capable to support the adaptive routing system. The preferred signalling system of choice is the CCITT SS No.7, as detailed in Section 5.7.

5.3.1.3 The Access Network

The access network will comprise the access switches, and remote access units. The access switches will provide circuit switched connections for the user population n the access area. Packet/message users in sparsely populated areas will be directly connected to a packet/message switch at a node through a dedicated circuit switched connection, alternately densely populated user areas will be served by a direct packet/message switch.

As far as traffic generating characteristics is concerned, two types of subscribers will be differentiated in the access area, namely, Direct WIS Subscribers (DWS) and Indirect WIS Subscribers (IWS). DWSs will have their terminals connected directly to an access switch (circuit or packet/message). The IWSs will use their normal PABX terminals, and ill have access to WIS by using the outward dialing facility of their PABX. The differences between DWS and IWS are explained in Table 5.2.

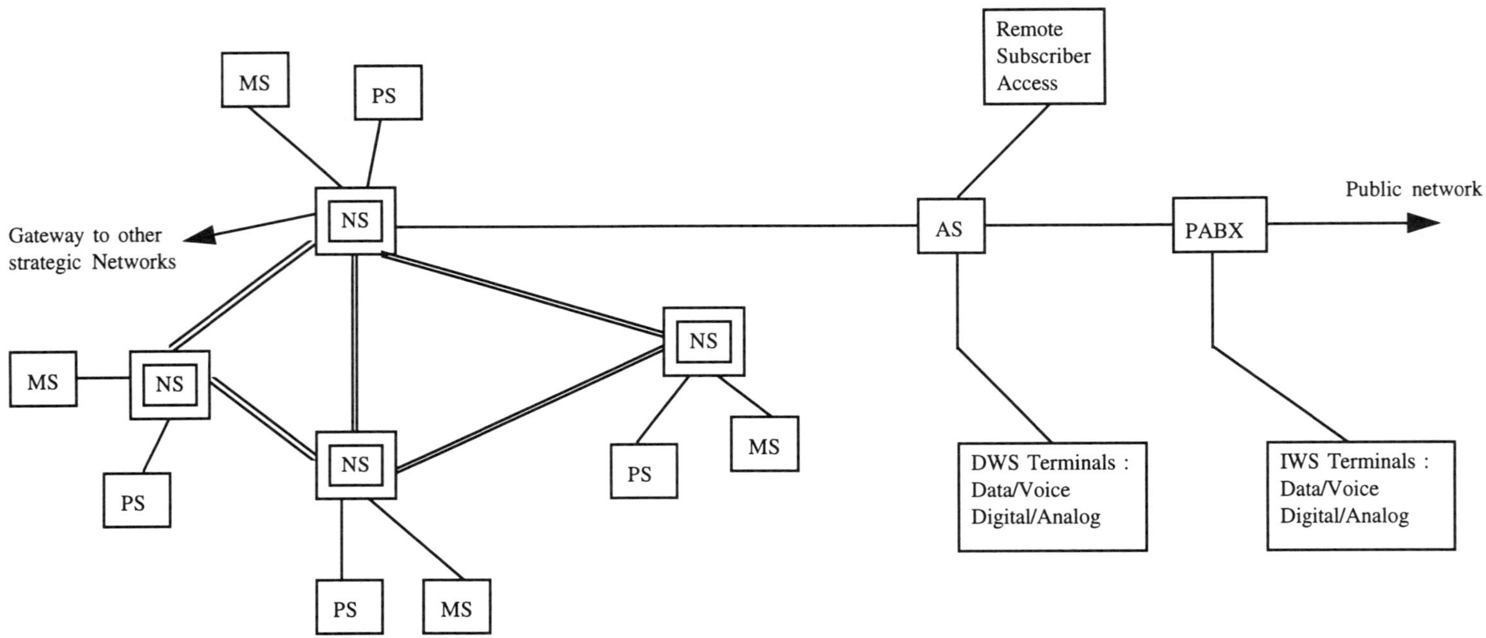

Figure 5.2 WIS Switched Network Architecture (Double Lines Denote the Nodal Network)

Table 5.2 Comparison of DWS and IWS Features

Feature	DWS	IWS
Network Connection	Connected to an access switch either directly or via a remote subscriber accesss unit	Connected to an access switch via a PABX or LAN
Security	A small percentage of DWS can be allotted end-to-end encryption facility	No specific security features will be available to IWS
Numbering	Each DWS will be identified through a unique WIS number	Not every IWS may be allocated a number. Certain sets of IWSs may be reached through a group number
Precedence/Preemption	DWSs will be assigned one of the four level of precedence and will have preemption commensurate to their precedence level	An IWS will only be assigned routine precedence. Preemption applies only to incoming calls.
Supplementary Services	DWSs will have the full range of supplementary services that do not compromise security	Some supplementary services will be offered only locally i.e., CUG, call transfer

5.3.2 Circuit Switching

The choice of switching methods used in a telecommunications network serving the various traffic classes is a major design decision. In the following subsections, the recommendations for:

a) Circuit switched network architecture with an integrated packet/message network and
b) Byte-oriented TDM switching method according to CCITT/CEPT standards will be justified.

5.3.2.1 Circuit Switching Architecture

With regard to the integration of voice and data the following possibilities can be considered:

a) Circuit switching of data as well as of voice,
b) Packet switching of voice as well as of data,
c) Combination of circuit and packet switching in an arrangement where voice channels are circuit switched while interactive data could be packet switched.

Bulk data could be either circuit or packet switched depending on which is more cost efficient.

The larger percentage of the subscribers will be expected to be communicating in the voice 0mode despite an increasing proportion of data communication. Most of the traffic volume quoted in Erlangs, as given in Chapter 2, can be transported adequately via circuit switching. WSS can in principle be based on the circuit switching technique. The circuit switch can accommodate various data services in the same switch matrix as digitized voice circuits, despite their widely varying characteristics. The class-of-service marks would allow the switch to separate various service types and ensure user to user compatibility. They would also enable services provided to be in accordance with the requirements of the individual subscribers.

However there are distinct advantages of providing packet/message switching services that cannot be provided with circuit switching such as burstiness of the traffic, reliability, bandwidth efficiency etc. Thus the integrated provision of packet and circuit switching in a network becomes an imperative necessity.

5.3.2.2 Circuit Switching Method

There are both economic and technical advantages to digital time division multiplexing (TDM) switching. Economically, in TDM switches, there are fewer equivalent cross-points for a given number of lines and trunks than in an analog space division or analog time division switching, and it has full availability (i.e. essentially non-blocking), in general leading to a better traffic performance for less expense. TDM switching is technically advantageous because it is regenerative, i.e., the switch does not distort the signal, the switching method is noise resistant, and the message format (binary) is compatible with data transmission and also with signalling. In conclusion digital TDM switching will be adopted for the WIS switching. The candidate TDM switching techniques are:

- CCITT TDM 64 kbit/s PCM switching,
- Asynchronous TDM fast packet switching (or ATM),
- Bit-oriented TDM switching.

In comparing time division switching methods which are based on byte-oriented switching at basic channel rate and bit-oriented switching one should note that bit switching techniques do not support word-structured channels such as PCM and since most of the digital traffic will be speech or data encoded using a PCM coder, 64 kbit/s byte-oriented TDM switching should be the preferred switching method. Other factors in favor of the byte oriented method are as follows:

Byte-oriented TDM switching is a state-of-the-art and proven technology. The 64 kbit/s byte-oriented time division switching is becoming overwhelmingly commonplace in both civilian and military networks. The ensuing logistic, procurement, and maintenance advantages should be taken into consideration.

Fast Packet Switching (FPS) is a new concept for providing integrated voice and data switching services as part of the integrated Services Digital Network. ISDN based on fast packet switching is expected to provide greater flexibility than the circuit-switching method which is used in today's public ISDN neworks [17]. Consequently this switching method runs as a close competitor to the byte oriented TDM switching. It is expected that ATM/FPS will be the switching method of the future Broadband ISDN networks.

It is desired that the basic channel rate in WIS switching satisfy two requirements: On one hand it should be the lowest rate in the time division multiplexing hierarchy to be used in transmission so that the multiplexing rate becomes n times the basic rate. On the other hand the CCITT defined user rates in the X.1 recommendations should possibly be a submultiple of the basic rate, (i.e., the data rate is the basic rate divided by n), for ease of submultiplexing and rate adaptation.

Integrated Services Digital Network (ISDN) is being defined by CCITT based on the 64 kbit/s channel rate, called the B channel. The basic access consists of 2B+D channels and the primary rate access consists of 30B+D channel, where D is respectively, a 16 or 64 kbit/s service channel. The choice of the 64 kbit/s rate can guarantee the future compatibility of the WIS network with integrated services and integrated network architectures. Accordingly the 64 kbit/s basic channel rate satisfies the subrate and multiplexing hierarchy requirements. Thus a standard CCITT type byte-oriented time division switching at 64 kbit/s basic channel rate should be the preferred switching method. The exchanges will be also switching the channels in the first level PCM multiplexing hierarchy at 2048 kbit/s as specified in the CCITT G.732 recommendation.

5.3.3 Packet Switching

5.3.3.1 General

Packet switching involves the transmission of messages using standard sized information packets processed and routed from node to node in a store-and-forward mode, enabling efficient use of the transmission resources. There are two general methods for routing packets through networks, that is datagrams and virtual circuits. In datagram mode, packets are routed independently and may therefore arrive at the destination with loss of order. In this mode (connectionless-mode) there is no time lost for setting up a route, but every packet must carry more overhead for routing information. In virtual circuit (VC) mode, a "VC request" packet sets up a logical connection between the source and destination.

Thus in this connection-oriented communication the routing decision is not performed for every packet and hence processing is simpler for multi-packet messages. Routing by virtual circuit is more vulnerable to node failures and less adaptive to changing traffic conditions. In CCITT recommendations packet services are based on Rec.X.25 for user to network interface and Rec X.75 for interfaces between networks. These recommendations define the first three layers in OSI/RM for these interfaces.

Packet switching will be suitable in such applications as interactive computer usage, electronic mail, teleprocessing, alarms, telemonitoring, transportation of message handling services, and data base services etc.. The advantages providing interactive data services through packet switching as compared to circuit switching re as follows:

Compatibility: For circuit switched connections the terminal devices must be strictly compatible in their signalling rate and and access protocols. Packet switching becomes increasingly important in the face of terminals that continue evolving and being diversified. For both message and packet switching networks, which are transaction oriented networks, the terminal devices need not be fully compatible to communicate as rate, protocol, and format conversions can be realized by the packet or message network itself.

Bandwidth Efficiency: Many data terminal devices are operated in the interactive mode. For such applications the bandwidth resource is much better utilized when it is dynamically shared among several users via packet switching, a packet switching network being more responsive in resource allocation. Furthermore packet switching suitably enhanced with store-and-forward functions will form a cost-effective transport medium for the message services.

Reliability: The standard interface for the packet mode terminal equipment is the X.25 protocol. The inherent error protection offered by the frame and packet layers of the X.25 interface makes the packet switching a more reliable transport mechanism for messages as well as for interactive data communication.

5.3.3.2 Packet Switched Network Architecture

The proposed packet switched architecture will be a one-level network with packet switches collocated with the nodal switches as shown in Fig. 5.3. The packet switches will be interconnected through dedicated circuits,with capacity n x 64 kbit/s, with automatic restoration capability.

The packet switched network will employ adaptive routing with flow control to attain survivability and load balancing. The access to the packet switches will be through switched and/or semi-switched connections from the access switches to the nodal switches. However, in the goal WIS architecture both the access and nodal switches will have integrated switching capabilities and a much wider deployment of packet switching capabilities will then be possible. Switching capabilities in an ISDN node including packet switching are illustrated in Fig. 5.3.

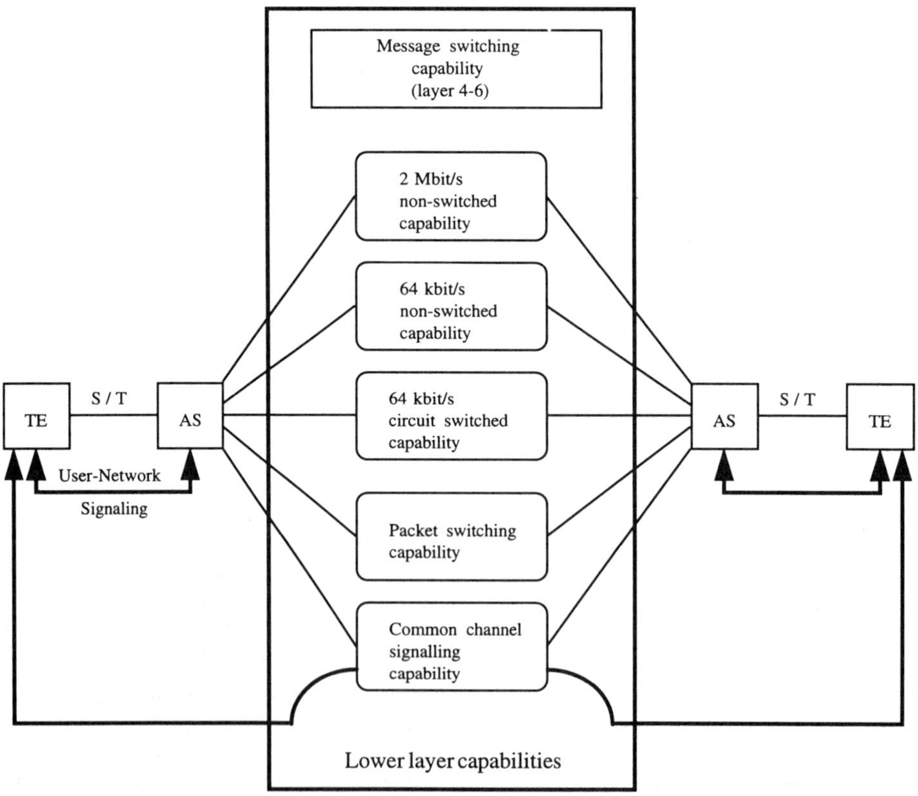

Figure 5.3 Switching capability in an ISDN node. AS: Access Switch; TE: Terminal Element; S/T: Interface Points

5.3.4 Message Switching

Message switching in the WIS architecture will mostly provide services belonging to the higher OSI layers. More specifically the functions and features of the message switches will be as follows:

- Message switches are store-and-forward switches and as such as they receive and deliver the entire message to the other user locations directly served by them or forward it to other switches for delivery to other user locations addressed in the message.
- Message switching centers will keep track of all messages received and offer guaranteed delivery and long term reproducibility despite misrouting and corruption of the message.
- Message switches will offer multi-level security and prioritization.
- Message switches will have collocated CMPDS (Computerized Message Processing and Distribution) capabilities as detailed in Chapter 9.
- On the network side message switches will perform OSI 4-6 layers functions, i.e., Transport Layer Session Layer, and Presentation Layer. Typical of Transport Layer functions are:

  a) optimization of the use of the available resource, e.g., the decision to use circuit switched or packet switched medium for message transportation,
  b) provision of end-to-end error detection and recovery,
  c) blocking and segmentation of Session Layer messages,
  d) provision of end-to-end flow control on each transport connection,
  e) multiplexing transport connections onto network connections.

The Session Layer functions are:

a) establishing and terminating connections between two presentation entities
b) provision for session synchronization and recovery,
c) provision for normal and expedited data exchange,

OSI layered architecture of WIS message switches is illustrated in Fig. 5.4.

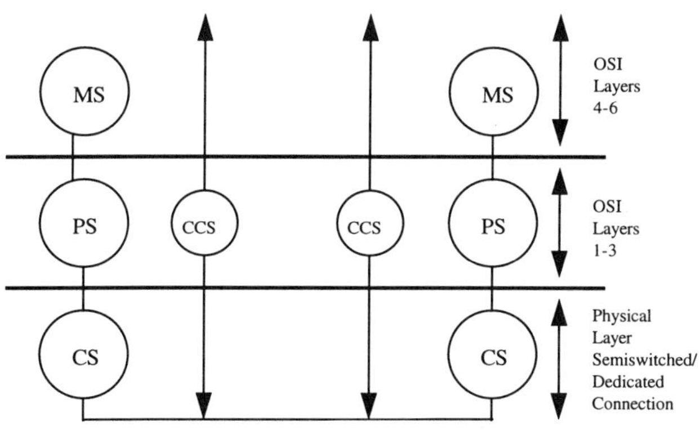

Legent :

CCS	Common channel signalling
PS	Packet switching
CS	Circuit switching
MS	Message switching

Figure 5.4 OSI Architecture of Message Switches When the Packet Switched Network is Uused as the Transport Medium

5.4 CIRCUIT SWITCHING SYSTEM

5.4.1 Nodal Circuit Switches

WIS Level 1 network is the backbone, common user, circuit switched network. This forms a nationwide mesh grid network, consisting of the nodal switches and of the internodal links interconnecting them. This circuit switched nodal network will also serve as a bearer for the packet switched and message switched services (overlay networks) and for the Common Channel Signalling (CCS) messages. Note that these various switching types can be integrated in a single switch (ISDN switch) or they can be implemented as separate subsystems, i.e., sharing the same installation and physical facilities or built in completely separate units and sites.

5.4.1.1 Nodal Switch Functions and Features

The nodal circuit switches will perform the following functions:

* The nodal circuit switches will be capable of making connections between isochronous channel time slots with the basic rate of 64 kbit/s, and where the channels are grouped in accordance with the primary frame structure (G.703/G.732). They may also be required to switch nx64 kbit/s groups, 1< n <32, in particular for the case of n=32, to switch 2048 kbit/s groups. The functional block diagram of nodal circuit switch is shown in Fig. 5.5. The connections will be used to transport the following types of information:

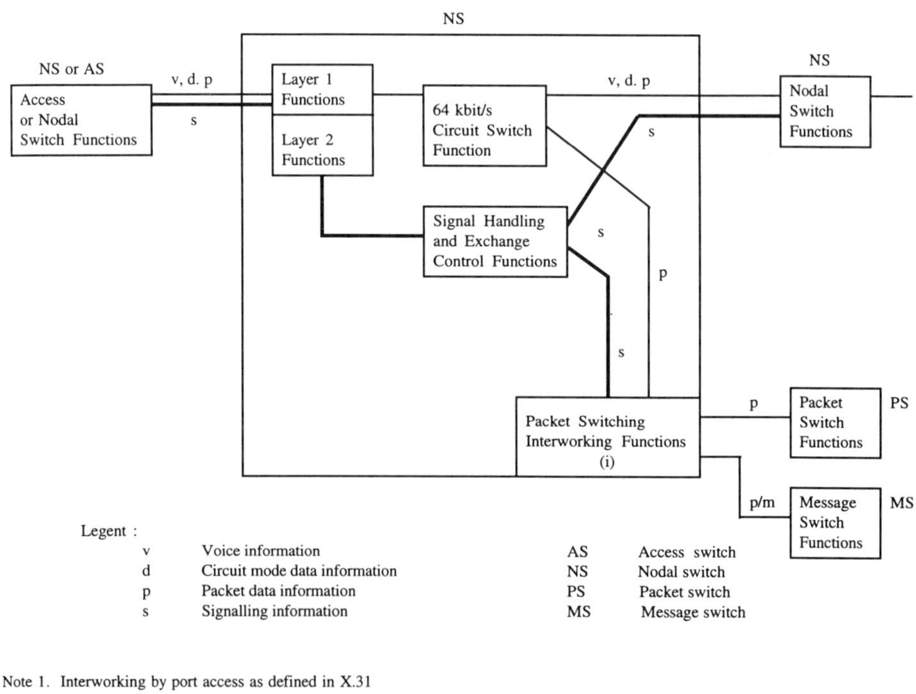

Figure 5.5 Functional Block Diagram of a Nodal Switch

- Telephony and associated voice-band information,
- Circuit switched data information (at 64 kbit/s or rate adapted to 64 kbit/s),
- Packetized data information: This involves the setting up of a circuit switched/semiswitched connection between the user terminal and the access port to the packet switch. The interworking function in the nodal switch is activated by an interchange of information between the user and the packet switching interworking function using packet transfer procedures (CCITT X.25).

- Message data information: This may involve the setting up of a circuit switched/semi-switched connection between the message switches. Alternatively message switches may be using the packet switched medium for their basic transport requirements.

  * Four types of interconnection will be handled by the nodal circuit switches as shown in Fig.5.6, namely:

  1) connections between nodes,
  2) connections between access switches connected to the same node,
  3) connections between access switches and the node to which they are directly connected,
  4) connections to gateway interfaces.

In addition the nodal switch must perform the following functions:

* Multihomed connections of access switches on an automatic basis.
* Perform adaptive routing functions for the purposes of call routing in a manner that is responsive to operational requirements in normal and damaged network conditions.
* Participate in the Network Timing and Synchronization functions as well as in the Crypto Key Distribution and Management
* Participate in the rendition of the supplementary services for circuit switched calls, such as Closed User Group, Call Interception, Precedence, Preemption, and Secure/Non-Secure Link Monitoring.
* Provide maintenance operations and network management functions.

Furthermore the nodal circuit switches should possess the following features:

* They should be operating with a maximum amount of autonomy and adaptability to enhance survivability of the network. These qualifications should bear significantly on the design of the routing method as well as on the organization and the design of the Routing, Synchronization, Surveillance and Control subsystems.
* The nodal switches can be located at specific transmission sites or authority centers, such as military headquarters themselves in order to benefit from their infrastructure and safety. The selection of the nodal switch locations wil be carried out based on balanced survivability considerations. Sites equipped with the nodal switches may or may not have collocated with them an access switch, a message distribution centre, or a network control centre.

To allow for flexibility and ample future growth all internodal transmission links will have capacity assignments as calculated on the basis of the following data for WIS [12]:

Number of DWS	:	N_{DWS}
Number of IWS	:	N_{IWS}
Number of access switches	:	S_{ACCESS}
Number of remote access units	:	S_{REMOTE}
Number of nodal switches	:	S_{NODE}
Traffic per DWS	:	λ_{DWS} Erlang
Traffic per IWS	:	λ_{IWS} Erlang
Traffic per access link	:	λ_{ACCESS} Erlang/channel
Traffic per internodal link	:	λ_{NODE} Erlang/channel
Average number of internodal links per call	:	L_{NODE}
Ratio of internodal traffic to total traffic	:	ρ

Then the following formulae can be used:

Traffic generated	$= T_{traffic}$	$= (N_{DWS}\lambda_{DWS}+N_{IWS}\lambda_{IWS})/2$
Total internodal traffic	$= T_{internodal}$	$= T_{traffic}L_{NODE}\,\rho$
Internodal channels	$= N_{channel}$	$= T_{internodal}/\lambda_{NODE}$
Average number of channels/link	$= N_{channel}$ /(Total number of links)	

the latter quantity resulting from network dimensioning study. The figure of "average number of channels per link" can be used to select the multiplexing hierarchy.

* Digital Access and Crossconnect Systems (DACS): Reliance for survivability only on a grid network and independent dual homing for access by the users to the network may result in an expensive network. The use of DACS can, however, provide higher flexibility, dynamic and better utilization of the transmission network, as well as drop and insert capability for the access switches as shown in Fig. 5.7.

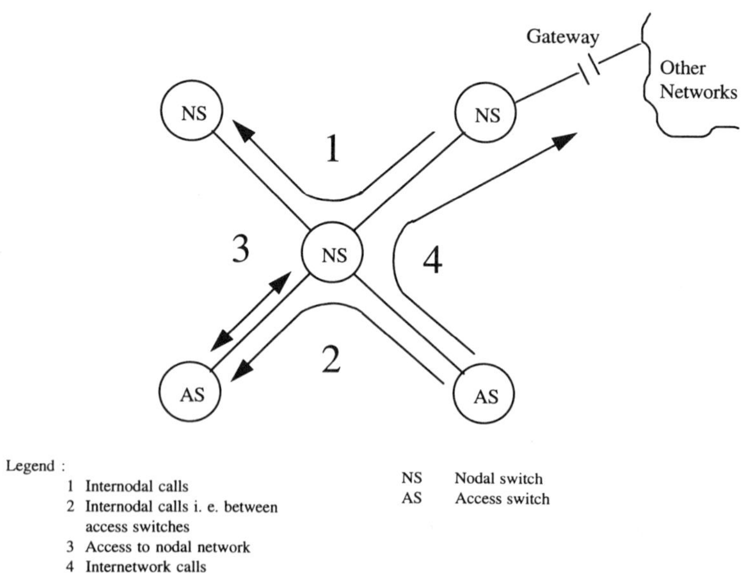

Figure 5.6 Types of Calls Handled by the Nodal Switches

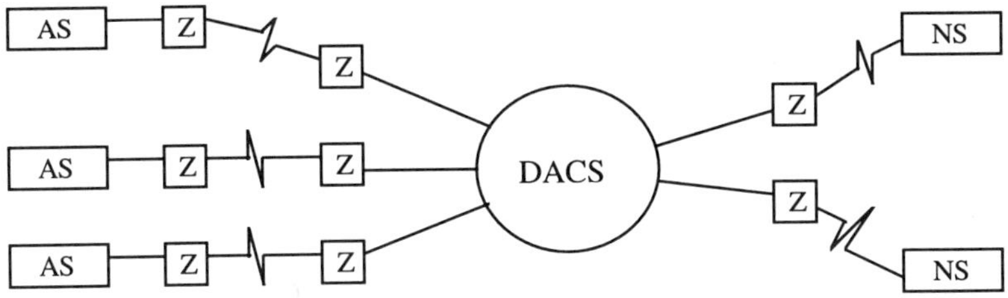

Z : Bulk Encryption Device

Figure 5.7 Deployment of a DACS in the Nodal Circuit Network

5.4.1.2 Nodal Interfaces and Bearer Service

Nodal switches will handle channels carrying A-law encoded voice signals, audio (e.g., analog modem) signals and data signals. However since WIS is expected to be a purely digital network, i.e., without any analog sections, signals will not be subjected to tandem A/D conversions, and speech compressors or interpolators will not be employed. Thus the following bearer services will be distinguished at the nodal circuit switches:

1) Circuit-mode, 64 kbit/s, 8 kHz, structured unrestricted, (channel B),
2) Circuit-mode, 2048 kbit/s, 8 kHz structured,unrestricted clock transparent channel (channel H12).

The nodal circuit switches will have the following type of equipment connected to them, remotely or collocated ,as in Fig. 5.8:

a) Digital Access Switches,
b) Message Switches,
c) Packet Switches,
d) Gateway units for interworking with other networks.
e) Other nodal switches.

The internodal and access links in general will conform to the CCITT Recommendations G.703 - G.705, G.732, G.735, G.745, G.53.

Interface a in Fig 5.8 is a digital interface as described in CCITT G.703, G.704, G.705 with a multiplex structure as in G.732, that is a nominal bit rate of 2048 kbit/s, 8 bits per channel time slot, 32 channel time slots per frame and with timing in the transmitting direction derived from the nodal switch.

Interface e is a digital interface as described in CCITT G.703, G.704, G.705 with a multiplex structure as in G.732, G.744 or G.753 that is with nominal bit rates, respectively, of 2, 8 or 34 Mbit/s , 8 bits per channel time slot, and with timing in the transmitting direction obtained from the nodal switch. Normally digital interface at nominal bit rate of 2048 kbit/s will be used in the nodal switches. Interfaces a and e will also support such functions a signalling insertion and extraction, code conversion, frame alignment, alarms and fault indications.

Interfaces d are gateway interfaces for other networks such as IVSN, CZCS, NSN etc. networks. They have the same characteristics as Interface e but they differ in interface functions such as signalling conversion, mapping of supplementary services, directory number translation, precedence conversion, maintenance procedure (see Chapter on Interoperability).

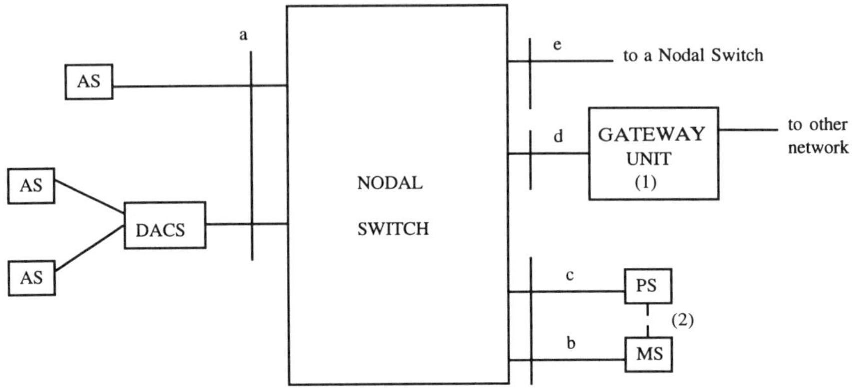

Notes :

1. Each network will have a different unit
2. Message switches may use the packet switching subnetwork for interswitch connections.

Legend :

AS	Access Switch
NS	Nodal Switch
PS	Packet Switch
MS	Message Switch
a,b,c,d,e	Interfaces

Figure 5.8 Interfaces Supported in a Nodal Switch

Interface b and c are X.75 type interfaces to a packet or a message switch. This is an interworking by port access where the nodal switch carries the interworking function unit (IWF) and the packet switch operates all mandatory elements of the OSI connection mode network service, as described in CCITT Rec.X.300.

5.4.2 Access Switches

WIS Level 2 network is a circuit switched network providing access functions, consisting of digital access switches and connected access equipment such as PABXs, concentrators, and multiplexers. An access switch serves WIS users in its area by providing switching among them and through connections to the nodal network.

5.4.2.1 Access Switch Functions and Features

An access switch provides originating, terminating, internal and through connections among users connected to it and access circuits. The access switch will establish circuit switched and semipermanent connections between channel time slots with the basic rate of 64 kbit/s. Services requiring less than 64 kbit/s for a connection will also be switched as 64 kbit/s connections. The connections will carry voice and data information as described in Fig. 5.9.

Bit integrity will be maintained to support secure voice and data services. The access switch will have means to disable digital processing devices such as A-law converters, digital PADs etc. The access switches will also perform the following basic functions:

* Perform interface functions to support the various user and network interfaces described in Section 5.3.1.3.
* Extract synchronizing information from one or more incoming bit streams in the access links.
* Interwork with nodal switches using WIS internodal signalling system, i.e., SS No.7 and with digital user equipment using access signalling procedures based on CCITT Recs. I.430, I.431, Q.920 (I.441), Q.930 (I.451), Q.701, Q.702 and Q.703 and with analog user equipments using DTMF, DP and E&M subscriber loop signalling procedures.
* Support various supplementary services as detailed in Section 5.2 and in Table 5.1.
* Limited transit routing capability: Some access switches may be required to perform limited transit routing function. Because of their dually homed connections, they may perform a "last resort" switching between the two nodal switches if all the internodal connections between them has become severed.

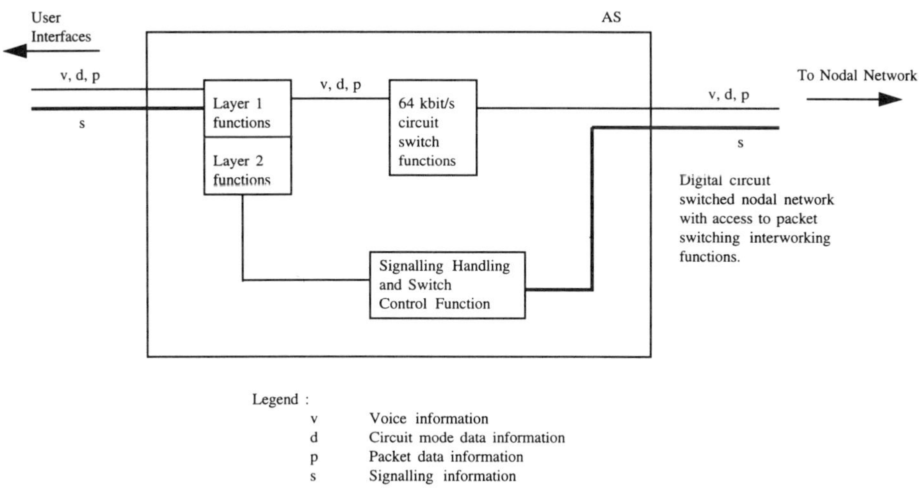

Figure 5.9 Access Switch Functions

5.4.2.2 Access Interfaces and Bearer Services

An access switch will provide 64 kbit/s circuit mode, 8 kHz structured, unrestricted channel bearer service where any bit pattern is allowed and where octet integrity is provided. This is the basic bearer capacity and various scenarios for its use are shown in Fig.5.1. For packet connections this channel is transparent to the access switch in that it connects a packet subscriber to the packet switching interworking function incorporated in the nodal switch.

Access Interfaces: The interfaces associated with the access switches are shown in Fig.5.10.

* Interface A: Standard 2048 kbit/s, 30/32 channel digital interface based on CCITT/CEPT Rec. G.703, G.704, G.705 and where the frame structure is according to Rec. G.732. This constitutes the standard interface to a nodal switch or to a DACS.
* Interface B: Analog subscriber line interface that is used to connect individual analog user terminals (analog telephone or data modem) or analog PABXs. The

output of the V.24 type data circuit terminating equipment will be converted to PCM as in the case of voice. The transmission characteristics of this interface are described in CCITT Rec.Q.517.

* Interface C: This interface handles data in 10 bit envelope structure as described in CCITT X.51, where bit 1 is a status bit conveying call control information, bit 2 is an envelope alignment bit, and bits 3-10 are information bits. This interface will be used to connect X-type data terminals (user classes 3-7) which will be rate adapted in the switch and then circuit switched. Packet mode data terminals will use this data interface where only level 1 interface functions (X.21) are performed. Level 2 and 3 functions are carried out between the packet switch and the user.

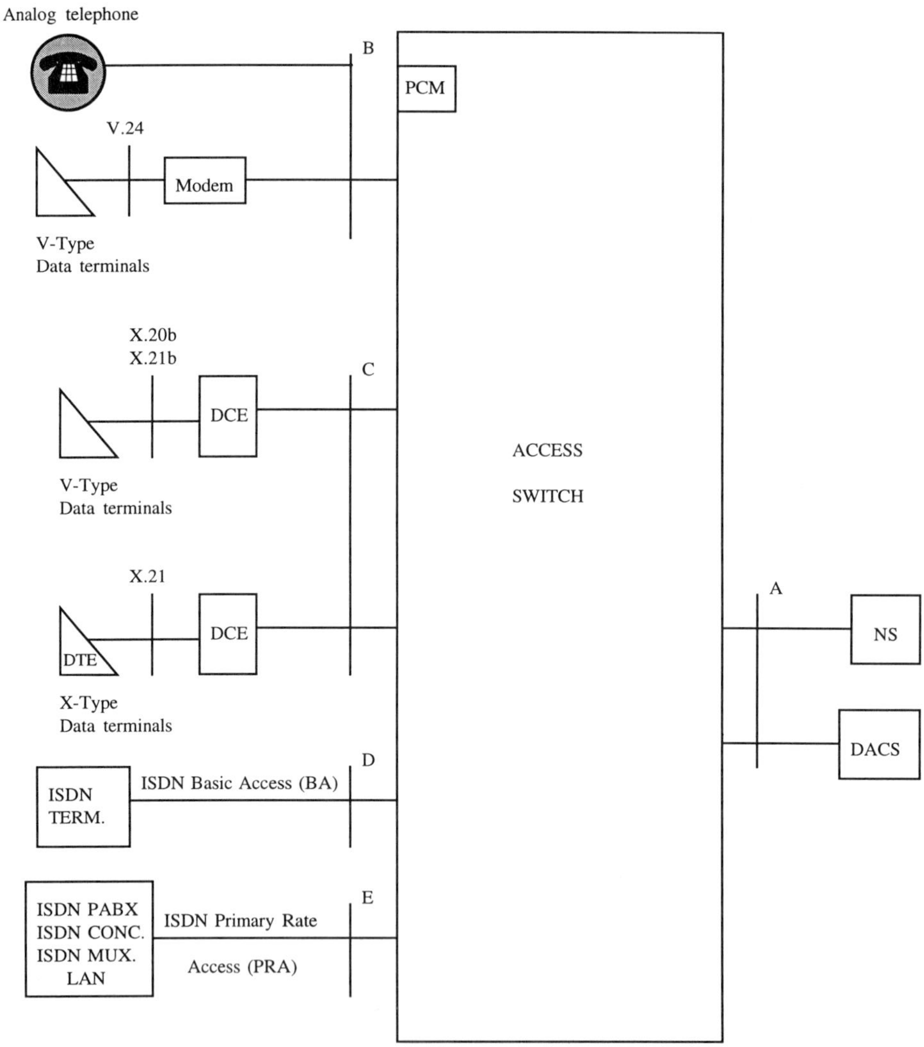

DCE : Data Communications Terminating Equipment

Figure 5.10 Interfaces Supported on The Access Switch

* Interface D: This is an ISDN basic access interface. The basic interface is composed of two B channels and a D channel, that is the 2B+D structure. The bit rates of the B and D channels are 64 kbit/s and 16 kbit/s, respectively. The specifications of interface D are contained in CCITT Recommendations I.430, I.440, I.441, I.450 and I.451. This interface will be used to connect ISDN terminals, non-ISDN terminals via terminal adaptor devices (TA).
* Interface E: This is the ISDN Primary Rate Access interface. The interface structure is composed of 30 B channels and a D service channel, each having 64 kbit/s rate. This interface will be used to connect ISDN PABXs, ISDN concentrators, LANs or ISDN multiplexers.

5.4.3 Access Components

Various access configuration will be used to connect WIS users to the switched network. Typical of the access components are PABXs, concentrators, Local Area Networks (LANs), and multiplexers. Their configuration will depend on the operational requirements, traffic statistics user density and cost considerations. LANs will be discussed in detail in Chapter 9.

5.4.3.1 PABXs

PABXs are used to provide voice and data communication services inside user locations and access into the public network. There are a number of situations in which a PABX is not suitable for data, an example being the case of mini or mainframe computers where a high volume of data of bursty nature may be exchanged between their systems. In this example, a LAN using a CSMA/CD or token passing access scheme would offer more flexible bandwidth allocation (Chapter 9). However if most WIS data applications are text riented (telematic services) then the requirements for data transfer between data processing systems and terminals both for distribution inside the user location and wide area networking can be meet by modern digital PABXs.

One can also argue whether WIS PABXs can be deployed as access switches. In this case some shortcomings may exist in the following areas: communication security; common channel signalling; precedence-preemption; multihoming and adaptive routing; conference; network surveillance and control.

Security requirements have been of minor importance in commercial PABXs. This situation could change with the introduction into PABXs of electronic mail features, which require security in case the information is misrouted.

In PABXs secure speech transmission can be achieved by means of external units. The need to have four-wire connections for secure voice speech communication is easily satisfied in digital PABXs, because the switching matrices and trunk interfaces already operate on a four wire basis. However electronic key distribution may create some problems if the key distribution is based on a common channel signalling scheme and the PABX does not handle it.

PABXs do not normally support common channel signalling. This will create difficulties in establishing network-wide supplementary services, for example, adaptive routing or CUGs that go beyond the user location.

Commercial PABXs usually incorporate a form of precedence and preemption handling through class-of-service and terminal marking procedures or by operator assistance. However it is doubtful if this capability will meet the precedence and preemption requirements of WIS. Multihoming and adaptive routing (limited) requirements may be meet by network-based PABXs, but the software effort to achieve this could be very high.

Conference calls may be achieved within PABXs. Network based conference calls may not be possible. Commercial PABXs offer internal and remote operation and maintenance facilities, but they may not meet the requirements of WNSC.

In conclusion, certain networking and military features can be incorporated in PABXs to the benefit of its users but full implementation of military features in present commercial PABXs would require a very high software effort and major hardware modifications and it is not considered to be practicable in the foreseeable future.

Features of WIS PABXs can be listed as follows:

- PABXs will support most of the user interface provided by an access switch. WIS PABX interfaces are shown in Fig 5.11. PABXs may also be required to provide protocol transparent services between office automation products by means of gateways and protocol converters.

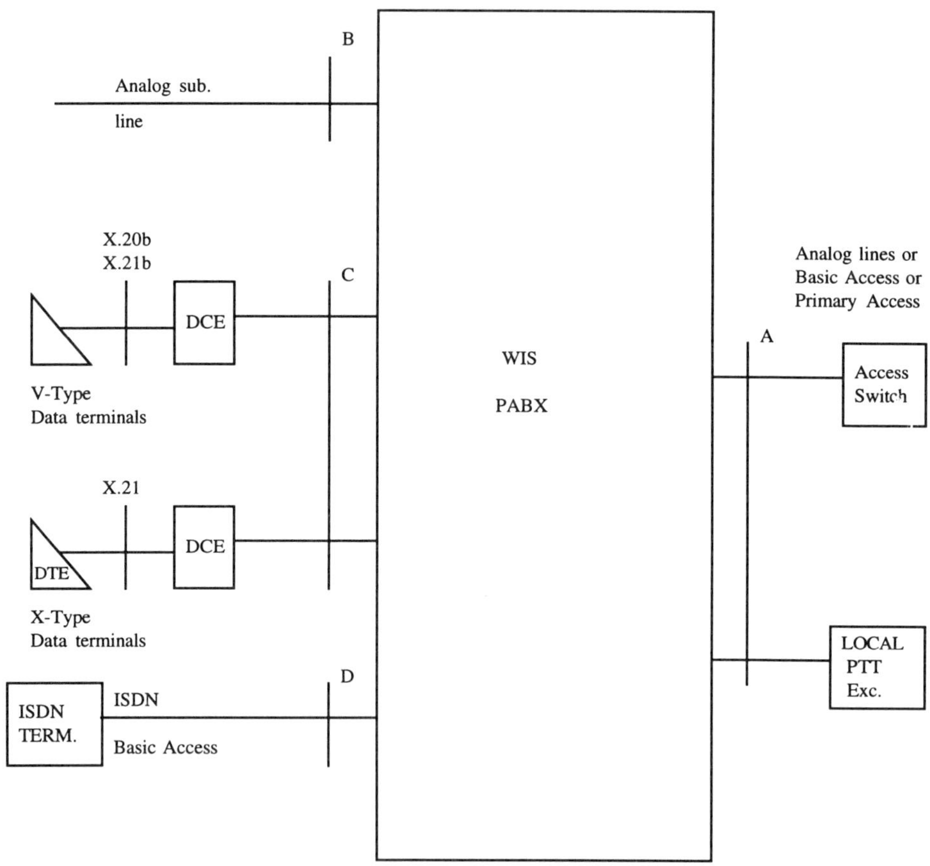

Figure 5.11 Interfaces Supported on WIS PABXs.

- Terminal devices will be linked to the PABXs through voice grade media (VGM), like twisted wire pairs. Since most of the data devices are collocated with voice communication devices the wiring plan of PABXs originally intended for voice only can as well serve data communication needs, up to 1 Mbit/s rates and covering distances up to 1 to 2 kilometers (Chapter 9).
- Voice and data facilities of the PABX should be coordinated through man-machine interface and/or be remotely configured.
- The centralized control in the PABX will aid in error detection, trouble reporting, trouble shooting, fault diagnosis, and in such other duties as traffic monitoring and resource allocation optimization at least to the level demanded by the WNSC (Chapter 11).
- PABXs should be apable of evolving into functionally integrated voice and data switches, hence they should be conceived from the start as ISDN PABXs.
- WIS PABXs should be highly maintainable and modular in design to allow for easy expansion and upgrades.

The following Table 5.3 and Table 5.4 show on a comparative basis the deployment of subscriber interfaces, network interfaces, and supplementary services across WIS nodal switches, access switches and PABXs.

5.4.3.2 Remote Access Units

WIS will provide services in several user locations, some of which will be equipped with access switches and the remaining sites will be equipped with remote access arrangements. In general the latter sites do not possess a sufficient number of subscribers to justify a complete autonomous switch. With the deployment of Remote Access Units (RAUs) it will be possible both to make efficient use of transmission capacity in the access area and also to make the connection of remote subscribers economical in sparsely populated sites. Typical of RAU equipment are Digital Subscriber Multiplexing Systems (DSMS) and Remote Switching Units (RSU).

Digital Subscriber Multiplexing Systems (DSMS): A DSMS is regarded as a system which provides connections from a local switch to more than one subscriber per wire pair, without traffic concentration. This means that each subscriber has always an available time slot. A WIS DSMS should have the following features:

- provide a suitable range of subscriber interfaces to meet the required WIS services, and accept any mix of subscriber interfaces,
- flexible operation as the multiplexer may be located either near the subscriber cluster (thus reducing line system cost) or at an intermediate point between the subscribers and the access switch,
- direct connection to the standard 2 Mbit/s interface of access switches,
- capability to interchange time slots to allow for "grooming' and "drop and insert".

Terminal equipment at the subscriber side of a DSMS will perform the following functions:

- Multiplexing function,
- BORSHT functions for the individual subscriber lines, i.e. battery feed, overvoltage protection, ringing, supervision and signalling, hybrid and testing,
- Monitoring and control functions which are appropriate for remote interrogation and control from a WNSC centre.

DSMS allows for any mixture of digital 64 kbit/s interfaces and analog voice interfaces on all 30/31 channels, and allows for digital voice connection and common channel signalling to be effected. Otherwise it is characterized by all standard features of a first order PCM multiplexer, summarized as: G.703 Digital interfaces, G.711 A-law PCM encoding, G.712 Voice channel transmission quality, G.732/734 Multiplex format and signalling. It is expected that in future DSMS units with ISDN basic access interfaces i.e. 2B+D will also be available. A DSMS associated with an access switch is illustrated in Fig. 5.12a.

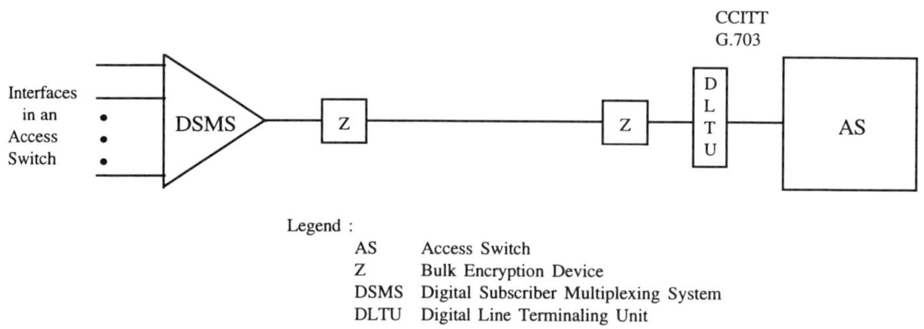

Figure 5.12a PCM DSMS Connected to an Access Switch

Digital Remote Switching Units (DRSU): A Digital Remote Subscriber Unit (Fig. 5.12b) is also referred to as as a Concentrator, a Remote Subscriber Stage, a Remote Line Unit etc. The remote switching unit implies that the traffic concentration is performed at the subscriber site by a unit that is controlled by the parent access switch. In this respect the RDSU can be visualized as the access switch interfaces moved to the subscriber location. This arrangement enables the remote switching unit to take advantage of the operation and maintenance capabilities of the control system in the parent access switch. Subscribers connected to a RDSU will have the same facility range as subscribers connected directly to the parent access switch.

There may be some other equipment used in the access area to enable the efficient use of transmission capacity in the access area or to make the connection of remote voice and data subscribers. examples of other access equipment are:

1) data multiplexer,
2) subscriber modem,
3) data multiplexer modem.

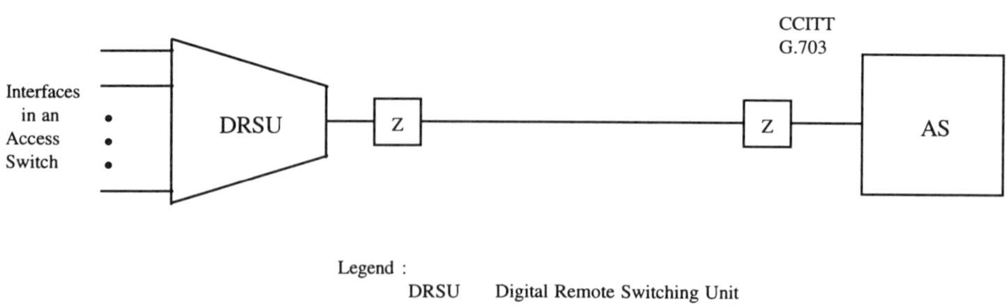

Figure 5.12b DRSU connected to an access switch.

Table 5.3 Deployment of Supplementary Services

Supplementary Services	PABX	AS	NS
Non-secure Conference	YES	YES	NO
Secure Conference	NO	YES	NO
CUG	Local	YES	YES
Announcements	YES	YES	YES
Operator Positions	YES	YES	NO
Precedence	YES	YES	YES
Preemption	Local	YES	NO
Delayed Hot Line	YES	YES	NO
Camp-on Busy	YES	YES	NO
Call Interception	YES	YES	YES

Table 5.4 Deployment of Interfaces

Subsciber Interfaces	PABX	AS	NS
Voice Frequency (Analog)	YES	YES	NO
Modem (V.26, V.27 etc.)	YES	YES	NO
X.20b, X.21/X21b	YES	YES	NO
ISDN Basic Access	YES	YES	NO
Remote Digital Concentrator	NO	YES	NO
Digital PABX	NO	YES	NO
Remote Digital Muldex	NO	YES	NO
PCM Muldex for Analog	NO	YES	NO
ISDN Primary Access	NO	YES	NO
Network Interfaces	PABX	AS	NS
Digital Interface (2 Mbit/s)	NO	YES	YES
Digital Interface (8 Mbit/s)	NO	NO	YES
Gateway Interfaces	NO	NO	YES
Multihomed Interface	NO	YES	NO
E&M Analog Trunk	YES	YES	NO
Analog Trunk with			
DC Loop Signalling	YES	YES	NO
City Line	YES	NO	NO
PAD Interface (X.25/X.3)	YES	YES	NO

5.5 CHARACTERISTICS OF CIRCUIT SWITCHES

5.5.1 Synchronization

Network timing is of vital importance for the disturbance-free operation of the WIS switching system. WIS synchronization subsystem shall provide the permitted timing relationships of the WIS switching, transmission, and terminal elements. Requirements to the WIS Synchronization Subsystem imposed by the WSS can be listed as follows:

* To meet the Bit Count Integrity of "1 slip per day in the end-to-end connections " as detailed in Chapter 6, the long term accuracy of the timing sources must be better than "1 part in 10E-10 ". This figure assumes a worst case Hypothetical Reference Connection (HRX) consisting of 10 consecutive independently timed Hypothetical Reference Digital Paths (HRDP) [10]. The quoted accuracy can be achieved at the central nodes equipped with caesium clocks which have a long term frequency stability of 10E-12 and at the other nodes equipped by means of high quality quartz clocks controlled with a mechanism having the desired long-term regulation stability.
* Maintain acceptable synchronization operation in the event of disruption or partial physical destruction, component failures, enemy interference, and jamming. Furthermore the timing and synchronization subsystem should not present an easier target than "network segmentation".
* Permit orderly growth, extension, reconfiguration, and provision of future services. This is an important requirement given the "phased" growth of WIS and the fact that different synchronization disciplines will be implemented during these phases (Chapter 6).
* Achieve an availability objective better than the WIS transmission and switching system availability.
* Provide synchronized timing to WIS access functions.

According to the WIS synchronization plan, a number of WIS nodes will be equipped with primary timing sources, i.e., high stability and high accuracy sources. The rest of the nodal and access switches will be equipped with secondary sources, i.e., medium stability and accuracy sources which will be controlled by the timing information derived from the incoming links. Typical primary sources are caesium clocks and typical of secondary sources are quartz clocks. The timing and synchronization issues for WIS are discussed in detail in Chapter 6.

5.5.2 Performance

The performance and availability design objectives relate to the technical capabilities and reliability of the switches to satisfy the stipulated network grades of service. ForIS an average Grade of Service of 5% for end-to-end connections is assumed as network dimensioning criterion [Chapters 2 and 4]. The contributions to the grade of service from WIS nodal and access switches are assumed to be negligible, since the internal blocking in the WIS switches should be less than 10E-5, when subjected to the call rates and traffic loads used for dimensioning the network. The performance requirements to the nodal and access switches are detailed in the sequel.

Performance of Nodal Circuit Switches

* Call Handling Capacity can be determined either in terms of

1) the number of circuits which it switches in lines,
2) the volume of traffic which it switches in Erlangs,
3) the number of busy hour call attempts which it can handle.

Due to the critical nature of traffic which WIS handles, it is preferable to dimension the switches according to the BCHA criterion. In addition. the following statistical mean values of delay should be satisfied in compliance with CCITT Rec. Q.504 for 95% of the calls:

Switch call set-up delay	< 250 ms
Through-connection delay	< 250 ms
Switch call release delay	< 250 ms
Switch signal transfer delay	< 100 ms

During overload conditions defined as double of the dimensioned load the above delay figures must in no case increase to more than 5 times the value defined for normal load. The following performance figures assuming at least 10,000 calls measurement period, are recommended as requirements for each nodal switch:

Probability of call failure general	< 5x10E-3
Probability of premature release	< 2x10E-5
Probability of release failure	< 2x10E-5
Probability of misrouting	< 10E-4
Probability of other failures	< 10E-4

Performance of Access Circuit Switches: The types of access units used to connect to WIS are basically remote access units nd access switches.

* The following mean delay parameters of the access switches should be satisfied in compliance with the CCITT Q.514 recommendations:

Dial tone sending delay	< 400 ms
Exchange call setup delay	< 250 ms
Through connection delay	< 300 ms
Ringing signal sending delay	< 650 ms
Exchange call release delay	< 250 ms.

* When handling the specified load at the specified channel capacity the processor should have at least 30% spare memory capacity and 30% spare processing capacity. The switch should not have its traffic capacity and call handling ability excessively deteriorated during periods of overload. The overload condition is quantified as 90% excess of the normal load at which point the switch performance should not fall below 50% of its throughput.

* The call processing performance objectives for WIS access switches should be in accordance with CCITT Rec. Q.514:

Call failure rate	< 1 x 10E-3
Premature release probability	< 2 x 10E-5
Release failure probability	< 2 x 10E-5
Misrouting probability	< 10E-4
No tone probability	< 10E-4
Probability of other failures	< 10E-4.

5.5.3 Availability

WIS switches should be designed for continuous operation with a mean down time (total system outage) less than 2 hours per year during a period of 20 years (i.e. switch life time). Assuming that repairs are done by simple module replacement and that a typical repair time (MTTR) is 0.5 hour, the outage time per year per switch yields a failure rate of 4 failures per year. This then yields a MTBF figure of 3 months and a switch availability figure of 0.99977.

Various redundancy schemes will be implemented in WIS switch design to obtain this availability figure. Examples of planned redundancy are multiple switching planes and duplication of common control equipment and control units, of line and trunk modules etc.. The MTBF figures of the above critical components can be calculated based on the overall MTBF figure of the switch.

There should be o single feature that could cause a complete switch outage. Furthermore the repairs should be executed with simple module replacements, which in turn impacts on the spare parts holding and the knowledge and skill of the personnel operating the WIS switches, on the organizational architecture of the WNSC, and on the hardware/ software resources for the switches.

5.5.4 Operations

WIS switches, in conjunction with the WNSC, will be implemented within the context of a Telecommunications Management Network (TMN) and should have the capabilities needed to allow the exchange to be operated and administered efficiently while providing services in accordance with WIS operational requirements. Types of operations and maintenance activities to be performed at input/output terminals using CCITT Man Machine Language (MML) are as described in CCITT Rec.Z.331.

5.5.4.1 Operations Features

Service provisioning and record: There should be efficient means of establishing service, testing, discontinuing service and maintaining accurate records for subscriber lines and services in the access switches and for interswitch circuits.

Translation and routing information: There should be efficient means of establishing, testing and changing call processing information, such as translation and routing information.

Resource utilization: There should be efficient means of measuring performance and traffic flows, to arrange equipment configurations as required to ensure efficient use of system resources, and to provide the best possible grade of service to all subscribers (e.g. load balancing). Note, however, that the resource utilization may not be optimized in a "democratic" fashion in view of operational requirements and priorities.

5.5.4.2 Switch Functions In Telecommunications Management Network

The Telecommunications Management Network (TMN) architecture, considers the switch to be a Network Element (NE) which can interact with Operations Systems (OS) within WNSC. Operations systems will be used to improve operating efficiencies and services by centralizing and mechanizing operations, as well as administrative and maintenance functions. Functions related to Operations Systems via TMN are:

- Subscriber administration
- Routing administration
- Network management
- Maintenance of subscriber links
- Maintenance of circuits between switches
- Switch maintenance.

5.5.5 Maintenance

WIS switches shall be capable of providing all information necessary for the identification of trouble conditions and the direction of repair activities.

5.5.5.1 Status

WIS switches shall provide information to WIS Surveillance and Control Subsystem (WNSC) personnel so that they can quickly ascertain:

- equipment/system status
- critical load levels
- trouble conditions
- network management controls in effect.

5.5.5.2 Inputs and Outputs

The switch shall be able to transmit and receive maintenance information and respond to commands from on-site and from remote WNSC control centres over the recommended interfaces. The switch shall use CCITT MML at its input/output terminals as covered in the CCITT Z.300 series of recommendations.

5.5.5.3 Tests

WIS switches shall have facilities for performing routine and diagnostic tests on its component parts, on the interfacing equipment or systems to make it possible to assess the switch performance, diagnose and locate faults within the switch and generate appropriate alarms.

5.5.5.4 Fault and Alarm Signals

WIS switches shall interact with access and internodal links as required to detect fault and alarm signals and take appropriate actions. At transmission interfaces, WIS switches will detect the following faults and alarm signals:

Faults:

- loss of frame alignment
- excessive error ratio
- loss of incoming signal.

Alarm signals:

- Alarm indication (remote alarm) received from the remote end
- AIS (Alarm Indication Signal). The presence of the AIS should be detectable even in the presence of an error ratio of 1 in 10E3.

Upon detection of faults and alarm signals, WIS switches will take appropriate actions and generate and transmit alarm messages as described in CCITT Recommendation.542. WIS switches shall monitor the error performances at the switch/transmission interface. The switches shall derive the following information from monitoring, details of which are given in CCITT Rec.G.821.

- degraded minutes (DM)
- severely errored seconds (SES)
- error free seconds (EFS).

5.5.6 Network Management

Network management is the function of supervising the performance of WIS network elements and taking actions to control the flow of traffic, when necessary, in order to promote the maximum utilization of the network capacity.

5.5.6.1 Sources of Information

WIS switches shall provide information on the status, availability, performance and configurationof:

- circuit groups
- call process statistics
- common channel signalling link sets
- other switches with direct links to this switch
- destination switches.

Status information is generated to indicate the service state of a switch or one of its subsystems by comparing the current value of event indicators with appropriate threshold values and/or detecting abnormal conditions. Such type of information assumes discrete values and it can be used, without other processing, together with traffic measurements data to activate traffic control routines. Status data will be collected according to a schedule or whenever requested by network operators.

In case of a fault, status information should be sent spontaneously to the WNSC control center. Performance information together with traffic measurements can be used for centralized processing or for network supervision in a WNSC centre. Configuration information is used for a network management data base at switch level. This information can typically include:

- threshold values actually used
- list of supervised circuit groups
- list of supervised signalling circuits
- list of supervised processors
- list of supervised destination codes
- list of primary and alternate routes for specified destinations.

Information collected at a switch for network management purposes may be sorted and preprocessed in the switch before being used for network management. Network management information may be sent on a scheduled basis. When triggered by abnormal situations (e.g. overload conditions, alarms, etc.) the information will be sent spontaneously. In addition information may be sent on demand, e.g. in response to an external request.

5.5.6.2 Switch Effected Controls

Network management controls provide the means to alter the flow of traffic in the network, in support of network objectives, that is, endurability (restoration, reconstitution) and load balancing. Controls in a switch can be activated or deactivated by input from a network management operations system or by direct input from a switch man-machine interface terminal. Network management control actions can be activated by:

a) pre-established logic responding to preset thresholds being exceeded,
b) manual overrides by external request.

In general, however, when automatic control operation is provided, means for human override should also be provided. WIS nodal switches and some major access switches shall be capable of applying a range of network management controls such as :

a) code blocking control, that is barring or restricting routing for a specific destination code.

b) circuit reservation, reservation of the last few idle circuits for high priority calls.

c) restriction of direct routing, that is limiting the amount f direct routed traffic accessing a route.

d) skip route, that is a control that allows traffic to bypass a specific route and advance instead to the next route in its normal routing pattern.

e) circuit directionalization, that is changing both way operated circuits to one way operated circuits.

However, the applicability of the above traffic controls will depend on the adaptive routing scheme implemented for WIS.

5.5.6.3 Automatic Control of Traffic

Automatic and/or dynamic network management controls can respond automatically to conditions internally detected by the switch or status signals from other switches and can be promptly removed when no longer required. Following types of controls may be provided in the WIS nodal switches depending on the routing system employed for WIS Signalling System.

Automatic Congestion Control System (ACC): ACC system allows a congested switch to send an overload indicator in a backward direction to the preceding switch. The switch receiving the overload indication should respond by reducing the amount of traffic offered o the congested switch. Congestion of a switch implies in this context the inability to find a trunk circuit.

A switch should establish a critical operating system benchmark, e.g.,the time required to perform a complete basic cycle of operations. The switch should continuously monitor this benchmark and, when continued levels of nominal performance are not achieved, a state of overload is declared.

Switches receiving an ACC indication from an affected switch or from the network operations centre should have the capability to institute the appropriate ACC controls and to notify its network management support system of the receipt of an ACC indication.

A switch should have the capability of assigning an ACC response category to individual circuit groups. There should be several categories available from which to choose and each category should specify how much traffic should be controlled in response to each of the received ACC indicators. The categories should be structured so as to present a wide range of response options to received ACC indicators. The control action options may be to skip a call to alternate route, to the next circuit group in the routing pattern or the cancel action which simply blocks the call.

Circuit Reservation Control: The Selective Circuit Reservation Network Management Control enables a digital switch to automatically give preference to a specific type of traffic over others (e.g., higher priority calls, direct routed calls over alternate routed calls) when circuit congestion is present or imminent.

A Selective Circuit Reservation Control may be defined, for a given circuit group, by the following parameters:

1) a reservation threshold,
2) a control response.

The reservation threshold defines how many circuits should be reserved for those traffic types to be given preferred access to the circuit group. The control response defines which traffic types should be given a lesser preference in accessing the circuit group, the quantity of each type of traffic to control, and how those calls denied access to the circuit group should be handled. Possible traffic types are Direct Routed (DR), Alternate Routed To (ART), Hard-to-Reach (HTR) and their combinations. The control action options may be SKIP or CANCEL.

Hard To Reach (HTR) Traffic : The Hard-To-Reach system for network management allows switches to automatically make more efficient resources during periods of network congestion. Improved performance with this system is derived from the ability to distinguish between traffic that is easy to reach (ETR) and traffic that is hard to reach (HTR). Hard-to-reach traffic (HTR), i.e., traffic with a low answer bid ratio should be applied heavier control. However this will not be in conflict with the priorities associated on the vital calls.

5.6 DATA COMMUNICATION

5.6.1 Traffic Sources

The present and future data traffic sources in WIS are either of messaging type (i.e., message handling system, facsimile or fax, document interchange and database access, electronic data interchange (EDI), telex/teletex, transaction processing, electronic mail or e-mail) or of teleaction type (i.e., network surveillance and control, telemeasurements, sensor data, alarms etc.) The characteristics of data traffic ensuing from messaging services have been outlined in Section 5.1.1.7. The features of the data traffic resulting from network surveillance and control activities are described below.

The design of the WIS Network Surveillance and Control (WNSC) subsytem is based on CCITT "Telecommunications Management Network" (TMN) concepts and principles. Data Communications Network (DCN) of WNSC/TMN provides communications channels between the network elements, i.e., witches, transmission equipments and the WNSC elements such as workstations and computer based operations systems. The DCN may be implemented using the capacity of the WIS packet switched or circuit switched data networks, or using the WIS common channel signalling network.

The volume of data generated in WNSC will not be very large. However the delay requirements are rather strict in that Operations System data must be delivered in near real time. Most of the information involved is not of perishable nature and data must be protected.

Amongst the options for the DCN, the common channel signalling No. 7 network is for the WNSC [5]. The CCITT No. 7 signalling system will provide connection-oriented and connectionless communications channels with distribution and file transfer capabilities and additional provisions for reliability. The other alternative for DCN is the X.25/X.75 option.

5.6.2 Comparison of Data Switching Methods

These data traffic sources as described above vary widely in their characteristics and requirements. These characteristics and requirements are burstiness, need for error protection, volume of data, necessity for a short delay, simplex - half duplex - full duplex communications, importance of security, reliability, data with archival value or with temporary status, need for multiaddressing capability and connectivity requirements. Although in principle either circuit, message or packet switching could be made to handle all these traffic types, no one methodology can be prescribed to be as the most cost-effective and flexible one. In conclusion these three types of switching methodology and possibly others will coexist and interoperate in WIS.

5.6.3 PACKET SWITCHING IN WIS

5.6.3.1 Architecture

Alternatives for packet switching architectures in WIS are as follows:

Option 1: Packet Switches in the Nodes Only

Packet switches are deployed collocated with the nodal circuit switches. These packet switches then perform both local and transit packet switching functions. The circuit switched backbone network provides the interconnections between the packet switches through 64 kbit/s semi-switched or permanent circuits or circuit groups with automatic restoration capability. The circuit switched network also provides access channels to packet mode data terminals which can be of switched, semi-switched or permanent variety (Fig.5.6.1). WIS circuit switched network will offer the following physical layer interfaces for the packet mode data terminals:

a) Access based on CCITT Recs. X.21, X.21b,
b) Access based on CCITT V-series Recs,
c) ISDN access in accordance to CCITT Rec. X.31.

The first two access schemes will be implemented in accordance with CCITT X.32, X.21, X.21b, and relevant V-series recommendations. In the third option, the access path between a packet mode DTE and a packet switch will be established, maintained, and disestablished in accordance with CCITT Rec. X.31 within minimum integration scenario.

At the link layer, the LAPB link access procedure of X.25 is used over a single switched physical circuits. At the packet layer, the formats and procedures are in accordance with Rec. X.25. The link and packet layer functions in X.25 are performed between the user terminals and the packet switches, while the circuit switches are only providing transparent X.21 connections.

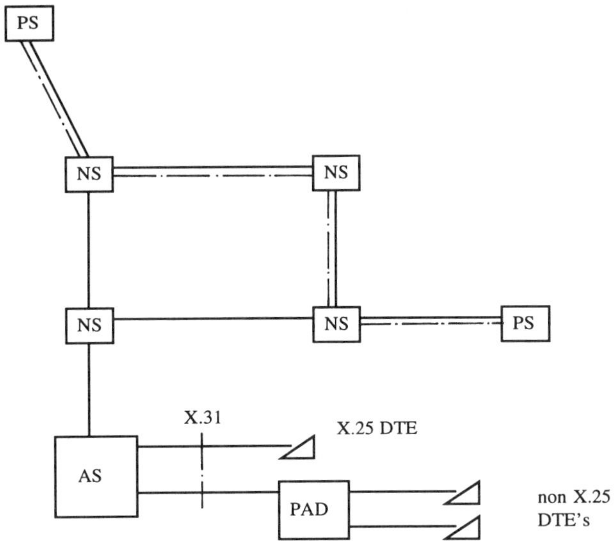

Figure 5.13 Option 1 : Architecture for Packet Switching; Remarks: NS: Nodal Switch; AS: Access Switch; PS: Packet Switch; PAD: Packet Assembly Disassembly Unit. Dashed Lines Denote Packet Mode Bearer Service.

Non-X.25 type data terminals: There will exist, for a long time to come, non-X.25 type DTEs that want to be connected to the packet switched network; these subscribers will be accomodated through PAD (Packet Assembly Disassembly) ports according to CCITT Recs. X.3, X.28, and X.29. PADS will be deployed at user terminals as shown in Fig.5.13 or pooled at the packet switches as shown in Fig. 5.14

Minimum Integration Scenario: This scenario refers to the transparent handling of packet calls through the circuit switched network according to CCITT Rec. X.31 and I.462 as shown in Fig. 5.14. The access/nodal switches provide only a physical 64 kbit/s switched or non-switched transparent connection between the packet mode terminals or PAD units and the access ports at the packet switch. The DCEs perform X.21/X.21bis or V-series procedures.

In the case of semi-permanent access, which is represented by the continuous line in the lower part of each connection in Fig. 5.14, the packet mode terminal is connected to the circuit switched network port on the packet switch called the "interworking port (IP)". The terminal adaptor serves to perform only the rate adaptation between the user rate at reference point R at the user terminal and the 64 kbit/s B-channel rate.

In the case of switched access, as illustrate in the upper part of each connection in Fig. 5.14, the packet mode user terminal is connected to the circuit switched network port on the packet switch (IP). In this type of connection, originating calls will be set up over the B-channel towards the packet switch using the ISDN signalling procedure prior to starting X.25 level 2 and level 3 functions. This can be realized by using either the hot line or complete selection features. For calls originated by the switch the same considerations as above apply. The circuit switched network port of the packet switch includes both rateadaption and call establishment functions. Packet mode terminals can also be first multiplexed through an X.51 scheme. According to CCITT Rec. X.51, while the transmission bit rate is 64 kbit/s, the multiplex structure gross bit rate is 60 kbit/s and padding techniques are utilized. The allowed sub-64 kbit/s bearer channel rates are 600, 1200, 2400, 4800 and 9600 bit/s. At the receiving sites these bearer channels are demultiplexed and onnected to the IPs (ISDN Interworking Ports).

A variant of this option is the case where packet switches are deployed at the access sites as well. Thus in those access areas where there is a significant concentration of packet mode terminals, the "access" packet switches can serve both to concentrate the packet traffic and to switch them locally, providing a more cost-effective connection means for the messages of the individual terminals to the backbone network as compared to semi-switched option as in Option 1 above.

Option 2 Separate Packet Network:

While Option 1 and its variant represent packet switched networks overlayed on the circuit switched network, in Option 2 a completely separate packet switched data network is envisaged. However the interconnections between the packet switches are provided via permanent connections in the circuit switched network. This packet switched network will also be a two-level network. The first level is the access level where the packet concentrators are used to concentrate the access area traffic towards the nodal packet switch. However, unlike the access circuit switches, they do not perform any local switching functions. The second level is the nodal packet network where the packet switches perform the local and transit switching functions, as well as networkwide functions such as routing and flow control. The packet switches are interconnected by dedicated n x 64 kbit/s channels offered from the backbone circuit switched network. Interworking between the circuit and packet switched networks are provided via interworking functions (IWF) located in certain nodal circuit switches which allow for packet subscribers not collocated with a packet switch to access the packet switched network through "port accessing" method as shown in Fig 5.15.

Option 3 Full Integration

In this option ISDN switches with circuit and packet switching capabilities will be deployed in both access and nodal sites. Data terminals of the circuit variety (X.21/X21b) and packet variety (X.25) will be connected to the network via terminal adaptor units.

Maximum Integration Scenario: This scenario refers to the case where a packet handling (PH) function is provided within the nodal circuit switch. The PH carries out the complete handling of the X.25 call. An access to a PH port can be established through the D-channel. The D-channel can also support packet data transmission rates up to 9.6 kbit/s, corresponding roughly to 7 kbit/s actual data rate. If implemented, this type of packet access would be adequate for the majority of packet data users, e.g., interactive users, message terminals, data base updates and queries. For data rates higher than 9.6 kbit/s the PH can be accessed through one of the B-channels using D-channel signalling. For both the B-channel and D-channel cases, it is possible to operate multiple terminals through Data Link Layer and Network Layer multiplexing procedures.

Comparison of the options: The choice among the architectural options will be based upon such factors as the concentration of packet type traffic, cost, and operational requirements. Option 2, i.e., a separate network, is a costly solution and it does not lead easily to future service and network integrations. Options 1 and its variant are viable only for those networks where predictions indicate that packet mode traffic will persist in having a low proportion within the total traffic. Option 3 is perceived as the eventual WIS architecture since it will provide the benefits of lower cost, enhanced services, and more flexibility.

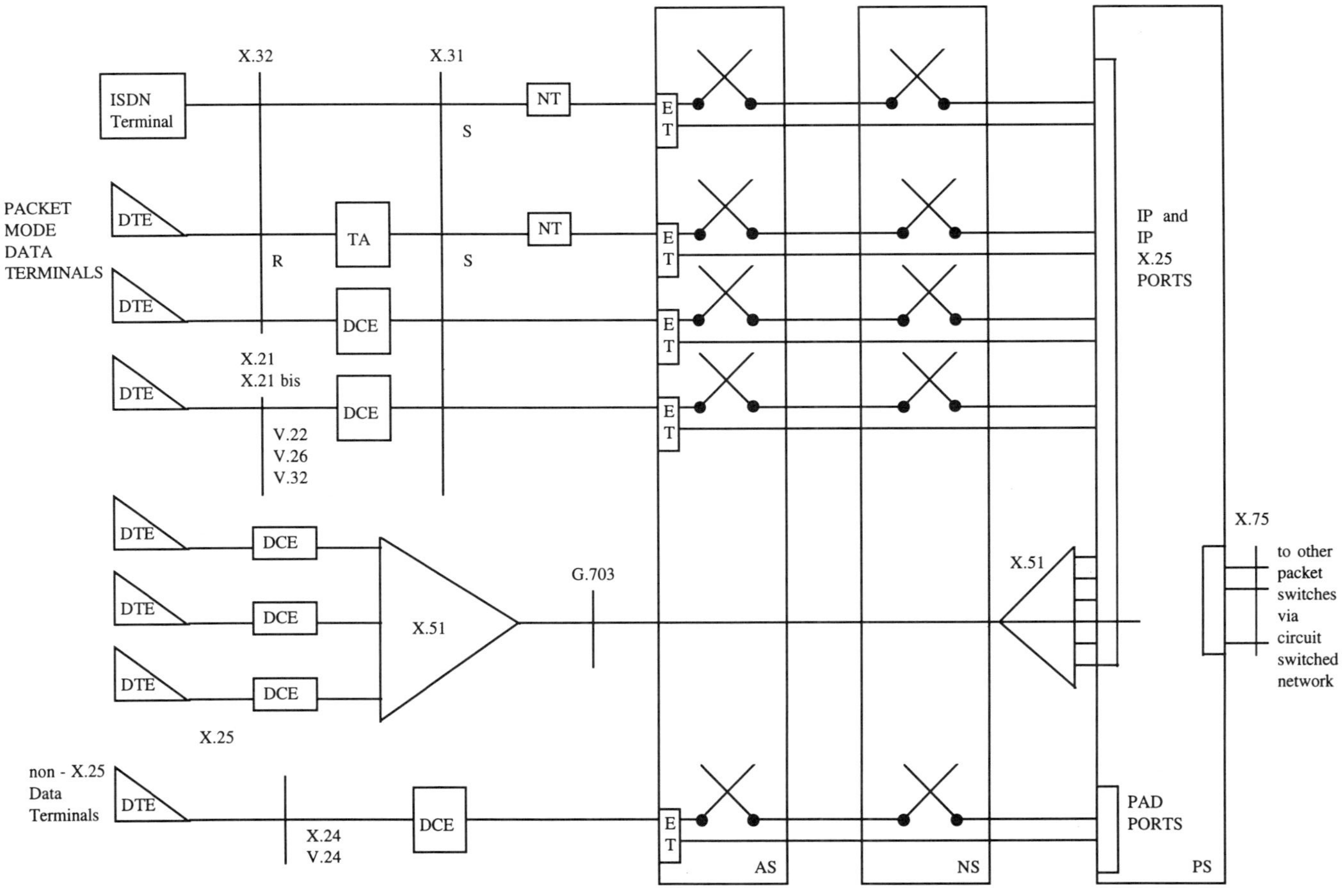

Figure 5.14 Access Methods for Packet Terminals in Option 1. IP , IP' : ISDN Interworking Ports

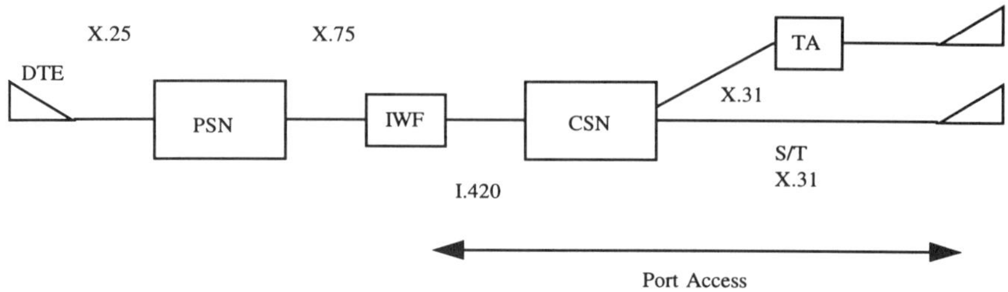

Figure 5.15 **Interworking Function Between Circuit and Packet Switched Networks in WIS. PSN:Packet Switched Network CSN: Circuit Switched Network IWF: Interworking Function**

5.6.3.2 Functions and Features

The functions of the WIS packet switches are as follows:

a) Establishment and maintenance of permanent virtual circuits (PVC).

b) Establishment and disconnection of virtual circuits (VC)

c) Selected packet switches should be able to establish interconnections to other packet switched networks ia gateway functions supporting X.75 protocol.

d) Control and maintenance of the packet network, such as activities for the collection and processing of traffic statistics, operations and maintenance functions.

e) Execution of adaptive routing and flow control functions.

The main features of the packet switches are as follows:

a) Adaptive routing capability in order to maximize the throughput while minimizing call set-up and packet transfer delays and to cope with link and node failures. Furthermore the routing algorithm must be able to control the dual-homing without any user intervention, and be able to use the secure and non-secure links in a controlled manner.
In the case of minimum integration scenario, an independent adaptive routing algorithm based on packet delay-throughput criteria can be implemented on the network formed from the semi-permanent circuit switched connections. This approach would be required especially in the case of datagram applications. On the other hand in order to set up virtual circuit connections, the packet network may simply rely on the adaptive routing capacity of the circuit switched network. Finally in the maximum integration scenario, a common routing scheme would be supported for the packet and circuit switched services.

b) The switch must be able to handle all recommended packet sizes in X.1., i.e for packet switched data transmission/service for DTE operating in synchronous mode using X.25 interface, and for start-stop mode terminals.

5.6.3.3 Service Features

The following minimum set of service features should be made available for selectable time periods:

a) Incoming calls barred (VC) which prevents incoming virtual calls from being presented to the DTE while the DTE may originate outgoing virtual calls.

b) Outgoing calls barred (VC) which prevents the DCE from accepting outgoing virtual calls from the DTE while the DTE may receive incoming virtual calls.

c) Non-standard default packet sizing, (VC, PVC), non-standard default window sizing (VC, PVC) where both features can be selected by means of the flow control parameter negotiation facility.

d) Default throughput classes assignment (VC).

e) Flow control parameter negotiation.

5.6.4 Message Switching in WIS

5.6.4.1 Message Switching Architecture

The alternatives for message switching architecture in WIS are basically the same as those for the packet switching. These alternatives are:

a) Message switches in the nodes only,
b) Message switches in the access and nodal switches,
c) Separate (non-overayed) message switching network,
d) Full integration.

Following the rationale used for the packet switches, Option (d) is perceived as the eventual WIS message switching architecture. Message switching encompasses functions of the higher OSI layers ,i.e., layers 4-6. On the other hand, for the lower layer functions, that is the basic transport, routing and call establishment functions, there exist two alternatives:

1) Use of the circuit switched network,
2) Use of the packet switched network.

In alternative 1 link level functions such as error protection, recovery of lost messages and flow control and Network Layer functions such as routing must then be performed by the message switches themselves. The message switches will be viewed by the backbone circuit switched network as just another access switch associated with nodal switches. The call establishment procedure for the message terminal is then the same as for a circuit mode terminal.

In alternative 2 the message switch is dedicated to higher layer functions, such as Transport Layer and Session Layer, while the communication functions (Layers 1-3) are performed by the packet switch.

5.7 ROUTING

The operational and system requirements of the WIS routing system is dealt with in detail in Chapter 7. Therefore in this Section the discussion will be limited to its implications on the switching subsystem.

* Signalling/Processor Capacity: The signalling and storage capacities and the switch processing capacities for routing related information should be properly dimensioned so as not lead to undesired delays.
* Overflow/Alternative Routing/ARR: The switches should be capable of managing:

  1) circuit group overflow, by routing the call via another group of circuits to the same destination switch;

2) alternative routing, where the group of circuits over which the overflow traffic is routed involves at least one exchange not involved in the previous choice route.;
3) automatic repeat attempt (ARR), that is a second attempt to set up a connection for a call when a difficulty is encountered in the setting up of this connection.

* Circuits in tandem: For reasons of transmission quality as well as the minimization of fast dialing and answer signal delays and the avoidance of signalling time-outs, it is desired to limit the number of circuits in tandem in an overall connection. However, survivability and resilience requirements may override these considerations in military networks.
* The choice of a free circuit in an outgoing route should be made in such a way as to minimize the probability of the same outgoing circuit being selected on ecutive calls. The selection procedure should ensure that all circuits comprising an outgoing route are traffic loaded as uniformly as possible.
* The system should allow for greatest flexibility in changes in routing schemes ,e.g., changes in spanning trees. By means of operator and/or WIS NSC it will be possible to create new routes, delete existing routes, modify the assigned trunk groups, allocate or remove trunk groups to that route, display current route information, update route dependent signalling and numbering parameters, change direction of scanning for both way trunk group hunting.
* The system should be able to provide routing to decentralized service points for special service calls. Provision should also be made for routing to the appropriate recorded announcement or tone source for calls to non-existing number group or code, spare or suspended subscriber number, secure/non-secure connection, etc.
* Satellite circuits: In WIS the following non-dedicated satellite connections are expected to provide capacity for access circuits to WIS nodal switches from fixed users in remote areas and from itinerant/mobile users. In WIS satellite routing is to provide access connections to afloat/airborne units, or ability for node skipping overlay connections for enhanced survivability. If satellite links will provide a node skipping capability, that is as an alternative to an all-terrestrial route within the switched network, then the routing system should limit the satellite paths to one in all but exceptional cases to preclude excessive propagation delays. In cases where access links connecting users to the nodal switches are provided via satellite communication using, for instance, low-cost or transportable earth stations, to satisfy the transmission requirement for not more than one satellite hop, the onward routing of calls by the parent node would be wholly terrestrial in such cases.
* Preemption: The capability to seize a circuit on which a call is in progress by a call of higher priority. In WIS the shortest path route is to be preempted.

5.8 SIGNALLING SYSTEM

5.8.1 Requirements

WIS signalling system will meet requirements of call control signalling for WIS voice and data services. It will also be used to transport other types of information between switches and specialized centres. In particular WIS signalling system will provide, when required, data transport facilities for:

- WIS users (user-to user data),
- WIS Network Surveillance and Control Subsystem (WNSC),
- WIS Timing and Synchronization Subsystem (for distribution of control and monitoring information)

- WIS Routing Subsystem.
- WIS COMSEC, for distribution of crypto keys.

The WIS signalling system must be optimized for operation over 64 kbit/s digital channels. However it should also be able to operate over analogue channels and at lower speeds and it should be suitable for use on point-to-point terrestrial and satellite links.

A common channel signalling system will be used for WIS, since it allows considerable economies of signalling equipment, especially with digital stored program controlled switches utilizing 64 kbit/s channels and is suitable for the transport of other types at information ,i.e., in multi-application systems. Beyond being fast and reliable, the WIS signalling system must satisfy the following operational and strategic requirements:

- Able to later for mobile and itinerant users,
- Able to use satellite communications paths effectively,
- Interwork with the signalling systems of other networks with which WIS has to interoperate
- Must have a level of survivability at least equal to that of WIS network it is supporting
- Must handle information for echo control (if satellite links are to be used)
- Support adaptive routing scheme employed for WIS,

5.8.2 Overview of Available Signalling Systems

There are over 300 national signalling systems in operation around the world for telephony and data, developed either by individual authorities or suppliers or as variants of international standards. In the international arena CCITT has specified six signalling systems (SS) which are in use today.

CCITT signalling systems No.1, No.2, and No.3 were developed for manual telephony and they are now obsolete. CCITT SS No.4 uses a code made up of two frequecies and pulse duration information to supervise lines and to carry address information. Despite its current widespread deployment, it is rapidly becoming obsolete. R1 and R2 are North American and European regional variants of an analog multifrequency signalling systems. These systems use the voice frequency telephone channel for transfer of information and the modulation technique is pulsed tones in various combinations. It should be noted that R2 was the first signalling system to provide "backward" signalling, which, for example, could give a more precise picture of where congestion was occurring, rather than merely advising that a route was busy as used in spill forward schemes. It could therefore support more intelligent routing schemes and have a reasonable capacity to support network management in this context. R1 and R2 are analog and in-band channel associated signalling methods where information is directly related to the physical channel carrying the signalling. However there are also digital versions for R1 and R2. In these analog signalling systems, switches can distinguish between tones of many different frequencies which allow for many possible combinations of signalling elements creating a larger vocabulary.

CCITT SS No.5 was developed from R1 and R2 for intercontinental routes, where handshaking is kept to a minimum to avoid delays in the call set-up over long terrestrial links or satellite links. It is also the current intercontinental signalling system.

CCITT SS No.6 is the first common channel signalling system, which can provide a somewhat extended range of network management signals in addition to call set up signals. The channel used is again a VF channel but the modulation technique is based on data modem standards (normally 2400 bits/s). Furthermore it has led to operational experience in common channel signalling. With the introduction of digital transmission, the development of CCITT SS No.7 had started, however, almost as soon as SS No.6 was specified, so only the most advanced countries have implemented SS No.6.

CCITT SS No.7 is a common channel signalling system capitalizing on the vast increase of capacity and peed offered by the 64 kbits/s traffic channel. Using highly efficient, reliable and flexible data communication protocols, a vastly increased capability of transferring signalling data between the switches as well as between the switches and the specialized centres becomes available. The possibility also emerges to produce a separate signalling data network in addition to point-to-point signalling. This allows intelligent switches to communicate with their surrounding switches in a much more general way. With these capabilitiesh owever, the management of signalling traffic and routes becomes much more important and results in extensive additional system complexity.

Signalling system No.7 provides extensive backward signalling. The much higher information capacity can be used for network management, intelligent routing, and, if needed, crypto key variable distribution. This signalling system is designed to meet present and future requirements of information transfer (circuit and non-circuit related) within telecommunication networks for call control, remote control, network data base access, management and maintenance signalling. Furthermore the inherent error protection mechanisms (i.e, HDLC checksum and ARQ retransmission protocol) provide a reliable means for the transfer of information in the correct sequence and without loss, duplication or mutilation of the No.7 message frames.

Signalling system No.6 has a limited capability of multiapplication in data transfer and a smaller signalling repertoire than SS No.7 and does not meet some of WIS signalling requirements. Other signalling schemes that could be considered for WIS could be CCITT X.75 signalling and ISDN PRA (primary access) LAPD signalling (Q-sig)

The X.75 protocol is used for packet switched networks. In the X.75 the message transfer part is more OSI oriented as compared to that of SS No.7 and it also does not have all the complicating STP (signal transfer point) functions as in the SS No.7. On the other hand CCITT X.75 protocol lacks the functions of the telephone and data user parts. Recently the LAPD (link access procedure in the D channel as in CCITT Q.920-Q.940) in ISDN Primary Access (PRA) has been developed by ETSI as the standard multilink signalling procedure between switches. In order to allow ISDN PABX's to interoperate with their full spectrum of supplementary services, this signalling system has in fact more user service functionality than CCITT No.7. While the No.7 system has a lot of its focus on the PTT's requirements for charging and O&M, Q-sig has been conceived to allow multivendor PABX's to interoperate over dedicated groups or through a public ISDN network. The sequel discusses the characteristics of CCITT No.7, while it should be kept in mind that either this or Q-sig could fill the needs of WIS.

In conclusion CCITT No.7 or Q-sig appropriately modified to satisfy the requirements of WIS could both be candidates for WIS signalling, while X.75 is reserved for packet nodes.

5.8.3 No.7 Signaling System

CCITT SS No.7 is an internationally standardized general purpose common channel signalling system conceived for use with stored program control exchanges. This signalling system is expected to provide the basis for signalling in connection with ISDN services. It is almost certain that ISDN user part (ISUP) will be able to encompass and handle all types of services. In addition to CCITT defined user/application parts, WIS specific user/application parts can also be defined and patched for network management, communication security etc..

5.8.3.1 Common Channel Signalling Features

Common channel signalling is a signalling method in which a single channel conveys, by means of labelled messages, signalling information relating to a multiplicity of circuits or processes, and other types of information such as that used for network management. This signalling system uses signalling links for transfer of signalling messages between points within a network. These points could be parts of a switch, an operations center or a database. Arrangements are provided to ensure reliable transfer of signalling information in the presence of transmission disturbances or network failures. These include error detection and correction on each signalling link. The system is normally mplemented with redundancy of signalling links and it includes functions for automatic diversion of signalling traffic to alternative paths in case of link failures.

5.8.3.2 The Structure of the SS No.7

The SS No.7 as froseen in WIS will a layered structure as illustrated in Fig. 5.16. To ensure flexibility for diverse applications and for open-ended evolution, the signalling protocol consists of functional modules patterned after the OSI/RM. The Message Transfer Part (MTP), which corresponds to the first three layers in the OSI/RM, serves as a transport system providing reliable transfer of signalling messages between the locations9 of communicating user functions. Various user parts can be attached to the MTP as shown in Figs. 5.16 and 5.17. The three MTP functional levels are:

Level 1: Signalling Data Link, defines the physical, electrical and functional characteristics of a signalling data link and the means to access it. WIS will provide 64 kbits/s digital paths (e.g., channel 16 in the 2 Mbits/s hierarchy) to be used for the signalling data link which is based on the CCITT Rec. G.703 and G.732. Note that Signalling Data Link consists of transmission and switching functions.

Level 2: Signalling Link Control functions provide measures for delimitation, error recovery, sequencing, error rate monitoring and link initialization with the purpose of reliable transfer of signalling messages between two points. If a satellite link is to be used, then forward error correction (FEC) can be implemented to avoid excessive delays instead of the retransmission technique (ARQ) as is commonly done in terrestrial links.

Level 3: Signalling Network Functions consist of such tasks as

a) message routing, i.e. selection of signalling links,
b) message discrimination that determines if that signalling point is the destination of the message,
c) message distribution that determines to which user part the message is to be delivered.

Level 3 also includes network management functions with the purpose of flow control, topology updates, reconfiguration, and restoration of signalling links.

Various User Parts shown in Fig.5.17 represent layer 4 functions. These User Parts provide procedures, interexchange signalling information, and functions enabling various services and controls to be realized between access points. The user parts in WIS would be:

- TUP: Telephone User Part which defines the necessary telephone signalling functions in the signalling system No.7 (CCITT Q.721)

- DUP: Data User Part is considered to be already "extinct" in the international data swicthing community and hence will not be realized in all possibility in WIS (CCITT Rec.Q.741 and X.61)
- ISUP: ISDN user part encompasses the signalling functions required to provide switched services and user facilities for voice and non-voice applications in an integrated services digital network. The ISDN User Part should meet requirements of service features, user facilities and network capabilities defined by WIS for internal and internetwork automatic and semi-automatic telephone traffic and for data transmission services. (CCITT Q.761)
- OMAP: The Operations and Maintenance Applications Part contains procedures for controlling, supervising, testing the signalling network, and carrying managements commands from the main network.
- MAP: The Mobile Application Part uses the services of TCAP to handle the messages and data which are exchanged between the mobile switching centers, user locations, and visiting location registers.

The network layer in SS No.7 consists of the SCCP (Signalling Connection Control Part) and of the Signalling Network Functions of the MTP. The SCCP provides services for connectionless communication, e.g., references for sequence numbering, circuit identities, and SCCP management itself.

The TCAP (Transaction Capability Application Part) represents Layers 5 and 6 in the OSI/RM. On top of TCAP, the application processes for the following WIS subsystems could be used:

* WIS Network Surveillance and Control Subsystem: WIS NSC may utilize functions equivalent to OMAP (Operations, Maintenance and Administration Part) and to some extent SCCP (Signalling Control and Communication Part) in the CCITT No.7 for its own information transfers. SCCP functions are required to transfer signalling information to computers at the maintenance entities and WIS NSC processors.
* Timing and Synchronization: If a disciplined synchronization method is assumed for WIS (e.g., master-slave, hierarchical master-slave, or mutual) then the signalling system must also carry the information related to the timing sources and their control. Typical of this information are the buffer resets, assignment of the master clocks, reconfiguration of the master-slave chains etc. In addition Mobile Application Part (MOAP), the set of procedures that deal with the data basis for the ocation of mobile and itinerant users could be implemented.
* COMSEC: Cryptographic application part (CAP) would be a newly defined user part for the allocation and distribution of cryptographic keys, for periodic changes in the link encipherment or for encipherment on a call-by-call basis. This part will have a transit-attribute so that crypto variables will be transmitted without encipherment and decipherment at each node.
* Other possible uses of SCCP are updating of vehicle location registers in mobile radio networks, user-to-user signalling that is not related to calls and which is handled by ISUP or the more specialized application parts.

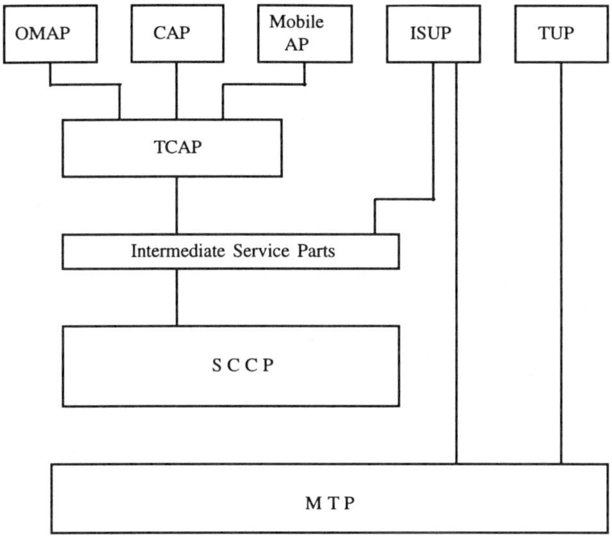

Legend :

AP	: Mobile Application Part
CAP	: Cryptographic Application Part
ISUP	: ISDN User Part
MTP	: Message Transfer Part
OMAP	: Operations Management Application Part
SCCP	: Signalling Connection Control Part
TCAP	: Transaction Capability Application Part
TUP	: Telephone User Part

Figure 5.16 Structure of SS No.7 as proposed for WIS

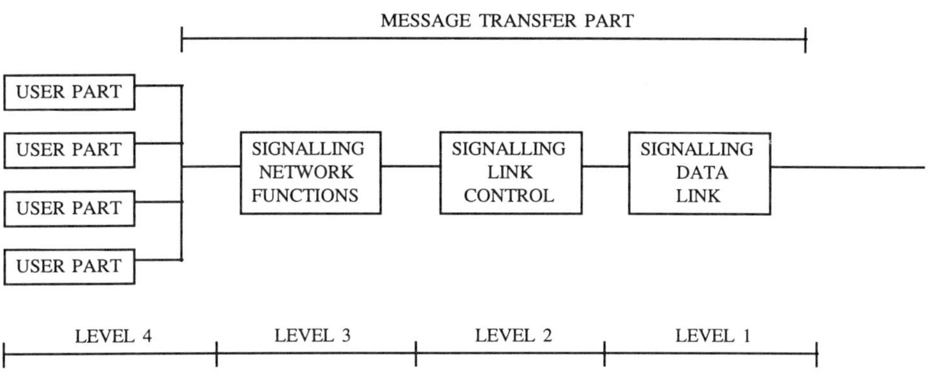

Figure 5.17 Functional levels of SS No.7

5.8.3.3 SS No.7 Error Protection

The error detection and correction mechanism in SS No.7 is more powerful than in any other existing signalling system. The error detection function is performed by means of 16 check bits provided at the end of each signal unit. The transmitting and receiving signalling points generate and check, respectively, the error detection and correction bits. If consistency is not found between the received check bits and the preceding bits of the signal unit, according to the algorithm, then the presence of errors is indicated and the signal unit is discarded.

The basic method is a non-compelled, positive/negative acknowledgement, retransmission error correction system. A signal unit which has been transmitted is retained at the transmitting signalling link terminal until a positive acknowledgement for that signal unit is received. If a negative acknowledgement is received, then the transmission of new signal units is interrupted and those signal units which have been transmitted but not yet positively acknowledged, starting with that indicated by the negative acknowledgement, will be retransmitted once more, in the order in which they were first transmitted.

The forced retransmission procedure is defined to ensure that forward error correction occurs in adverse conditions (e.g. high error rate and/or high traffic loading). When a predetermined number of retained, unacknowledged signal exists, the transmission of new signal units is interrupted and the retained signal units are retransmitted cyclically until the number of unacknowledged signal units is reduced.

5.8.3.4 CCS No.7 over Satellite Links

In WIS, satellite links will be used for national/international connections, to provide a node skipping overlay to the terrestrial network for enhanced survivability, as access circuits for afloat units and aircraft, and in certain cases as access circuits to WIS nodal switches from fixed users in remote areas and from itinerant/mobile users. The problems in the se of SS No.7 over satellite links are as follows:

a) One earth station (ES) serving two or more switching nodes: If associated mode signalling is used, the signalling channels of different nodes could either be multiplexed onto the same common signalling channel or each incoming signalling channel could be made to occupy separate satellite channel, at the cost of additional satellite capacity. To allow the nodes using one ES to share a single satellite signalling channel to the nodes using the distant ES, a signal multiplexing arrangement would be required at each ES.

b) Echo suppression and Call history: SS No.7 should contain messages for echo suppressor control. In addition messages indicating whether a satellite link has already been used or not (i.e., call history) should be included in order to avoid double satellite hops

5.8.4 X.25/X.75 interface in WIS

CCITT Rec. X.25 is a packet mode DTE-DCE interface whereas X.75 defines the characteristics and operation of an interswitch signalling system for international packet switched data transmission services. Although CCITT SS No. 7 has been indicated as the appropriate signalling system for WIS, X.25-X.75 standards remain as a potential alternative as this interface is broad enough in scope that it can be adapted to this purpose. The following modifications are suggested for this interface to become a viable candidate for WIS signalling system:

a) As part of the Higher Level Functions, the relevant User Parts (i.e., TUP and eventually ISUP) should be adapted into the X.25/X.75 protocol.

b) As this protocol will only be used to carry signalling messages but not the user messages, some simplification is possible.

c) The X.75 protocol is said to possess a "miopic" view of the network as compared to the CCITT SS No.7. This protocol must be enhanced and imparted with networkwide information if it is to carry management and control messages and can have restoration and adaptive routing capability.

Otherwise this signalling system possesses advantages similar to those of SS No.7, i.e., in more detail:

- Speed: The interswitch transmission speed will be 64 kbit/s. Furthermore there exist switched and permanent virtual circuits, datagram or fast select facility alternatives for the switching of the signalling messages, which can be used optimally to satisfy WIS signalling speed requirements.
- Reliable: The inherent error recovery, sequencing and link initialization mechanisms in the X.25/X.75 frame layer provides a reliable transport medium for signalling messages.
- Growth potential: The X.25/X.75 interface corresponds to the first three layers of OSI/RM. However the packet layer of this interface can potentially support higher OSI layer protocols and hence appropriate "user parts" as in the No.7 signalling alternative can be developed.
- Integration: The X.25/X.75 interface will be used for signalling and data transfer between packet switching exchanges in the nodes. Thus the adoption of the same interface to support circuit switching signalling, network surveillance and control signalling, and COMSEC signalling, as well as for packet switching requirements per se seems a cost-effective solution for the WIS signalling system.

5.8.5 Access Signalling

The following types of access signalling will be supported in WIS:

a) Analog loop signalling where call control is achieved by a set of pulses and/or multifrequency tones. The following types of loop signalling can be distinguished:

   - Dial pulse line,
   - DTMF line,
   - 2W-DC loop trunk,
   - 4W E&M trunk.

b) Digital Data-Loop Signalling: Concerning circuit switched data, user signalling is obtained by 8-bit characters according to the international alphabet No.5 (IA5) together with the use of additional control information carried outband. This signalling protocol is standardized in CCITT Rec. X.21.

c) Digital Access Signalling: This is a three level protocol as detailed in CCITT Recs. 1.430, I.440, and I.450 or similarly in Recs. Q.920 to Q.931. The physical layer should include both point-to-point and multipoint (passive bus) arrangements. The layer 2 protocol is the LAPD (Link Access Protocol in the D channel). Layer 3 protocol should allow signalling information (s-type), packet information (p-type) as well as telemetry information (t-type). The s-information procedures should have capabilities for common telephony features such as conferencing, call hold, call transfer etc., it should allow for easy mapping for circuit switched data according to CCITT Rec. X.21., and it should provide for easy interworking with the future ISDN user part of signalling system No.7. Finally this signalling procedure should cater for the distribution of crypto variables for end-to-end encryption applications.

5.9 WIS NUMBERING PLAN

This section the operational requirements of the WIS numbering plan and the related issues are considered. In the light of these requirements three numbering methodologies, namely, functional numbering, geographical numbering and ISDN numbering are discussed.

5.9.1 Requirements of the WIS Numbering Plan

Operational Requirements

* The numbering scheme should be as "pure" as possible across the services. Furthermore any numbering conversion should take place at the gateways.
* It should permit common procedures in the numbering plan for all users and services. As an example of common procedure the first n digits (n=3) could be used for numbering translation for routing purposes while the last m digits (m=4) could be used as an access code. All number translations for routing purposes should take place in the nodal and access switches.
* The numbering system should be sufficiently redundant for the ability to create a deducible numbering scheme, to allow for future expansions in the user population, in the number of nodal/access switches and in the types of services.
* It should provide access to multihomed static users users with alternate static locations mobile users without caller knowing current location service positions national strategic networks.

5.9.2 Numbering Methodologies

Possible numbering methodologies to be considered for WIS, are as follows:

a) Functional numbering
b) Geographical numbering
c) ISDN numbering
d) NATO numbering

5.9.2.1 Functional Numbering Plan

A numbering plan based on functional splitting identifies each subscriber according to his functional membership. A typical usage could be in military strategic networks, where the functional areas could be Land Force, Naval Force and Air Force Commands, Gendarmerie, Coastal Guard, General Staff, Ministries, the Security Council etc. These functional areas are further split into subareas and sub-subareas. The numbering code should identify the direct WIS subscribers in the access switch and the indirect WIS subscribers at the PABX level. A special prefix should be reserved for public networks, like PTT, so that after dialing this digit one will receive a dial tone denoting the dedicated access to the PTT network.

5.9.2.2 Geographical Numbering Plan

In this numbering plan the WIS network has been split into four regions as shown in Fig. 5.18. The first digit represents the region number while the second and third digits represent nodal and access switches, respectively. Furthermore one can use the prefixes 0,8,9 respectively for PTT, Mobile/Itenerant users, and other network accesses. In this figure

1<R<7	denotes the region number
0<N<9	denotes the transit node number in region R
0<A<9	denotes the access switch attached to the node
0<P<9	denotes the PABX attached to the access switch.

This method is advantageous in that there is no need form code translation in the routing functions except for the firsts digit (the region digit). In this sense there is less computational load in the switches as compared to any functional numbering plan. On the other hand geographical numbering is less deducible, hence more difficult to remember. A compromise would be that the subscribers could dial the functional number and at the access switch this number would be converted to the geographical number. This type of technique is called digit replacement method. This would, however, necessitate the management of the subscriber number data base at the access switches.

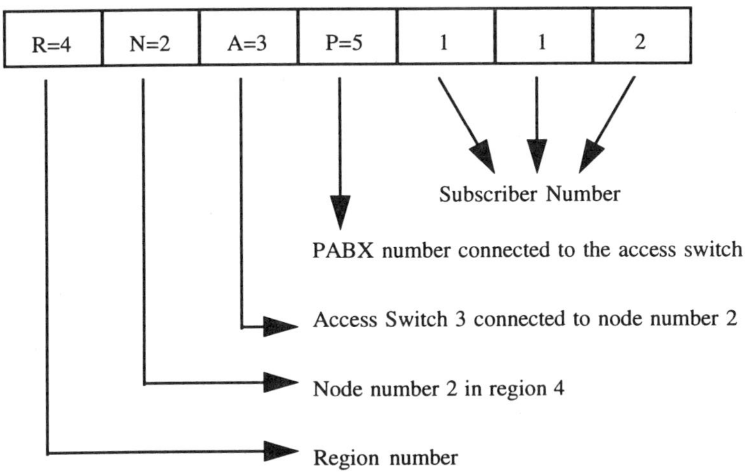

Figure 5.18 A Geographical Numbering Example

5.9.2.3 ISDN Numbering Plan

The structure of an ISDN address is shown in Fig. 5.19. The ISDN numbering plan mimics in essence the CCITT telephone numbering scheme where for example the country cods identifies the country or the geographical area. The national destination code in the case of WIS implementation would correspond to the geographical or functional number of the subscriber.

The advantage of this numbering scheme is the presence of the subaddress which bears information about the type of service, the type of terminal and it can be used to identify a terminal at a subscriber premise among several others having the same type and functionality.

Thus multiple services and terminals at a subscriber's premise can be reached with a single ISDN number, an important step towards attaining integration of services. CCITT has foreseen 40 digits for the subaddress. However this subaddress is not considered as part of the proper ISDN number, in the sense that it is not subjected to any processing at the switches and is treated only at the premise's network termination. In this sense the ISDN subaddress can be accommodated into any numbering scheme (e.g., functional, geographical).

5.9.2.4 NATO Numbering Plans

NATO Tactical Switched Network Numbering Plan (STANAG 4214) covers allocation of nationality identifiers, area codes, routing information to be exchanged between systems managers and also suggests routing rules. As this numbering plan is geared more the joint multinational operations in the combat zones it lacks the functionality needed for WIS and hence will not be considered further.

STANAG 5046 Type Numbering Plan is devised for military organizations of NATO member nations from the level of army headquarters downwards. This numbering plan provides a unique, deducible, constant length subscriber address. The directory structure accommodates international gateway codes and special system facility instructions and national codes. The directory matrix allows to to identify HQs above corps, corps, division and HQ below division, major units, independent minor units and communication installations.

In this numbering plan the subscriber number can be easily deduced using the knowledge of organization of the forces and applying the logical steps illustrated in the matrices and tables. The STANAG 5046 numbering scheme is functional, deducible and could form a basis for WIS numbering with appropriate modifications.

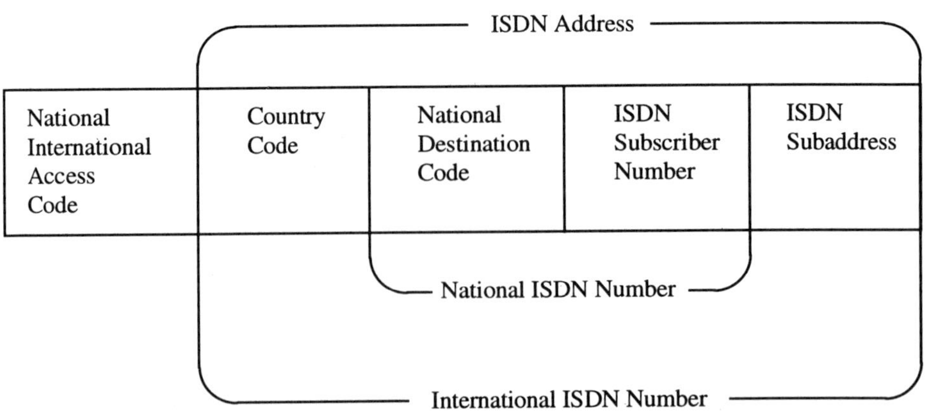

Figure 5.19 ISDN Numbering Structure

5.10 REFERENCES

[1] J. G. Gruber, N. H. Lee, "Performance Requirements for Integrated Voice/Data Networks", IEEE J. Select. Areas in Comm., SAC-1, 981-1005, Dec. 1983.

[2] M. Decina, A. Roveri, "ISDN (Integrated Services Digital Network) Architectures and Protocols, in Advanced Digital Communications", Ed. K. Feher, 1987, pp. 40-132.

[3] J. McDonald, (Ed), "Fundamentals of Digital Switching", Plenum Press, 1983.

[4] J. Bellamy, "Digital Telephony", J. Wiley & Sons, 1982.

[5] R. L. Freeman, "Telecommunication System Engineering", Wiley, 1980.

[6] A. Joel, "Electronic Switching: Digital Central Office Systems of the World", IEEE Press Book, 1982.

[7] M. Purser, "Computers and Telecommunication Networks", Blackwall Scientific Pub., 1987.

[8] J. W. Modestino et al. "Modeling and Analysis of Error Probability Performance for Digital Transmission oer the Two-wire Loop Plant", IEEE SAC-4, 1317-1330, 1986.

[9] D. I. Parnas, "Software Aspects of Strategic Defense Systems", Comm. ACM, 28, 1326-1335, Dec. 1985.

[10] M. J. Ross, "Military/Government Digital Switching Systems", IEEE Comm. Mag., 18-25, May 1983.

[11] M. J. Ross, "Circuit vs Packet Switching in Digital Switching Systems", Ed., A. Joel, Plenum Press, 1983.

[12] CCITT COM XI. No.R5E (Jan. 1982) and CEPT-T/CS(82)75 of Working Group T/No.11 and T/CS 62-04.

[13] STANAG-5046. "The NATO Military Communications Directory System".

[14] Amer. Nat. Stand. Inst. Proposal Standard X3.102, "Data Communication User-Oriented Performance Parameters", Dec. 15, 1982.

[15] Deutsch D., "Electronic Mail Systems in Digital Communication", T.C. Bartee (Ed.), Howard W. Sams Books, 1986.

[16] C. L. Heitmeyer and S.H. Wilson, "Military Message Systems: Current Status and Future Directions", IEEE Trans. Communications, 28, 1645-1654, Sept. 1980.

[17] J. S. Turner, "Design of an Integrated Service Packet Network", IEEE J. Selected Areas in Communications, Vol. SAC-4, pp. 1373-1380, Nov. 1986.

[18] "NATO Software Quality Control System Requirements", AQAP-13, Aug. 1981, NU.

[19] AC/317(WG/1)(SG/1) N/47, "NATO Use of National Switched Networks".

[20] MC 225 (Military Decision), "NATO Communications Security Policy CCITT Recommendations".

Relevant CCITT Recommendations

- CCITT Rec.E.123. Notation for national and international telephone numbers.
- CCITT Rec.E.183. Guiding principles for telephone announcements.
- CCITT Rec. E.541. Overall grade of service for international connections (subscriber-to-subscriber).
- CCITT Rec. E.543. Grade of Service in digital international telephone exchanges.
- CCITT Rec.E.427. Collection and statistical analysis of special quality of service observation data for measurements of customer difficulties in the international automatic service
- CCITT Rec.G.703. Physical/electrical characteristics of hierarchical digital interfaces.
- CCITT Rec.G.704. Functional characteristics of interfaces associated with network nodes.
- CCITT Rec.G.705. Characteristics required to terminate digital links on a digital exchange.
- CCITT Rec.G.711. Pulse Code modulation (PCM) of voice frequencies.
- CCITT Rec.G.721. 32 kbit/s adaptive differential pulse code modulation (ADPCM).
- CCITT Rec.731. Primary PCM multiplex equipment for voice frequencies.
- CCITT Rec.732. Characteristics of primary PCM multiplex equipment operating at 2048 kbit/s.
- CCITT G.735. Like above but with digital access at 384 kbit/s.
- CCITT Rec.745. Second order digital multiplex equipment operating at 8448 kbit/s and using positive/zero/negative justification.
- CCITT Rec.751. Third order digital multiplex equipment operating at 34368 kbit/s and using positive/zero/negative justification.
    - CCITT G.703. Electrical and physical characteristics of digital interfaces.
    - CCITT G.704. Functional characteristics of digital interfaces, both at the 64 kbit/s channel rate and at hierarchical bit rates.
    - CCITT G.705. Characteristics to terminate digital links on a digital exchange and in particular they will conform to.
    - CCITT G.732. Characteristics of Primary PCM Multiplex Equipment at 2048 kbit/s.
    - CCITT G.745. Second order digital multiplex equipment operating at 8448 kbit/s and using positive/zero/negative justification.
    - CCITT G.753. Third order digital multiplex equipment operating at 34368 kbit/s and using positive/zero/negative justification.
- CCITT Rec.G.821. Error performance of an international digital connection forming part of an integrated services digital network.

- CCITT Rec.I.320. ISDN protocol reference model.
- CCITT Rec.I.412. ISDN user-network interfaces, interface structures and access capabilities.
- CCITT Rec.I.431. Primary rate user-network interface layer 1 specification.
- CCITT Rec.I.440 (Q.920) ISDN user-network interface data link layer-General aspects.
- CCITT Rec.I.460. Multiplexing, rate adaptation and support of existing interfaces.
- CCITT Rec.Q.504. Performance and availibility design objectives (digital transit exchanges in integrated digital networks and mixed analogue digital networks).
- CCITT Rec.Q.507 Transmission characteristics (transit exchanges).
- CCITT Rec.Q.513. Connections, signalling, control, call handling and ancillary functions (local exchanges).
- CCITT Rec.Q.514. Performance and availability design objectives (local exchanges).
- CCITT Rec.Q.517. Transmission characteristics (local exchanges).
- CCITT Recs. Q.601-Q.685. Interworking of Signalling Systems.
- CCITT Recs. Q.721-Q.725. SS No.7: Telephone User Part.
- CCITT Recs. Q.761-Q.766. SS No.7: ISDN User Part.
- CCITT Rec.Q.920. ISDN user-network interface data link layer-General aspects.
- CCITT Rec.Q.930. ISDN user-network interface layer 3 specification.
- CCITT Rec.T.73. Document interchange protocol for the telematic services.
- CCITT Rec.V.21. 300 bits per second duplex modem standardized for use in the general switched telephone network.
- CCITT Rec.V.22. 1200 bits per second duplex modem standardized for use in the general switched telephone network and on point-to-point 2-wire leased telephone-type circuits.
- CCITT Rec.V.24. List of definitions for interchange circuits between data terminal equipment and data circuit-terminating equipment.
- CCITT Rec.V.26. 2400 bits per second modem standardized for use on 4-wire leased telephone-type circuits.
- CCITT Rec.V.27. 4800 bits per second modem with manual equalizer standardized for use on leased telephone-type circuits.
- CCITT Rec.V.29. 9600 bits per second modem standardized for use on point-to-point 4-wire leased telephone-type circuits.
- CCITT Rec.V.35. Data transmission at 48 kilobits per second using 60-108 kHz group band circuits.
- CCITT Rec.V.36. Modems for synchronous data transmission using 60-108 kHz group band circuits.

- CCITT Rec.X.1. International user classes of service in public data networks and integrated services digital networks (ISDN)
- CCITT Rec.X.20. Interface between data terminal equipment (DTE) and data circuit-terminating equipment (DCE) for start-stop transmission services on public data networks.
- CCITT Rec.X.20 bis. Use on public data networks of data terminal equipment (DTE) which is designed for interfacing to asynchronous duplex V-series modems.
- CCITT Rec.X.21. Interface between data terminal equipment (DTE) and data circuit-terminating equipment (DCE) for synchronous operation on public data networks.
- CCITT Rec.X.21 bis. Use on public data networks of data terminal equipment (DTE) which is designed for interfacing to synchronous V-series modems.
- CCITT Rec.X.22. Multiplex DTE/DCE interfaces for user classes 3-6.
- CCITT Rec.X.24. List of definitions for interchange circuits between data terminal equipment (DTE) and data circuit-terminating equipment (DCE) on public data networks.
- CCITT Rec.X.25. Interface between data terminal equipment (DTE) and data circuit-terminating equipment (DCE) for terminals operating in the packet mode and connected to public data networks by dedicated circuit.
- CCITT Rec.X.28. DTE/DCE interface for a start-stop mode data terminal equipment accessing the packet assembly/disassembly facility (PAD) in a public data network situated in the same country.
- CCITT Rec.X.31. Support of packet mode terminal equipment by an ISDN.
- CCITT Rec. X.32. Interface between data terminal equipment and data circuit-terminating equipment for terminals operating in the packet mode and accessing a packet switched public data network through a public switched telephone network or an integrated services digital network or a circuit switched public data network.
- CCITT Rec.X.51. Fundamental parameters of a multiplexing scheme for the international interface between synchronous data networks using 10-bit envelope structure.
- CCITT Rec. X.61. Signalling system No.7. Data user parts.
- CCITT Rec.X.75. Terminal and transit call control procedures and data transfer system on international circuits between packet-switched data networks.
- CCITT Rec.X.121. International numbering plan for public data networks.
- CCITT Rec. X.130. Call processing delays in public data networks when providing international synchronous circuit-switched data services.
- CCITT Rec. X.131. Call blocking in public data networks when providing international synchronous circuit-switched data services.
- CCITT Rec. X.135. Delay aspects of grade of service for public data networks when providing international packet switched data services.

- CCITT Rec. X.136. Blocking aspects of grade of service for public data networks when providing international packet-switched data networks.
- CCITT Rec.X.140. General quality of service parameters for communication via public data networks.
- CCITT Rec.X.150. Principles of maintanence testing for public data networks using data terminal equipment (DTE) and data circuit-terminating equipment (DCE) test loops.
- CCITT Rec.X.400. Message handling systems: system model-service elements.
- CCITT Rec.Z.331. Introduction to specification of the man-machine interface.

CHAPTER 6

TIMING AND SYNCHRONIZATION

6.1 INTRODUCTION

The WIS network will consist of digital access and nodal switches interconnected by time division multiplexed and bulk encrypted links. Timing and Synchronization (T & S) functions are therefore of vital importance for WIS. The synchronization concept must define the technical criteria to distribute and maintain permitted timing relationships at the WIS witching, transmission and terminal elements.

6.1.1 WIS Synchronization Objective

The synchronization subsystem must permit orderly automatic service to all users during peacetime whilst degrading gracefully to maintain network services to the maximum extent if WIS is disrupted or damaged. Disruption to individual segments of the synchronization subsystem must not promulgate unduly into neighboring segments of WIS.

The T&S requirements for WIS are as follows:

a) Bit Count Integrity (BCI) and Delay: To meet slip rate and delay performance objectives for various services that will be transported over end-to-end and/or link encrypted connections, and to satisfy BCI loss requirements for the HRC (Hypothetical Reference Circuit).

b) Maintain acceptable T&S operation even in the event of disruption or partial physical destruction, component failures, enemy interference, jamming etc. In fact WIS T&S subsystem should not present an easier target than "network segmentation" via destruction of links.

c) Permit orderly growth, extension, reconfiguration and provision of future services.

d) Achieve an availability objective better than the WIS transmission and switching system availability.

e) Maintain agreed synchronization plans with other national military communication systems as required.

f) To provide synchronization to WIS access functions.

g) Allow for control and monitoring functions from the WIS Network Surveillance and Control (WNSC). When WIS is damaged, the synchronization scheme should preferably be able to readjust so that as many nodes as possible remain usable for service without the intervention of the WNSC.

h) Be able to accommodate existing and future crypto equipment deployment with minimal changes.

i) To enable interconnection with other users and to provide possibly services to these users. These users will be referred to as the "NATO/NTTS/SATCOM connection"

One can state that the WIS switching nodes and transmission media will be subjected to disruptive influences and threats such as natural phenomena, e.g., atmospheric and temperature effects, noise; inadvertent and accidental effects, e.g., power failures, interference from other systems; intentional interference by an enemy force, e.g., jamming; and finally sabotage and destruction of WIS equipment. During hostilities the WIS network could become severely degraded and segmented for periods of time and it is expected that the timing in the segmented regions should be able to freewheel for periods of time sufficiently longer than the mean restoration time.

The T&S Planning Baseline for WIS is as follows:

a) The digital bearer channels will be isochronous transparent at the rate of 64 kbit/s with end-to-end timing. Channels may be sub-multiplexed and/or rate adapted for services requiring a bit rate below 64 kbit/s or combined to form higher order multiplex groups in accordance with CCITT Recommendations, e.g., G.735/G.737.

b) Both circuit switched and packet switched bearer services will be required as part of WIS ISDN concept based on CCITT I-series Recommendations.

c) The digital network will consist of WIS network elements such as switches, digital patch panels, transmission equipment terminals, gateway switches.

d) WIS network will have to interoperate with various other digital and analog networks.

e) All transmission links (internodal and node-to-access) in WIS will be link encrypted and a certain percentage of subscribers with DWS status will possess end-to-end encryption equipment.

6.1.2 Network Synchronization Criteria

An essential requirement of an integrated digital network is that the clocks at a multiplicity of diverse locations are maintained to within sufficiently close limits to ensure that the network of transmission and switching equipment is able to operate without unacceptable disruption. The network synchronization design is influenced by the following factors in a digital network:

- Survivability criteria,
- Reliability/Availability of the equipment,
- Performance objectives (e.g., delay, slip rate),
- Interoperability with other networks,
- Requirements of the crypto equipment,
- Error rates of trunk transmission.

6.1.3 Performance Criteria

Each node in a digital communications network derives waveforms from a local digital master clock to correctly receive digital groups, to perform switching functions and to transmit digital groups to other nodes. The objective of T&S is to maintain the frequencies of the master clock at each node of the network within a predetermined tolerance and accuracy such that data is neither inadvertently gained or lost.

Without T&S the nodal master clocks will certainly differ in frequency, possibly by indeterminate amounts depending on the type of frequency sources used (Appendix 6C). Bits in an incoming group can not then be correctly identified at the receiving node and they must therefore be stored in an elastic buffer and subsequently aligned with the local master clock. If there is a difference between the received frequency and the local clock frequency, excess/deficient bits will be stored / lost until the buffer is eventually exhausted / empty. At this point the receiving node must either lose or add bits into the received group, preferably in a determined manner known as controlled slip. The process of deletion or insertion of bits, or groups of bits within a digital group due to timing inequalities or imperfections is known as slip. Although slip cannot be eliminated, it can be controlled by appropriate design as to the moment at which it occurs and the magnitude of the slip, i.e., the number of bits that are either lost or inserted. The rate of occurrence of slips can be regulated by applying adequate control of the frequency of nodal clocks in the network and by using elastic buffers of adequate length in accordance with a T&S plan. However, although the rate of slip decreases as the buffer lengths increase, the transmission delay through each digital node increases and for this reason buffer lengths in practical networks are of limited size, typically one or two frame lengths.

The T&S plan must then state the synchronization performance objectives and the methods of achieving an acceptable slip rate. In more detail it must define the allocation of frequency accuracies in the clocks in the nodal switches, together with the method of synchronization used. The planning objective is to meet the requirement for the acceptable overall slip rate on a connection containing the maximum number of nodal switches allowed by the routing plan of the network.

Other characteristics of the transmission medium which have an impact on the T&S Plan are jitter, wander and bit errors.

Jitter: It is the short-term fluctuation of the frequency of a received bit stream, generally introduced by digital regenerators or higher-order bit-justification multiplexers. The CCITT Rec. G.823 considers the amplitude and rate of jitter within a digital network and is described in Appendix 6A. Provided that the specification of the transmission regenerators and multiplexers are suitable, the effects of jitter can be removed by a suitable elastic buffer to receive each digital group.

Wander: It is the long-term variation of the frequency of a bit stream, generally caused by path-length variations due to temperature changes in the case of terrestrial paths (Appendix 6F) or by orbital perturbations and variations of a satellite's nominal position in geostationary orbit for satellite links. CCITT Rec. G.824 provides the typical usable amplitudes and rates, as described in Appendix 6.B. If wander is not controlled sufficiently, it may be necessary to use excessively long elastic buffers, thus causing unacceptable transmission delay between users.

Bit errors: One of the most significant transmission impairments observed in digital networks is the occurrence of bit errors. These can be introduced by both predictable sources (e.g., thermal noise, crosstalk, etc.) and also by unpredictable sources (e.g., jamming, interference, malfunctioning of equipment etc.).

Because bit errors may occur anywhere within the digital bit stream, the CCITT Rec. G.732 defines a multiframe frame-alignment pattern such that, if the rate of errors is controlled within defined bounds, each elastic store is able to determine the correct frame boundaries within the bit stream and hence deduce the position of each time slot. The frequency of the bit stream is not affected and therefore the impact on the synchronization scheme is negligible in a CCITT Rec. G.732-based network

It follows then that the two direct end-to-end performance objectives for a T&S system are:

a) End-to-end delay,
b) Controlled slip rate.

6.1.3.1 Hypothetical Reference Connection

Standard hypothetical reference connections (HRX) are used to study the overall network performance objectives. The HRX reflects impairments (e.g., delay, slip, bit error etc.) due to different types of transmission elements (e.g., multiplexers, radio links, transcoders) and due to different types of switching elements (including digital circuit switches and digital patch fields). For WIS a worst case HRX is assumed, for which the path length measures 3000 km.

6.1.3.2 Delay Performance

Delay performance affects the network operations in the following aspects:

a) On voice and data connections incorporating 2-wire/4-wire hybrids echo control techniques may need to be applied to maintain acceptable quality. CCITT Rec.G.131 gives echo tolerance curves, which express minimum echo path reference equivalent for a given delay or similarly maximum delay for a given echo path reference equivalent. Note that even if all subscriber connections are made 4-wire, there would still be a need in certain voice connections for echo control due, e.g., to imperfect acoustic isolation. For connections involving satellite links, echo control must necessarily be used.

b) The design of protocols for error control, as in the packet switching protocols or as in common channel signalling systems, is affected by delay. As an example, the throughput achieved with ARR (Automatic Repeat Request) type of error correction scheme depend upon the end-to-end delay figure and the size of the sliding windows.

c) Call set-up times (which is rarely a limiting factor).

For WIS CCITT recommended delay figure of 50 ms (two-way path delay) will be assumed. To determine whether the WIS network satisfies these objectives, the following planning values can be obtained from CCITT Rec.G.114.

For WIS assuming that the longest path length is 3000 km, the number of A/D stages is 2, and finally the number of digitally switched exchanges is 10, then mean one-way delay should not exceed 0.004x3000+10x0.45+2x0.3 = 17.1 ms. Thus the worst case echo delay (two-way) of about 34 ms is well below the CCITT delay objective of 50 ms. Secure voice services in WIS networks will employ 4-wire connections and they would not in general require echo control devices. However if a secure voice call terminates in another distant network where there are 2-wire connections, then the WIS subscriber may encounter objectionable echo. Provision of echo cancelers or suppressors for this situation would be extremely difficult as digital suppression in the return path will cause the crypto units to lose synchronization. However such users with secure voice equipment who encounter objectionable echo can always revert to talk in the clear mode and than echo control devices can be enabled at the terminating switch.

6.1.3.3 WIS Bit Count Integrity Objective

Slips (bit, byte, frame) result invariably in the loss of bit count integrity (BCI), which in turn may be magnified by the loss of synchronization and its recovery process. How much the user is affected by the loss of BCI depends on the type of traffic (e.g. plain voice, secure voice, data, facsimile etc.) and on the connection methodology (circuit, packet or message switching). For WIS a slip rate objective of 24 hours for mean time to loss of BCI (MTTL BCI) can be adopted. This MTTL BCI figure is based on the logical assumption that crypto keys are normally changed once every day.

6.1.3.4 Interoperability with other networks

Since WIS has to provide communications services both to WIS users as well as non-WIS users, WIS must cater for both non-switched reconfigurable transmission connectivity and switched circuit interoperability.

Non-switched interoperability: WIS may supply 2.048 Mbit/s groups on a point-to-point basis through its transmission media. If the nodal frequency deviations are within the tolerance limits of CCITT Rec. G.732, this has no impact on WIS T&S.

Switched interoperability: It is reasonable to assume a digital HRX consisting of three intermediate (through) (IC) circuits, two terminating circuits (TC), and various crossborder links (CBL) (Fig. 6.1). The contribution of CBLs to the overall slip rate is rather small as they have a slip rate of 1 slip per 70 days. On the other hand since the local digital loops are more slip prone, the terminal segments are attached a weight of 1.25, as compared to the through segments. Therefore solving for the WIS slip rate S (assuming 4 CBLs):

$$3xS + 2.5xS + 4/70 = 1 \qquad \text{and therefore} \qquad S = 1 / 4.6.$$

In other words the end-to-end slip rate for the WIS HRX should be less frequent than once every five days, while this proposed objective applies for more than 98.9 % of the total time within at least one year observation periods.

In summary the objectives for controlled slip rate over international links and the slip rate apportioned to WIS are as follows:

End-to-end slip rate in	Not worse than 1 in 24 hours
WIS sliprate	Not worse than 1 in 5 days
Crossborder link slip rate	Not worse than 1 in 70 days

The practical implication of the above results can be explicated as follows. Assume that either a mutual hybrid or a master slave scheme for WIS T&S has been selected. These two options have in common the fact that there will be x master nodes equipped with primary standards and thus x regions in WIS that operate plesiochronously are created. Consider now a worst case HRX for WIS that crosses 9 such plesiochronous regions. To satisfy the above slip requirements, than the slip rate objectives between the plesiochronous regions should be "1 slip in 45 days". This objective can easily be satisfied if the master nodes contain caesium clocks. Actually the slip rate performance in the vast majority of calls will be significantly better.

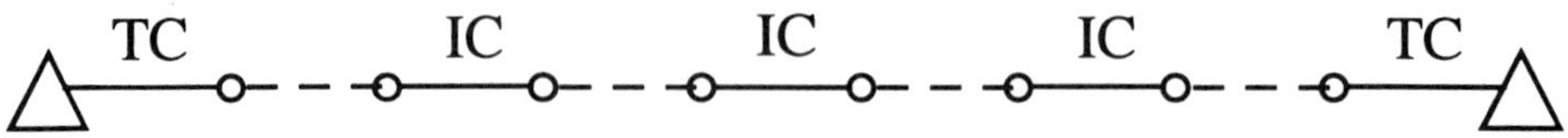

Figure 6.1 **A Digital HRX for Slip Rate Computation (CBL : Crossborder Link; TC: Terminating Circuit; IC: Intermediate (through) Circuit).**

6.2 NETWORK SYNCHRONIZATION METHODS

The options available for the synchronization of a digital network are:

a) the plesiochronous option,
b) the synchronous option.

A plesiochronous network is a network in which the control clocks at the nodal switches are independent of each other; however their frequency accuracy is kept within narrow specified limits.

A synchronous network is a network in which the clocks are controlled to run, ideally at identical rates, or at the same average rate with limited relative phase displacements.

6.2.1 Plesiochronous Operation

In the plesiochronous operation the network is not synchronized, but merely high accuracy clocks are used that are free running and adequate buffers are used to absorb the frequency differences. The buffer length depends on differential clock drift, path length variation, link data rate, and buffer reset period. These clocks are in effect running synchronously, but the frequency spread is confined enough to result in an acceptably low level of BCI loss. The clocks in each node should maintain their high accuracy during the lifetime of the equipment.
The CCITT has recommended plesiochronous operation between international networks. Plesiochronous operation is also a viable option for WIS because it is unconditionally stable for all the topologies, and because the nodal clocks operate independently, it is also the most survivable option.

If the accuracies of all the nodal clocks could be maintained sufficiently precise, the mean frequencies of oscillation of these clocks would be the same and identical to the mean frequency of the network. In practice, plesiochronous operations give rise to slips as a result of limited accuracy of the clocks. But if the slips occur infrequently enough (i.e., low slip rate), their influence on transmission performance will be marginal. Hence, clocks with sufficiently high frequency accuracy may serve the purpose of network synchronization. However, for some types of clocks (namely crystal controlled) the needed frequency accuracy can not be kept for long time periods. This is known as limited long term frequency stability. To compensate for limited long term stability, which increases slip rate, clocks in the plesiochronous network, from time to time, have to be checked against external frequency reference, or be based locally on a high stability standard (e.g., an atomic frequency).

The plesiochronous method is suitable for all kinds of network sizes. Its attraction lies in the ease of implementation, and absence of frequency stability problems. The slip performance is also easily quantifiable. The dominant disadvantage is the high procurement cost and also the high lifecycle costing due to their short lifetime and hence to high MTBF figure of these clocks. In fact atomic clocks cost about a hundred times with respect to Voltage Controlled Crystal Oscillators (VCCO), and the lifetime of caesium clocks is four years. The shorter lifetime results in an increased maintenance operation and lifetime costing. Consequently, the T & S reliability would not be adequate unless replicated clocks were used at each node.

6.2.2 Synchronous Network Operation

When the network is operated in the synchronous mode, a controlled relationship between all the clocks is required throughout the network, so that virtually a fixed relationship between digit time slots is established. At first sight, the simplest solution would appear to be the establishment of a single master clock controlling the whole of a digital network. Such an arrangement is known as despotic synchronization, that is where a unique clock has full power of control over all other clocks in the network. Are advantages in having a measure of related control between the clocks at the important nodes of a network. Instances of the latter principle are hierarchically synchronized and mutually synchronized networks.

The goal of the synchronous approach is to avoid slips by using some method of frequency or phase control throughout the digital network. Options that have equal or better performance and that are more economical with respect to the plesiochronous operation are:

- Master Slave,
- External Reference,
- Hierarchical Master-Slave,
- Network-wide Justification,
- Mutual Synchronization with single-ended and double-ended options.

A brief discussion of each of these techniques with their relative merits in the context of WIS is given in the following section [2,3,4,12].

6.2.2.1 Master - Slave Method

In the network master method a single master clock is transmitted to all other nodes enabling them to lock onto a common frequency. The phase-lock principle is used to keep the phase difference between the master clock and the slave clock constant or even to make it zero. In this manner the frequencies of master and slave clocks will be the same. Information on the instantaneous phase difference between the clocks is contained in the level of the elastic buffer at the slave exchange terminal. This information is then used in a control circuit, a phase-locked loop, to adjust the frequency of the slave clock. The slave clocks thus follow the master, and if the buffer size and control circuit are suitably designed, then no slips should occur. The master-slave concept is easy to implement as shown in Fig. 6.2 and is stable. However the reliability is low due to dependence upon a single master clock. Reliability and survivability considerations preclude this technique as a candidate synchronization method for WIS.

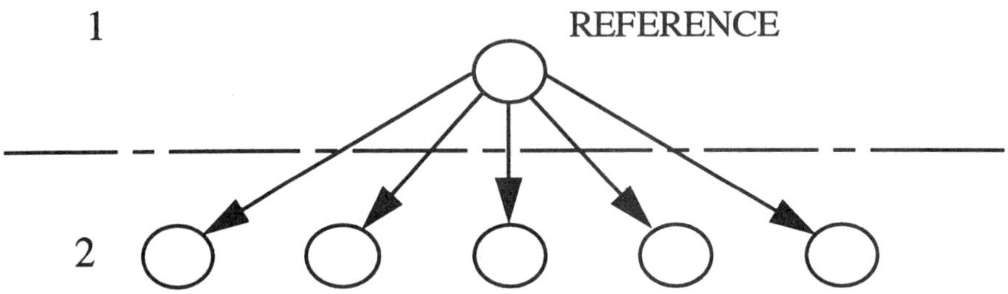

Figure 6.2 Master-Slave Method of Synchronization. The Circles Represent Clocks

6.2.2.2 Hierarchical Master-Slave Method

In the hierarchical master-slave method, all exchange clocks are arranged in a hierarchy and every clock is assigned an identification label, i.e., a rank, according to its place in the hierarchy. The hierarchical master-slave concept has better reliability, is less sensitive to link failures and is suitable for any network structure. This is achieved at the expense of increased complexity, because information about the hierarchical status of the clocks and possibly about the quality of the links has to be continuously distributed to and evaluated at every node by the control system. Hierarchical structure of a synchronization network is shown in Fig. 6.3

A network reference frequency is transmitted to a few selected higher-level switching nodes. After these nodes synchronize their clocks to the reference and hence remove the transmission link induced jitter, the reference is passed on to lower level switches by way of existing digital links. The next lower level switches, in turn synchronize to the incoming link from the higher level and pass timing on to another level of switches by way of their outgoing links. Thus all switching nodes become effectively synchronized either directly or indirectly to the same reference and hence they all run at the same nominal clock rate.

In the case of the failure of the master clock a new master having the next highest rank is automatically selected. The master clock(s) is usually a primary standard (e.g., caesium or rubidium) while the slave clocks are moderately stable crystal clocks.

In a non-trivial network not all sites will have a direct connection to the master clock and as shown in Fig. 6.4 the nodes will be divided into a number of levels. Level 1 is designated as the master reference site and those sites which have direct links to this reference are in level 2. Those sites which have direct links to level 2 will be placed in level 3 and so on. This simple arrangement is not very satisfactory since failure of links or site clocks and in particular the failure of the master reference clock will cause loss of synchronism at a number of dependent sites. Standby reference sites and standby reference links are generally introduced to provide better availability. This then requires fault detection circuitry to determine failures and to cause the changeover to the standby links or equipment. It is desired that the changeover takes place without manual intervention, i.e., in a self-organizing fashion. Typically then three kinds of control signals are transmitted:

- Designation of the node used as the master reference for the local clock
- The number and quality of links that have been encountered in the path from the reference clock to the local clock
- The rank of the local clock.

Each node monitors the frequency of the clock coming from a number of sites in a higher level and some weighted average of these frequencies is used to control the local clock. In either case mechanisms must be employed to prevent faulty links from affecting the clocks adversely. In large networks, the control chain (depth of the hierarchical tree) may be undesirably long. In this case, deployment of more than one master clock, such that each node will be connected through certain control chains, to more than one such reference would then improve the survivability and reliability of the network synchronization. At any one time a region composed of a master-clock and the nodes it controls directly will be operating plesiochronously with respect to similar such regions. Every node is then provided with the following components:

a) local quartz clocks,
b) selector for synchronizing master links,
c) frequency averager,
d) management control interface.

6.2.2.3 External Reference Method

a) Radio Beacons and Satellite Transmissions

Many radio beacons have been established throughout the world for a number of reasons, most of them being navigational. These beacons are all referenced back to a primary standard. Some of these systems provide excellent frequency accuracy as shown in Table 6.1. Each transmission and switching node could, therefore, be equipped with a receiver which would enable the local oscillator clock to be locked to one of these radio references. The advantages here are that cheaper crystal oscillators could be used and no specific network topology would be needed for the timing and synchronization plan.

On the other hand there exist several disadvantages such as difficulties encountered in radio propagation due to fading and path length variations. It may not even be possible for certain nodes to detect the radio signal, particularly the nodal switches which may be sited in valleys, in contrast to transmission nodes located on hill tops. From the military survivability point of view, the beacon would be very susceptible to jamming and also not very survivable unless several beacons are used. Further these beacons would not be directly controlled by the WIS network surveillance and control center. In conclusion, synchronization through radio beacons is not an acceptable T&S solution for WIS.

b) Pilot Carriers

In frequency division multiplex systems, reference frequencies are transmitted to the various repeater station sites in the form of pilot tones. This enables the local oscillators to be locked to a reference frequency source via this pilot carrier. In an all digital network, however, be it a radio or a cable network, this technique cannot easily be adapted for a number of operational and technical difficulties.

It would need to be added on a per channel basis and therefore would present a large overhead, not only in the increased bit rate capacity, but also in the cost and complexity of the equipment. This would especially be so in the case of digital switches when the justification control bits would need to be separated onto different carriers. Furthermore fades experienced on radio links could result in loss of BCI on individual 64 kbit/s paths with consequential problems of synchronizing encryption equipment. No organizations are known to have made any serious proposals for such a technique to be adopted in a digital switch or multiplexer for these reasons, and therefore it is not recommended for WIS as well.

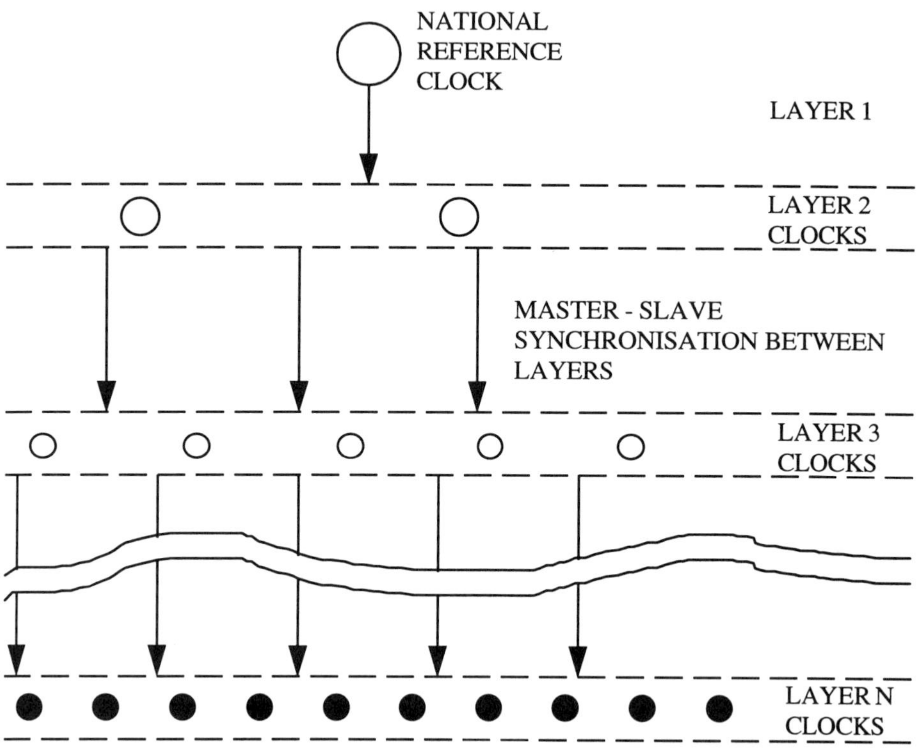

Figure 6.3 Hierarchical Structure of a Synchronization Network

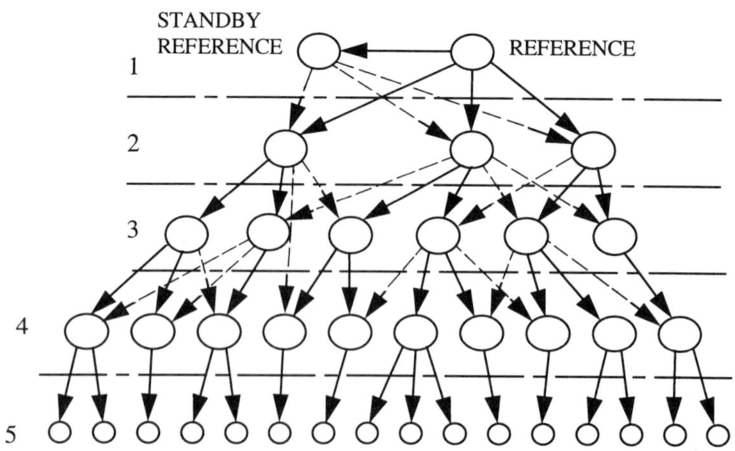

Figure 6.4 **A Master-Slave Synchronization Network With Many Levels and Standby Synchronization Links.**

Table 6.1 Comparison of Frequency Sources

Dissemination Techniques	Accuracy For Frequency Comparison or Calibration	Accuracy For Time Transfer	Ambiguity	Coverage	Reference
VLF radio OMEGA	$1x10^{-11}$	Envelope 1^{-10} ms	1 cycle	Nearly global	UTC
LF radio LORAN		1 μs (ground) 50 μs (sky)	30-50 ms	Most of nothern hemisphere	NBS master clock
HF/MF radio WWV and WWVH	$1\text{-}10^{-9}$	500 μs	Code: year Voice: 1 day Tick: 1 s	Hemisphere	NBS master clock
TV (VHF/UHF radio) TV Line-10 Color Subcarrier	$1x10^{-11}$ $1x10^{-11}$	1 μs N.A	N.A N.A	Network Coverage	None
Satellite (SHF radio) GOES TRANSIT	$3x10^{-10}$ $3x10^{-10}$	30 μs 30 μs	1 year 15 min	Western hemisphere Nearly Global	UTC UTC
Global Positioning System	$5x10^{-12}$	1 μs	N.A	Global	UTC
Defence Satellite Communications System	$5x10^{-12}$	1 μs	N.A	Nearly Global	UTC

6.2.2.4 Mutual Synchronization

This is a concept of achieving synchronism in a highly interconnected digital network without any one master [18-20]. In mutual synchronization, a common network clock frequency is established by having all nodes in the network exchange frequency references according to a distributed control algorithm. Each node averages the incoming references and uses this for its local and transmitted clock. The main attractiveness of a mutually synchronized network is its ability to remain operational in spite of a clock failure at a node, hence the survivability and the simplicity of meshed interconnectivity since no single clock or transmission path is then essential. The disadvantages of this method are however.

- The eventual system frequency is difficult to predict, since it is a function of oscillator frequencies, network topology, link delays, and weighing coefficients.
- Changes in link delay or nodal dropout can cause significant perturbations in nodal frequencies and a permanent change in system frequency.
- Lack of a fixed reference results in offset with respect to any external network or other external sources of time or frequency, making mutual synchronization incompatible for interworking with other networks.

The primary objective of the clock control is to ensure that the clock frequencies have the same long-term mean value and limited instantaneous deviations from that value. Short-term fluctuations are smoothed by elastic buffers, which allow a system with varying delays about a central (delay) value without causing slips. Two methods of control are available for mutual synchronization schemes:

a) single-ended control,
b) double-ended control.

Single ended control is suitable for use in a network of an arbitrary structure. Reliability, survivability and stability of the clock may be lower than those for the methods already described. However the stability of the system is affected directly by the ambient temperature. In single-ended control, the phase of the switch clock is the average of phase offsets between the local clock and all incoming clocks as illustrated in Fig. 6.5. The weakness of this method lies in its inability to cope with the effects of transmission delay variations caused by the temperature changes.
The double-ended control improves the synchronization system further by making it independent of delay variations. The somewhat more complex control scheme is particularly beneficial in networks containing long links. In this case, the input to the control circuits consists of the difference between the "single-ended" information and phase offsets measured at all cooperating nodes. The schematics of the double-ended control algorithm is shown in Fig. 6.6.

Variants of mutually synchronized networks are shown in Figs.6.7-6.8. A totally mutually synchronized network as in Fig. 6.7 will gradually drift in frequency, relative to some absolute frequency, as the individual oscillators age. For interworking with other networks it is necessary that this network be referenced to a primary standard. Generally one site is nominated as a reference as shown in Figs. 6.8 where the mutual synchronization with hierarchical levels is illustrated, and the other sites are mutually synchronized to each other and then a number of them are slaved to the reference. If the reference fails, the remaining sites will remain mutually synchronized. For survivable interoperability with other networks multiple reference nodes must be provided, consistent with the threat, using either "hot-standby" or "multiple-master" references, as shown in Figs. 6.9.

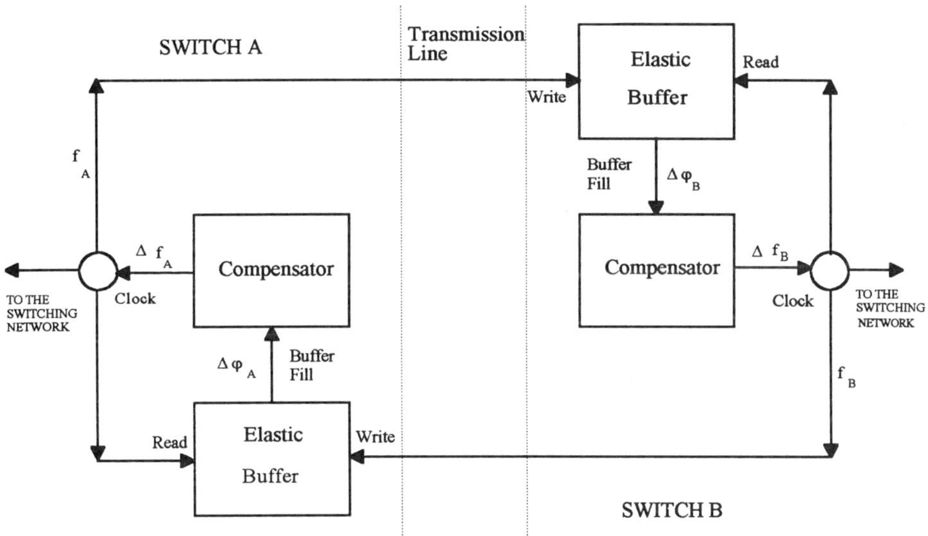

$\Delta\varphi_A$, $\Delta\varphi_B$	Instantenous phase difference measured at A and B respectively.
Δf_A , Δf_B	Frequency correction applied to the clock A and B respectively.
f_A , f_B	Frequency output from the clock A and B respectively.

Figure 6.5 Single-Ended Control in the Mutual Synchronization Scheme

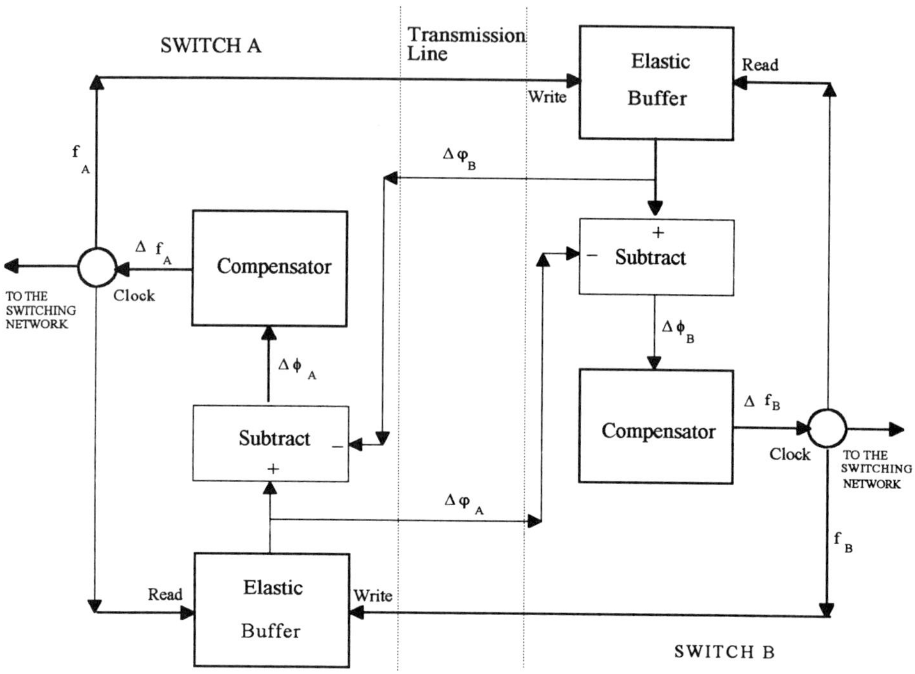

$\Delta\ \varphi_A$, $\Delta\ \varphi_B$ Instantenous phase difference measured at A and B respectively.

$\Delta\ \phi_A$, $\Delta\ \phi_B$ Instantenous corrected phase difference calculated at A and B respectively.

$\Delta\ f_A$, $\Delta\ f_B$ Frequency correction applied to the clock A and B respectively.

f_A , f_B Frequency output from the clock A and B respectively.

Figure 6.6 Double-Ended Control in Mutual Synchronization

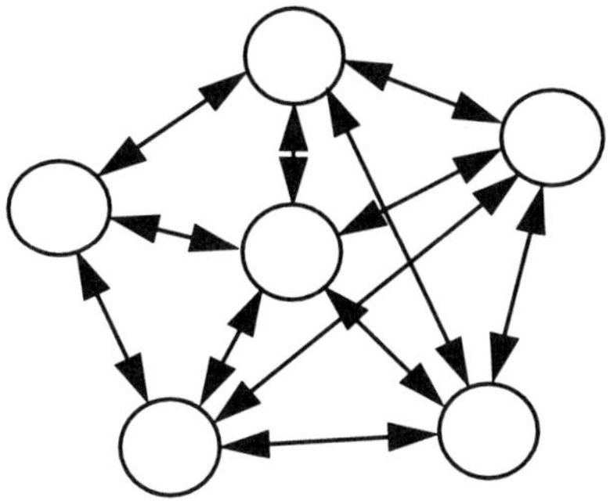

Figure 6.7 Mutual Synchronization Without a Reference

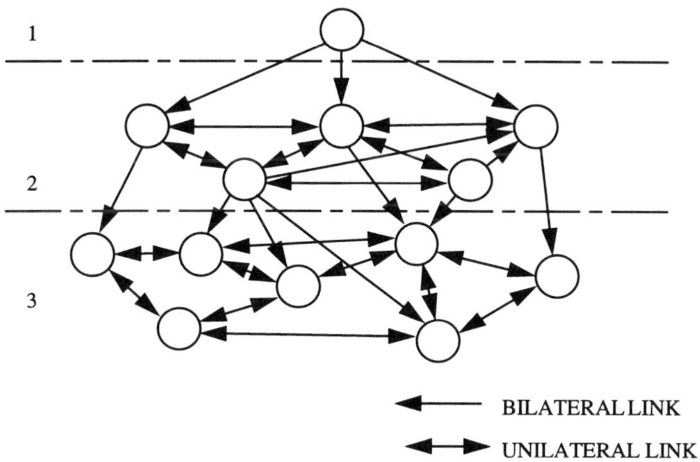

Figure 6.8 Mutual Synchronization With a Reference And Hierarchical Levels

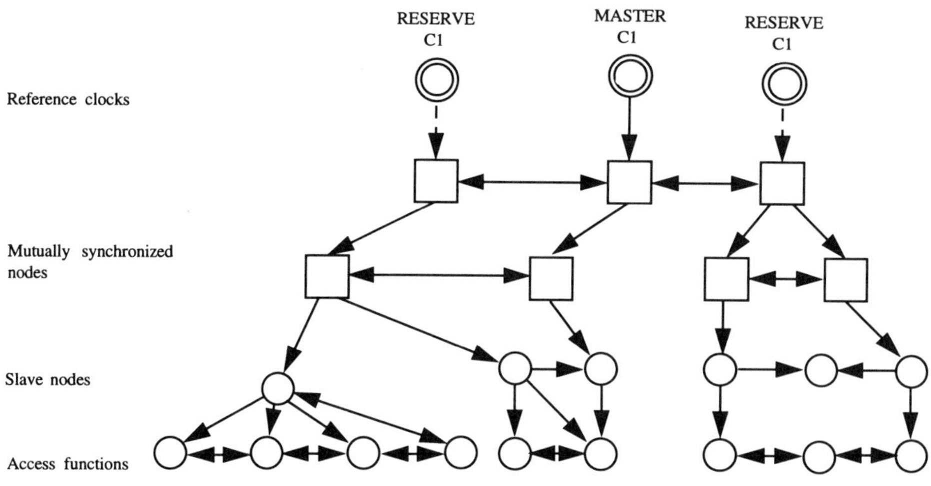

Figure 6.9 Mutual Synchronization Using a Hot Standby.

6.2.2.5 Network Wide Justification

In the networkwide justification method, asynchronous streams of 64 kbit/s are justified to a common higher bit rate and then assembled either in a switch or multiplexed onto a higher digital carrier of 2048 + kbit/s. Thus TDM channels are provided at the user level through which user data flows at lower and variable rates and the differences of user clocks are absorbed by internal justification, which effectively overcomes any BCI loss problems. Consequently the network reference clocks do not need to be very accurate and low grade oscillators can be used at the asynchronous nodes. However it also means that justification is needed on a per channel basis, which in turn presents a large overhead, not only in the increased bit rate capacity, but also in the cost and complexity of the equipment. For example in the case of digital switches the justification control bits would need to be separated onto different carriers. Furthermore fades experienced on radio links could result in loss of BCI on individual 64 kbit/s paths with consequential problems of synchronizing encryption equipment. The pulse stuffing techniques are becoming obsolete, even in higher order multiplexers, except in certain instances like digital satellite links. For these reasons the networkwide justification method is not considered to be an acceptable option for WIS.

6.2.3 Packetization

The synchronization discussions of the preceding sections have implicitly assumed that a synchronous, circuit-switched network was being considered, since prevailing digital voice networks operate in that manner. On the other hand, packet-switched networks are designed such that asynchronous techniques can be used for the transmission of packets between nodes. Therefore a networkwide synchronization scheme would not be needed. Currently the available packet transmission rate is too slow to support 64 kbit/s digital services and the processing delay in each node of 30 ms could cause unacceptable cumulative delays. This option eliminates the need for a synchronization function, but the network would not conform to international standards and would require special development. However research into fast packet switching, frame relay and ATM techniques could lead to practicable techniques for implementation in a nodal network in the near future and therefore are worth considering

6.2.4 Comparative Evaluation

All the synchronization concepts described above have certain distinct advantages while at the same burdened by some disadvantages. Pros and cons depend on the network size, its topology, distances between nodes, network connectivity, types of transmission media, cost of synchronization equipment, system complexity, reliability, maintenance, and management effort, survivability against threats, interoperability, life cycle, flexibility and, ease of network reconfiguration. A comparison of network synchronization methods on the basis of above mentioned criteria is shown in Table 6.2.

The plesiochronous approach offers advantages of being relatively simple to implement, and due to the autonomous nature of the nodal clocks it is highly robust against physical electronic threats. It is not susceptible to jamming and spoofing since it does not depend on information transfer between the station clocks for its operation. Its major drawbacks are that a larger number of relatively expensive caesium clocks with somewhat poor reliability characteristics are required and periodic timing slips due to the resetting of buffers must be accepted. With atomic clocks the system could run for extended periods without buffer resets. The effect of buffer resets would be minimized by coinciding them with other planned traffic interruptions such as key variable change on the link, preplanned resetting at low traffic periods or off-loading of switched traffic prior to the reset. The external-reference method, compared to the other despotic methods, is easier to implement and may take advantage of some already existing precise frequency standard, although the system reliability and survivability vis-a-vis threats are severe problems.

The hierarchical master-slave method improves reliability, is less sensitive to link failures, and is suitable for any network structure. This is gained at the expense of increased complexity. The enhanced master-slave technique scores equally high as the plesiochronous technique on all major criteria but it has a potentially lower cost. One of the main features of this technique is the adaptive reorganization capability. Among several possible adaptive schemes the one where the best available clock is selected as the master and the slaved nodes are adaptively constructed seems to be the most viable one. Relative to the plesiochronous approach the main drawbacks of the enhanced master-slave technique are greater implementation complexity, requirements for clock error measurements and coordination, and somewhat greater vulnerability to sophisticated jamming. The additional complexity of the enhanced master-slave technique can be quantified in terms of information processed at each node and exchanged between nodes to achieve synchronization and the related hardware and/or software. This additional traffic will be proportional to the sampling rates for measuring timing errors and data transfer associated with adaptive reorganization.

The mutual synchronization method has the advantages of system reliability and high survivability. Interoperability problems can be overcome by designing a hybrid mutual synchronization scheme whereby one or more masters are assigned and the other nodes are mutually synchronized while being at the same time slaved to the master nodes. One should, however, note that mutual synchronization schemes and in particular, the double-ended control version, involves a significant amount of system complexity. The double-ended mutual synchronization method is desirable to make the synchronization subsystem independent of environmental effects such as temperature variations, especially if there exists hops with large path lengths. One additional disadvantage of the mutual system is that in a system where every clock controls all the others, the addition of a new clock requires that the phase of the clock be initially in clock alignment with the phase of the rest of the system. If then the original phase difference propagates through a part of the system it may cause some buffers to slip.

It is observed that the network master method gets a poor to fair score on most of the selection criteria and will not be further considered for WIS. From these qualitative analyses, the following methods, namely.

- plesiochronous
- multiple master-slave
- hybrid double ended mutual synchronization

are viable candidates for WIS. The final selection between these methods as detailed in the next Section, will rest upon such considerations as life cycle costing, interoperability etc..

Table 6.2 Characteristics of Network Synchronization Concepts

Method	Main disadvantages	Main advantages	Network Size	Network topology	Distances between exchanges	System complexity and technical risk	Survivability	Interoperability	Management effort	50 year life-cycle cost US $000	15 year life-cycle cost US $000
Plesiochronous	Cost	No network stability problems Easy to implement	International		Long	Very low complexity and risk	Very good	Fair.Interworking may increase buffer reset rates	Low	5,000	2,000
Network Master	Capacity needed for extra signalling	No network stability problems	Small (*)	Star	Short (**)	Low complexity and risk	Poor	Fair to good. Buffers may have to be reset at interfaces to other network	Very high		
Hierarchical Master-slave	System complexity Extra signalling capacity is needed	System reliability	Medium	Star or mesh	Short-medium	Medium complexity and low technical risk	Fair with automatic nodal reordering	Fair to good	Medium to high	2,000	872
Multiple Master-slave	Complexity of clock average	Reliability		Mesh	Medium	High complexity and some technical risk	Fair with automatic nodal reordering	Fair to good	Medium to high	2,000	872
Single-ended Control MUTUAL	Network stability problemsDependence on delay variations	System reliability Champ Clocks	Medium-big	Mesh	Medium	High complexity and some technical risk	Good but affected by transients	Fair.Network frequency difficult to predict	Low to medium		
Double-ended control MUTUAL	Network stability problems. Extra signalling capacity is needed	Advantages of single-ended and independence of delay variations	Medium-big	Mesh	Long (**)	Very high complexity and some technical risk	Good	Fair to good	Low	2,000	1252

Synchronization is derived from inter-nodal groups, G
(*) small networks have 5-7 nodes
(**) short distance < 20 km: long distance < 100 k

6.3 TIMING AND SYNCHRONIZATION SUBSYSTEM

Various network synchronization methods have been discussed in Section 6.2. The criteria for the selection of a network synchronization method for WIS are as follows:

1) The WIS users' requirements, especially the influence of cryptographic equipment,
2) Requirements for survivability of the WIS network,
3) Cost,
4) Requirements from non WIS users,
5) Interoperability with PTT for interchange of transmission capacity,
6) Interoperability requirements for tactical area communication networks and other national strategic networks,
7) Management issues and technical risks,

6.3.1 The Impact of Interchange of Transmission Capacity

It is anticipated that WIS will exchange transmission capacity with PTT network. This section discusses the impact of transmission interchange on the synchronization function.

PTT Transmission Media: PTTs are installing digital networks which is based on the CCITT G.702 hierarchy and for most PTTs their paths would be geographically similar to WIS transmission paths. The issue to be resolved is whether this can be used to provide transmission capacity for groups of WIS channels. There are two possibilities vis-a-vis the second-order multiplexing structure of the PTT transmission systems. Either the PTT uses:

1) Synchronous second-order multiplexes, i.e., CCITT G.744, or
2) Asynchronous second-order multiplexes using justification, i.e., CCITT G.745.

Synchronous Multiplexes: If the PTT uses synchronous second-order multiplexing equipment, as per CCITT G.744, then the WIS groups must be transferred synchronously with the PTT synchronization scheme. This can be achieved either by

a) Using intermediate elastic buffers at each of the interchange points between the PTT and WIS as in Fig 6.10a. The WIS and PTT synchronization function would then interoperate plesiochronously. However, extra transmission delay of 128 us per buffer would be introduced at each interchange of transmission capacity. The cumulative delay of several interchange buffers would adversely affect the transmission performance for user service;

b) Using synchronous interoperation, as in Fig. 6.10b, where direct interchange of groups could then be achieved if the WIS and PTT were fully synchronized. No intermediate elastic buffer would then be needed.

c) Using justification multiplexing: Using plesiochronous interoperation, as in Fig.6.10c direct interchange of groups could be achieved. Both the PTT and WIS should then use higher order justifying multiplexes according to CCITT G.745.

Limits for plesiochronous operation: The limits stated by the CCITT G.745 gives the maximum justification rate as 8 kbit/s for each group of 2048 kbit/s. In principle this could be used as the limit for plesiochronous operation of the two networks i.e., a maximum frequency deviation of 1 in 256, for each 32 channel group (G.703). However, the tolerance stated for the nominal bit rate for G.745 is +30 parts per million. The length of the elastic buffer in Fig.6.10a must be considered since it contributes to both the slip rate and the path delay. For a maximum slip rate of 1 per day, a clock accuracy of 30 ppm, the buffer length can be calculated as follows.

Clock accuracy	= 30 x 10 E-6
Phase drift in bits/sec	= (30 x 10 E-6)(2.048 x 10 E6)= 61
Bits accumulated in one day	= 61 x 24 x 60 x 60
Frames to be buffered	= 61 x 24 x 60 x 60/256 = 205875
Time delay	= 205875 x 125 x 10 E-6 = 26 sec

This amount of storage delay is of course unacceptably large. For an acceptable delay of 1 frame, i.e. 128 us, the frequency deviation should be better than 10E-11, for which the stability of caesium clocks is necessary.

In conclusion one can state that,

1) If both PTT and WIS use justification multiplexing, Rec. G.745, there be, minimal impact on the synchronization schemes, provided that plesiochronous operation is achievable, i.e., each side referenced to separate atomic caesium standards;
2) If one of the PTT or WIS networks uses synchronous multiplexing, then either (a) synchronous operation with a reference to the same master clock standard or (b) plesiochronous operation with separate caesium clocks, should be used.
3) In each of the above cases the WIS synchronization function must be referenced to a caesium clock (or equivalent). In the case 2(a) above this could be provided by the PTT.
4) Using synchronous interoperation, as in Fig.6.10b direct interchange of groups could then be achieved if the WIS and PTT were fully synchronized. No intermediate elastic buffer would be needed.

6.3.2 Survivability Considerations

The network synchronization function must be survivable to a degree consistent with the WIS nodal grid. The threat models to assess the network survivability consist in segmentation of the network into isolated regions, for which, N or more nodes must be destroyed. When the network is sparsely connected, synchronization distribution must be feasible across the remaining interconnecting links, i.e.. whenever connectivity between trunk nodes remains, and timing distribution must also be achieved to maintain service.

Survivability for NTTS: Cross-border links (CBL) for NTTS will be provided at $n_{terrestrial}$ terrestrial sites and n_{satcom} SATCOM sites. The NTTS timing information must be maintained at each CBL until that CBL is destroyed. When all the CBL's are destroyed, NTTS timing is no longer usable. It can therefore be concluded that the NTTS synchronization scheme need be no more survivable than the group of CBL's.

Survivability comparison of the T & S options:

- Plesiochronous Mode: Of no special measures were taken and the nodes were allowed to operate plesiochronously using internal quartz clocks, then slippages would occur over time. A worst case analysis can be done by considering an HRDP (Hypothetical Reference Digital Path) of several links, with each adjacent node drifting in opposite directions of frequency change. The drift is cumulative and the effect on the total frame slip rate of the HRDP will increase with time. For example let us assume a drift rate of + 1x10 E-10 per day and an HRDP consisting of 10 links, then one would have 0.7 total slips/day, which would accumulate to 21 slips/day over a period of one month. Therefore, clocks with low drift rate are used for plesiochronous operation e.g., atomic clocks such as caesium standards. Levels of survivability in the plesiochronous operation and the number of atomic clocks deployed can be estimated from possible segmentations of the network under threat scenarios.

- Mutual Synchronization Mode: The mutual synchronization mode can maintain synchronization both in the grid-connected network and when the network is sparsely-connected due to damage. When the network is segmented the segments will operate their separate and individual mutual synchronization, but there will not be synchronization between disconnected regions. The options available are:
    - The whole of the nodal network is mutually synchronized
    - A core of nodal network is mutually synchronized with master/slave distribution to the peripheral nodes.

If the whole of the nodal network is mutually synchronized, all of the nodal clocks operate nearly-synchronously since the significant instants of separated clocks may not always coincide due to the response function of the multi-node synchronization function. The mean frequency of the whole network may also drift with time. If quartz oscillators are used then the drift will not affect service to WIS users and the WIS users requirements will be satisfied.

For reliability reasons (the MTBF of a caesium clock is about 3 years), several caesium clocks must be used, either collocated or distributed over several locations. For an interoperating network, if each CBL is to be as survivable as the international connectivity, then it is reasonable to equip each CBL with a caesium derived reference clock signal. At this stage more than one option appears for deployment of the reference clocks:

In the first option, SATCOM distributes the caesium standard reference signal to the CBLs in WIS. Double-ended control would be necessary because of the long path lengths used. With this option WIS operates nearly-synchronously with the SATCOM. In the second option a caesium-derived reference signal is distributed to the CBL, from WIS. Because the WIS nodes would be mutually-synchronized, this implies that a caesium clock could be located at a node in the network which is remote from the CBL. In this way one caesium clock could synchronize two or more CBL's for plesiochronous operation. In the third option a caesium reference clock could be placed at each of CBL's such that plesiochronous operation with other users remains possible even when there is no remaining connectivity with the WIS network. If a clock averaging circuit is used at each node, then multiple caesium reference sources may be connected simultaneously to the mutually-synchronized WIS network.

Hierarchical Master-Slave: A synchronization distribution graph is constructed for each master node such that each master controls one region of the network. Care must be taken of the number of plesiochronous boundaries that the worst case HRDP traverses, and that the calculated maximum rate of slip remains within the bound set by the operation of bulk crypto devices. The availability objective leads to a requirement for replication of master clocks, either by triplication at each master node or by the provision of alternative synchronization paths from each master clock to each master node. The survivability requirement necessitates the provision of alternative paths for synchronization such that when a master clock is damaged, other routes can provide stable waveforms from alternative masters. Synchronization distribution must be available from at least two alternative sources at the same or higher hierarchical layer.

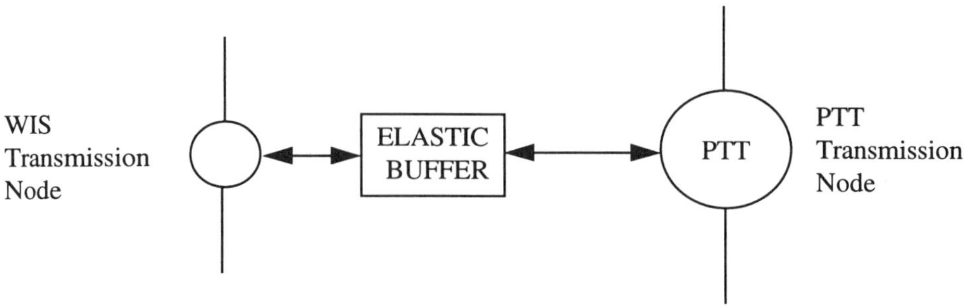

Figure 6.10a. Plesiochronous Interoperation of the PTT and WIS Synchronization Schemes.

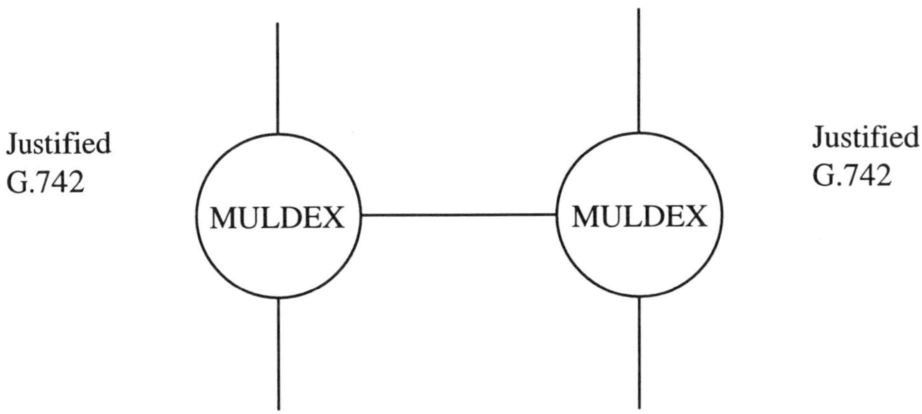

Figure 6.10b WIS-PTT Synchronous Interoperation

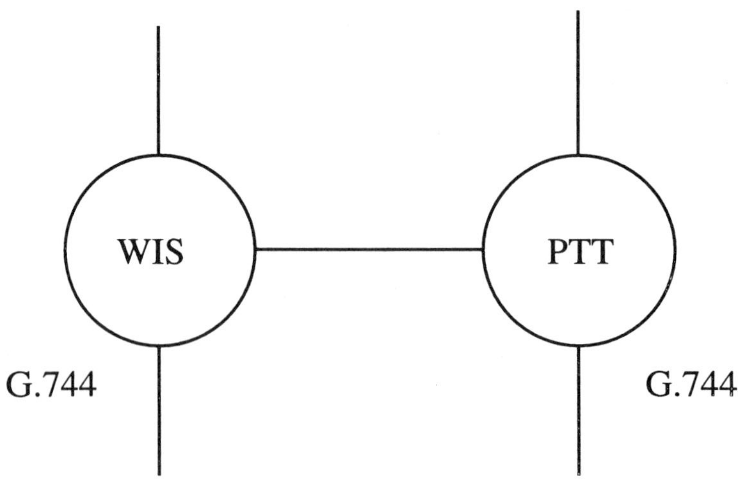

Figure 6.10c Plesiochronous Interoperation of WIS and PTT With Asynchronous Multiplexing

6.3.3 Influence of Cryptographic Equipment

The COMSEC plan for the WIS implies that the internodal trunks will be protected by encryption, and that each link uses two crypto devices, one at each end. Error-free groups can only be transmitted whilst the crypto devices are in synchronization and normally they are synchronized when a key is loaded. Bit errors caused by poor transmission will not normally desynchronize the cryptos (although the data will be corrupted). However BCI failures could cause desynchronization of the cryptographic devices, depending on the method of timing distribution.

Timing Distribution: Three methods are available in practice to provide timing to the crypto devices at each node of an internodal link, namely:

1) Internal timing from an asynchronous clock source as in Fig. 6.11a,
2) Co-directional timing distribution whereby the nominated master crypto supplies timing to the remote slave crypto as in Fig. 6.11b,
3) Local timing whereby each crypto derives timing waveforms from the local node as in Fig. 6.11c.

For cases 1 and 2 the influence of loss of BCI on the cryptographic devices should be minimal. However, if option 3, where local timing, is used, the cryptographic synchronization will be adversely affected, depending on the method of crypto synchronization which is used Cryptographic synchronization methods are discussed below.

Cryptographic Synchronization: Two principles of crypto synchronization are available, namely:

1) Cipher text feedback, 2)Autokeying.

With cipher text feedback (CTAK) the cryptos resynchronize continuously as cipher text is transmitted. Each bit error causes error extension (of about 100 bits, depending on the logic used). A frame slip will lead to immediate resynchronization of the cryptos, with an error extension of about 100 bits.

With autokeying (KAK) the cryptos are synchronized once only and can only be disturbed by a loss of BCI. A bit error in the crypto text will cause one bit error in the clear text. A frame slip (or loss of cipher bits) will cause the cryptos to lose synchronization completely until an external event occurs to initiate the synchronization process which can use about 100 bytes of data.

This indicates that different control criteria must be used for CTAK or KAK cryptographic devices and these will affect the unavailability objectives of the HRDP when local timing is supplied to the cryptos.

The unavailability criteria of the HRDP when local timing is supplied to the cryptos are as follows. For CTAK cryptos no external control procedure is necessary and each bit error or slip will cause an error extension of 100 bits. The effect will be to worsen the overall BER by a factor of 100 over the objectives stated in Chapter 8. For KAK cryptos external control is necessary to resynchronise the cryptos when slip occurs. Possible schemes to achieve this are:

a) manual resynchronization after an audible alarm;
b) resynchronization from the nodal switch control;
c) automatic resynchronization from a local synchronization monitor, as depicted in Figure 6.11b. Of these, the automatic resynchronization scheme is likely to respond most rapidly.

It is informative to derive a lower bound of the performance in terms of lost bits during resynchronization:

- The criterion to detect loss of frame alignment according to CCITT Rec. G.732 is that when 3 consecutive frame alignment signals have been received with an error.
- The number of missing frames before the alignment error is registered is 3 x 2 = 6 frames
- The times T to detect the desynchronization situation is T = 6 x 125 us = 750 us
- The number of frames used to complete the resynch is about 1000/256 = 4
- The number of frames actually lost due to slip is: N = 6 + 4 = 10
- The minimum error extension of the resynchronization process is: 256 x 10 = 2560 bits per group or 8 x 10 = 80 bits per channel.

Comparison of CTAK and KAK on the Quality Objective of the HRX: This section compares the impact of the two types of crypto synchronization on the bit error performance of the HRX. CTAK causes an error extension of approximately 100 bits in a 32 channel group for every bit error and every loss of BCI. This worsens the performance of the HRDP for each 64 kbit/s circuit by a factor of 100 with the consequences that either the worsened BER is accepted (Chapter 8), i.e.,

a) BER worse than 10 E-4 (instead of 10 E-6) for 0.54 % of 1 minute intervals in a month.

b) BER worse than 10 E-1 (instead of 10 E-3) for 0.0734 % of 1 seconds in a month.

c) The errored seconds figure 0.435 % in 1 month would not change significantly.

or the design parameters of the transmission system must be made more stringent, i.e., approximately 100 times better to satisfy the overall HRDP objective..

The KAK mode does not cause error extensions for bit errors, but with every frame slip and the consequent loss of synchronization leads to a minimum error extension of 2560 bit/group or 80 bit/channel. For example, for a representative slip rate of 1 in 24 hours, one has, considering the fact that each slip causes 80 bit errors, and therefore it can be shown that the resulting total number of errored seconds due to slip in a month is 30. Hence when KAK type crypto equipment is being used slips will have only a marginal effect on the HRDP quality objectives. In conclusion the use of crypto equipment of the KAK variety does not impact significantly on the transmission quality objectives.

6.3.4 Life Cycle Costing

Consider the life cycle costing of clock sources over Y years, where subscripted P's denote the costs of the related items:

1. Caesium clock

Purchase:	$P_{caesium}$(typically 60.000 US dollars)
Replacements	Y / MTBF (typically 2,5 years, so that over 50 years 20 replacements would be needed) and unit price $P_{replace}$
Labor	P_{labour} (e.g., 3000 $)
Non-Indexed life cycle cost	$P_{caesium} + Y / MTBF\ (P_{replace} + P_{labour})$

2. Quartz Oscillators

Purchase P_{quartz} (typically 6000 US dollars)

Memory / Flywheel P_{memory}

Replacement $P_{replace}$ (MTBF of 50 years)

Non-Indexed life cycle cost $P_{quartz} + (P_{replace} + P_{memory})$

The non-indexed life-cycle costs of the two clock sources, for a lifetime of 50 years, would be, 320.000 \$ and 13.000 \$, respectively. Similarly the life cycle costing for the whole network can be estimated by multiplying these unit costs with the numbers deployed. For example in a 25 node network the life cycle cost for Y = 50 years would be 8.000.000 \$ if plesiochronous operation with caesium clocks is opted, while the cost drops to 305.000 \$ with quartz sources. For mutual synchronization, on the other hand, in addition to the cost of clocks the cost of averaging circuits must be added.

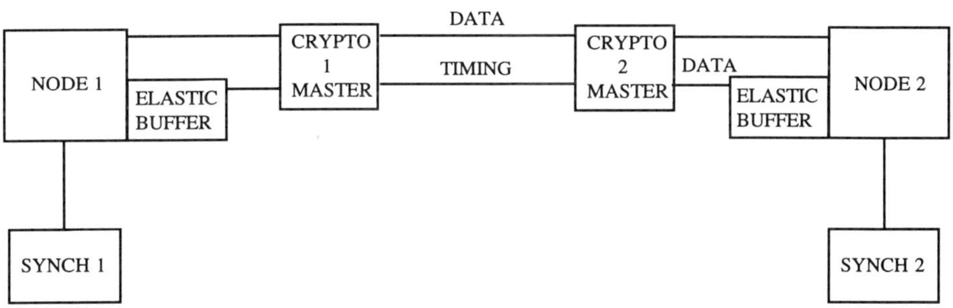

Figure 6.11a KAK Crypto Auto Resynchronization Function

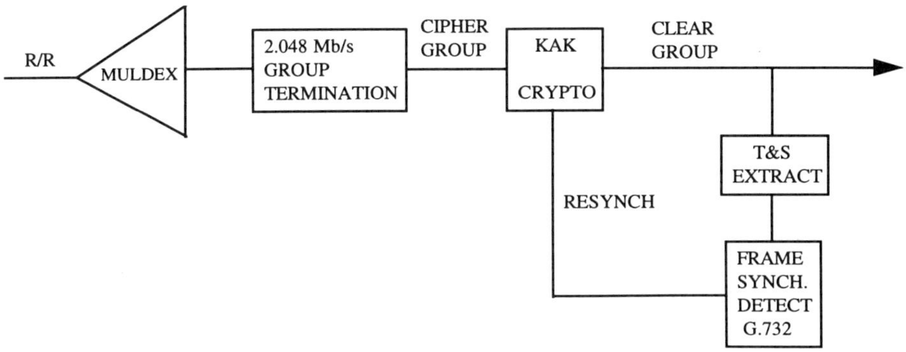

Figure 6.11b. Codirectional Timing Distribution

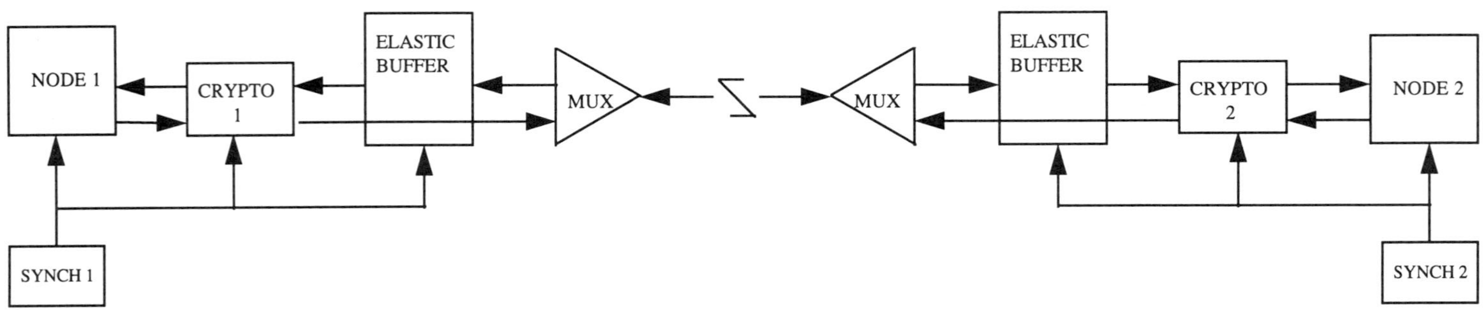

Figure 6.11c Local Timing Distribution of the Cryptographic Devices

6.3.5 Management Issues

Management is concerned with the continuity of the synchronization scheme, even when the network is damaged or when a failure of a clock has occurred. This is concerned with both the availability and the survivability issues. The issue to be addressed is how the direction of distribution of timing information is changed so that, even under network damage The available connectivity is used to deliver synchronization waveforms to all operational nodes using available links.

The complexity of the link management and monitor depends on the synchronization scheme chosen. These are summarized below:

1. Plesiochronous: Alternate timing to be taken automatically from two or more adjacent master clocks. The implication is that it needs an automatic changeover device and a clock averager.
2. Master-Slave: Alternate timing to be taken automatically from designated alternate masters. The main implications are:
   a) Designated standby plan;
   b) Automatic changeover;
   c) A clock averager is needed;
   d) "Flywheel" stabilization of quartz clocks is needed to prevent transient effects during the changeover period.
3. Hybrid Mutual Synchronization: (hot-standby) Network adjusts automatically, but standby master must be connected to the network when required. The implication: Management must instruct the new master to attach the new reference.
4. Mutual Synchronization: Network adjusts automatically. The are no implications

In all cases there will be an extra management duty to repair the failed clock, for which a maintenance and repair organization will be needed. However the mutual synchronization and plesiochronous schemes seem to have the least impact on management issues.

6.3.6 Conclusions on Factors Analysis

The three synchronization methods namely plesiochronous, multiple master-slave and hybrid mutual with double ended control all satisfy WIS user requirements, NTTS requirements and interoperability requirements.

The plesiochronous method has a simple management but is handicapped due to high life cycle cost and needs a continuous procurement of spare primary standards due to their short MTBF figure.

The other two synchronization methods in this respect seem more advantageous because of their significantly lower life cycle cost. Another advantage may accrue due to the fact that in the disciplined method the synchronization monitoring and control functions provide a second source information on the status of links and nodes.

Thus both the mutual synchronization method with double-ended control and the multiple master-slave technique are likely to be the best long term objectives. In both schemes the network would be equipped with a small number of caesium clocks to satisfy the requirements for interoperability with other networks. All other nodes should be equipped with one or more reliable (MTBF = 30 years) double-oven temperature controlled quartz clocks. The frequency of the clocks should be capable of external electrical control via a local electronic "flywheel" averaging circuit to smooth the effects of any transient phase discontinuities when the control mode is altered.

6.4 REFERENCES

[1] B.N.Kearsey, R.W. McLintock, "Jitter in Digital Telecommunication Networks", British Telecom. Eng., 3, pp.108-116, July 1984.

[2] R. Smith, L.J. Millott, "Synchronization and Slip Performance in a Digital Network", British Telecom. Engineering, 3, pp.89-107, July 1984.

[3] R.A. Boulter, W. Bunn, "Network Synchronization", British Telecom. Engineering, 3,1984.

[4] J. Bellamy, "Network Synchronization, Control and Management" in Digital Telephony ,Wiley 1982.

[5] D.R. Smith, "Network Timing and Synchronization", in Digital Transmission Systems, Van Nostrand Reinhold Co., 1985.

[6] R.A. Boulter et al., Timing and Synchronization of the NICS II Digital Network, STC TM-652, NATO Rest., May 1981.

[7] CCITT Rec. G.811: Timing requirements at the outputs of reference clocks and network nodes suitable for plesiochronous operation of international digital links.

[8] CCITT Rec. G.821: Error performance of an international digital connection forming part of an integrated services digital network.

[9] CCITT Rec. G.822 : Controlled slip rate objectives on an international digital connection.

[10] CCITT Rec. G.823: The control of jitter and wander within digital networks which are based on the 2048 kbit/s hierarchy.

[11] CCITT Rec. G.921 :Digital sections based on the 2048 kbit/s hierarchy.

[12] M.C. Thompson, H.B.Jones, "Radio Path Length Stability of Ground-to-Ground Microwave Links", NBS Technical Note 219, Washington Government Printing Office, 1964.22.J.J. Spilker, "Digital Communications by Satellite", Prentice Hall, 1977.

[13] R.Freeman, "Standard Time and Frequency" in Reference Manual for Telecommunications Engineering, Wiley, 1985.

[14] D.L.Duttweiler, "The Jitter Performance of Phase-Locked Loops Extracting Timing from Baseband Data Waveforms", Bell Sys.Tech. Journal, pp. 37-58, Jan. 1976.

[15] D.L.Duttweiler, "Waiting Time Jitter", Bell Syst. Tech.Journal, pp.165-207, Jan. 1972

[16] M.B.Brilliant,"The Determination of Frequency in Systems of Mutually Synchronized Oscillators", Bell Sys. Tech. Journal, pp.1737-1748, Dec. 1966.

[17] M.W.Williard, H.R.Dean, "Dynamic Behaviour of a System of Mutually Synchronized Oscillators", IEEE Trans. Comm., COM-19, pp.373-395., Aug.1971.

[18] J.Yomoto, "Stability of a Synchronization Control System for an Integrated Telephone Network", IEEE Trans. Comm., COM-22, pp. 1818-1953, Nov.1974

APPENDIX 6 A

TIMING INSTABILITIES

If the internal and external operations of a digital network are timed from a single master, a particularly stable source is not needed since all commonly clocked elements would then experience timing variations in common. However in digital communication systems most equipment extract their timing information from the incoming digital links or channels. These timing sources are invariably subject to long and short term phase variations. Indeed the recovered timing signals will appear to be randomly phase modulated with respect to a stable clock reference.

The short term variations of the significant instants of a digital signal from their ideal positions in time is called "jitter". On the other hand very low frequency jitter is often termed "wander", e.g., jitter with a bandwidth of less than 0.01 Hz. If jitter and wander are not removed they tend to have a cumulative effect in some systems, causing a digital exchange to eventually make slips which invariably will lead to loss of BCI. Loss of BCI may result in the disruption of end-to-end encrypted connections, if they do not have inherently a resynchronization mechanism.

6A.1 Types of Jitter

Timing Jitter: Certain amounts of timing instability called "timing jitter" inevitably builds up in a digital transmission system. The main sources of timing jitter are as follows:

a) Noise and Interference : Receiver clocks are synchronized to transmitter clocks through some timing recovery circuit like a phase locked loop. The local clock maintains the desired average frequency but produces inherently a certain amount of phase jitter as it hunts continuously the underlying clock frequency of the source. At low signal to noise ratios these fluctuations are more pronounced, and in fact the local clock may loose synchronism altogether, in which case the transmission link is considered to be broken. Another important consideration is that jitter tends to accumulate in tandem clock recovery circuits, as each regenerative repeater outputs a jittered bit stream due to timing uncertainties accumulated at previous stages. Such accumulated jitter may cause subsequent timing circuits to have difficulty in tracking the receive timing as well as may lead it to sampling errors and even loss of lock.

b) Changes in Path Length: Communication path length changes occur as a result of atmospheric bending of radio paths or contraction and expansion of metallic guided media with temperature variations. Changes in the path length signify a decrease or increase in the number of bits stored in the medium. Jitter in this case is due to the rate of the change of the length of the transmission medium. A similar effect is also produced by the changes in the velocity of propagation, which itself may be due to variations in temperature and humidity or may be induced by Doppler shifts, e.g. in the case of signals transmitted and received to and from airplanes or satellites.

Systematic Jitter (Pattern Dependent Jitter) : Owing to implementation imperfections (like intersymbol interference or variations in the density of timing marks) in the timing recovery circuits, the jitter produced by repeaters is dependent on data patterns. For example, a pattern may produce phase ramps that tend to accumulate and get amplified in a repeatered line section. The systematic nature of this jitter makes it to be the most significant source of accumulated line clock jitter. This type of jitter becomes of concern usually at timing interfaces of higher-level multiplexers or in switches, that do not have a local clock with adjustment capability.

The main sources of pattern dependent jitter are as follows:

a) Intersymbol interference : Imperfect equalization results in residual intersymbol interference, the extent of which depends on the pattern content of the signal. This intersymbol interference causes a positional displacement of each pulse.

b) Finite Pulse Width Effects : This type of jitter is dependent on both the pulse shape and the pattern content of the signal used to excite the timing recovery circuit.

c) Amplitude -to- Phase Conversion : Ideally, the limiting amplifier which follows the tuned circuit should produce a squared-up timing signal that is completely independent of the amplitude of the applied input signal. However most practical circuits exhibit voltage offsets, due to aging and temperature effects, and the output phase and pulse width are affected to a small extent by the amplitude of the applied input signal, which itself can vary according to the pulse density of the input digital signal. Operational experience indicates that pattern-dependent jitter for a single regenerator falls typically in the range of 0.4 to 1.5 % of a unit interval RMS, for all digit rates [1].

<u>Random Jitter :</u> Jitter sources within a regenerator that are not strongly dependent on the transmitted signal are often termed random jitter sources. This type of jitter is not highly correlated at the regenerators and hence the rate of accumulation is less than for pattern-dependent jitter. The main sources of this jitter are as follows :

a) Differential pulse delay : The output drive circuitry usually produces a positional asymmetry between the positive and negative pulses of the regenerator. However this type of jitter is typically high frequency and is therefore considerably reduced by subsequent timing circuits.

b) Crostalk coupling in cable systems is different from one regenerator to another, the jitter produced within each regenerator is uncorrelated and is, consequently, considered to be of little significance.

<u>Waiting Time Jitter :</u> Higher order asynchronous multiplexers employ a technique called justification or pulse stuffing. The asynchronous multiplex brings the input tributaries to a common digit rate by the controlled addition of justification bits. These extraneous justification bits are identifiable so that they are removed at the demultiplex stage to recover the original data stream. However derivation of synchronous subchannel clocks becomes complicated by the insertion and removal into TDM data streams of these overhead bits which create irregularities in the data arrival rate. These irregularities cause "waiting time jitter". The timing recovery circuit, e.g., PLL in each tributary regenerates a continuous clock signal with a rate nominally equivalent to the original tributary timing information. Typically the clock generated by the PLL will gradually increase in frequency and then abruptly slow down again due to the justification process. Although most of the high frequency components of this jitter are removed by the PLL, a low frequency component always remains, called "waiting time jitter".

6A.2 Jitter Reduction and Control

If proper control is not exercised in limiting the amplitude of jitter in a digital network then the following degradations may arise :

a) an increase in the probability of introducing errors in digital signals at points of signal regeneration,

b) a degradation of digitally-encoded analog information as a result of the phase modulation of the reconstructed samples in the D/A conversion,

c) the introduction of slips into digital signals resulting from the overflow or underflow of storage devices, or the overload of phase comparators that are used in the synchronization subsystems of digital switches, multiplex equipment, digital patch fields, gateways etc.

In digital transmission systems timing recovery circuits such as PLLs in regenerative repeaters or muldex equipment will filter out most of the high frequency components of the transmission induced jitter. However the remaining low frequency components will continue to have long-term and short-term effects on the network components and end-users.

The most effective way to combat jitter is by means of buffers called elastic stores. Digital symbols are written into a store at the input clock rate but are read out with a different and less "jittery" clock. If the read clock is jitter free as it would be in a switch or digital patch field with an independent clock, then the transmission induced jitter is effectively removed from the digital stream. In some elastic store implementations there exists a buffer occupancy monitor to control the transmit clock on a long term basis as shown in Fig.6A.1

Providing that the buffer storage is sufficient to accommodate the misalignment between the input bit stream and the read timing signal, the jitter reduction can be achieved without introducing uncontrolled slips. However there will be also the long term jitter effects and no matter how long the elastic stores are made, there will eventually be overflows, or underflows causing slips. The elastic store size is an important parameter in jitter reduction and prevention of uncontrolled slips. The rate of controlled slips depends on the other hand on the accuracy of the clocks. The waiting time jitter in asynchronous muldexes was stated to be also an important source of timing disturbance.

6A.3 Effects of Jitter

Jitter affects both the network components and the user services.

User Services: Jitter present in a digital stream will cause analogue samples to be displaced from their correct positions at the D/A conversion stage. This random phase modulation of analog samples results only in a slight distortion of the reconstructed analog samples [4]. Previous studies indicate that an rms jitter of 3 UI (unit interval) in the incoming 2 Mbit/s stream results in a signal to distortion ratio of 33 dB in the reconstructed signal from 8-bit A-law PCM code. One can also observe that for a 2 kHz tone (e.g., used for control or signalling purposes), this amount of jitter corresponds to about 1 degree of phase uncertainty, where 3 x UI x 360 / 500 us = 1^{0}, where UI = 1/2.048 us and 500 us is the period of the 2 kHz tone. Phase variations of this magnitude will not have any significant bearing on the voice as well as non-voice services. For example modem outputs up to 9.6 kbit/s can be transported through such channels without much detriment.

Network Digital Equipment: Jitter affects the operation of digital equipment in a network by introducing uncontrolled slips or by introducing sampling errors. Uncontrolled slips occur when the buffer capacity associated with asynchronous muldexes is exceeded. On the other hand sampling errors occur as the sampling instants are displaced from their optimal position (maximum eye opening position) in the bit detection process. In synchronous muldexes uncontrolled slips are quite unlikely to occur as their buffer storage is sufficiently large so that jitter requirements will not demand additional storage. On the other hand asynchronous muldexes have a smaller buffer storage (e.g., 10 bits) since most of the mean frequency differences are compensated by the inherent justification process. In this case allowance must be made for jitter (e.g., 2-3 more bit spaces) to avoid uncontrolled slips.

Calculation of path delay variation, T: Delay variations of line-of-sight microwave links as due to meteorological parameters such as atmospheric pressure, temperature, relative humidity, wind velocity, and solar radiation amount to 0.1 to 0.2 ns/km over a 24 hour period [12]. Assuming then a hop-length of 100 km, the delay variation amounts to 20 ns which for the 2 Mbit/s stream is equivalent to 0.04 bits. This quantity is truly negligible. For geostationary satellite paths, on the other hand experienced path delay variations due to satellite orbit inclination, satellite orbit eccentricity, atmospheric and ionospheric variations may be as much as $\pm$ 1.4 ms which for a 2 Mbit/s stream amounts to 2800 bits of buffering. However satellite ground terminals will be located only in specific gateway nodes.

6A.4 Jitter Specification

To ensure full interconnection compatibility between the equipment forming a digital network, it is necessary that the output jitter at any interface in a digital path does not exceed the jitter that is tolerable at the input of the succeeding equipment. This has led CCITT to define input and output jitter tolerance masks and recommendations on the control of jitter and wander (Rec. G.823). The basic philosophy of jitter control in digital networks is based on the needs to recommend a maximum network limit that should not be exceeded at hierarchical interfaces, and to give the ability to apportion jitter to individual digital equipment. For individual digital equipment, the jitter is specified in the three following ways:

a) Jitter tolerance of digital input ports,

b) Maximum output jitter in the absence of input jitter,

c) The jitter transfer characteristics measured between input and output ports.

The limits given in Table 6A.1 (Table 1 in Rec. G.823) represent the maximum permissible levels of output jitter at hierarchical interfaces within a digital network. These limits are to be met for all operating conditions and regardless of the amount of equipment preceding the interface.

These limits for input and output jitter apply for all network nodes, that is elements in the digital network equipped with a reference clock. Examples of such network nodes are digital exchanges, synchronous digital multiplexers, digital patch fields. CCITT Rec.823 has a detailed specification on jitter test and measurement procedure

Table 6A.1: Maximum Permissible Output Jitter at an Hierarchical Interface (G.823).

Digit rate (kbit/s)	Parameter value	Network Limit		Measurement filter bandwidth		
		B1 unit interval peak-peak	B2 unit interval peak-peak	Band pass filer having a lover cut off frequency f1 or f3 and an upper cut off		
				f1	f3	f4
64		0.25	0.05	20 Hz	3 kHz	20 kHz
2.048		1.5	0.2	20 Hz	18 kHz (700 Hz)	100 kHz
8.448		1.5	0.2	20 Hz	3 kHz (80 kHz)	400 kHz
34.368		1.5	0.15	100 Hz	10 kHz	800 kHz
139.264		1.5	0.075	200 kHz	10 kHz	3500 kHz

APPENDIX 6B

TIMING INACCURACIES - SLIPS

6B.1 Slip Behavior

The process of deletion or insertion of bits (or groups of bits) within a digital stream due to timing inequalities or imperfections is known as slip. Slips cannot be eliminated but can be controlled by appropriate design and be reduced to acceptable levels. Its frequency can be regulated either by the use of highly accurate clocks, which minimize slip frequency or by control timing information dissemination so that all nodes are made synchronous and slips are eliminated.

Elastic stores at digital switches are written into by the covered line clock, but read from at the local rate (the rate of the reference clock in the switch) as shown in Fig 6A.1 This method eliminates the line induced jitter. The elastic stores, however, will eventually overflow or underflow, depending upon whether the write or read clock, respectively, is faster. An overflow signifies loss of some data units and an underflow amounts to a double read, causing extraneous bits to be inserted into the bit stream.

Uncontrolled slips are very detrimental to the operation of a digital network as they cause loss of frame synchronization and consequently several more frames would usually be lost in the resynchronization process. Therefore slips are allowed to occur only in prescribed manners that do not upset framing, which is called controlled slips. The capacity of a buffer store determines the maximum number of bits that constitute a slip and, in a multiservice digital network it is usual to design buffers so that only one or two types of slip occur:

a) A single slip of 8 bits from a 64 kbit/s digital stream is known as an octet slip and represents the slip of one sample in a pulse code modulation (PCM) encoded analogue signal.

b) A slip of one frame from a primary level digital stream is known as a frame slip. For networks operating with a primary rate of 2048 kbit/s, a frame slip represents simultaneous slips of one sample (that is, octet slips) in 30 channel PCM encoded signals.

The choice between these two types of slip can be made by equipment designers without consideration of digital network topologies or services because the average impact of both types of slip, in terms of bits gained or lost per unit time in any 64 kbit/s channel, is identical.

As long as slips are controlled so that they do not disrupt higher level synchronization processes, their only effect is an infrequent repetition or deletion of information within affected TDM channels. For example the audible effect of slips on a digitized voice signal s an occasional " click" and only one slip in 25 is said to cause an audible click. Such slip rate objectives can be very easily achieved for voice. Assuming that one click in 5 minutes is an acceptable performance, than clock deviations of 1 part in 10E5 would suffice. On the other hand both voiceband data and encrypted traffic (voice or data) are much more susceptible to slips. Encryption/decryption processes usually rely on bit synchronous scramblers and unscramblers. When the bit count is altered by insertion or deletion of bits in a time slot, counters at the source and destination lose synchronization. Unintelligible speech results until the scrambler-unscrambler become synchronized again; worse yet the connection may be lost. For digitized channel carrying voiceband data the effect is quite detrimental as well; for example one slip corresponds to a time shift of 125 us which for an 1800 Hz carrier amounts to 81 degrees of phase shift; high speed data modems employing phase shift modulation with coherent detection will than be severely affected and the operation of the voiceband modem can be upset for several seconds [4].

The main sources of timing inaccuracies are imperfect clocks, jitter and transmission delay variations. However as discussed in Appendix 6.A, jitter and transmission delay variations are filtered out at the timing recovery circuits and they seldom cause a controlled slip. The dominant source of controlled slip is then clock inaccuracies.

Limited frequency accuracy and stability of clocks result in the accumulation of the phase difference between nodes, eventually producing a slip. For example consider two centers interchanging bits at a rate of 2 Mbit/s and clocks controlling these centers of having a frequency accuracy of the order of between 10 E-7 for a crystal oscillator, and 10 E-12 for an atomic clock. The phase difference, expressed in bits, accumulated during one day of operation and the number of slips using 8-bit buffer would then be:

Frequency accuracy	Bits accumulated per day	Slip per day (8-bit buffer)
10 E-7	17280	2160
10 E-12	0.17	0.02

6B.2 CCITT Slip Rate Objectives

Controlled slip rate objectives on an international digital connection are given in CCITT Rec. G.822. This Recommendation deals with end-to-end controlled octet slip rate objectives for 64 kbit/s international digital connections. These slip rate objectives have been established to satisfy requirements for telephone and non-telephone services in an ISDN network.

The slip rate objectives for an international end-to-end connection refer to a standard digital Hypothetical Reference Circuit (HRX) of 27500 km in length. Furthermore it is assumed that international links are operating plesiochronously, with reference clocks as specified in Rec. G.811. The clock accuracies specified in G.811 cause one slip per 70 day in an international link and therefore for the standard digital hypothetical reference circuit consisting of 13 nodes, the plesiochronous operation allows one slip per 5.8 days.

The CCITT slip performance objectives for an international digital connection of 27500 km in length are shown in Table 6.4 and these impairments are allocated as follows:

Each local portion	:	40 %
Each national transit portion	:	6 %
International transit portion	:	8 %

Table 6A.2 Acceptable Controlled Slip Performance on a 64 kbit/s International Connection of a Bearer Channel

Performance Category	Mean Slip Rate	Proportion of time (> 1 year)
Acceptable	< 5 slips in 24 hours	> 98.9 %
Degraded	< 5 slips in 24 hours & < 30 slips in 1 hour	< 1.0 %
Unacceptable	> 30 slips in 1 hour	< 0.1 %

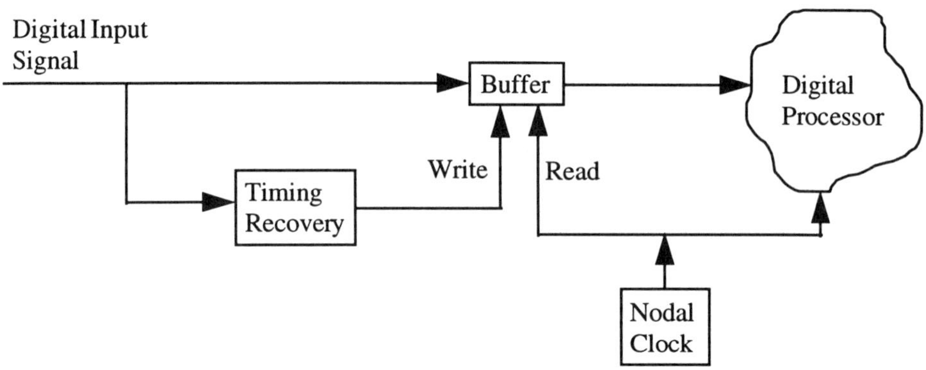

Figure 6A.1 Buffer Store at the Input of a Digital Processor.

APPENDIX 6C

FREQUENCY SOURCES AND CLOCKS

6C.1 Frequency Sources

Frequency determining devices can be grouped into three classes: mechanical, electrical and atomic. Mechanical and electrical resonators are of little importance in today's telecommunication networks. Therefore atomic frequency sources are used in telecommunication network synchronization. Comparisons of the characteristics of various frequency sources are given in Tables 6A.3 and 6A.4.

Quartz Oscillators : Crystalline quartz has great mechanical and chemical stability, which is a very useful characteristic in a frequency source. Drift or aging and dependence on temperature are common traits of such oscillators. The aging rate can be considered to start after the initial warm-up time and remains in one direction. Thus periodic frequency checks and corrections are needed to maintain a quartz crystal frequency standard. The most stable crystal oscillators exhibit a drift characteristic as good as 10 E-11 per day). The good short-term stability characteristic and frequency control capability make crystal oscillators ideal for use as slave clocks with, e.g., atomic clocks.

Caesium Oscillators : Caesium oscillators are the only type of commercially available equipment that do not exhibit any systematic long term drift and therefore would be able to meet the 10-20 year lifetime without adjustment. It has a long term accuracy of ± 1 part in 10 E-11 and is a true primary standard requiring no other reference for calibration. However it is expensive and has quite a short time between failures (between 2 to 4 years) and hence requires replication to meet the MTBF requirement. A further difficulty with the caesium standard could be the long mean-time-to-repair (MTTR). Since the instrument is so complex, it typically takes about 4 weeks for repairs and testing by the manufacturers. Caesium standards are furthermore able to monitor their own various control mechanisms and bring-up an alarm immediately if any defect is detected. This ensures that in case of faults, automatic changeover is initiated.

To reduce the cost and maintenance effort of the replicated unit, it would be possible to provide only one caesium standard together with two high quality quartz crystal oscillators for use as standbys. The greater reliability of these secondary standards would also increase the overall MTBF. The crystal oscillators would normally be locked to the output of the caesium standard and would have a memory associated with each one such that when the caesium failed, the standby oscillator is able, at the instant of failure, to continue operating at the frequency of the caesium standard. With time it would drift in frequency at a rate determined by the quality of the oscillator. The quality would need to be such that the oscillator remained within the frequency requirement during the repair or replacement time of the caesium standard.

Rubidium Oscillators : Rubidium oscillators vary in their resonance frequency by as much as 1 part in 10 E9 because of differences in gas composition, temperature, pressure and in the intensity of light. Therefore rubidium oscillators require initial calibration and also recalibration because they exhibit a frequency drift or aging like crystal oscillators. On the other hand the medium term stability of rubidium oscillators (i.e. minutes to a day) is excellent.

If the rubidium standards are to be maintained to within ± 1 part in 10 E10 of the network frequency, then they must be recalibrated. This could be achieved either by some planned sequence of manually calibrating the rubidium standards to incoming links starting at the nodes adjacent to a reference node containing a caesium standard, or by physically taking a primary standard to each site once a year and recalibrating the instruments. The failure rate of rubidium standards is not significantly better than that of caesium standards and so, to provide the requisite reliability, three rubidiums would be required.

6C.2 Clock Reliability

Synchronization may fail either due to clock failures or due to link equipment failure without the clock failing. If the latter occurs, the clock will continue to function and drive the node, though at a slightly deviated frequency as compared to the network. In this case the end user will be subjected to occasional loss of BCI at a rate dependent upon the clock frequency inaccuracy. On the other hand failure of a clock is of prime importance since all traffic passing through that node will be lost. Its overall reliability must therefore be comparable with that of other subsystems which cause complete failure at that node, e.g., the power supply. The reliability figure quoted for nodal clocks, is a failure rate of 0.01 failures per year or an MTBF figure of 100 years [6]. Since single oscillators cannot meet this figure, the basic oscillator must be replicated. Normally this is achieved by using one oscillator as the "worker" and one or two "standby" oscillator. The variants of which are shown in Fig 6.A2. A changeover switch then selects the worker oscillator and feeds the output to a waveform generator which distributes the waveforms to other parts of the site. The oscillators are not normally phase-locked in any way, and so if the outputs of the changeover switches are to be used directly, then care has to be taken in their design to ensure that the phase discontinuities on the output waveform introduced during the changeover period, are within the limits set by CCITT Rec. G.811.

Table 6A.3 Comparison of Frequency Sources [5]

Characteristic	Quartz	Quartz (Temperature Compensated)	Frequency Source Quartz (Single Oven)	Cesium	Rubidium
Basic resonator frequency		10 kHz to 100 MHz		9,192,631,770 Hz	6,834,682,613 Hz
Output frequencies provided	$1x10^{-9}$ typical	10 kHz to 100 MHz	$1x10^{-9}$ to 1x10-10	1,5,10 MHz typical	1,5,,10 MHz typical
Relative frequency drift, short term. 1 s	$1x10^{-7}$ typical	$1x10^{-9}$ typical	$1x10^{-7}$ to $1x10^{-9}$	$5x10^{-11}$ to $5x10^{-13}$	$2x10^{-11}$ to $5x10^{-12}$
Relative frequency drift, long term, 1 day	$5x10^{-6}$ per year	$1x10^{-8}$ typical	$1x10^{-9}$ to $5x10^{-11}$ per year	$1x10^{-13}$ to $1x10^{-14}$	$5x10^{-12}$ to $5x10^{-13}$
Relative frequency drift, longer term		$1x10^{-9}$ to $5x10^{-7}$ per year		$<5x10^{-13}$ per year	$1x10^{-11}$ per month
Principal environmental effects	Motion, temperature, crystal drive level			Magnetic field, accelerations temperature change	Magnetic field, temperature change. atmospheric pressure
Principal causes of long-term instability	Aging of crystal, aging of electronic components, environmental effects			Components aging	Light source aging filter and gas cell aging, environmental effects

Table 6A.4 Characteristics of Various Frequency Sources [13]

CHARACTERISTICS	KIND OF OSCILLATOR			
	CRYSTAL	CAESIUM	RUBIDIUM GAS CELL	HYDROGE N HASER
Basic Resonator Frequency	10 kHz to 100 MHz	9.192.631.770 Hz	6,834.682.608 Hz	1.420.405.75 2 Hz
Output Frequencies Provided	10 kHz to 100 MHz	1. 5. 10 MHz Typical	1. 5. 10 MHz Typical	1. 5. 10 MHz Typical
Resonator o	10^4 to 10^6	10^7 to 10^8	10^7	10^9
Relative frequency stability, Short-Term, 1 second	10^6 to 10^{-12}	$5x10^{-11}$ to $5x10^{-13}$	$2x10^{-11}$ to $5x10^{-12}$	$5x10^{-13}$
Relative frequency stability, Long-Term instability	10^{-6} to 10^{-12}	10^{-13} to 10^{-14}	$5x10^{-12}$ to $3x10^{-13}$	10^{-13} to 10^{-4}
Principal frequency stability Long-Term ýnstability	Aging of cyrstal, aging of electronic components, environmental effects.	Components aging	Light source aging, filter&gas cell aging, environmental effects	Cavity pulling, environmenta l effects
Time for clock to be in error, I microsecond	1 sec. to 10 days	1 week to 1 month	1 to 10 day	1 week to 1 month
Fractional frequency reproducibility	Not applicable. Must calibrate	$1x10^{-11}$ to $1x10^{-12}$	$1x10^{-10}$	$5x10^{-13}$
Fractional frequency drift	$1x10^{-9}$ to $1x10^{-11}$ per day	$<5x10^{-13}$ per day	$<5x10^{-13}$ per month	$<5x10^{-13}$ per year
Principal environmental effects	Motion, temperature, cyrstal drive level	Magnetic field accelerations, temperature change	Magnetic field, temperature change, atmospheric pressure.	Magnetic field, temperature change
Start-up after being off	Seconds to hours (may have systematic offset)	30 to 60 minutes	10 to 60 minute	Few hours
Resonator reliability, time between replacements	Very reliable	3 years	3 years	No data
Typical size, cubic centimeters	10 to 10.000	10.000 to 700.000	1.000 to 20.000	300.000
Typical weight in kilograms (pounds)	0.1 to 10 (2 OZ to 22 LB)	16 tý 400 (35 to 70 LB)	1 to 20 (3 to 45 LB)	200 (460 LB)
Power consumed watts	0.1 to 15	30 to 200	12 to 15	40 to 100
Estimated price	(LOW)	(MEDIUM)	(MEDIUM)	(HIGH)

Alternatively, if three oscillators are to be provided instead of a changeover switch, a simple majority decision gate can be used, but this has the disadvantage that the three oscillators have to be phase locked. This then introduces problems of interaction when one fails and presents difficulties in retiming and maintenance. Furthermore the links between the oscillators reduce the reliability. Triplicated oscillators with triplicated changeover switch, therefore, give the greatest reliability. This configuration also gives the ability to compare the output frequency of each oscillator with the other two oscillator outputs in order to detect faults which have caused any one of the oscillators that have gone outside the frequency specification. This inherent control mechanism is particularly significant when the transmission or switching nodes are operating without any external reference. Furthermore the configuration in Fig.6.A2, especially if the site is large, and there are long cabling runs to different equipment, then enables the waveform distribution to be secured by taking an output from each waveform generator and performing a simple majority decision at the input to each equipment. This protects against a single cable failure or failure of a waveform generator or changeover switch.

6C.3 CCITT Specifications

CCITT specifications for clock errors as defined in Rec.G.811 are as follows:

Time Interval Error (TIE): This is the variation T, of the time delay of a given timing signal with respect to an ideal timing signal. The TIE over a period of s seconds is defined to be the magnitude of the difference between to be the magnitude of the between the time delay values measured at the end and at the beginning of this period, i.e., TIE(s) = T(t+s) - T(t).

Frequency Departure: The quantity f/f is the TIE divided by the duration of the period (e.g., s seconds). The short time statistics derived from these curves imply that the characteristics of the frequency source should be such that over any 1 ms observation interval, the time interval error should be less than 61 ns on any 2048 kbit/s stream. This figure can be used, e.g., in the specification of the transient behavior of the changeover switches and phase locked loop oscillators. On the other hand the long term behavior implies that for observation periods beyond 10 days, the required frequency accuracy can only be satisfied by a caesium standard.

This recommendation contains also specifications and/or information on the degradation and unavailability of reference clocks, on the forms of aligning equipment and the reliability of reference clocks.

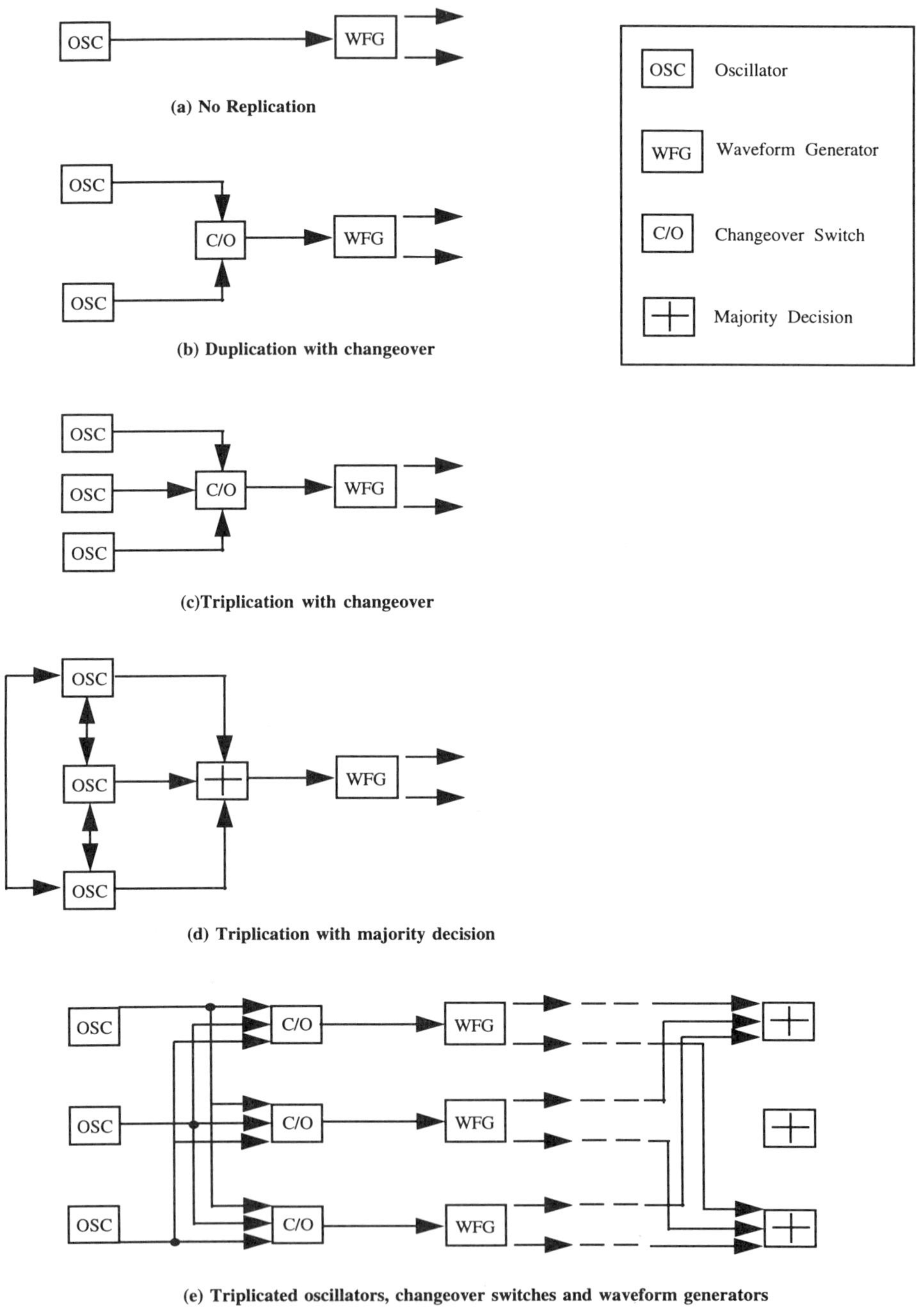

(a) No Replication

(b) Duplication with changeover

(c)Triplication with changeover

(d) Triplication with majority decision

(e) Triplicated oscillators, changeover switches and waveform generators

Figure 6A.2 Clock Configurations

APPENDIX 6D

ECHO CONTROL

6D.1 Echo Suppressors

In a digital network it is recommended to use digital cho suppressors. Thus speech detection is implemented by sampling the bit stream in the two directions of transmission and the loss is implemented by digital attenuation. The digital suppressors are less costly and more stable in operation as compared to analog suppressors are discussed in CCITT Rec.G.164.

6.D.2 Echo Cancelers

Echo cancelers suppress solely the echo in a contaminated signal in the receive direction. Since there is no voice switching action full duplex transmission becomes then feasible. The cancelers are designed to accept standard CCITT PCM input signals (G.732 or G.733) and the nonlinear PCM companding characterictics is utilized as an integral part of canceler processor algorithms. Recommended characteristics of echo cancelers are given in CCITT G.165 Rec. Echo cancelers improve the signal to echo ratio significantly more as compared to echo suppressors, they are finding extensive application in ISDN subscriber loops; on the other hand not enough experience has been gained so far, as they are not yet widely deployed in PTT networks, and their operation is quite sensitive to phase roll effects in channels. Both echo suppressors and echo cancelers can be used on connections in which the one way delay of the echo does not exceed the threshold values for the successful operation of these echo control devices. For the digital echo suppressors this limit is 25 ms while for echo cancelers a figure of 32 ms is quoted.

6.D.3 Balance Network

Improving the balance return loss of 4-wire/2-wire hybrids has the advantage that it is quite cost-effective and the echo signal becomes attenuated without the overall reference equivalent of the end-to-end connection being changed. For nailed-up connections this technique can be used because the impedance connected to the 2-wire part of the hybrid can be known and is relatively constant, so that a suitable balance network is designed and implemented. For example, from CCITT echo tolerance curves one can determine that if the return loss in increased by 6 dB, then the tolerable round trip echo delay could also be increased by up to 30 ms.

CHAPTER 7

ROUTING METHODS

7.1 INTRODUCTION

To remind the reader in this juncture that WIS will be an integrated digital network which will provide various teleservices such as plain and secure voice, facsimile and data communications. Network architecture will be based on a common user, primarily circuit switched, area grid network. The area grid network is also called the nodal network consisting of about 25 nodal switches at the network nodes and some 50 internodal links connecting these nodes. This network forms level 1. Level 2 is the access network, comprising the access equipment through which the users are connected to the network.

Three types of connections are envisaged: semi-permanent connections, circuit switched connections and packet and/or message switched connections. A major part of the traffic carried by WIS will be over switched circuits. It is envisaged that the functions of routing the calls will be done at the nodal switches (level 1), by the routing system software.

This Chapter discusses the subject of the requirements to a routing scheme for WIS. WIS nodal network is described elsewhere in this book. The Switching subsystem is the subject of Chapter 5. The objective of routing is to establish a succesful connection between any two nodes in the network. "Routing Scheme" defines how a set of routes is made available for calls between a pair of nodes. In view of the nature of the network and the need to distinguish from numbering schemes, in this report, the term "routing" is used to describe also the process of locating the called subcriber inthe network.

This Chapter begins in Section 7.2, with a statement of the operational and technical requirements. In Section 7.3 some factors influencing the selection of a routing scheme are discussed. Section 7.4 describes some candidate routing schemes for the system. Section 7.5 is devoted to a discussion of requirements and methods, leading to the conclusions and a tentative routing scheme proposal.

The network structure which is of a grid type is designed with the intention of meeting many of the military operational requirements, however those requirements directly influencing the selection of a routing system are discussed in Section 7.2. Aspects of survivability, redundancy, distributed operations and the ability of the routing system to operate in a damaged and fragmented network are highlighted in this section. A fundamental operational requirement is that the routing system will be able to locate itinerant users wherever they may be in the network.

Additionally, 5 levels of priority are defined for WIS traffic. This leads to the possibility of pre-emption of calls of lower priority during congestion conditions.

A fundamental objective of any routing scheme is for the network to handle the largest possible amount of traffic under any given condition. A measure of this is the Grade of Service (GOS). In a network offering a variety of services with 4 levels of priority as mentioned above, a modified definition of GOS is required. These factors are discussed in Sectioon 7.3. Another topic discussed in this section is the access procedures of the subscribers to access the WIS nodal network.

After the discussion of operational and technical requirements and factors in Section 7.2 and 7.3, a survey of routing methods which are considered candidate methods are presented in Section 7.4. This section begins with the CCITT definitions, principles and a classification of routing methods.

Direct routing, hierarchical alternate routing, dynamic (adaptive) routing are described; the effect of routing scheme parameters is discussed. Salient features of two adaptive and one non-adaptive method which may be adopted for WIS implementation and which are generic are briefly described:

- Own Exchange Only Routing (OEO)
- Right Through Control Routing (RTC)
- Saturation Search (SS)

Following these descriptions some derived and hybrid techniques based on these techniques are described and discussed. These techniques include Right Through Control with Saturation, Delegated Right Through Control, and Intelligent Own Exchange Only Routing all of which were subject of detailed evaluation during NICS studies in NATO under the direction of one of the authors of this book.

For network performance evaluations a method which is a fast and easy to use is modelling the network and the routing method on a computer. "Event-driven" simulation models where each call is represented and followed through in time are particularly useful in this regard. Use of such simulations makes it possible to study the behaviour of the network with different routing schemes for different configurations, different damage scenarios, different priority systems, etc. In NICS Architecture Studies the routing schemes were extensively simulated.

Conclusions from the simulations carried out for the NICS studies are presented in Appendix 7A. For these studies a set of especially designed software were used. Simulations carried out on the WIS network using the same suite of programs with scenarios developed for WIS yielded some results which are also used in the discussions in Section 7.5. These latter simulations are described in Appendix 7B.

In Section 7.5 an evaluation of the "basic", "candidate" techniques against the requirements is presented. This leads to a set of properties for the routing scheme which must be used in the WIS network. To meet varying requirements and conditions the scheme must be a hybrid, combining the basic properties of being adaptive, dynamic and intelligent. A proposal for the WIS circuit switched routing functions is then presented and shown on flowcharts. In the proposal the basic route establishment functions are carried out by a system similar to the Delegated Right Through Control method of NICS. In this system the nodes have connectivity and status information. The originating node controls the establishment of a route to the destination node; however the intermediate nodes on the proposed route have also the possibility of recalculating the route in case of blocking. This system can also handle satellite links if these are used. It is also envisaged that saturation search can be invoked at any node.

Routing scheme will be implemented in software. Issues in the implementation of this large and complex software are discussed in Section 7.6 and Appendix 7C.

Conclusions and Recommendations are given in Section 7.7 of this Chapter.

7.2 REQUIREMENTS

7.2.1 Operational requirements

As stated in Section 7.1 WIS Nodal Switches will carry out the circuit routing functions. The routing scheme to be used in WIS must

- Find a route from the originating subscriber to the destination subscriber, if such a route exists,
- Use the network resources efficiently,
- Permit the resulting call set-up to conform to the CCITT requirements as regards capacity and delay,
- Contribute to the overall communications security performance of the network.

Operational requirements which directly influence the choice of a routing system for WIS are discribed in the following section.

7.2.1.1 Survivability

The grid type network of WIS is designed to provide alternate transmission paths with the objective of increasing survivability. In order that this objective is realized:

- All Nodal Switching equipment must be hierarchically equal and tasked with similar functions, for automatic back up when required.
- Routing functions (and data bases) should be geographically distributed.
- Redundancy and the capability to efficiently utilise it must be available.
- A compromise of a Nodal Switch must not give the enemy access to information on other parts of the network.

These will ensure availability of resources under damage conditions as well as COMSEC performance as required. Additionally, adaptation and reacting to changes and graceful degradation of performance under stress/damage conditions will be achieved.

7.2.1.2 Zonal operation

Under operational conditions routing scheme must be able to operate in zones and in sub-network constituents, which would be self contained as regards

- operations and
- data.

This means a distribution of functions and data over the network, as mentioned in connection with the survivability considerations above. It also means that the routing scheme should be able to work without the complete information on the network being available in any one node.

7.2.1.3 Multi-level Precedence and Pre-emption

The routing scheme for WIS must be able to distinguish at least 5 levels of precedence on calls. These will have different, specified grades of service.

The Grades of Service and input traffic percentages for various precedence levels are as follows:

Priority level	GOS [Chapter 5]	Traffic (Assumed)
Flash-override	0.0 %	very small
Flash	0.01 %	15 %
Immediate	0.1 %	25 %
Priority	1.0 %	30 %
Routine	4.0 % for DWS and 10 % for IWS	30 %

The required GOS for the Flash-override traffic is assured by routing it over permanent or hotline circuits.

Although multi-level precedence and pre-emption is a switch function routing scheme must provide priority marking of routes and calls. Seizure of links must ensure a low rate of non-successful seizures; links must not be seized unnecessarily.

7.2.1.4 Services to Itinerant and mobile users

Some subcribers should be able to move within the entire WIS network or within a limited part of the network keeping their own directory number.

Access to such a subscriber should be possible after a move using his number even if the old location is inaccessible. Further, the same facilities should be available to a group of subscribers.

7.2.1.5 Numbering Plan

The routing scheme must be able to operate with a fixed numbering scheme. Such a scheme may include,

- a free numbering plan,
- a functional numbering plan (Chapter 5) or
- geographically oriented numbering plan.

The logic of the routing scheme must also be able to accommodate combinations of above alternatives.

7.2.1.6 Additional Considerations

In addition to the requirements above, the routing scheme to be adopted for WIS should be designed so as to provide;

- The possibility of handling secure as well as non-secure links, combined with subscriber and call categories defining requirements for secure circuits.
- Interoperability with other networks,
- Routing effectively calls requiring different types of circuits, viz. all digital or

mixed analogue and digital, during the transition period, and be able to set up calls taking into account the transmission constraints effecting the permissible make-up different types of connections.

- Routing calls to access switches homed on two, or more nodes.
- Routing calls over a satellite system while observing the configuration presented by such a system and the transmission characteristics of satellite circuits (delay and bandwidth).

7.2.2 Technical requirements

In addition to the operational requirements stated in section 7.2.1., technical requirements to be considered in the choice of a routing scheme include the following:

- Provision of the required capacity and grade-of-service should be cost- effective.
- Equipment must be used effeciently and call tromboning must be avoided. All connections should be set up using the best route available: according to given criteria, for example, path length, number of hops, blocking probability, path reliability etc.
- Routing algorithms should be reliable in that they should provide protection against entering deadlock states and looping.
- Routing information should be transported over a common channel signalling system.
- The routing scheme must support subscriber group services e.g. line groups, broadcasting, conferencing, call transfer and closed user groups.
- The software for the routing scheme must be secure, robust and well tested (See Appendix 7C and Chapter 5).

7.3 FACTORS AFFECTING REQUIREMENTS TO WIS ROUTING

7.3.1 General

In this section some factors of importance to the choice of WIS routing scheme are discussed: Call priorities and the possibility of pre-emption of calls and the definition of grade of service are interrelated subjects and are reviewed in this section. Pre-emption is a switch function yet, its mode of application is determined by the routing parameters which in turn influences the performance. Another subject which is important for the selection of a routing scheme is the access of subscribers to the network. Before proceeding to a survey of routing techniques available in Section 7.4, these topics are considered.

7.3.2 Pre-Emption

An important property of military networks is the possibility of pre-emption, i.e. the seizure of a communications path on which a call is in progress, by a call of higher priority. It is recognized that in times of stress there will be a higher demand for services. This can be taken into account to a certain extent by dimensioning the network. But there will always be instances where the capacity will not meet the demand. In such a situation, it is necessary to differentiate between the essential and non-essential traffic. With multi-level precedence and pre-emption it can be assumed that non-essential users will not cause degradation of the service given to the vital users. Thus, in effect, pre-emption is a special service introduced to permit the lower precedence traffic to use the network during periods of light load and limited or no damage conditions. Pre-emption applies on a link by link basis, with two or more priority levels enabling the system to differentiate between different calls. In WIS, for the switched traffic 5 levels of priority is envisaged, with the FLASH-Override traffic being regarded as part of FLASH traffic using hot-lines.

In order to make maximum use of the available network resources, it is essential that calls are not unnecessarily pre-empted by attempting to set up high precedence calls without first checking that the called subscriber can be found or that his circuit is not already engaged on a call of equal or higher precedence. This implies the use of a common channel signalling system and a routing procedure which can check the availability of the connection end-to-end up to the called subscriber before pre-emption is initiated.

It should be borne in mind that a pre-empted call is not necessarily valueless. Thus in planning studies consideration should be given to the value of pre-empted traffic.

Several different implementations of pre-emption in routing systems are possible [15], some of these will be described in connection with routing schemes in later sections.

7.3.3 A "GRADE OF SERVICE" Definition for WIS

Grade of Service (GOS) is a value normally used to describe the ability of the network to provide successful call set-up at given load levels. "Grade of Service" means effects like blocking and/or delay for bids which are offered to a network/network component. These effects are caused by the fact that traffic handling capacity of a network/network component is finite and demand traffic has stochastic nature. To emphasize that main interest is focused on traffic-dependent effects the term "traffic Grade of Service" is sometimes used.

"Parameters that measure these effects are called GOS parameters. The GOS parameters provide a measure of adequacy of network/network component under specified conditions and, therefore, they are used for dimensioning of network components and resources" (CCITT Recommendation E. 720 ISDN Grade of Service Concept (Draft)).

In order to select the parameters when defining GOS in WIS the following factors should be considered:

- WIS provide integrated access to a wide variety of telecommunication services through a small set of standardized user-network interfaces.
- Services have heterogeneous traffic demand profiles (calling intensity, service interest and call holding times) and diverse performance requirements, (delay, success rate and service quality).
- The traffic streams generated by user demands may in a varying degree use the same network resources.
- The configuration and implementation of a user's terminal and its man-machine interface may vary from one service to another service and one user to another user.
- End-to end out-of-band signalling and control capability are provided, allowing the network the capabilityof efficient coordination of resources with respect to a full visualised, instantaneous picture of user demands.

In the case of WIS, because different levels of priority are defined for the users and pre-emption is allowed as described above, the connections that are established are not necessarily allowed to be completed and terminated normally. Therefore a performance measure which is the average probability of success, POS (A), of the calls may be used instead of the normal GOS. This is the probability that it is not terminated by pre-emption.

This definition was used in [15]. The numerical value of POS(A) can be estimated from the fraction:

$$POS(A) = \frac{\text{Number of 'calls' successfully terminated}}{\text{Total number of routing attempts}}$$

In the formula the successful routing attempts do not include those which were subsequently pre-empted.

The causes of failure or non-completion of a call may be:

- Failure to find a free transmission path.
- Unavailability of other call related/service required (signalling registers, conference equipments etc.)
- Overloaded processors of the switching system.
- Pre-emption invoked by a call of higher precedence.

7.3.4 Subscriber Access

Subscribers are connected to WIS network using Access Switches. Each Access Switch (AS) in turn will be connected to two Nodal Switches (NS) (i.e. parent nodes) for reasons of survivability. This raises two routing issues; namely, those relating to originating and to terminating calls respectively. The relationship between these routing issues and the internodal network routing scheme is discussed in this section. It should be noted that the relationship described below is what could be proposed if the aim is to minimize the routing intelligence in the AS. This may not be an optimal concept, however, if the potential contribution of the AS's in the case where they could perform transit functions is considered.

7.3.4.1 Terminating Calls

The originating node will make several routing attemps to reach (nearest) one of these destination parent nodes. On each attempt, three results are possible;

- An internodal route is found and an access link is free or pre-emptable. In this case the call will be connected,
- A route is found but the access circuits are blocked and cannot be pre- empted.
- It is not possible to reach the chosen destination node at the given precedence level, although the access circuits may be free.

In the second and the third cases, the call should be directed to the alternate parent originating node. The originating AS will then re-attempt a call set up using a route to an alternative parent Nodal Switch.

7.3.4.2 Originating Calls

The handling of outgoing calls impacts on the control at the access switch but does not place any additional requirements on the internodal routing system. The AS will have at least two separate groups of circuits connected to two different Nodal Switches. For an outgoing call the AS may seize any of these circuits. The AS may have a table indicating the preferred access link to be used for calls to particular destinations. The preference will be based on making the best use of the internodal network. For example, if the destination switch is connected to one of the parent nodes (local) but not to the other, then that node is the obvious best choice for routing the call.

These tables may contain:

- Codes of the other access switches connected to the parent nodes of the Access Switch holding the table.
- Codes of other Access Switches for which there is both a high volume of traffic and a significant difference in internodal network efficiency depending on parent node choice.

7.4 ROUTING

7.4.1 General

The objective of routing is to establish a successful connection between any two nodes in the network. A "routing scheme" defines how a set of routes is made available for calls between a pair of nodes (CCITT E. 170 Draft).

The routing function in a communications system directly influences several important system characteristics such as grade of service, cost, reliability and in military systems it contributes to survivability.

Routing is extensively studied and documented in the literature [2,3,4,5]. For different applications such as voice, data, etc. and different network togologies and requirements different routing schemes have been developed. The use of common channel signalling systems and stored program controlled (SPC) exchanges has made this possible. A routing scheme in this context consists of an algorithm (or method) which is applied to a real physical network.

In WIS, subscribers will access the network through an access switch which in turn is connected to two nodal switches (for survivability reasons), or through a PABX. The task of routing calls will thus be delegated to the nodal switches. "WIS routing", therefore, refers to routing over the backbone network (switch node to switch node) switched traffic.

In the following sections two distinct processes i.e.

- locating the called subscriber,
- selecting a path to the subscriber,

which constitute the WIS routing function are considered together.

7.4.2 Logic of routing

7.4.2.1 Classification

A routing scheme for WIS must meet the operational requirements stated in Section 7.2.

In addition;

- It must be computationally efficient and inexpensive to apply.
- The algorithm must be realistic, and similar to the one actually implemented in the final operating network.

There are several view points to classify the traffic routing schemes. The various types are classified as fixed (non-adaptive, unintelligent) or dynamic (adaptive, intelligent) according to the network information used by the scheme; hierarchical and non-hierarchical according to the structure [2,3,6,7, CCITT Recommendation E.170 (draft)]. A diagram showing the relationship of routing schemes is given in Fig 7.1, the discussion below is based on this figure.

The simplest form is "direct routing" in which the number dialled by the calling subscriber determines the exchanges through which the call will be routed. The digits of the called subscriber's number are used to select an outgoing trunk at an exchange; with each exchange removing one or more digits from the number and progressively establishing connection. This simple method is impractical for calls going through a significant number of exchanges and uses transmission facilities inefficiently. This form of routing assumes a fixed topology and known traffic patterns.

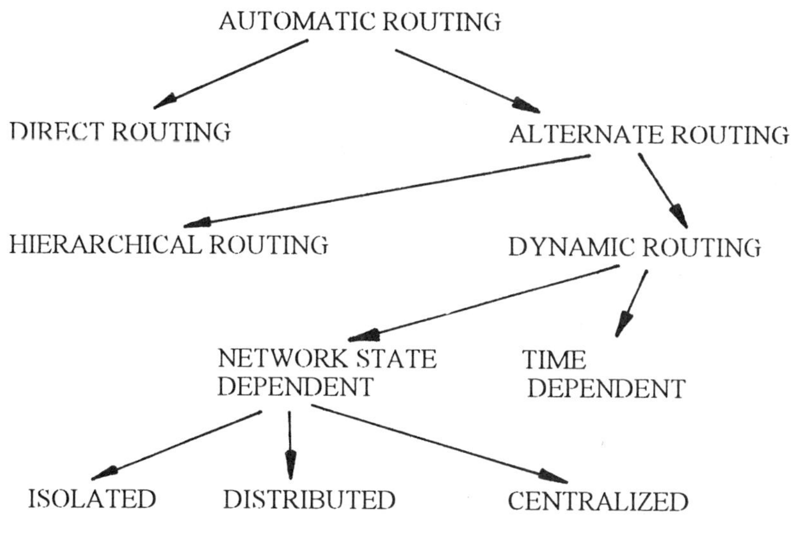

Figure 7.1 Classification of Routing Schemes

"Alternate Routing" provides the possibility of selecting trunk groups based on loading. In the first form of this method, in addition to direct trunks between exchanges, tandem exchanges were used and exchanges had overflow trunks to the tandem (Fig 7.2). This scheme significantly increases the efficiency of operation.

Alternate routing is accomplished through the use of routing tables stored at the switch that identify the primary and alternate routes from each switch to every other switch in the network. Routing tables at a given switch are unique to that switch and are fixed.

7.4.2.2 Hierarchical Schemes

Using "Alternate Routing" in a network is more complicated as there will be multiple choices of routes, giving rise to call looping. Adapting a hierarcy of switching units would alleviate this problem (e.g. local, area, trunk etc.). However, "it is important to note that the concept of "hierarchical routing" need not be directly related to the concept of a hierarchy of switching centres".

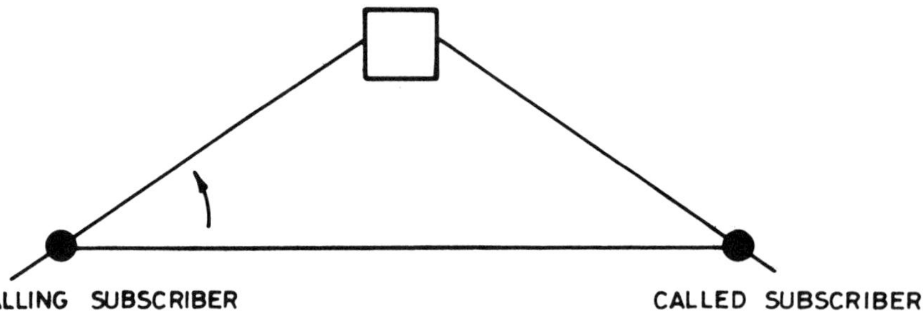

Figure 7.2 Overflow Onto a Tandem Switch

"A routing structure is hierarchical if, for all streams, all calls offered to a given route, at a specific node, overflow to the same set of routes irrespective of the routes already tested. The routes in the set will always be tested in the same sequence although some routes may not be available for certain call types. The last choice route is final in the sense that no traffic streams using this route may overflow further". (CCITT Recommendation E. 170 (draft)).

Thus hierarchical routing is based on overflow onto a set of fixed sequence of alternative routes (See Fig 7.3). This possibility of re-routing blocked calls increases network efficiency.

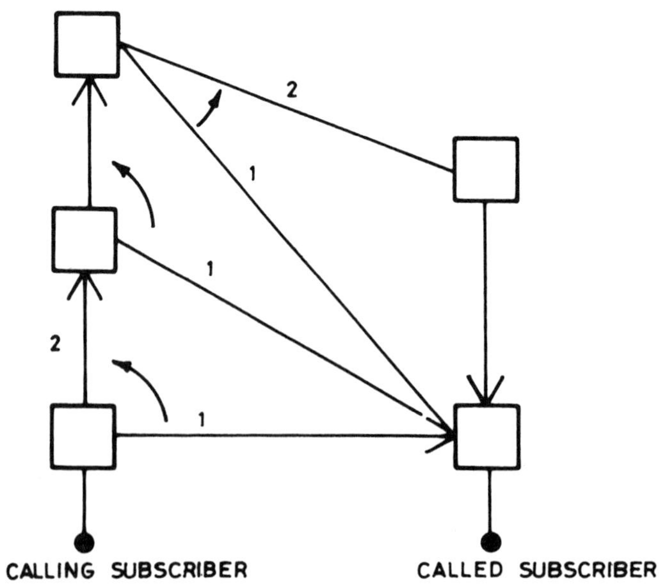

Figure 7.3 Hierarchical Routing in a Non-Hierarchical Network

However, the hierarcy is designed on the basis of the network being fully operational. The fixed sequence of alternatives do not allow for

- changes in traffic patterns
- failures and damage to network,

both of these cases leading to congestion. In other words with a hierarchical routing scheme network resiliency is low. Manual correction in the form of re-routing, blocking traffic for certain destinations, reserving trunks etc. may be applied; at the expense of losing traffic and delays.

The technical constraints which were applicable during the development of hierarchical schemes are no longer applicable. Dynamic routing schemes co-ordinate the actions of all the switches in the network and make better use of the facilities as discussed in the next section.

7.4.2.3 Dynamic Routing

A "dynamic routing scheme" incorporates frequent automatic variations of the routes using two basic mechanisms, namely: time and state [8,9].

a) Time dependent routing takes advantage of noncoincident busy periods across the network, usually on a preplanned basis, but has limited ability to accommodate unplanned traffic fluctuations.

b) State dependent routing which senses, by various methods and to varying degrees, network connectivity, occupancy and routes the traffic accordingly.

State dependent schemes are generally referred to as being "adaptive" and appear to offer increases in network efficiency and resilience especially during crisis and damage conditions in a military environment, in which case information on the topology is also an input for the routing decisions.

An adaptive routing scheme must perform a number of functions:

- Measurement of network parameters pertinent to routing strategy,
- Forwarding of the measured information to the point (s) (Network Control Center or nodes) at which routing computations take place,
- Computation of routing tables,
- Conversion of routing table information to routing decisions.

A classification of adaptive schemes can be made according to the quality (detail) of information and where the routing decisions are made.

Data collection and subsequent route selections can be done centrally by a routing control center or in a distributed manner by the individual network nodes. Adaptive algorithms collect and evaluate information on the status of the links and nodes and the traffic. The collection of information is transported through the use of common channel signalling.

The schemes may be broadly grouped into three categories based on how large portion of the network is taken into account in performing routing calculations.

a) Centralized methods make network-wide routing calculations. A centralized system with all messages flowing inward to some central processing facility is essentially a star configuration with links radiating from a single node. It is the simplest form of network topology and requires a link to be dedicated between the central node and each terminal.

With a centralized approach, processing and storage at network nodess are greatly reduced. Global information about the network required to run the algorithm (current topology, line capacity, conditions of links and nodes etc.) need only be kept by the central supervisor. Path setup is then accomplished through routing messages sent to each node. The reliability of the central node greatly affects the overall reliability of the centralized network. Its failure causes the failure of the network whereas an individual link failure will only affect a single device per link. The routing operation is vurnerable, since, in the case of supervisor node failure the network or some of its components may be left with no supervision. To increase reliability, the central node is duplicated.

Clearly this approach is undersirable due to survivability considerations, as discussed in Section 7.2.

b) Isolated methods consider only the state of the individual switch (processor loading, link group status, etc.) and the patterns of recently received calls. In other words, each node adapts to the changing conditions of the network based only on information it has gathered itself. The nodes do not transmit routing information to their neighbours. Basically, the difference is only in the organization of the switching function. From the graph theory point of view, an isolated network is described as a mixture of star and mesh components. An isolated network is more reliable than a centralized one due to the additional nodes and corresponding connecting links which permit some paths to be duplicated.

c) Distributed methods perform calculations relating to a limited section of the network, either the origin neighbourhood, the destination neighbourhood, approximate through routes, or some combination of these. The routing calculations may be performed in the neighbourhood routing centers or in the switches. The distributed network thus consists of a set of mesh subnetworks in which each node is connected to at least two other nodes and this provides a topology which is inherently reliable.

With a distributed approach, more processing and storage are needed at each network node than the isolated case. The required information is exchanged among nodes in the network. This implies some means of disseminating changes in topology (nodes and links going down), congestion etc.

In the distributed system case, two alternatives are possible for forwarding the pertinent information to the various nodes:

i) Forward only a limited amount of network information to each node,
ii) Forward "global" network information to all nodes.

The design of a distributed network must be concerned with the properties of nodes as well as the network's topological structure. Performance criteria like response time, throughput, and network reliability have to be considered. The evaluation of the properties of a network of nodes is concerned with nodal characteristics such as message handling and buffering, error contorl, flow control, and reliability.

The routing operation in a distributed system is, in general, more resilient when network components fail and inherently more complex than in the centralized system (See Section 7.2.1).

Another classification of dynamic routing methods is made on the basis of the frequency with which routing propositions are updated. Updating can be made:

- for each call;
- periodically.

In the latter case the set of "fixed" alternate routes which are offered are updated at regular intervals.

This classification differentiates between "spanning tree" and other adaptive routing methods all of which find and use shortest paths (Section 7.4.3.).

7.4.2.4 Hybrid Methods

The above discussion classified routing methods according to their salient features. In the implementation of a routing scheme the inclusion of features from different adaptive and non-adaptive methods is feasible [10,3]. Such a "hybrid" method may mix different methods:

- in time, or
- in geography,

applying different algorithms sequentially and achieving greater flexibility. For example tables can be maintained for frequently called subscribers, or following the establishment of a route into a network zone saturation search can be invoked etc. Such approaches are discussed further in this and the following section of this chapter.

7.4.2.5 Parameters

A "routing principle" includes the search method and the algorithmic aspects of a "routing scheme" which are neccessary for determining a route; such principles are described by such terms as Own Exchange Only (OEO), Right Through Control (RTC), Saturation Search (SAT), etc. as described in Section 7.4.3.

A routing scheme on the other hand, also includes a number of parameters which the designer is free to alter. Such parameters can greatly influence the performance of the routing algorithm to the extent that differences can be as great as those due to different principles (See Conclusions in Appendix 7A). These parameters can be incorporated into the switch software:

- Pre-emption rules,
- Updating frequency if link status information is updated periodically,
- Limits on route lengths, with or without the use of satellites,
- The amount of traffic status information available to the switch processor and any limits placed on the maximum number of links (L) in a route.

On the last point, traffic status information is the "free/available at precedence N" status of each link. If this is repeated to adjacent switches and by them to others in the network up to L-1 times, each nodal switch knows the traffic status of the network up to L links away at each precedence level. Since all switches always know the status of the links connected directly to them, the size of the traffic status reporting area, L is always at least 1.

It is clear that the applicable parameters should be identified and set after experience with the system.

7.4.3 Routing schemes to be considered for WIS

7.4.3.1 General

Circuit switched public networks are normally, organized in a hierarchical structure. Most routing schemes, including adaptive/dynamic ones, are based on such a structure [11,12]. For grid structured military networks a number of non-hierarchical, adaptive routing strategies have been developed. These developments are based upon, among others,

- Operational requirements,
- Development of digital switches,
- Use of common channel signalling systems,
- Functional numbering plans.

Example of such strategies and their applications can be found in Autovon (US), NICS studies, NDDN (Norwegian Digital Defence Network), the saturation search algorithm etc. [1,13].

In these routing schemes the call control procedures are of two main types:

a) "Step by step" which distributes the decision of selecting a route to the nodes involved. Supervisory control signals pass from one node to the next (Progressive call control).

b) "Right-Through" in which the originating node, maintains control of the call set-up until a connection has been completed (Originating call control).

In either case from all the possible paths between the source node and the destination node a particular path is selected. A sequence of calls set up from a particular origin node to destination nodes form a "tree" rooted on the origin node; some routing schemes are based on this approach.

At each nodal switch a directory of all the WIS subscribers, indicating the access (and nodal) switch each subscriber is connected can be maintained. However due to considerations of security for some users and to deal with mobile users,

- the inclusion of a search activity in the routing scheme for WIS,
- the use of routing zones in the WIS network

are necessary.

To discuss these and other topics within the context of available methods some generic methods and their realizations are presented below:

7.4.3.2 Generic Methods

The following methods which are considered generic are presented in this section

- Own-Exchange - Only routing
- Right Through Control routing
- Saturation Search

First two routing methods are adaptive and are applied on a per call basis. They display the scope of the routing information used at a node. The third method has a very narrow view of the network, due to its robustness it is already widely incorporated into hybrid routing methods for military networks.

During the description of these "generic" methods the application of pre-emption will also be discussed in order to define terminology. In the following section methods derived from these will be reviewed.

a) Own-Exchange-Only Methods

In the own-exchange-only routing methods each nodal switch acts on the information it has about the links emanating from it. At each node fixed routing tables are used. These tables contain the identity of the next link to be chosen for routing to the destination subscriber in question. When the originating node determines the destination node for a call it consults this table and selects the first outgoing link and the node. An enquiry is sent to this "intermediate" node, which in turn performs the same action. When the destination node is reached a reply message is sent and route seizure is initiated starting from the originating nodal switch. Clearly if the routing tables at each node are designed to minimize the number of links and if the first choice is taken each subsequent node then the path-length will be a minimum.

In the case of a blocked enquiry a reply message is sent to the origin node, which re-attempts the call, deleting its earlier choice of outgoing link.

If the first choice link is busy at the originating node or at any intermediate node pre-emption may be applied according to pre-set rules.

If in the tables the original, undamaged network connectivity information is used, this form of the method is called the "unintelligent OEO". If the tables reflect current damaged state of the network it is said that the method is "intelligent", in other words the method is adaptive.

In both cases in addition to the pre-emption, the rules applicable to the route selection process at each switch are:

- Soft pre-emption :The route involving the pre-emption of the lowest precedence call is selected; if there is more than one such link, that occurring first in the routing table is used.
- Hard pre-emption :If any link is free, the route selected is the first free link based on the routing tables. If no free link is available, the first pre-emptable link is chosen.
- Ruthless pre-emption :The links are tested in the order given by the routing tables, and the first link found to be free or pre- emptable is chosen.

Clearly, in the OEO method soft pre-emption is not possible as the switch currently taking routing decisions can not see which route involves the pre-emption of the lowest precedence call. However in the Right-Through Control method (which is discussed in the next paragraphs) such is possible due to the network information available at a node.

b) Right Through Control

In this system a map of all network links is kept at every node. Also, in the generic form, for each called subscriber the nodal switch it is connected is known. In the Right Through Control principle, the originating node calculates, using all these maps, and the information on the subscribers a route to the destination node; in such a way that the route with least number of links is selected. The route to be adopted is decided on the basis of a real-time calculation of the permitted route to the destination switch, taking into account the traffic status of the links, the call precedence and the pre-emption rule in force. A routing enquiry is then sent along this shortest route, which is seized if free or pre-emptable throughout. Any links found to be blocked are discarded from subsequent calculations to find an alternative route. The RTC concept is intended to optimize the use of network resources by the use up-to-the minute network information (See Fig. 7.6).

Route seizure is carried out on a link-by-link basis in a backward direction from the destination subscriber.

The present status of the internodal network is used to determine the route, in other words the maps are updated and a route is searched for each call. In order to reduce the amount of information involved RTC principle may be applied on a zonal basis which reduces the perception of a node of the whole network.

To estimate the time available to route a call in the complete WIS network the following calculations can be made:

- Total assumed traffic in the TAFICS network	2000 Erlang
- Less Local traffic	500 Erlang
- Inter-nodal traffic	1500 Erlang

If this traffic is equally divided among nodes we have 1500/25 = 60 Erlang/Node originating traffic.

Assuming average hold time of 180 secs this is equal to

$$\frac{60}{180/3600} = 1200 \textit{ originatingcalls/hr}$$

in other words 3 seconds/call is available for RTC routing for the undamaged network under normal traffic load.

c) Saturation Search

Saturation search (known also as flooding) is a method commonly used in military networks [14,12]. In Saturation Search routing, a node receiving from one of its connected neighbours a call request to a distant terminal broadcasts a search signal over all outgoing links which have one or more circuits free or pre-emptable at that moment. Because of its inherent ability to locate all possible alternative routes, it offers robust performance and is simple to implement (Fig. 7.4 and 7.7).

This method usually uses a common channel signalling system. Without seizing circuits the location of the subscriber and all possible routes to the subscriber can be found simultaneously. Clearly this method is very useful in the case of itinerant users and damage condititons. It is also an advantage that the method is of a distributed nature, without central control.

An implementation of this method would be as follows:

Using the common channel signalling the calling subscriber's node (originating node) sends a Search Message (enquiry) on all outgoing links containing a free or pre-emptable circuit. The search signal contains all relevant details of the call and a tagnumber that uniquely defines the call request during the search procedures. The tag number consists of the originating node identity and a serial number. Each adjacent node upon receipt of the search signal checks if the called subscriber is connected to that node. If not, the tag

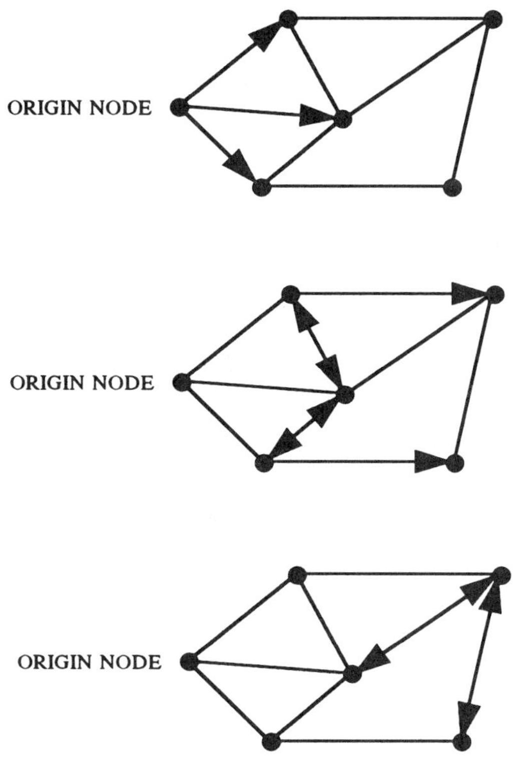

Figure 7.4 Saturation Search Steps

number is entered into a tag table and the Search Message is relayed on all outgoing links which are free or pre-emptable, except that on which the node received the message. To go on unnecessarily, each node receiving a Search checks it against its tag table to see if that message has been received before. If it has, the node terminates the enquiry and does not relay the Search Message again. In this way the search signal is propogated as a wave front until it reaches all attainable nodes if there is a free or pre-emptable path between the originating and destination nodes.

When the called subscriber is located, the destination node with which it is connected sends a reply signal back over the link that first passed the Search Message to the node. The next node receiving the reply signal consults its tag table to determine the link over which it first received the Search Message and relays the reply signal over this link, and so until the reply signal is received by the originating node. It is easily seen that the reply message will arrive at the originating node over the shortest path which is available at the time of the search. It must be pointed out here that of the possible routes one which is best is selected; "best" depending on various factors (including least number of pre-emptions). It should also be noted that this route found with saturation may then be attempted for subsequent calls to the same subscriber.

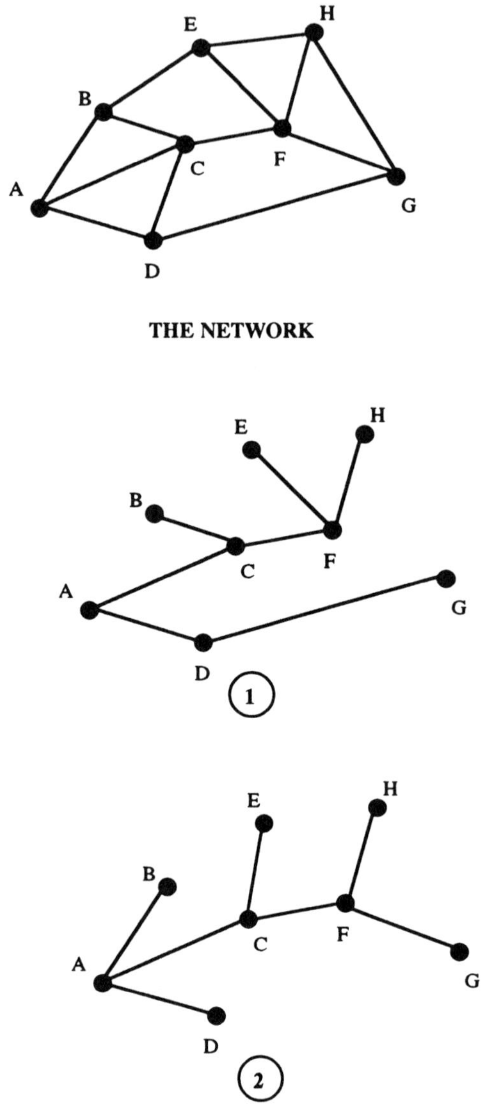

TWO SPANNING TREES FOR THIS NETWORK WITH ROOT A

DISTRIBUTED REPRESANTATION
AT NODE F

(1) Root A to E and H
(2) Root A to H and G

Figure 7.5 Spanning Trees

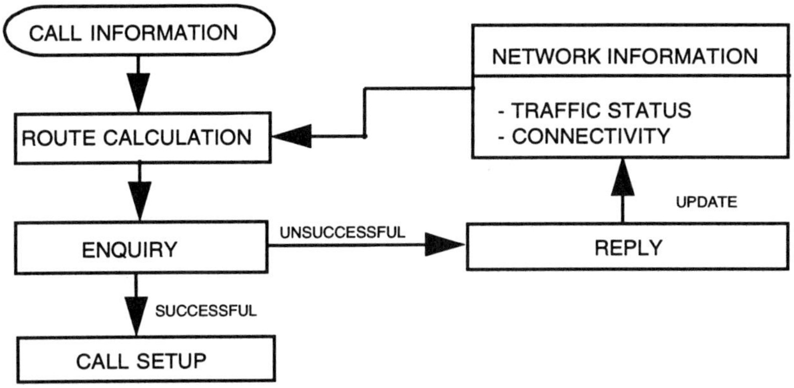

Figure 7.6 RTC-Sequence of Operations

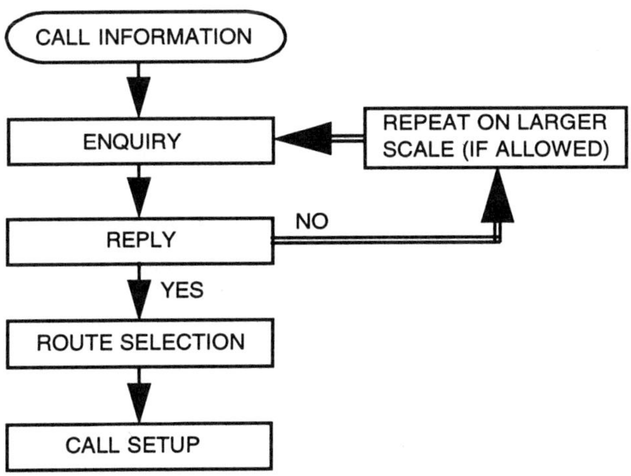

Figure 7.7 SAT-Sequence of Operations

In saturation routing, a Search Message is sent over every available link in the network each time a call is to be routed. The signalling load on a link is therefore in principle independent of the number of links in the network but proportional to the number of call attempts. Although common channel signalling such as CCITT Signalling System No 7 is sufficient, the signalling load associated with saturation search may be unduly heavy when the network is large [1]. It would therefore be advantageous, based on prior knowledge, to search in a limited area only. In this case search signal is broadcast inside a pre-determined zone, and would not produce long paths. If however the zonal search is not successful, several approaches may be adopted for he search process, including:

i) Zonal saturation in the neighbouring nodes.

ii) Total saturation in the whole network.

The disadvantages of saturation search are:

- Message traffic increases.
- The switches have to perform a lot of work in the searching process.
- All candidate channels on the route are reserved.
- The method does not view the whole network, route found may not be the optimum.

7.4.3.3 Derived Methods and Hybrids

a) General

The three methods discussed in section7.4.3.2 indicate the increasing use of information and adaptiveness from no-information saturation search, through Own-Exchange-Only up to Right Through Control. Some other methods designed and implemented for military networks, if examined, display an affinity to these basic, generic methods. Some are hybrids as defined in section 7.4.2.4. In this section some methods are described briefly.

b) Spanning Trees

A "spanning tree" of a graph is a tree which has the same nodes as the graph, in other words it "spans" the graph. This concept is also applied to routing in a grid network. For a given graph there may be more than one tree which accepts one node as root. A well known property of spanning threes is that if the graph is connected i.e. if there exists a path between any pair of nodes then there exists at least one tree which spans the whole graph. In a spanning tree with root at the originating node paths to all destination nodes are available (Fig 7.5).

Several algorithms to find spanning trees of a connected graph are available and puslished in the literature. These algorithms also cover the cases where each link or node has a weight or a cost associated with it.

In spanning tree routing two approaches are possible:

- If the node to which the called subscriber is connected is identified, the tree with this particular node as the root is used.
- If the node to which the called subscriber is connected is not known at the calling (originating) node then a search begins at this node and goes out to all nodes; the called subscriber will eventually be found.

In the case of congestion and/or damage the tree which was determined earlier is updated. Efficient methods of representing and updating spanning trees are available [1]. Updating a spanning tree starts from the root the search to update is sent to all neighbours of the root. If all the links are intact then there is no change on the branches recursively for each neighbour, in case a node can not send a search to another node this information is returned to the root node immediately. After evaluation of this information, the root node modifies the configuration of the spanning tree for itself. Time necessary for this process is dependent on the processing capabilities of the switches.

An updated tree will reflect the current state of the network. Therefore a routing scheme based on spanning trees is effectively an adaptive one depending on the frequency of updates.

The following observations can be made on such schemes:

- A spanning tree gives only one available path between source and destination. Not having any alternative path may be considered as a disadvantage.
- Each switch has been spared of the work involved with each search occurrence,
- If updates are not made with sufficient frequency the network's robustness will be reduced somewhat since not all paths will be made available for each search.

Spanning tree updates:

If a spanning tree method is used for routing, trees will have to be updated at frequent intervals to reflect network conditions. Time between two updates will depend on

- The size of the network
- Complexity of the decision criteria for link selection,
- Traffic load.

Considering these factors with reference to WIS, the first two factors will not present any problems as the total network has 26 nodes. Even if the link selection criteria is complex, time to update will not be significant using modern processors.

To evaluate the effect of the third factor similar computations as those carried out in section 7.4.2.3(b) can be made. It was found that there are on the average 3 seconds between call attempts at a node. This means that if the spanning trees are updated every 30 seconds, 9 calls from a given node will use the same spanning tree. If for any reason (including damage, congestion) calls can not be established on the average about 5 calls will wait 15 seconds for the establishment on the new tree rooted on this particular node.

As will be discussed in the next section for RTC, the use of a periodically updated spanning tree method can be combined with saturation search method, should the old tree be broken since the last update.

A route found in a spanning tree is one found by RTC or OEO methods according to the algorithm constructing the tree. The difference in approach being the periodic updating of the spanning trees as opposed to route computation for each call.

c) RTC/SAT system with call transfer tables

The Right Through Control with Saturation Search (RTC/SAT) was discussed extensively during NICS Architecture development [See Appendix 7A].

This system is based on sequential use of two routing principles: Rigth Through Control (RTC) and Saturation Search (SAT). Most calls are handled by RTC, SAT being used only for important calls after the RTC attempts to establish a connection have failed. If satellite links are available calls are routed over these links after reference to fixed look-up tables, Call Transfer Tables (CTTs), which are stored at nodes.

When a call is initiated, the origin node first checks whether or not the called subscriber is also attached to the same node, (a local call). In this case if a free or available access link exists, then the route is established without investigation of the internodal network. In this connection a "free" link has circuits not in use, a "blocked" link has all circuits busy at a priority equal to or higher than that of the call in question. An "available" link has all circuits busy, but at least one is available to the call in question by pre-emption. If the called subscriber is connected to a distant node, the routing system attempts to find the "best" internodal path. Under lightly loaded conditions, the "best" internodal path is the shortest internodal path in terms of the number of internodal links used in the connection.

However, when the network is loaded, the routing system calculates a longer "free" path rather than employ pre-emption in the shortest path.

The origin node initially attempts to route the call by using RTC. The route calculations are carried out by the origin node and the results communicated to en-route nodes. The information used by an origin node for RTC route calculation consists of an internodal connectivity matrix, a traffic status matrix, a CTT, and a terminal affiliation list.

The internodal connectivity matrix contains information on the connections between nodes. It covers the whole network, except for satellite links, and is updated each time the connectivity changes.

The traffic status matrix contains information on the state of traffic carrying links (free, available, or blocked) in an area around each node defined in terms of the number of links distant. The size of the traffic status area is generally taken to be about two links, although there is no fundamental restriction on the size as for as this description of routing systems is concerned. Satellite links and terminal links are not included.

The CTTs contain fixed entries for certain terminals and direct calls via pre-defined Satellite Groud Terminal (SGT) nodes.

The terminal affilication list simply contains, for each terminal, a list of nodes to which it is attached (from one, up to and including three).

The sequence of events following the calculation of a route is shown is Fig. 7.6. An enquiry is sent along the route chosen, to the destination node. If no blocked links are encountered (including the destination terminal link), seizure commences towards the origin node. If blocked internodal and/or terminal links are encountered, a reply containing the relevant information is sent by the destination node. The contents of this reply are used to update the information used in any further automatic routing attempts (generally, up to three are allowed).

If this procedure fails to find a route and if the class of service and priority of the call allow it, saturation search is invoked. The sequence of operations is shown in Fig 7.7. The origin node first calculates the maximum distance that the search messages should propogate, this being based on the internodal connectivity. A search message containing the path length constraint is sent out, each node rebroadcasting the message that arrives first, on the free and available outgoing links, except satellite links. On receipt of an enquiry, the destination node sends a reply containing the state of the route. There can be up to two destination nodes, hence up to two separate replies can be received by the origin node. The origin node picks the "best" reply and commences a seizure sequence towards the destination node. If no reply is received and if the class of service and priority of the call allow, unlimited saturation search is invoked. The sequence of operations is identical to those described above with the exception that there is now no path length constraint [See Section 7.4.3.2 (c)].

Calls for which satellite links may offer the "best" route are treated rather differently. The origin node first consults a fixed look-up table termed CCT 1. Against the wanted terminal are listed up to three "upside" SGT/Nodes through which the call might be routed. These are treated as destination nodes and the RTC/SAT procedures described above are used. If and when an enquiry arrives at an SGT/Node, it scans another fixed look-up table called CCT 2. This contains a list of up to three satellite links over which the call might be routed. If one of these satellite links is free or available, it is seized and the route to the origin node is set up. The destination terminal number is transmitted over the satellite and the SGT/Node at the "downside" end again uses normal RTC/SAT procedures to route the call to a destination node (satellite links cannot, of course, be used in this attempt). If successfull, the remaining part of the route is seized. If unsuccessfull, the previously seized links are released.

d) Delegated RTC System

In this system, as in the RTC/SAT system, the origin node calculates a route right through to the destination terminal, selecting a destination node in the case of a multi-access destination terminal. The route calculation is similar to that used in RTC and described in the previous section. It is based on full connectivity information plus traffic status information from the traffic status information of from the traffic status area. In contrast to the RTC/SAT system, the calculated route may involve a satellite link. There is therefore, no "two stage" process (i.e., routing and seizure to the satellite ground terminal (SGT) Node followed by routing and seizure from next SGT/Node) concerning routing over the satellite.

Having determined what is considered to be the "best" route, the origin node dispatches an RTC enquiry message. Each node, en route, examines the list of nodes and call priority contained in the equiry message. If, to that node's knowledge, the remainder of the route is free or available, the enquiry message continues on its original path. If, however, a link on the originally calculated route is known to be blocked, the intermediate node attempts to recalculate the route from itself to destination node, using the path length constraints contained in the equiry message. It then reforms the equiry message and sends it over the modified route. If the intermediate node cannot calculate a suitable route, the equiry message is deleted and a reply sent to the origin node giving the intermediate node identification. This intermediate node is then deleted in any further route calculation. Reply or set up is initiated by the destination node over the path followed by the enquiry message, in a manner identical to the RTC part of the RTC/SAT system. If the route becomes blocked during seizure the origin node is informed. As this system has an improved capability for routing long distance calls, the saturation search element has been removed.

e) IVSN Routing

The IVSN routing system is based on Own-Exchange-Only principles using look-up tables, with route finding and seizure performed in a single step [13,15,16,17]. When an originating access switch initiates a call, the call information is transferred to the originating node and the originating node will then control the routing to the destination station. The call information contains the destination station number, the call type and the call priority.

The originating node determines the destination node by means of a look-up table, and inspects its routing table. For each destination node the routing table contains up to four choices of links to neighbouring nodes in an ordered list, each leading to the destination node over a different route. The first choice route is in general the shortest, and the later choices are alternative routes which may be longer. The links are tested in turn and if any of them has a free circuit, it is seized. Provided that an outgoing route is found, the originating node now delegates the call routing to the next node. It uses the common channel signalling system to forward a message containing the call information and the call history: call type, call prioty, originating station and node, and destination station and node, list of seized circuits, pre-emption allowed flag, pre-emption used flag and satellite used flag. The call history is used to prevent loops in the route and to restrict the use of satellite channels to only one per route.

A node on this path then performs the following routing functions: if pre-emption is not allowed, it looks for the first free circuit; if pre-emption is allowed, it looks for the first free or available circuit. If a circuit is found, it is seized, the call is again forwarded, and the next node repeats the process. Eventually the destination nodal switch is reached, and the connection to the destination station is completed if the access link can be seized.

When the originating node does not succeed in finding a free route to the destination, it sets the pre-emption allowed flag. In this case, each intermediate node tests its first choice link for a free circuit first and then a pre-emptable circuit. If none exists, it similarly tests its second and subsequent routing choices. If the search is successful, the chosen circuit is seized and the call is forwarded.

If on either a free or hostile search some intermediate node finds all its outgoing route choices can not be user, it sends a common channel signalling system message dropping the call back towards the originating node for a further routing attempt. All circuits seized in connection with the abortive routing attempt are released during the drop-back. If this occurs after per-emption of one or more circuits, some effect of unnecessary pre-emption will also arise if a call reaches the destination node by pre-emption, only to find the called station is busy. The sending of an "access status enquiry" message and an "access status reply" message over the signalling channels prior to a routing attempt could alleviate this particular cause of lost calls.

If an enquiry towards a destination node fails, the first link on that enquiry is deleted from subsequent attempts to find a route to the destination node. Up to three attempts may be made to route to the destination node. When these have been exhausted, the destination node is deleted from the list of destination nodes and further routing attempts are made (if the destination terminal is multi-access) up to a specified maximum.

There is a manual capability to update IVSN routing tables, but this requires local action at each switch. Therefore this can not be done remotely from a central network control point, instead it can be done from a local control point.

7.5 DISCUSSION

7.5.1 General

In the previous sections of this chapter requirements and factors affecting the selection of a routing scheme were outlined together with a set of "candidate" methods to be considered. These methods did not include those used for public networks for reasons to be stated below. On the other hand these methods include features which may form the basis of a routing scheme for WIS . In order to bring out these features a discussion is presented in this section. The approach taken is shown schematically on Fig. 7.8. The text elaborates the cause-effect relations shown on this figure. A method which deals with situations displayed on this figure will no doubt have to be adaptive.

7.5.2 Routing in public networks

In the public networks, today, circuit switched traffic forms the largest part of network usage.

Routing in such circuit-switched networks is basically hierarchical with the possibility of alternative routing. The user population is diversified (business, homes), with stable statistical behaviour as regards traffic generation [7,11].

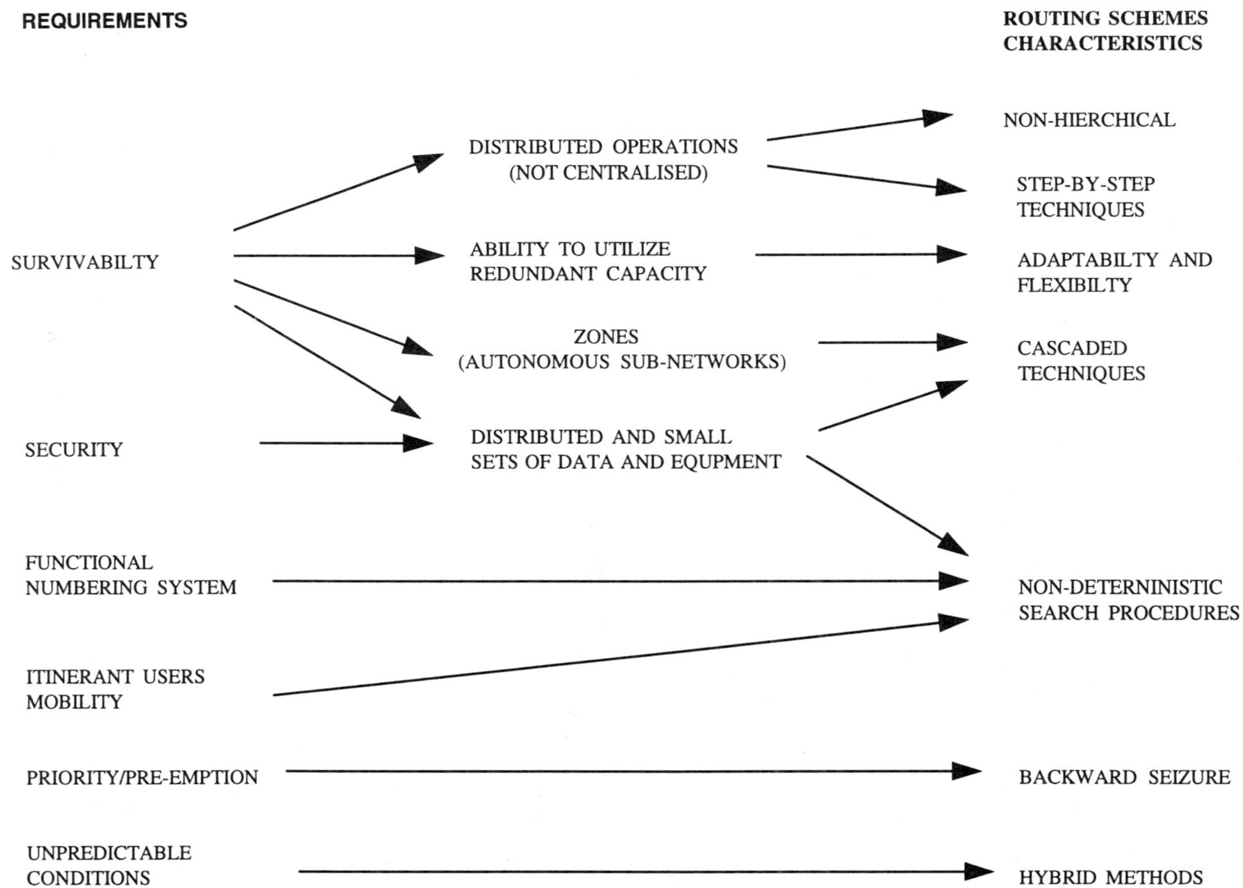

Figure 7.8 Cause-Effect Diagram for Routing

Routing is not usually dynamic, the system does not adapt itself to changes in the network and the traffic, in other words "network resilince" is low. Failures cause local congestion for which alternate routing over tandem paths is usually provided with restrictions on the number of tandems. In order to avoid the possibility of call looping in alternate routing a hierarchical routing network structure was found to be necessary.

Alternate routing techniques are unable to adapt the network to major perturbations, of the type which may be experienced in military networks in operational use. However, this short-coming is being remedied by the adaption of dynamic routing methods in the future systems using modern switching equipment.

Another fundamental property of the public networks is the uniformity of its users, in other words subscriber priorities are not defined or provided for. As a conclusion of this no pre-emption of calls is normally provided, a circuit set up for a particular call is kept over the duration of the call.

The following are a brief summary of the properties of Public Network Routing Systems:

1- Routing is hierarchical.
2- Hierarchy is dependent on all equipment being fully operational.
3- Failures cause local congestions.
4- Routing is not usually dynamic hence does not adopt itself to network and traffic changes.
5- No pre-emption is possible.
6- Numbering is fixed, not functional.
7- No provision for security is provided.
8- General restriction to at most one tandem per path.

Clearly, in military networks such as WIS, the requirements (outlined in Sec. 7.2.1) for survivability, adaptation to rapid topology changes, free numbering, mobility, pre-emption and priority will lead to different routing strategies from those used in public networks. WIS network must, therefore, have a specially designed adaptive routing system as discussed in the following section.

7.5.3 Routing in WIS

7.5.3.1 Survivability Considerations

One of the most important military requirement is the survivability. Survivability means generally the resilience of the communications system to:

- Loss of transmission and/or switching capacity,
- Traffic overload, both locally and network wide.

As a military network, the WIS needs to be survivable under all conditions of network stress, i.e., in times of political tension, crisis and war. The overriding operational requirement for the network design is to be provided to all of its users, or an appropriately selected subset thereof, an acceptable level of network performance.

The kinds of threat to which the WIS might be subjected can cause either dynamic network overloading due to increased traffic or jamming or static network damage due to failure or destruction of network elements. To predict the complete knowledge about the survivability of a military network under conditions of political tension, crisis and war is a very complicated procedure since it is difficult to determine the threat level, the amount of increased traffic, the resulting damage and the performance of the actual degraded network. To have some approximations in this subject, a network design where the analysis of the performance of the communication is provided should, in principle, identify the expected level of threat and construct damage scenarios. According to the performance analysis outputs, the network topology is revised.

Network damage scenarios may be differentiated between two different types of attack strategies;

- Random attack or failure,
- Intelligent attack.

Failures due to random attack occur on a probabilistic basis. Intelligent attact always aims at applying damage to the network in those locations where the impact on the overall network performance or on a selected communication needline can be maximized. Intelligent attack therefore covers the most prominent vulnerabilities of the network, i.e., those points whose destruction will produce maximum damage. Studies relating to WIS survivability design are reported in Chapter 13. From the point of view of routing the operational characteristics which contribute to survivability include (see Fig. 7.9):

a) Distributed equipment,
b) Distributed operations,
c) Distributed data,
d) Ability to ensure the use of redundant capacity,
e) Zonal operation on fragmented network.

a, b and c lead to the usage of a non-hierarchical method using an approach where the route is determined on a step-by-step basis.

Redundant capacity which must be available must also be usable, this means that the routing scheme must be adaptive, using the resources wherever and whenever available.

Under severe damage conditions where the network is fragmented each zone should be able to operate on its own. This again precludes a centrally controlled routing method. It also indicates that a method which operates in an "autonomous" zone and yet combines easily with the same method applied to a neighbouring zone is desirable.

Under damage conditions, zones will be formed and identified by the network control system when possible its contingency plans for reconstitution and reconfiguration of the network [see Chapter 11].

The contribution of the routing scheme to WIS survivability can be evaluated in terms of its ability to route calls of different levels of priority under several damage and overload conditions, using simulations as described in Section 7.5.3.6 below.

7.5.3.2 Security of Subscriber Information

In order that the network and subscriber data are not compromised such data should be kept in small sets, distributed over the whole network. The security of this information must be ensured and means must be provided for the erasure and destruction of this information when considered necessary.

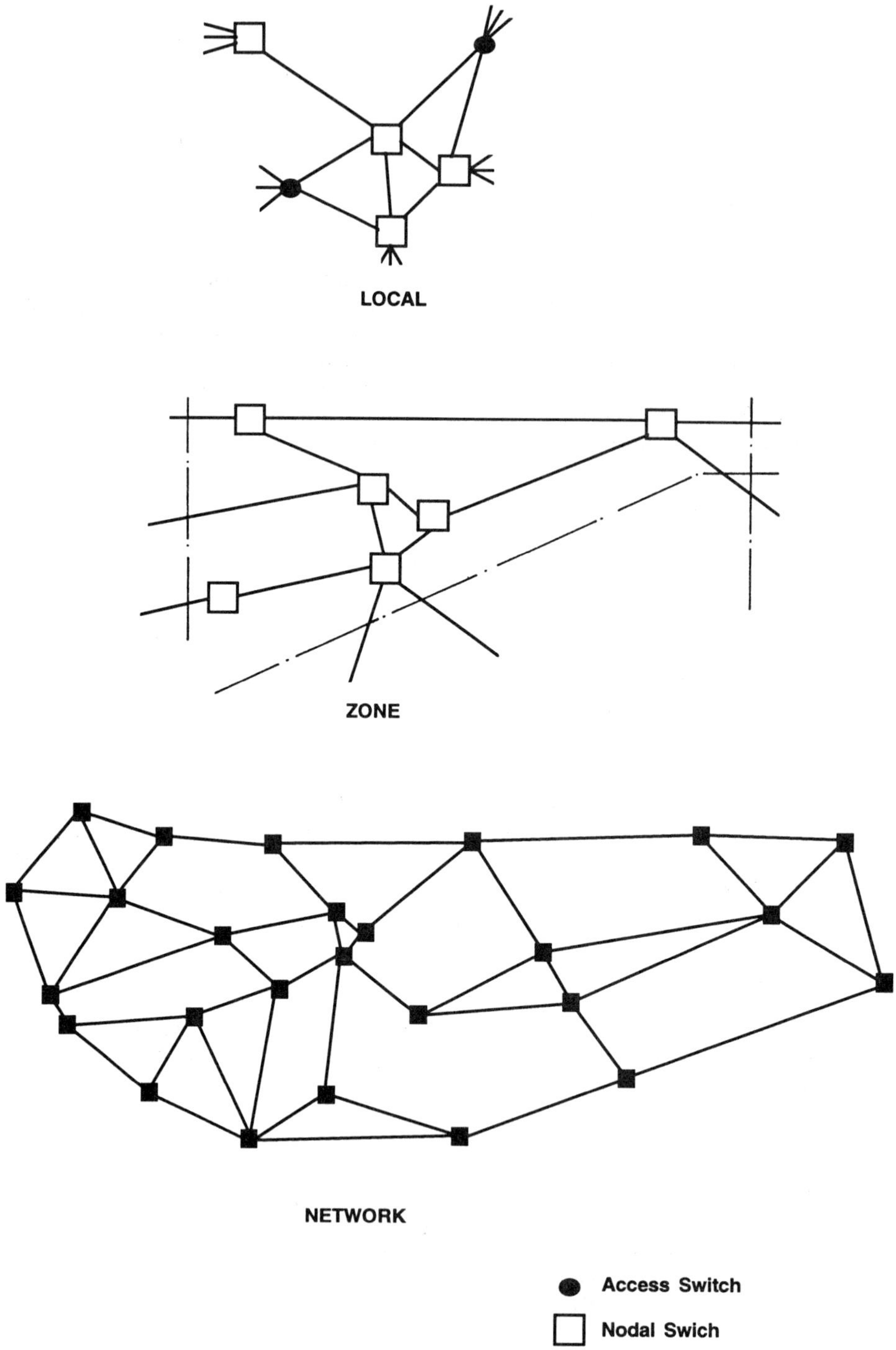

Figure 7.9 Local, Zonal and Network Routing

7.5.3.3 Mobility and Functional Numbering System

A very important requirement for WIS is its capability to serve mobile and itinerant users. Such users will keep their subscriber numbers wherever they may be in the network. The numbering system proposed for WIS makes provisions for this. In fact the numbering system is functional which again precludes a hierarchical routing scheme. These considerations lead to the use of non-deterministic search procedures. Not withstanding this conclusion if any information on the location (connected "parent" switch node) is available this information should be kept in a table of recently called numbers and used, for efficiency.

7.5.3.4 Priority, Pre-emption

In WIS, 5 levels of user priority is envisaged, with the possibility of pre-emption. Implementation of pre-emption using a step-by-step routing technique is not without its risks. If links are seized up the path towards the destination subscriber by pre-empting other calls, eventually a point may be reached from which no further progress is possible with some calls unnecessarily interrupted. To avoid this condition which directly influences the probability of success of a call, the seizure of links should be carried out in the backward direction. In other words seizing links should begin after tracing a path which is free or pre-emptable.

7.5.3.5 Unpredictable Conditions

The above paragraphs (Section 7.5.3.1 to Section 7.5.3.4) cover requirements affecting the selection of a routing scheme. It has been pointed out in Section 7.4.1 and elsewhere (Appendix 7B) that adaptive, non-hierarchical methods bear a substantial similitude to the extent that performance differences between them become comparable to parameter differences within a method. Thus during usage a scheme may be tuned for better performance by

- alternating rules for route calculations,
- extending data bases/tables,
- alternating decision rules on pre-emption,

To cater for conditions not envisioned at the beginning a further measure is the specification of a hybrid scheme; in other words the use of more than one method in a routing scheme. A well known example of this is the use of Right Through Control (RTC) with saturation routing, whereby RTC routing up to a search zone is followed by a saturation search within that zone or in the whole network when RTC fails to establish a successful connection. This was proposed in NICS studies. In the following proposal given in Section 7.5.4 a similar approach is taken.

7.5.3.6 Simulation Results

The comparative evaluation of different routing schemes require event driven simulators. For NICS studies at SHAPE Technical Centre (STC), a suite of programs were developed under the direction of Prof. Ince. Using these programs several routing schemes were tested under different conditions of traffic, network damage etc. Details and results can be found in STC documents [13,14,15,17]. In Appendix 7A a brief description of these simulations and conclusions are given.

In these simulations, methods described in detail in Section 7.4.3 of this chapter were tested on the NICS network, using various scenarios. It was observed that the following cases gave significant results for comparison.

- Normal conditions, design load.
- Double traffic, no damage.
- Double traffic with network damage.

Probability of success figures for double traffic with 8 % damage to nodes and 15 % damage to links are shown an Table 7.1. Since the ability of the routing system to perform will under conditions of network damage is of particular concern these figures are dicussed below:

The performance of the non-adaptive systems are significantly inferior and they would not meet operational requirements. The adaptive systems perform well for higher levels of priority under damage conditions. The RTC (or its DRTC variant) system and the IOEO systems are not statistically different, all are acceptable. In other words using an adaptive system, with full network information high priority calls go through, while lower priorities are considerably pre-empted. For a damaged network with overload, POS figures close to 100 % are observed for Precedence 1. The RTC system performance appears slightly better than IOEO because initial routing decisions take into account busy links at remote stations.

Additional tables covering also normal traffic and double traffic with no damage are given in the Appendix 7A (See tables 7A.1, 7A.2, 7A.3).

The same programs were used for a set of simulations runs on the WIS network. These runs produced results confirming the NICS studies. The high performance of adaptive RTC method (with saturation) can be observed on Tables 7B.1, 7B.2 and 7B.3 of Apendix 7B.

It is therefore concluded that a routing system based on RTC concept with additional facilities of saturation and delegation of the route calculations should be recommended for WIS. Such a system is described below.

7.5.4 A proposal for WIS automatic routing

7.5.4.1 Introduction

As a result of the discussion presented in Section 7.5.1 to 7.5.3.5 several alternative proposals for a routing system for WIS may be made. It has been noted that during the NICS routing studies [9] (See also Appendix 7A) the performance difference between routing systems is dependent largely on the parameters. With this in view a proposal is now offered in this section as a routing system concept for WIS. Note that this is used as an example of a system that has the primary characterictics to meet the WIS operational requirements. This proposal illustrates how these can be understood and implemented. Given the very high complexity of switching/routing software, it may well be that such a routing scheme will be constructed by incorperating parts of existing and well-tested software.

7.5.4.2 The Routing Scheme

The routing scheme proposed for WIS is based on the Delegated RTC concept which was briefly outlined in section 7.4.3.3(d), with the following features:

Table 7.1 Probabity of Success For Adaptive and Nonadaptive Routing Systems

ROUTING SYSTEM	POS (A) (voice and data) %			
	Prec.1	Prec.2	Prec.3	Prec.4
Adaptive systems:				
RTC/SAT	99	37	7	3
DRTC	98	33	6	3
RTC	96	35	7	3
IEOE	96	36	7	3
Non-adaptive systems:				
RTC/SAT + CCT	62	41	8	3
RTC + CCT	61	37	7	3
OEO	57	31	8	3

a) Intermediate nodes check the status of links ahead during the enquiry sequence, using their stored traffic status information. If blocking is encountered, the intermediate node recalculates the route to the destination.

b) Saturation search is invoked whenever necessary and for mobile users.

c) Satellite Links are included in the route calculation.

In this scheme, (See Fig. 7.10) as in the RTC/SAT system, the origin node calculates a route right through to the destination terminal, selecting a destination node in the case of a multi-access destination terminal. The route calculation is similar to that used in RTC and described in section 7.4.3.3(c). It is based on full connectivity information plus traffic status information from traffic status area.

Figure 7.11 illustrates the sequence of events following the origination of a call. Having determined what is considered to be the "best" route using a shortest path algorithm the origin node dispatches an RTC enquiry message over that route. Each node, en route, examines the list of nodes and call priority contained in the enquiry message. If, to that node's knowledge, the remainder of the route is free or available, the enquiry message continues on its original path. If, however, a link on the originally calculated route is known to be blocked, the intermediate node attempts to recalculate the route from itself to the destination node, using path length constraints contained in the enquiry message.

It then reforms the message and sends it over the modified route. If the intermediate node cannot calculate a suitable route the enquiry message is deleted and a reply sent to the origin node giving the intermediate node identification. This intermediate node is then

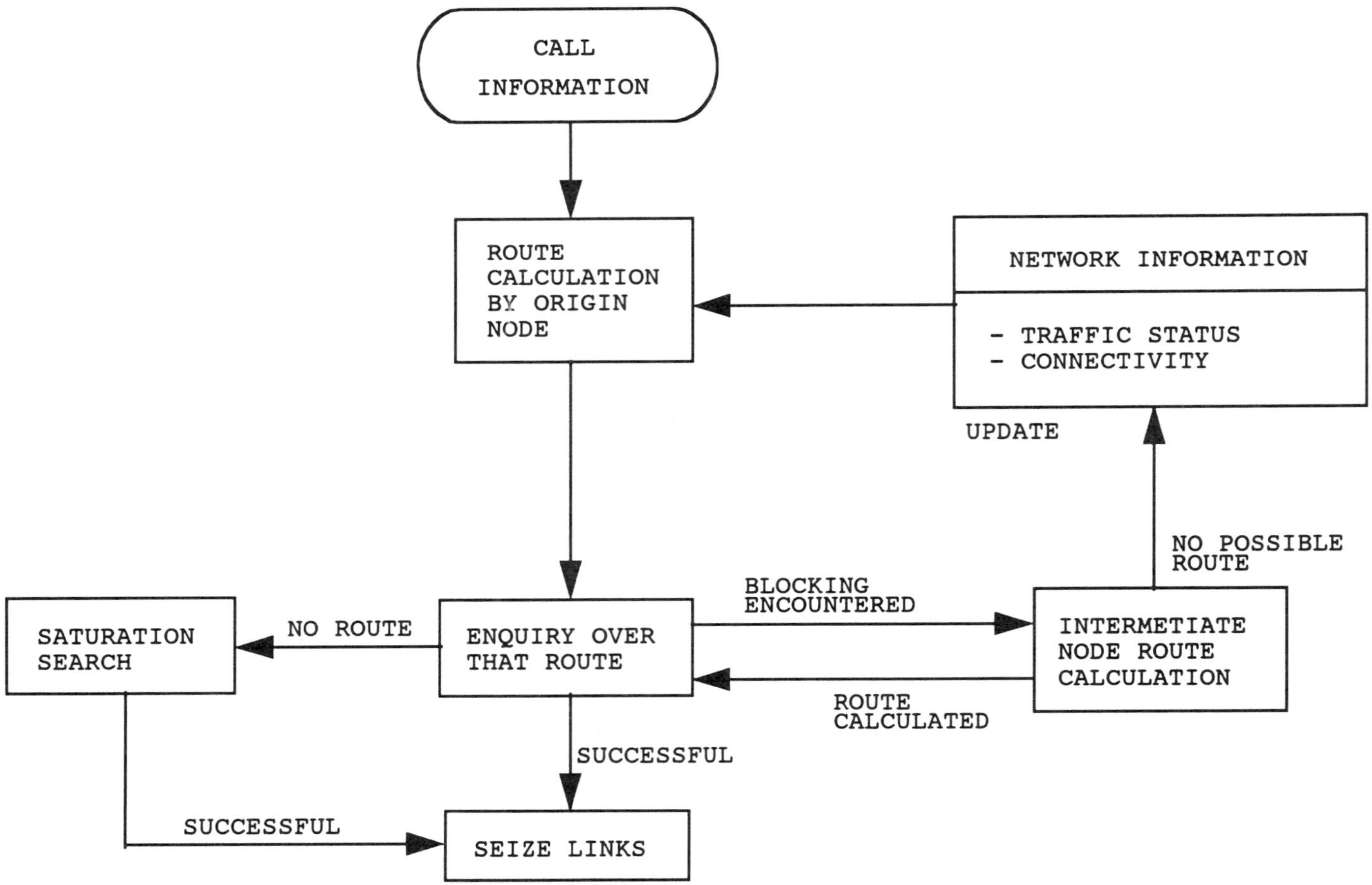

Figure 7.10 DRTC-Sequence of Operations

deleted in any further route calculation by the origin node. Reply or set up is initiated by the destination node over the path followed by the enquiry message as in the RTC system. If the route becomes blocked during seizure the origin node is informed. This system has an improved capability for routing long distance calls, however the saturation search is also included:

a) Nodal Processes

The following paragraphs describe the actions taken by a node in response to the arrival of various originating call requests. The figures are in flow diagram form as this is a convenient way to illustrate the logical processes (See Fig. 7.11).

b) Origin of Call

If no destination node can be found for the call or no terrestrial route (even assuming that all links are free) can be calculated the saturation search may be invoked. In other cases the initiation of an DRTC enquiry sequence will begin. The following paragraphs discuss each of the operations in more detail.

c) Route Calculation

The node calculates a route to the destination terminal by entering the destination node(s) into the routing algorithm and choosing the route of lowest priority within the path length constraints imposed. The operations are described below.

The node has two connectivity and traffic status matrices (or matrix sets when call priority is included), one for the terrestrial network alone, [T], and one for the terrestrial and satellite network, if satellite links are used [T+S], in which rows and columns with satellite links are identified. A minimum terrestrial route length in terms of the number of links, D_{min} (T), is calculated from [T] and this the following values are calculated [10].

$$D_{max}(T) = D_{min}(T) + N$$

$$D_{max}(T+S) = D_{min}(T) - M$$

where N and M are integer parameters.

It is possible that D_{max} (T+S) is negative or zero and in this case satellite routes are not considered further.

If D_{max} (T+S) is positive, the routing algorithm operates on [T+S] until routes of length up to D_{max} (T+S) have been investigated.

There are two possible outcomes to this:

i) Destination node not found within D_{max} (T+S) links: In this case no route is calculated. A new calculation starts, using [T] alone. This is identical to that used in the RTC/SAT route calculation.

ii) Routes found with one or more satellite links. In this case the route calculation is not completed but a more complex shortest path algorihm is used. This algorithm calculates the shortest available path between the source and the destination with only one satellite link. If no route can be found (either because of blocking or because no possible route exists), the lowest priority route is calculated from [T].

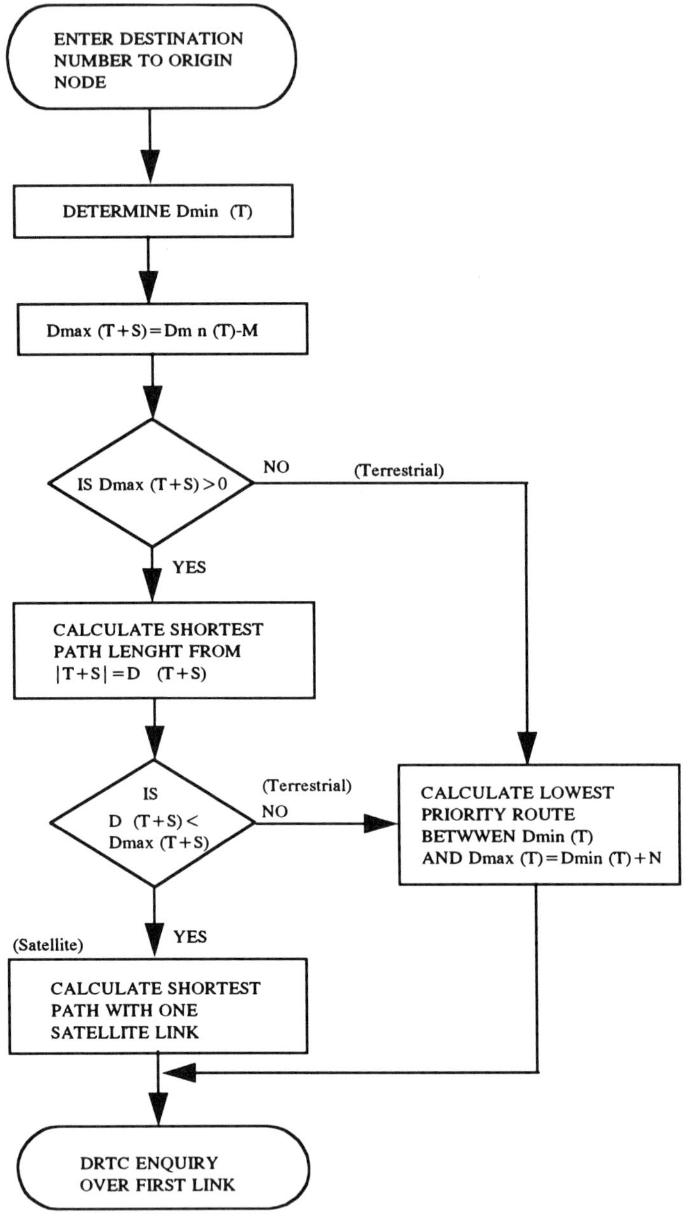

M : A constant which determines if a satellite link is to be used.
N : A constant which determines the pre-emption applicable.
T : Terrestrial.
T + S : Terrestrial and satellite links.
D_{min} : The number of links in the shortest possible path.
D_{max} : D_{min} + M, where M is normally 2.

Figure 7.11 Delegated RTC-Route Establishment Procedure

For the routing calculations described, traffic status information is available for the routing algorithm. It is worth emphasizing that for a route involving a satellite link the shortest available route is chosen (to minimize the effect of links, the route preference is the lowest priority route between D_{min} (T) and D_{max} (T).

Following the calculation of a route, the origin node sends an enquiry message containing the call identity, priority, maximum possible route length, and list of nodes along the route. The RTC enquiry message is then sent to the first node on the route.

d) DRTC enquiry message

Figure 7.12 illustrates the actions taken by a node on receipt of a DRTC enquiry message. The first check performed by the node is to determine whether it is the destination node. If so, it compares the call priority with the status of the terminal link to the wanted terminal, seizing the terminal circuits if possible and initiating backwards set-up over the internodal route taken by enquiry message (the first step being the composition of a set-up message and the transmission of this to the node from which the enquiry was received). If the destination terminal link is blocked, then a reply message containing this information is sent to the node from which the enquiry was received.

If the node receiving an enquiry message is not the destination node, then the actions taken by the node are more complex. The node compares the concepts of its traffic status table with the links yet to be enquired (these are contained in the enquiry message). If the links are free or available, the enquiry message in sent to the next node contained within the message. However, if any of the links are blocked the node will attempt to calculate a route to the enquiry message. The intermediate node uses the same route calculation procedure as described for an origin node (call origination) except that if a satellite link has already been traversed only the [T] matrix is used. In calculating the path length contsraint, the node subtracts, the number of links so far traversed and uses this value as its D_{max} (T) or D_{max} (S).

If a route is found, a new enquiry message is formed and transmitted to the next node on the new route. If the intermediate node is unable to calculate a free or available route within the path length constraint, it may invoke saturation search or delete the enquiry message and send a "call blocked during enquiry" message back to the origin node.

If the routing attempt subsequently fails, and a new routing attempt is made by the origin node, the node (s) at which the enquiry message was deleted should be avoided in the new routing attempt.

e) DRTC reply

A reply message is generated in three circumstances following an enquiry message:

- Free/available internodal route but blocked access link
- Original route blocked and intermediate node unable to recalculate
- A call starts seizure along an enquired free/available path and then encounters blocking.

If a reply message due to any of these causes arrives at an intermediate node, it is retransmitted over the link from which the original enquiry was received, for details see Ref. 8 and Fig. 7.13.

f) DRTC seizure

The seizure of links in DRTC is identical to the seizure in the RTC i.e. from the destination to the originating node.

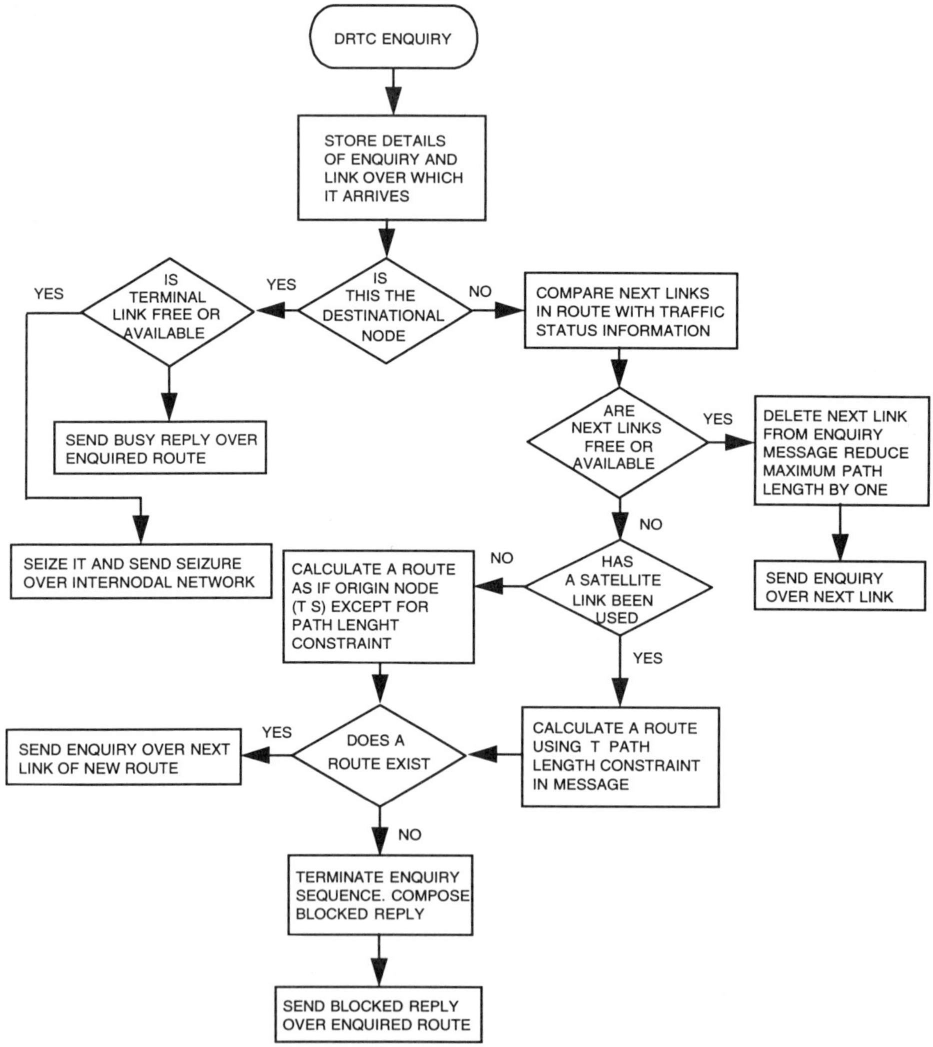

Figure 7.12 DRTC-Receipt of an Enquiry Message

g) Traffic status changes

A traffic status message is generated by a node if the traffic carried in any of its outgoing links changes in a way that affects the routing of calls over that link - i.e., when link changes beween free, blocked, and any of the available states. The nodes at each end of a link transmit the status change information (the identity of link and its new state). However, the concept of traffic status area is introduced. The status area of a given node, V, denotes the collection of all internodal links about which node V holds traffic status information. The extent of the status area is defined by the "radius", r_v, i.e. the area defined by nodes of distance r_v links.

The radius of a status area depends on the priority level, and generally increases with increasing priority.

It is desirable that the status change messages should be sent over the minimum number of links. This minimum number of links equals the number of nodes that have to be notified of the status change (excluding the two nodes interconnected by the link changing its status).

7.6 IMPLEMENTATION ISSUES

The routing system features proposed in this Chapter for adoption in WIS will enable the full operational requirements to be met, especially as regards survivability, connectivity, priority and pre-emption and independence from a particular numbering system.

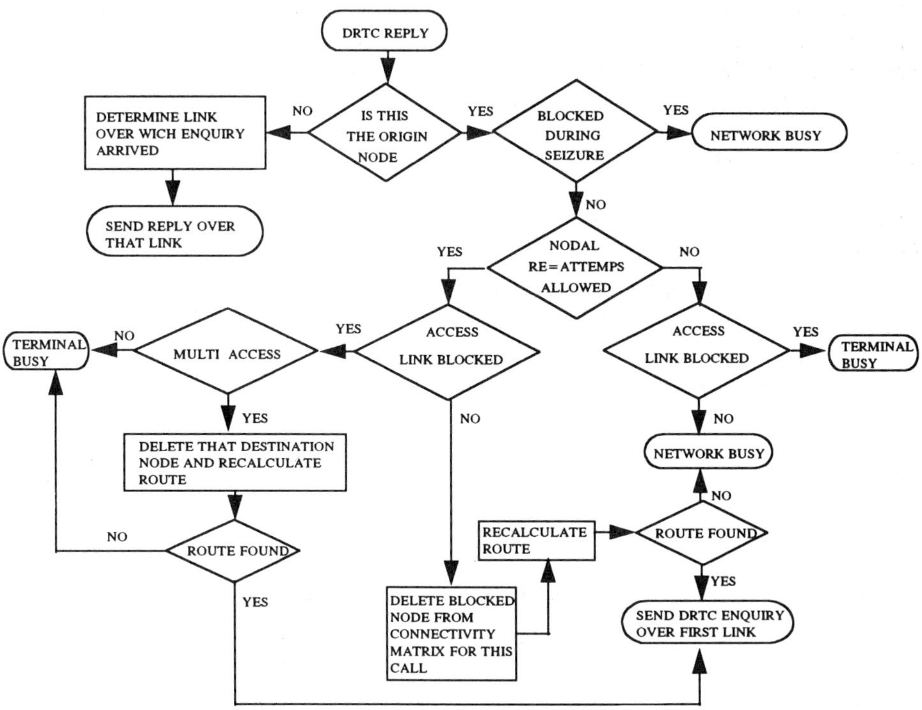

Figure 7.13 DRTC-Processing of Reply Message

The planned WIS switches [Chapter 5] and the use of CCS will make it possible to specify an adaptive, dynamic and intelligent routing system. In particular it is envisaged that a scheme with the properties described in Section 7.5 will be implemented. The routing scheme will be implemented in software and it is expected that there will be little or no equipment modification.

To reduce costs and the risks inherent in the production of complex software, WIS routing software, will have to be developed using existing software but with the addition of vendor's purpose built modules. The vendor should be selected on the basis of having implemented routing software for a similar system. Modules to meet specific WIS requirements will be designed by the vendor in close cooperation with the design authority.

The use of modern languages such as ADA, CHILL or C will contribute to this development consistent with software availability. It should be borne in mind that the performance and reliability of this highly complicated software will need to be tested using

- computer simulations, and wherever possible under,
- operational conditions.

A computer simulation model representing complete details of the routing algorithm and its variants is necessary for use during the design and development of the routing scheme to test vendor's existing system and additional features required in accordance with this Chapter.

The implementation of WIS as has previously been stated will be carried out in a phased procurement process which is planned to last several years. Phase I of this implementation is proposed to be a pilot system with a view to modelling the complete developed network. Therefore the routing scheme as well as the other features of WIS should be tested in this pilot system represents a reduced WIS network. During the following phases of WIS implementation, the impact of the system extensions on the routing scheme and the WIS network performance should again be examined using computer simulations.

7.7 CONCLUSIONS AND RECOMMENDATIONS

The following are the conclusions reached based on the foregoing Sections regarding the automatic routing system for WIS.

1. A routing scheme which is non-hierarchical, adaptive, dynamic and intelligent must be used for WIS. Schemes developed for NICS and similar strategic systems with these properties could be adapted for use in WIS.
2. Priority and pre-emption features must be incorporated into the scheme.
3. Each WIS switch should be able to invoke saturation search.
4. Routing scheme must be a hybrid of the available generic techniques both in time and in geography to be able to meet unforeseen operational conditions.
5. Routing scheme should not be based on a particular numbering plan.
6. Each node should be capable of handling calls "local" to that node.
7. The possibillity of setting routing parameters such as limiting route lengths, pre-emption method used, number of retries etc. should be available.
8. During the design, development and implementation particular attention should be paid to the system software design quality and software security.

9. The set of algorithms which makes up the routing software must be based on a system which has already been implemented and have additional custom-built features to meet the special WIS requirements specified in this book. The routing system design must me tested on "an event-driven simulator" which should be maintained at the design authority. The simulator must be used to during the design and development of the routing scheme and the system performance must be shown to be satisfactory under certain network damage conditions and at given traffic levels.

7.8 REFERENCES

[1] R. Bottheim: "A Method for Optimization of Large Mesh-Structured Circuit Switched Networks", Ph. D. Dissertion, The University of Trondheim, April 1987.

[2] B. R. Hurley, C. J. R. Seidl and W.F. Sewell: "A Survey of Dynamic Routing Methods for Circuit-Switched Traffic", IEEE Communications Magazine, Vol. 25, No.9, 1987, pp. 13-21.

[3] G. W. Bernas, D. M. Grieco: "A Comparison of Routing Techniques for Tactical Circuit-Switched Networks", ICC 78, IEEE, pp. 23-51, 1978.

[4] T. G. Yum and M. Schwartz: "Comparison of Routing Procedures for Circuit-Switched Networks", IEEE Transaction on Communications, Vol. Com-35, No.5, May 1987, pp. 535-544.

[5] D. A. Garbin and J. E. Knepley: "Marginal Cost Routing in Nonhierarchical Networks", Proc. NTC-81, Denver, 1981.

[6] W. N. Hsieh and I. Gitman: "Routing Strategies in Computer Network", IEEE Computer, 1984, pp. 46-56.

[7] I. M. Soi and K. K. Aggarwal: "A Review of Computer-Communication Network Classification Schemes", IEEE Communications Magazine, Vol.19, No.3, 1981, pp. 16-23.

[8] G. R. Ash, R. H. Cardwell and R. P. Murrey: "Design and Optimization of Networks with Dynamic Routing", The Bell System Technical Journal, Vol.60, No.8, 1981, pp. 1787-1891.

[9] Michael Purser: "Computers and Telecommunications Networks", Blackwell Scientific Publications, 1987.

[10] C. S. Warren: "The NICS Nodal Network: A Description of the Routing System Simulated", STC Technical Memorandum TM-453, The Hague, 1975 .

[11] P. R. Bell and K. Jabbour: "Review of point-to-point Network Routing Algorithms", IEEE Communications Magazine, Vol.24, No.1, 1986, pp. 34-38.

[12] M. J. Ross: "Military/Government Digital Switching Systems", IEEE Communications Magazine, Vol.21, No.5, May 1983, pp.1825.

[13] I. Q. Jensen, M. R. Miller and W. Roehr: "Comparison of Zonal Saturation Search and Right Through Control Methods of Routing for the Stage II NICS", STC Consultant Report CR-NICS-56, The Hague, 1980.

[14] I. Q. Jensen, M. R. Miller and W. Roehr: "A Simulation Study of the Effects of Routing Parameters for the NICS Nodal Network", STC Consultant Report CR-NICS-56, The Hague, 1980.

[15] M. R. Miller and I. Q. Jensen: "Performance Simulation Comparisons Between The IVSN, RTC and OEO Routing Systems", STC Consultant Report CR-NICS-40, The Hague, 1981.

[16] P. Henneberg: "Improved Tools for the Design of the Initial Voice Switched Network", STC Technical Memorandum TM-711, The Hague, 1984.

[17] F. Van Duppen: "IVSN Event Simulation Users Manual", The Hague, 1983.

APPENDIX 7A

NICS ROUTING SIMULATIONS

7A.1 SIMULATION METHODOLOGY

A powerful tool in network studies is modelling and simulation of the network and the routing scheme on a digital computer. During the design of the NICS network under the direction of Prof. Ince a suite of programs were produced and extensively used to simulate the specially designed non-hierarchical, adaptive routing schemes for NICS. The suite of programs and the studies conducted using these programs are documented in detail in STC reports and memoranda of the period [10,13].

The basic simulation method used is of the type known as "next event" or "event by event" simulation. In this type of simulation, calls are generated by sampling from their assumed distributions. According to the state of the network and the routing algorithm each event is followed through with its consequences in the computer of the system in operation.

The method is particularly suitable for the study of complex stochastic systems, with a large number of parameters and interactions. In such cases an analytical approach is not usually feasible. In the case of routing methods which has several levels of priority and the possibility of pre-emption the representation of detail available on computer models proves to be particularly effective.

7A.2 REPRESENTATION ON NETWORK AND EVENTS

In the suite of simulation programs produced for the NICS studies the network is specified by listing the terrestrial internodal links. Satellite links are treated in a similar manner. This information, "i.e. the network connectivity" is specified at the beginning of a simulation run and can not be changed afterwards.

The traffic load on a link is represented by calculating at any time, the number of calls of each priority on the link. Searching for a free channel or seizing a channel carrying a call of lower priority is effected on the basis of this information.

The simulation program allows the representation of the access links in the model in the form of a statistical evaluation as well as the simulation of Common Channel Signalling.

A matrix giving the traffic between each origin and destination node pair was used as input to the model. A "call generator" for each origin node generates the calls with the appropriate priority and destination. The model also allows data traffic and multiple addressing, however this facility was not used.

After a call for a certain origin-destination node pair is created the route for this call is determined, using the applicable routing algorithm.

7A.3 SCENARIOS

The simulation runs can be made in different conditions under which the network will operate, i.e., scenarios, covering for example:

i. Traffic load

   a. Normal design load
   b. Double load representing a crisis condition

ii. Damage to network

   a. Uniform damage of various levels
   b. Localized damage

It is also desirable to include other factors such as;

- ECM,
- Satellite connections, damage to ground terminals,
- Increases in error rates,
- Changes in priority percentages of traffic.

Taking into account all these possibilities in simulations would lead to a large number of cases to be run on the computer. In NICS studies various scenarios were developed and simulated. It was found that the following three scenarios gave significant results for comparison between different routing methods:

A- Normal conditions, with design traffic load, no damage to the network
B- Double traffic representing a crisis condition, no damage to the network
C- Double traffic load and the network subjected to damage of about 15 % to links and 8 % to nodes.

7A.4 OUTPUTS

During a simulation run the computer generates a record of all events. This is available on tape and can be treated afterwards, at three levels of detail:

- System level; giving overall statistics,
- Node/Link level; giving local statistics,
- Event level.

A large amount of output was obtained from the simulation runs for NICS. Of these the POS (A) i.e. probability of success figures are considered to be highly significant and therefore reproduced here . On the Tables A.1 to A.3 POS (A) figures obtained for the undamaged network (Scenario A of NICS studies), network with double traffic (Scenario B) and the damaged network with double traffic (Scenario C), for several routing systems are shown respectively.

7A.5 MAIN CONCLUSIONS OF NICS STUDIES

Simulation studies carried out on the NICS network have produced much experimental evidence on the behaviour of a grid network under different conditions and many conclusions for the selection of a routing scheme. These conclusions are in general, applicable not only to NICS network but also to networks of similar layout and purpose. Main conclusions of these studies were reported in the STC reports and those relevant to WIS are summarized below.

1- Adaptive routing methods (RTC/SAT IOEO) perform successfully and similarly to the extent that a selection between them depends largely on routing parameters. Under light traffic there is little difference between the performance of the routing systems, as seen in Table 7A.1. It was reported that most calls are established over the first choice routes [14].

2- It was observed that with heavy traffic and damage, high precedence calls are routed without problem by all the systems while traffic with lower levels of precedence shows serious degradation of performance (Table 7A.2 and Table 7A.3).

3- The use of connectivity information to automatically update routing decisions has appeared to be essential for a survivable routing system.

4- Use of call transfer tables for routing is not recommended.

5- The use of common channel signalling does not present undue risks.

6- Spill over from a "fat" route onto a "thin" one should be avoided.

7- Pre-emption rules should be parameterized.

In summary, the adaptive systems e.g. RTC and IOEO were the most suitable candidates for the NICS. It was observed that there is not much performance difference between them, both have the capability of automatically adapting to changes in network connectivity.

TABLE 7A.1 POS(A) Figures for NICS Scenario A

System	P 1		P 2		P 3		P 4	
	V	D	V	D	V	D	V	D
IVSN	100	100	100	100	99	100	94	97
(I) OEO	100	99	100	100	99	99	87	94
RTC S.	100	99	99	99	96	96	79	92
RTC H.	100	100	94	100	89	97	77	89

V = Voice, D = Data, H = Hard pre-emption, S = Soft pre-emption.

TABLE 7A.2 POS (A) Figures for NICS Scenario B

System	P 1		P 2		P 3		P 4	
	V	D	V	D	V	D	V	D
IVSN	100	100	81	94	25	62	4	12
(I) OEO H.	100	100	86	95	26	57	5	18
(I) OEO R.	100	100	77	92	34	67	8	24
RTC H.	100	99	80	90	34	62	5	19

V = Voice, D = Data, H = Hard pre-emption, R = Ruthless pre-emption.

TABLE 7A.3 POS (A) Figures for NICS scenario C

System	P 1		P 2		P 3		P 4	
	V	D	V	D	V	D	V	D
IVSN	87	82	19	44	3	9	1	4
IEOE H.	92	90	22	44	3	12	1	5
IOEO R.	93	90	25	48	4	14	2	6
RTC	94	92	24	45	3	13	1	5
OEO PLC=2	71	68	25	39	3	12	2	5
OEO PLC=15	72	69	21	34	3	11	1	5

V = Voice, D = Data, H = Hard pre-emption, R = Ruthless pre-emption,
PLC= Path Length Constraint.

APPENDIX 7B

WIS ROUTING SIMULATIONS

7B.1 ROUTING SYSTEM SIMULATED

For the WIS network some simulations were conducted using some of the programs prepared for NICS studies. Three network configurations produced int he early stages of WIS network design were used and run under different conditions, using the initial estimated of traffic load. The purpose of the tests was to check the dimensioning, to observe the behaviour of the network and to compare the three routing methods namely: adaptive RTC/SAT, nonadaptive OEO and partially adaptive IVSN. To our knowledge no other simulation packages of comparable ability in representing details were available, thus these tests used the NICS programs. This Appendix presents the results, along with a brief discussion.

7B.2 SCENARIOS

The schedule of runs for WIS which could be conducted was as follows:

I - Normal conditions, with design traffic load, no damage to the network

II - Double traffic representing a crisis condition, no damage to the network

III - Double traffic load and the network subjected to damage of 19 % to nodes and 19 %, 18 %, and 19 % to the links of 18, 26 and 33 node networks respectively.

IV - Double traffic load and the network subjected to damage of about 17 % to nodes and 17 %, 16 % and 16 % to the links of 18, 26, and 33 node network respectively.

As seen, damage details were specified for scenarios III and IV for network of 3 sizes namely 18, 26, and 33 Node Networks.

The table below shows the runs made on the STC Computer System.

Table 7B.1 Run Schedule

An "✓" indicates a successful run.

		Scenarios											
		I			II			III			IV		
		18	26	33	18	26	33	18	26	33	18	26	33
Routing Methods	RTC/ SAT	✓	✓	✓							✓		
	IVSN	✓	✓			✓			✓			✓	
	OEO												

7B.3 INPUT DATA

The user of the simulation model has a large number of options available to him. Input supplied to the model and the values of some critical parameters were as follows:

- Network configuration, following WIS studies. (No satellite links were included in the simulations)
- Traffic matrix, obtained from data supplied by the Authority
- Link capacities for each link, computed from a dimensioning program.
- Priority structure of calls:

Priority :	1	2	3	4
% :	10	15	10	65

The percentages being used are based on earlier assumptions and are similar to NICS.

The fifth priority, namely FLASH-Override, is very small and is included in Priority 1.

- Holding time : Negative exponential, average 3 minutes.
- Number of user re-attemps : 9
- Probability of re-entry : 0.8

Note: The matrix showing the traffic in Erlangs between all pairs of nodes was also available separately.

7B.4 RESULTS

The runs conducted as shown in Table 7B.1 gave numerical results of some significance. Those for 26 node network configuration are summarized below:

7B.5 CONCLUSIONS AND REMARKS ON WIS SIMULATION RUNS

From the set of computer runs conducted the following observations can be offered:

1- Probability of success:

POS (A) figures obtained and presented in Table 7B.1 for all the network configurations, with an adaptive algorithm is high. Particularly for priorities 3 and 4 of voice traffic no appreciable degradation is observed. This is due to an "overdimensioning" of the network; it appears that residual channel capacity is available on the links. This can be observed on the channel usage output.

In the case of a semi-adaptive method such as IVSN some reduction for priority 3 and 4 traffic may be observed; however results for this case is not available.

The POS (A) measurement may not be particularly useful when the grade of service is poor (i.e. low probability of success) [14]. Because it is then not possible to distinguish between a system carrying a low number of calls of varied duration and one carrying the same number of calls, but only of short duration because of pre-emption of the rest. Both will yield identical POS (A) result, although the traffic carried (in Erlangs) will be different in two cases.

On the other hand, when the channel availability for all priorities of traffic is high (see tables) and little or no pre-emption is observed the usual GOS figure becomes meaningful.

Table 7B.2 POS (A) % values for 26 Node Network

Scenario	System	P 1	P 2	P 3	P 4
I	RTC/SAT IVSN	98 100	98 98	98 99	97 74
II	IVSN	99	98	96	9
III	IVSN	87	78	25	4
IV	IVSN	76	46	25	4

Table 7B.3 Output counts of 26 Node Network under RTC/SAT runs.

Number of Calls Successfully Completed					
Sce.	P1	P2	P3	P4	Total
I	3451	4839	3238	21586	33114

Channel Capacity			
Sce.	Occupied	Total	%
I	3395.16	4022	84

Number of Calls Pre-empted					
Sce.	P1	P2	P3	P4	Total
I	0	0	0	325	325

Number of Calls Blocked					
Sce.	P1	P2	P3	P4	Total
I	67	100	79	1062	1308

Mean Path Length				
Sce.	Shortest	Accual	With Pre-emp	Without Pre-emp.
Sce.I	2.02	2.08	2.07	2.82

Table 7B.4 Output Counts of 26 Node Network Under IVSN Runs

Number of Calls Successfully Completed					
Sce.	P1	P2	P3	P4	Total
I	1887	1929	3834	28805	36455
II	7655	11475	7678	17772	44580
III	6142	9037	3990	7874	27052
IV	4024	5111	2580	5403	17118

Channel Capacity			
Sce.	Occupied	Total	%
I	3782.79	4022	94
II	3978.96	4022	99
III	2576.46	4022	64
IV	1652.44	4022	41

Number of Calls Pre-empted					
Sce.	P1	P2	P3	P4	Total
I	0	0	2	1557	1559
II	0	137	241	27134	27512
III	0	590	4746	26768	32104
IV	0	968	2342	18738	22048

Number of Calls Blocked					
Sce.	P1	P2	P3	P4	Total
I	0	0	2	8763	8765
II	0	133	234	149537	149904
III	855	2315	10520	151696	165386
IV	1212	5677	7227	101550	115666

Mean Path Length				
Sce.	Shortest	Actual	With Pre-emp.	Without Pre-emp.
I	2.01	2.36	2.37	2.11
II	1.91	2.24	2.28	2.07
III	1.90	2.34	2.26	2.69
IV	1.90	2.42	2.38	2.69

2- Average Number of Links in a call path:

The traffic between the nodes is defined by the traffic matrix. When the information on this matrix is applied to the network topology the average path length, in terms of the number of links, traversed by a call is obtained. This average is computed in the computer runs and tabulated results are available. A study of these results (presented in Tables 7B.2 and 7B.3) shows that this average is about 2.

3- Pre-emption:

It appears that (again due to large channel capacities) there has been a few instances of high priority calls pre-empting low priority ones (Tables 7B.2 and 7B.3). In fact in some cases no pre-emption of 2nd and 3rd priority was observed. This needs to be verified through other runs changing percentages of various priority levels.

APPENDIX 7C

ROUTING SCHEME SOFWARE REQUIREMENTS

To support and maintain the WIS System Software broad guidelines may be given for the software performance and quality. The most important requirement is the strict application of software security measures during design, development, and implementation. The same considerations would apply to the specific case of routing system software which is large and complex [See Chapter 5]. The following requirements are therefore to be classified as general requirements and requirements which are applicable to routing system.

General requirements:

1- Acceptable	:	Quality and efficiency of the software should be satisfactory for users.
2- Well documented	:	The software documentation must provide good communications between the users and the manufacturer.
3- Cohesive	:	There should be maximum interaction within each software module.
4- Economic	:	The software should be cost effective at procurement.
5- Efficient	:	Resource utilization should be optimal.
6- Fast Development Rate	:	Relatively less time should be needed for future development.
7- Feasible	:	The resulting system should satisfy all feasibility criteria.
8- Flexible	:	It must be easy to modify, add or delete software modules.
9- Hierarchical	:	Components of the system should have an hierarchical relationship.
10- Low coupling	:	There should be a minimum of interaction between components.
11- Modular	:	System should have relatively independent and single function parts that can be put together to make a complete system.

In addition the routing system software should satisfy the following requirements:

1- Well tested	:	Changes of future failure and/or user dissatisfaction should be minimized.
2- Simulation tested	:	Routing simulation for different conditions should be provided.

3- Compatible	:	The system should fit the total, integrated system even if it consist products of different manufacturers with minimum effort.
4- Transportable	:	Written in a common language such as ADA, Chill, C, etc.
5- Hardware independent	:	It should be able to run on several different vendors hardware.
6- Maintainable	:	Maintenance of the software and future enhancement times and efforts should be small.
7- Reliable	:	Error rate should be minimized, outputs should be consistent and correct.
8- Observable	:	Easy to perceive how and why actions occur; traceable.
9- Simple	:	Complexity should be low, and easy to understand; unambiguity should be minimized.
10- Robust	:	The system should operate effectively under normal, peak load, crisis and recovery conditions.
11- Uniform	:	Structure of components should be uniform.
12- User friendly	:	It should meet user needs; while being understandable and usable.
13- Easy adjustment	:	System parameter adjustements should be easily accomplished.
14- Manual intervention	:	Operator access should be possible.
15- Secure	:	System should be protected against unauthorized interventions.
16- Data protected	:	Security and availability of network data should be ensured under all conditions.

CHAPTER 8

TRANSMISSION ENGINEERING

8.1 INTRODUCTION

8.1.1 General

The WIS digital transmission planning studies cover the determination of transmission performance criteria for the user services. The depth of this chapter is intended to be sufficient to enable decisions to be taken on the key characteristics of the transmission Architecture of WIS.

The establishment of transmission availability and quality objective for WIS requires the studies of the key elements of the transmission system such as, digital coding techniques with related channel rates, digital multiplexing hierarchy, and digital network synchronization plan in the light of any relevant requirement which may be stated by the Authority for design purposes.

8.1.2 Structure of the Chapter

Requirements, stated implicitly and explicitly for WIS, will have to be taken into account when defining the WIS Transmission Subsystem (WTS) concept and standards. These are discussed in section 8.2 below.

The concept for the system outlined in Chapter 2 stipulates that standard A-law PCM according to CCITT Rec. G. 711 [1] operating at 64 kbit/s would be used as the basic voice encoding principle in WIS. Section 8.3 outlines the various signal encoding techniques that may be relevant to WIS and to the systems and access circuits that may be connected to it.

In Section 8.4, the multiplexing hierarchies are presented and the multiplexing scheme based on a primary level of 2048 kbit/s is described. The hierarchy defined conforms to CCITT Rec. G. 702, G. 703, and G. 751.

The evolutionary timing and synchronization strategy described in Chapter 6 "Timing and Synchronization" is briefly outlined in Section 8.5.

The transmission performance requirements are discussed in Section 8.6. For the specification of the transmission standards for WIS, a hypothetical reference digital path (HRDP) of 3400 km is assumed. Availability and quality requirements for this HRDP satisfy the limits given in both CCITT G. 821, CCIR Rec.557 and Technical Performance Criteria developed for similar defence networks, considering CCITT G.801 and CCIR Rec.556. Effect of multipath fading and selective fading is formulated following CCIR Rep. 338 and the improvements obtainable by frequency and space diversity are also presented. Outage time calculations have been carried out for a typical digital link to demonstrate the application of the methodology given.

Cumulative quality and availability performance figures have been calculated for a connection assumed to have about 30 hops with given path profile data which show, in this case, that the standards set for WIS are feasible.

The important aspect of a frequency allocation plan for WIS is treated in Section 8.7, considering the long-haul and short-haul networks, separately.

Finally, issues related to WIS implementation and procurement are discussed in Sections 8.8 and 8.9, respectively.

8.2 SYSTEM REQUIREMENTS

WIS is assumed to make use of a mix of radio relay and optical fibre cable systems. Radio relay systems to be established for WIS will be assumed to form the backbone of its transmission subsystem (internodal network) and will also provide links to connect the user locations (access network).

The transmission media, together with the switching system, and the analog-to-digital conversion method used must meet the performance requirements given, explicitly or implicitly, by the Authority.

The WIS network will be required to interoperate with the public network using leased links during the entire implementation phase and perhaps also during the operational phase of WIS. It is therefore a requirement that the transmission standards are consistent with those used by the public network operator.

The hypothetical WIS network configuration, given in Fig 8.1 is determined on the minimum cost-basis as outlined in Chapter 4 taking into account the user locations and traffic volumes and relations between them as assumed in this book. As can be seen from Fig 8.1 there are 14 transmission nodes and 25 switching nodes with a total of about 250 hops within the WIS internodal network. The dimensioning of the network done using both analytic and event-driven simulation programs has shown that 34 Mbit/s transmission rate for the internodal links would be quite adequate. The access connections, however, do not have to be of necessarily this rate. It is proposed to utilize a single transmission rate for the radio system in order to reduce the complexity of the equipment. The optical cable systems will use the range of rates specified by CCITT.

The transmission media in WIS will consist of:

a) LOS radio relay digital links.
b) Optical cable with digital transmission.
c) Rented digital channels and groups.
d) Digital satellite links.

8.2.1 Integration

The WIS transmission system is fully integrated on primary group level (2048 kb/s). At this level and above, all subnetworks share the same transmission recources.

8.2.2 Survivability

WIS specific transmission routes for the internodal network will generally follow the existing public network routes, but additionally access links will be established to connect the WIS users to the network.

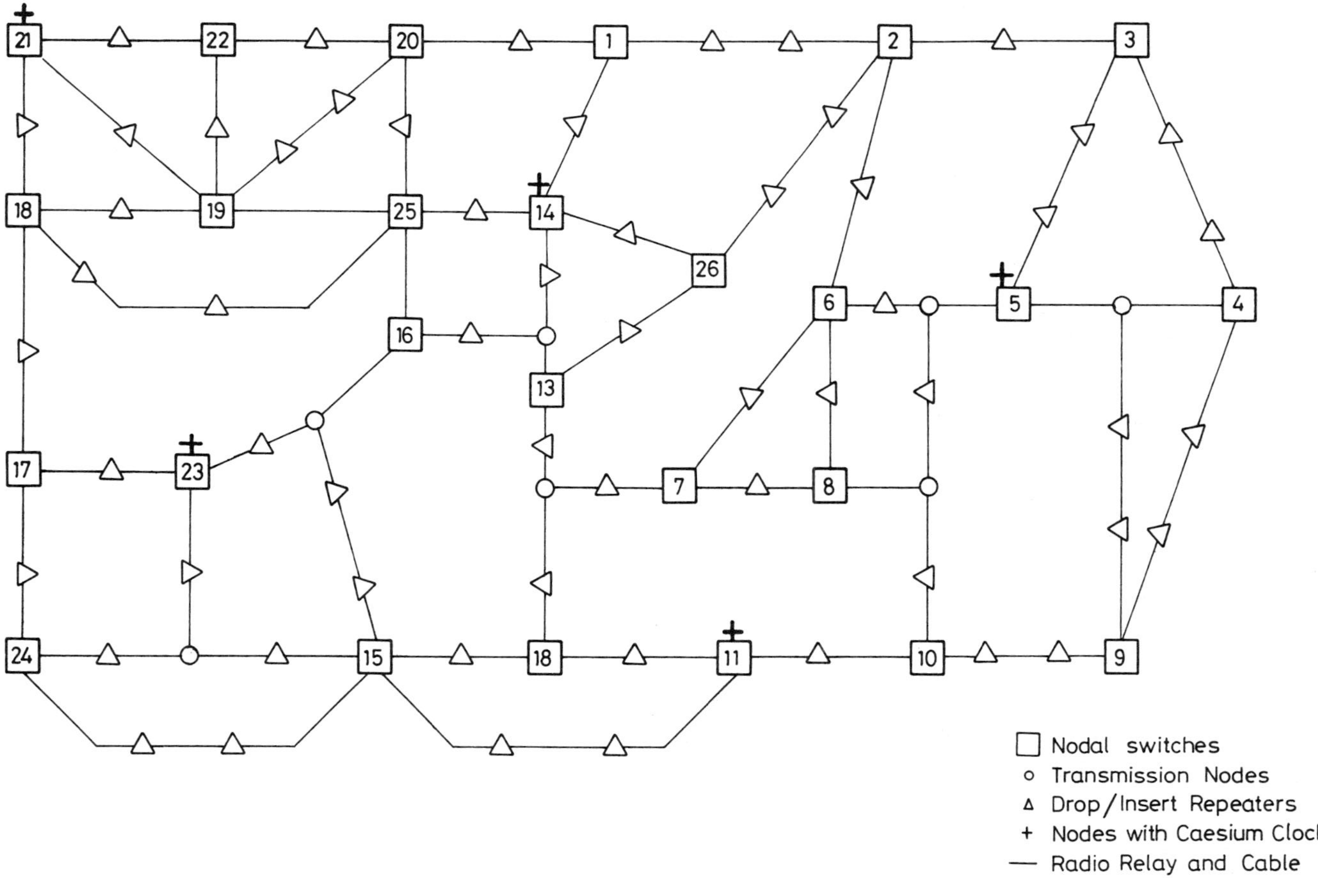

Figure 8.1 WIS Nodal Network Transmission Structure

The users are connected to the switching nodes with at least two different transmission links, preferably using different media; LOS radio and optical fibre.

The network topology together with transmission media diversity as described above is the most important survivability feature of the transmission system.

In addition, the transmission sites will be protected against EMP and provided with Physical Site Protection and Hardening as specified in Chapter 13. With respect to Electronic Warfare (EW), no specific threat has been defined. When and if the threat and required degree of ECCM are specified, they can be included in the design calculation for the radio links. In general terms, measures that may be taken to protect the radio system against ECM include nulling antennas, low side-lobes, and natural and artificial screening of the antennas.

8.2.3 Communications Security

Communications security is basically catered for by the bulk encryption of the transmission links and by the use of end-to-end crypto equipment as described in Chapter 12. The transmission system is designed to support the COMSEC subsystem by having a performance in terms of Bit Error Rate (BER) and slip rates which are in compliance with the requipments of these crypto devices.

8.2.4 Readiness

The allocation of transmission resources for readiness purposes will be part of an overall readiness plan. This will include the use of transportable replacement equipment such as antennas and complete LOS radio stations.

8.2.5 Interoperability and Interconnection

The transmission system will be a key element to achieve interoperability and interconnection to other networks, in particular in the early stages of WIS when the switching subsystem is not in full operation. Interconnection functions supported by the transmission system are typically:

- Physical and electrical termination
- Voice encoding
- Code conversion
- Rate adaption
- Group and channel routing

8.2.6 Standardization

Transmission equipment is mainly based on the CCITT G-series of recommenations.

8.2.7 Reliability

Because WIS is organized as a grid network and will have an adaptive routing system [see Chapter 7], transmission equipment failures will, in very few cases, affect the end users. Besides, transmission equipment generally have a very high degree of reliability and duplication of equipment purely to protect against equipment failures will therefore only be used in exceptional cases.

8.2.8 Availability, Flexibility and Responsiveness

The transmission system will be engineered to meet the requirements set by the Authority.

8.2.9 Maintenance Support

All transmission equipment units will provide local and remote alarms for equipment failures and provide local and remote alarms for equipment failure and failure of incoming signals. Loop-test, self diagnostic and local test access will enable rapid fault location. Modularity of the equipment will ensure easy replacement of faulty units.

8.2.10 WIS Users

All static WIS users will be provided access to the network by the WIS transmission system. Land-mobile WIS users will be served through appropriately located WIS pick-up points with adequate access capacity as defined in Chapter 2. The users will be provided with appropriate access links to be interconnected with the pick-up point by transmission means organic to their organization. Afloat WIS users will be served through a gateway (or a ship-shore buffer) using shipborne satellite terminals or/and by HF radio.

8.2.11 WIS Services

The services defined in Chapter 2 can be supported by the WIS switching system (WSS). However, the functionality of the Transmission System is essential for service requirements such as

- Support of telegraph and data equipment with a variety of interface standards and bit rates.
- Circuit quality requirements (voice intelligibility, bit error rate etc.)
- Rerouting of point-to point circuits in case of failures.

8.2.12 SUB-SYSTEM Requirements (SATCOM Links)

It is assumed that SATCOM links, either leased from international systems (INTELSAT, EUTELSAT) or obtained from a national system which may be procured by the Authority, may be used (either as a supplementary or an integral part of the WIS backbone network) for the following purposes:

- Interswitch circuits in the WIS providing a node skipping overlay to the terrestrial network to enhance survivability.
- Access circuits to WIS nodal switches from fixed users in remote areas and from itinerant/mobile users.
- Circuits between ships/aircraft and users in fixed locations.
- Broadcast and emergency nets.
- Dedicated circuits between individual itinerant/mobile users.
- Dedicated circuits between individual users in fixed locations.

As far as switched networks are concerned the main problems with the use of SATCOM are associated with signalling, routing [see Chapter 5], and transition from peacetime to operations under wartime stressed conditions [2]. Frequency allocations for mobile applications may also present problems.

SATCOM links can be designed to meet any transmission standards (See Section 8.6) with respect to error-rate and slip. Assuming direct digital interface (DDI) and TDMA being used, TDMA would have its own frame synchronization which would be independent of the terrestrial system and this would require synchronization between the two. There are different methods which can be used: fully synchronous interface, interface based on frame slips and interface based on justification.

8.2.13 Network Control

Transmission equipment will interface the WIS Network Surveillance and Control (WNSC) System for collection of alarms and status information, and for remote control of the system [see Chapter 11]. The transmission system will also provide a separate communication channel to the WNSC by offering the use of the service channel on the radio links as a connection between control units in adjacent nodes. This channel will not depend on Nodal Switches nor on the multiplexers for its operation.

8.3 SIGNAL ENCODING

8.3.1 Requirements

The WIS transmission system will serve different functions:

a) Transparent through-connect of 2048 kbit/s groups for tactical network applications.

b) Provision of transmission facilities for other communications systems. For this application, the requirement for WIS transmission is generally to provide transparent channels of 64 kbit/s between pairs of terminations in the network.

c) Interconnection of WIS switching nodes to form an Integrated Digital Network.

For a), the internal structure of the 2048 kbit/s signal is of no interest to the network. The network will offer a bit-transparent, clock transparent channel between pairs of terminations.

For b), the transparent channels through the network are the 64 kbit/s channels, carried in primary groups with frame structure according to CCITT Rec. G. 732. The channels may be terminated as analog or digital user interfaces, or as timeslots (as in DACS)* in a primary group. The network is required to offer cross-connection of 64 kbit/s channels as a transmission function. It is therefore a requirement to use the same representation of the user signal within the 64 kbit/s timeslot for groups of terminals that may be cross-connected with each other.

For c), the compatibility of terminals within a user group is required as for b). In addition, the signal encoding must be compatible with the switching system and with signal processing and conversion equipment within the network.

It should be noted that it is assumed there is no requirement for interconnection between (b) and (c). The signal encoding in the two networks need not therefore necessarily be identical. This indicates that ISDN interfaces, basically defined in a switched circuit environment, is of little relevance to (b), while it is vitally important for (c) to ensure future enhancements of the system.

---

* DACS = Digital Access and Cross - Connect System

8.3.2 Channel Transparency

The basic idea of an Integrated Digital Network is that all signals should be converted to bit streams; and all bit streams should be switched and processed in the same way by the network. Although this should definitely be the basic guideline for the network design there are aspecsts associated with traffic types that, in some cases, call for a different processing by the network of the different traffic types:

- -Voice
- -Non-voice analog
- -Digital

The three categories differ mainly in the transparency required for the connection.

8.3.2.1 Voice

The quality of a voice circuit is characterized by its intelligibility. The signal waveform itself need not necessarily be reproduced in every detail. Furthermore, it is in particular situations convenient to apply nonlinear processing such as speech interpolation, echo suppression, echo cancelling, and bandwidth-compression techniques. This can be used for the benefit of either cost reduction, or to remedy transmission impairments, even if the transparency of the channel is violated.

8.3.2.2 Non-Voice Analog

This category covers typically modem signals from low and medium rate data modems. The quality is characterized by the accuracy in the reproduction of the waveform. Voice-processing devices such as echo control and speech interpolation may have to be controlled separately in the case of non-voice analogue traffic, because the signal waveform is not always carried faithfully through such devices.

8.3.2.3 Digital

For a digital connection, the bit-by-bit channel transparency is essential. The quality of this circuit is characterized by the bit error rate and the slip rate.

In the general case, the user bit rate is carried on a network channel with a higher bit rate. The characteristics of this conversion will also determine the relationship between the network channel characteristics and the user channel characteristics.

8.3.3 Encoding of Voice Signals

The following sections contain some important aspects of various analog-digital conversion techniques [3].

8.3.3.1 PCM

In line with the principles incorporated in the concept of WIS, voice signals in WIS will be encoded using PCM and A-law companding in accordance with CCITT Rec. G.711.

PCM is the predominantly used principle for A/D conversion of voice for civilian networks. The signal quality is good, even after several A/D and D/A conversions in cascade. This encoding will therefore form the general basis for the WIS system.

However, some other voice encoding systems are of interest to WIS. One area of interest is interface to networks and terminals where other encoding methods are used. The other area is the dedicated circuits where submultiplexing of channels within the 64 kbit/s timeslot potentially can be of interest to the users of WIS to improve bandwidth utilizations.

8.3.3.2 CVSD

CVSD (Continuously Variable Slope Delta Modulation) is the encoding specified for use in Tactical Communication Networks with which WIS will have to interoperate. WIS will therefore have to be provided with an interface supporting CVSD.

CVSD is a waveform encoding system where the bit rate is equal to the sampling rate. The advantages of CVSD are acceptable speech quality at medium bit rates, low complexity, and low sensitivity to bit errors.

The quality of speech reproduction is such that if 16 kbit/s is used, only one A/D/A conversion should be permitted on a connection. A tandem connection of a 16 kbit/s CVSD section and another section with significant signal distortion should also be avoided. Thus, a 16 kbit/s CVSD section used as access to an HF or VHF base-station should be avoided. At 32 kbit/s, CVSD is good enough to allow a few conversions in sequence. For interconnection between tactical networks and WIS, with CVSD/PCM conversion on the gateway, the signal quality will be dominated by the CVSD quantization noise. If the connection is terminated in a tactical network in one end and WIS in the other end, the voice at both ends will sound like a CVSD connection.

If two CVSD networks are interconnected through a PCM network, there are two options. The first possibility is that the CVSD signal is converted to PCM (code conversion) at one gateway and converted back to CVSD on the other gateway. In this case, the quantization noise of the CVSD encoding process will be added twice. According to observations above, only 32 kbit/s CVSD would give an acceptable signal quality in this case.

The second option is to carry the CVSD encoded signal at 16 or 32 kbit/s transparently through the PCM network. In this case, a 16 kbit/s connection would give the same quality as if it were set up within one CVSD network. To enable this type of transition, the intermediate PCM network must be able to treat the signal as a 16/32 kbits/s data channel.

8.3.3.3 LPC 10

LPC 10 (Linear Predictive Coding with 10 tap coefficients) is a source coding of voice that enables intelligible speech at a very low bit rate to be obtained. For military applications, 2400 bit/s is used [3].

The application of LPC 10 is mainly to offer digitally encrypted voice on an ordinary analog voice channel.

For WIS, LPC 10 will appear as an analog non-voice signal. A pure PCM network will easily carry this type of signals. Since, however, the advantages of LPC-10 at a low bit rate is most evident for non-homogenous connections through different networks, the interoperability aspects must be carefully assessed [see Chapter 10].

8.3.3.4 ADPCM

ADPCM (Adaptive Differential PCM) is a waveform encoder where the difference from the previous signal sample is encoded [3]. The ADPCM transcoding technique is used to convert a 64 kbit/s PCM channel to and from a 32 kbit/s channel. The band width is utilized more efficiently than for PCM. This technique allows the use of a 2048 kbit/s group to carry 60 voice channels (CCITT Rec. G.721) or to increase the bandwidth on the voice channel to 7 kHz. CCITT Rec. G 721 recommends a transcoder that transforms 2x30 PCM channel to 60 ADPCM channels at 2048 kbit/s.

The drawback associated with ADPCM transcoders is that bit transparency on the PCM channel is not preserved. The transcoding will work properly for voice and for PCM encoded modem signals up to 2400 kbit/s. For rate adapted data traffic, the most commonly used rate adaption methods can not be used together with a PCM-ADPCM-PCM conversion.

When used in a digital network carrying both voice and data traffic, the different traffic types will need different routing and/or processing when ADPCM is involved. Although this kind of traffic-type dependent processing cannot always be avoided, it is a strong argument against the introduction of ADPCM in the general 64 kbit/s switched part of WIS.

In the part of WIS providing semi-permenant channel routing end-to-end (ref. 8.3.1b above), ADPCM can be utilized to increase the number of analog voice channels that can be carried on a given bit rate capacity [3]. It may be feasible if a large number of voice channels terminate on the same location at both ends. Possible ADPCM transcoders should be part of the user system. The facility offered by WIS will still be to provide 64 kbit/s digital channels through the network.

8.3.4 Rate Adaption

Although WIS is a digital network, signals from digital terminals can not normally be connected directly into the network. Most digital terminals especially from the pre-ISDN era, have a bit rate that is different from the network rate of 64 kbit/s. Additionally, digital terminals require a variety of electrical, mechanical and signalling characteristics of the interface.

The difference in bit rate can be regarded as an obstacle, but it does also give an opportunity to utilize the extra bandwidth for something useful. Possible benefits can be;

- Better utilization of the 64 kbit/s bearer channel. This means submultiplexing or exchange of status and control information,
- Reduced sensitivity to bearer channel degradation. The rate adaption may be designed to offer a lower bit error rate on the user channel than on the bearer channel,
- Clock transparency.

In some cases, the terminal is not slaved to the network frequency, although the two frequencies may have a fixed nominal relationship. The redundant capacity can be utilized to carry the true user bit rate between the terminals.

- Interoperability.

8.4 MULTIPLEXING

8.4.1 General

The WIS multiplexing system can be split into three subsystems:

a) Multiplexing of channels up to the primary rate of 2048 kbit/s. This function is taken care of by the access switches for the switched part of WIS. For the point-to-point circuits primary multiplexing is in some cases a WIS network function, while in other cases, the 2 Mbit/s signals are carried transparently through the network.

b) Higher order multiples to 8.448, 34.368 and 139.264 Mbit/s for transmission on LOS radio links or other transmission media.

c) Routing of TDM groups for optimum utilization of the transmission grid network. Routing will basically be on primary group (2 Mbit/s) level.

Fig. 8.2 gives an overview of multiplexing arrangements in an Access Node, While Fig. 8.3 shows a transmission node housing a nodal switch. Figures 8.4, 8.5, and 8.6 show the multiplexing arrangements at:

- group drop/insert repeater,
- channel drop/insert repeater/transmission node,
- through repeater.

8.4.2 Multiplexing Requirements

8.4.2.1 Primary Rate Multiplexing

For WIS Switched Subsystem, (WSS), termination and multiplexing of user signals are regarded as an integral part of the Access Switch function.

For the WIS service of semi-permmant channel routing, primary rate multiplexing is used to multiplex 30 user channels with various characteristics into a uniform TDM frame at 2048 kbit/s, according to CCITT Rec. G.732 The user interfaces to be supported are:

- 64 kbit/s codirectional/contradirectional interface according to CCITT Rec. G.703.
- 4-wire V.F. interface with E&M signalling.
- V.24 data interface at various bit rates.
- 50-300 bit/s Telegraph interface.

The WSS will, in addition to the above interfaces, support:

- ISDN interfaces (Basic Access, Primary rate Access)
- 2-wire V.F exchange line interface for connection to public switching system.
- 2-wire V.F channel unit for LB (local battery) telephones.
- 2-wire and 4-wire V.F channel units for CB (central battery)telephones.
- Data subscriber interface for connection of data terminals to WIS switched system.
- X.25 interface unit for packet mode data terminals.

There will be other interfaces in WIS requiring special gateway arrengements, such as the interface to tactical networks. This interface is regarded as part of the switching system.

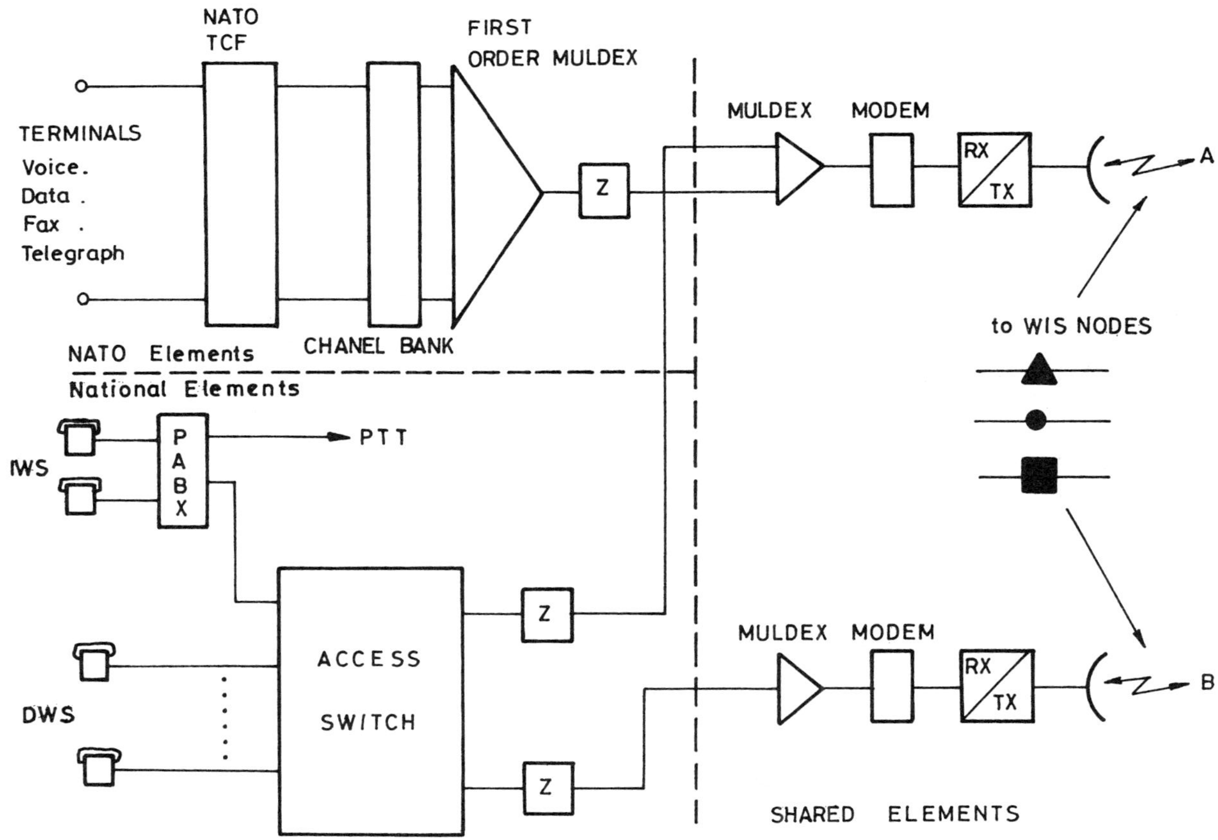

Figure 8.2 Multiplexing Arrangements in an Access Node

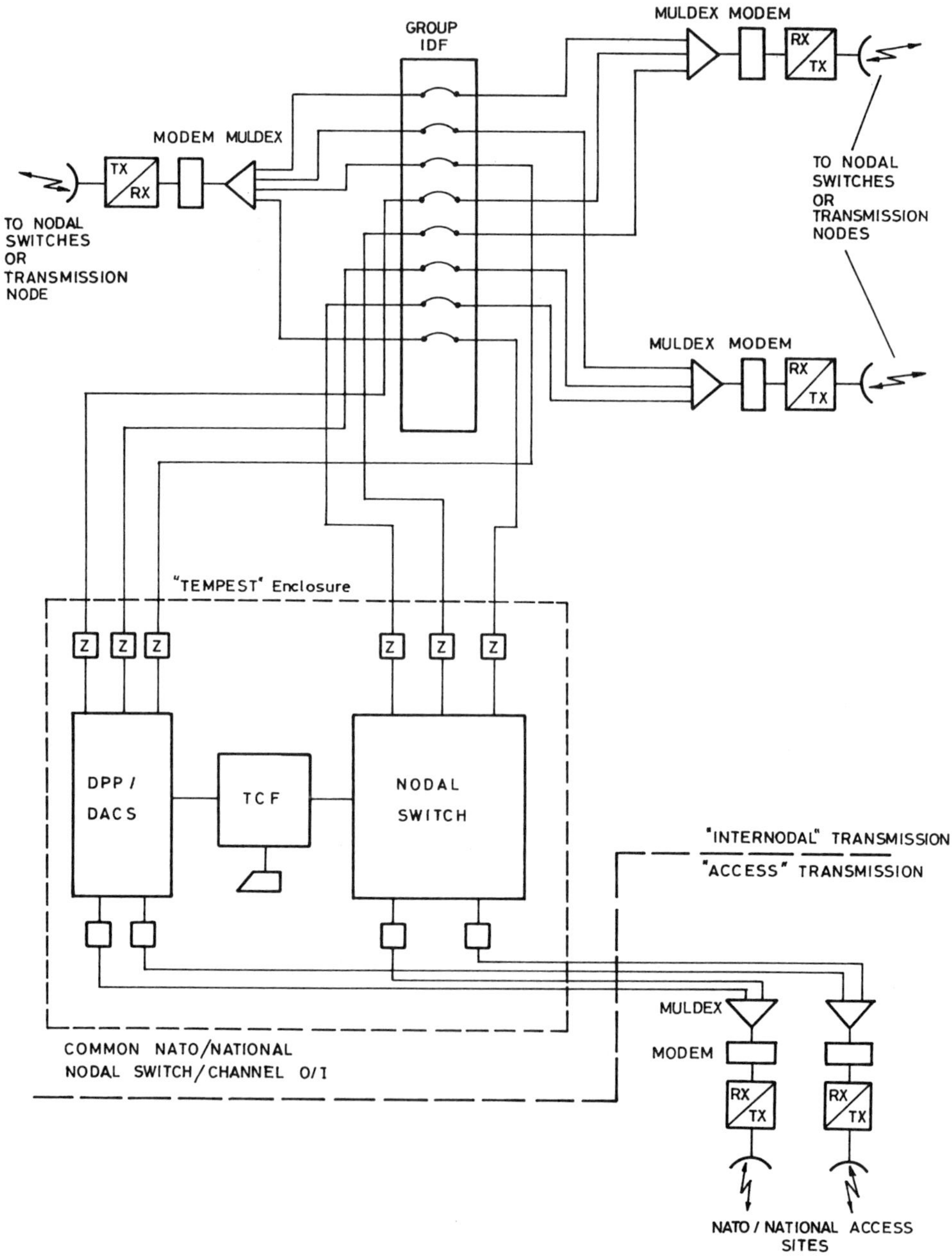

Figure 8.3 A Typical Nodal Switch/Transmission Node

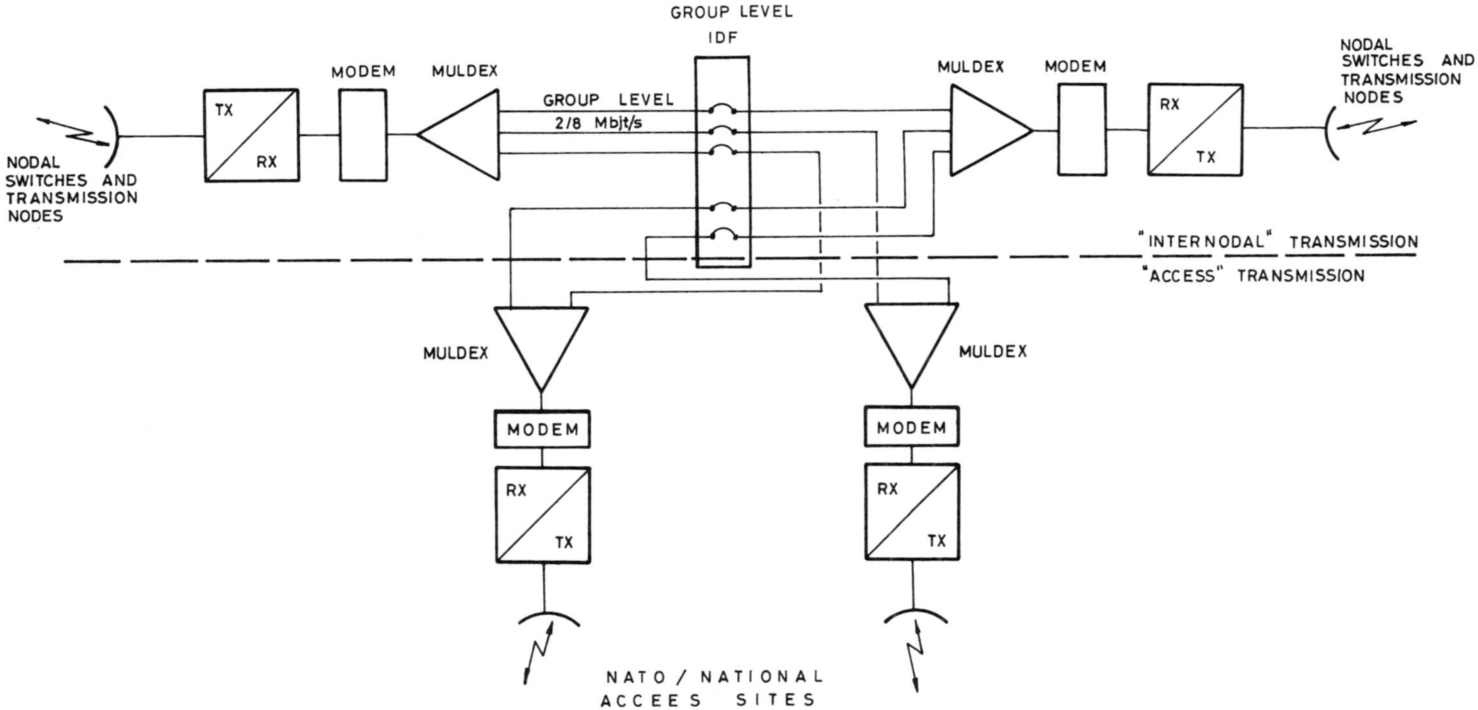

Figure 8.4 Configuration Group Drop/Insert Repeater Installation

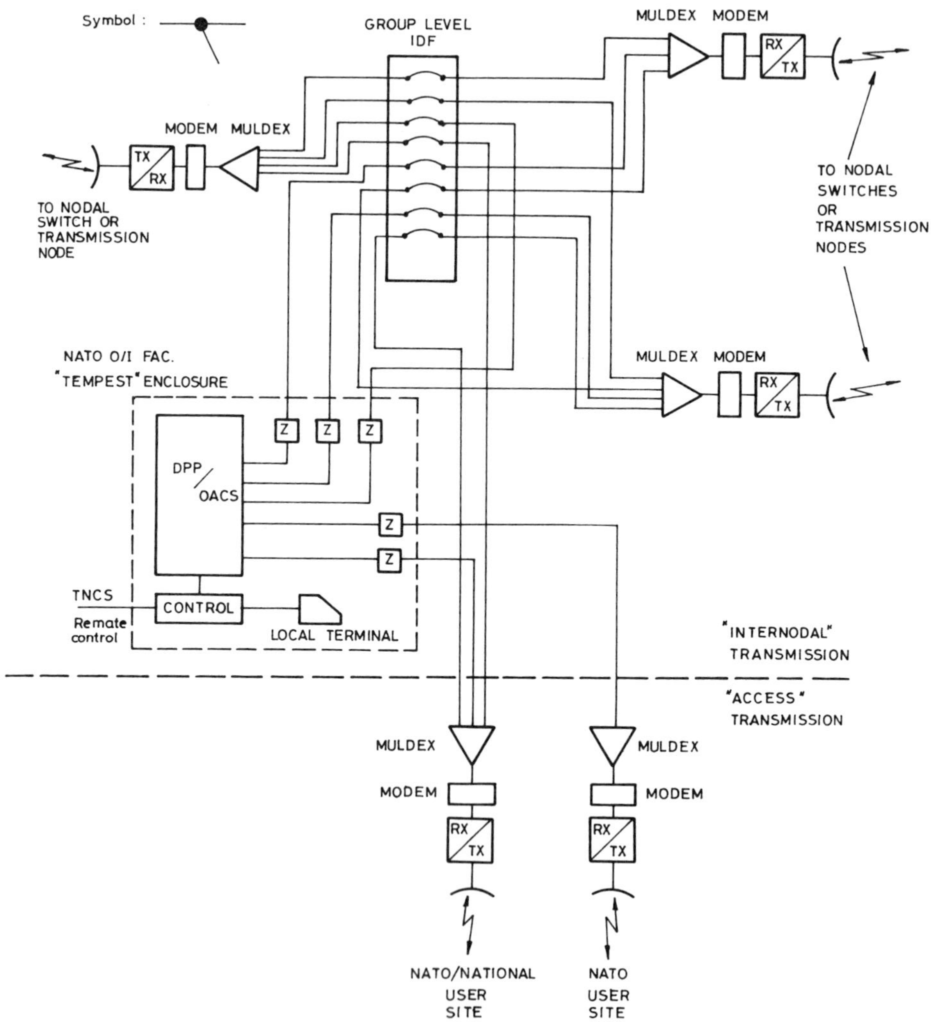

Figure 8.5 Configuration Channel Group/Insert REPEATER

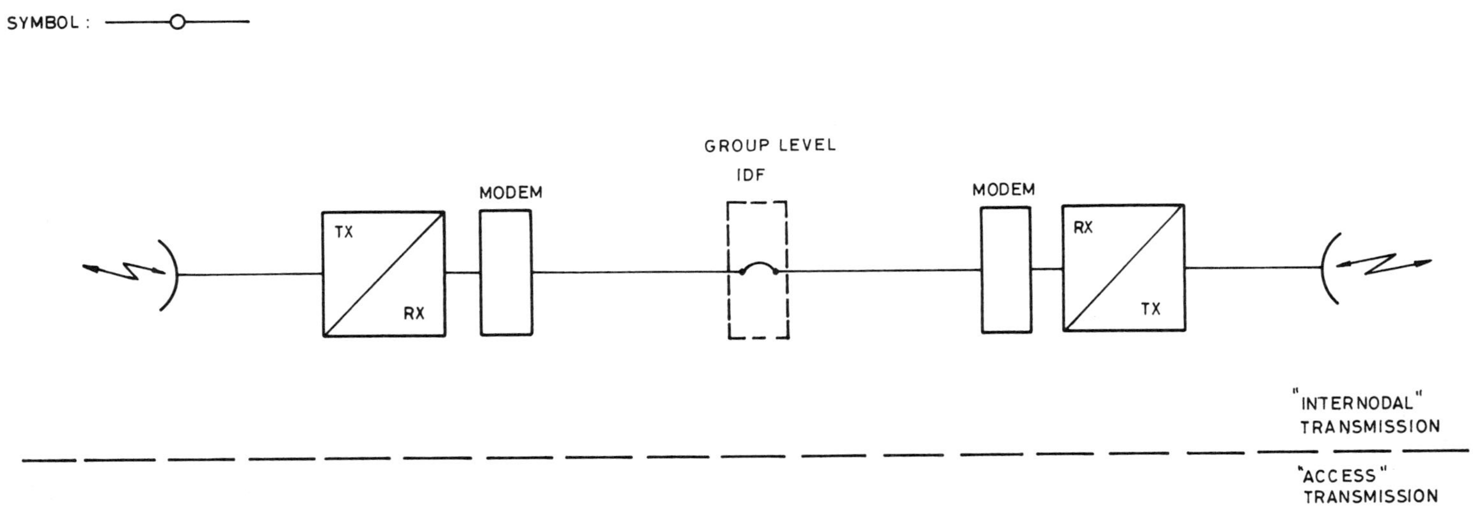

Figure 8.6 Configuration Through Repeater Installation

For through connection of TDM groups between tactical networks, ordinary 2048 kbit/s interfaces are offered. Multiplexing of TDM groups at 512 and 1024 to 2048 kbit/s is a WIS external matter. The framing pattern on this 2048 kbit/s group will not be according to CCITT Rec. G.732. However, this framing pattern will pass transparently through the WIS primary group network.

8.4.2.2 Primary Group Routing

Each primary group has two terminations in the network. Each termination may be either:

- an external primary group interface
- A primary rate multiplexer
- A TDM port in a Nodal Switch
- A Digital Access and Cross-connect System (DACS)

Between the two primary rate terminations, the primary group can be routed in different ways through the grid network, depending on capacity, availability, route length etc. Therefore, at each transmission node there must be facilities for patching the primary groups through the network in accordance with an overall transmission facility allocation plan (Figs. 8.3, 8.4, 8.5).

Bulk encryption will be used on primary rate level. Consequently, the primary rate terminations must be equipped with pairs of Bulk Encryption Devices (BEDs).

8.4.2.3 Higher Order Multiplexing

Between transmission nodes (TN), there are transmission facilities, mainly LOS radio links working at 34 Mbit/s, carrying 480 channels of 64 kbit/s rate. The primary groups must be multiplexed in higher order multiplexers to form a 480 channel group suitable for transmission.

On a fully equipped TN, there will be a complete higher order multiplexing function for each higher order transmission link connected to the TN. However, in some cases where there are only 2 transmission links terminating on the node, only a few primary groups will be available for termination or routing. The majority of groups are directly through connected. This is called a drop-insert node.

8.4.2.4 Circuit Routing

The user channel requirement at the various terminations in the network will generally not be a multiple of 30 channels. In a large portion of the terminating primary groups, the number of occupied channels will be rather low. It is therefore advantageous to "fill up" primary groups in the heavy parts of the network by patching timeslots (by using DACS) from primary groups. With a low channel occupancy, into fully occupied primary groups. In principle, this patching must be done separately for the different sub-networks defined in Section 8.3.1 above.

8.4.3 Main elements in the WIS multiplexing system

8.4.3.1 Primary Rate Multiplexers

The primary rate multiplexing function is basically performed by multiplexing equipment in accordance with CCITT Rec. G.732, but extended to support various voice and non-voice interfaces as defined in Section 8.4.2.1

For point-to-point circuits each user-rate interface can be the aggregate signal from a sub-multiplexer. This is in particular relevant for telegraph channels, where a TDM multiplexer (CCITT Rec. R.101, R.102, R.111 or another format) or an VFT multiplexer (CCITT Rec. R.35) can be used. R.111 is the preferred standard, because this offers the full utilization of the 64 kbit/s channel.

Such sub-rate multiplexing is considered to be a part of the user terminal area rather than an integral part of WIS.

For circuit switched connections, sub-rate multiplexing is generally not feasible, because the WIS switching system can not switch sub-rate channels independently.

8.4.3.2 Higher Order Multiplexers

The multiplexing of primary rate signals generally follow the standard hierarchy based on 2048 kbit/s as defined in CCITT Rec. G.702.

The basic hierarchy is to use second order multiplexers according to CCITT Rec.G.742 to multiplex up to 4 primary groups to one secondary group at 8448 kbit/s. Four secondary groups are multiplexed in a third order multiplexer to form a 34368 kbit/s group according to CCITT Rec.G.751. The WIS was designed at a time when the plesiochronous digital hierarchy was predominant. It is not considered relevant to change this to a synchronous SHD hierarchy as an internal function in WIS, since there is no need for the increased capacity of SHD and bandwidth available 2 Mbit/s groups from WIS may, however, easily be carried inside a public (PTT) SHD infrastructure where this is available.

CCITT recommends different frame structures, both at 8448 kbit/s and 34368 kbit/s. The options are:

a) G.742/G.751 working with positive justification.
b) G.745/G.753 working with positive/zero/negative justification.

The functional differences between the two standards are not essential. The choice is mostly a question of availability of equipment.

The primary groups within WIS will generally be bulk-encrypted. It is therefore essential that the higher-order multiplexers are truly frame-independent, bit tranparent at 2048 kbit/s. Alarm indication signals generated by a primary rate multiplexer is only available at the opposite multiplexer at the red side of primary group termination. The alarm hierarchy can therefore not utilize frame alarm on the primary groups on a link by link basis.

8.4.3.3 Primary Group Routing

Routing of primary groups should be carefully optimized on the basis of:

- Capacity,
- Route length,
- Route occupancy,
- Survivability,
- and other operational factors.

The allocation of primary groups should not be altered very often, because of the complex impact of change, to all the users of the common primary rate network . It is therefore considered adequate to implement this cross-connect by means of a cable patching field, DDF (Digital Distribution Frame).

8.5 TIMING AND SYNCHRONIZATION

8.5.1 Requirements

The purpose of the Timing and Synchronization subsystem is mainly to keep the rate of bitslip below a specified level. A bitslip is an event where bits are deleted or inserted in an end-to-end digital connection. Difference in the nodal clock rates of interconnected nodes causes loss of data by slip. Each slip causes the loss of 8 bits of data for each 64 kbit/s channel affected and the greater the difference in clock rates, the greater the rate of slip.

The general target slip rate for international connections is given in CCITT Rec. G.811. For WIS, the slip rate objective is taken to be less than 1 slip in 14 days, degrading to a maximum rate of 2 slips per day if the network is damaged.

The requirement is further discussed in Chapter 6. The argument used there, is that because the bulk crypto equipment will resynchronize once every 24 hours and thereby cause an interruption in the primary group connection there is no point in specifiying a lower slip rate than 1 slip/24 hours. The two requirements can be considered consistent if:

- 1 slip/24 hour applies to the maximum length connection within WIS (10 Nodal Switches)
- 1 slip/14 days applies to a connection between any 2 nodes in the network, regardless of the synchronization system in use (plesiochronous, mutual, master-slave)

Acceptability of different controlled slip rates is given in Table 8.1 below:

Table 8.1 **Acceptability of different controlled slip performance on a 64 kbit/s international connection or a bearer channel.**

Performance Category	Mean slip rate	Proportion of time (Note 1)
Acceptable (Note 2)	< 5 slips in 24 hours	> 98.9 %
Degraded	> 5 slips in 24 hours and < 30 slips in 1 hour	< 1.0 %
Unacceptable	> 30 slips in 1 hour	< 0.1 %

Note 1 -Total time > 1 year
Note 2 -The nominal slip performance due to plesiochronous operation alone is not expected to exceed 1 slip in 5.8 days.

For a 2048 kbit/s system above threshold values are 5 slips of 1 octet on 32 channels of 64 kbit/s.

$$5 \times 8 \times 32 = 1280 \text{ single bits in 24 hours}$$
$$30 \times 8 \times 32 = 7680 \text{ single bits in 1 hour}$$

8.5.2 Recommended timing and synchronization subsystem for WIS

The Timing and Synchronization (T&S) aspects of WIS are dealt with in depth in Chapter 6. On the basis of a 4-phase implementation of the WIS the main conclusions are;

- Phase I: It is recommended to operate the network in a plesiochronous mode. Each of the nodes is equipped with a high-stability reference clock (Fig. 8.1), used to control the phase and frequency of the outgoing 2 Mbit/s groups from the node. The slip rate is a function of buffer size and clock accuracy.
- Phase II (the expansion phase), where additional nodes are being added to the network. For this phase, a multiple masterslave nodes method is recommended. The original nodes are used as master-nodes. The remaining nodes are either master-slave nodes, receiving and distributing timing information, or slave nodes, synchronized to higher rank nodes.
- Phase III is the final stage where Hybrid Mutual Synchronization is foreseen. This is not a result of network growth, but of the expected matureness of this method. If the method proves to be reliable and efficient at an early stage, phase II can be omitted.

8.6 TRANSMISSION PERFORMANCE CRITERIA AND REQUIREMENTS

8.6.1 General Aspects of Digital Transmission Modeling

Digital transmission network models are hypothetical entities of a defined length and composition for use in the study of digital transmission impairments (e.g. bit error, jitter and wander, transmission delay, availability, slip, etc.). The diversity of possible network situations implies that individual models can only represent a small portion of typical real entities. However, a limited number of such models together may be sufficiently representative to provide a useful tool upon which studies may be based.

Three such models are given by CCITT Rec. G. 801.

i) Hypothetical Reference Connection (HRX),
ii) Hypothetical Reference Digital Link (HRDL),
iii) Hypothetical Reference Digital Section(HRDS).

Brief description of these models are presented in the following subsections.

8.6.1.1 <u>Standard Hypothetical Reference Connection (HRX)</u>

Since the overall network performance objectives for any parameter need to be consistent with the user requirements, such objectives in general, should relate to a network model which is representative of a very long connection, such as the standard Hypothetical Reference Connection (HRX) as given in CCITT Rec. G. 801.

Figure 8.7 shows the HRX for a length of 27.500 km. The difficulty of identifying every conceivable practical implementation of a connection and the undesirability of producing too many options naturally require that this "Standard HRX" may need to be appropriately modified in composition to suit the particular task in hand. A situation can be envisaged where many similar HRX's exist to serve specific functions, but in all cases they are derivatives of the "Standard HRX". This approach is adopted in discussing the performance requirements for WIS, in the following sections.

8.6.1.2 Standard Hypothetical Reference Digital Link (HRDL)

To study the digital transmission impairments, a network model, called Hypothetical Reference Digital Link (HRDL) comprising a combination of different types of transmission elements is suggested. HRDL can be considered as a constituent element of a HRX, thus permitting the apportionment of overall performance objectives to a shorter model.

CCITT Rec. G. 801 (Also CCIR Rec. 556) defines a standard HRDL of 2500 km with 9 hops (Fig. 8.8).

8.6.1.3 Standard Hypothetical Reference Digital Section (HRDS)

The "Hypothetical Reference Digital Section" model is constructed to accommodate the performance specificaton of transmission (i.e. digital line and radio systems). Such a model (Fig 8.9) is defined for each level in the digital hierarchies. HRDS can be regarded as a constituent element of a HRDL. The value of Y is dependent on the network application and the lengths of 50 km and 280 km have been identified by the CCITT (Rec. G.921). Digital equipments such as multiplexers/demultiplexers are not included in the HRDS model. HRDS is particularly useful in studying the propagation and performance of real digital radio links.

8.6.1.4 Transmission Plan

The subjective quality of a telephone connection is governed by a set of transmission parameters such as:

- -Signal level,
- -Distortion,
- -Delay,
- -Echo,
- -Noise.

In a digital network such as WIS, these parameters are more easily controlled than for an analogue network. Nevertheless, performance objectives should be specified also for a digital network. It is of particular importance to establish these parameters for connections between terminals in WIS and terminals in other networks (PTT, etc.)

a) Signal Levels

First of all, reference circuits defining levels, reference equivalent (SRE, RRE, CRE)* and 2/4 wire points should be established, typically for all WIS circuit level interfaces, internal and external.

---

* SRE : Sending Reference Equivalent
RRE : Receiving Reference Equivalent
CRE : Corrected Reference Equivalent

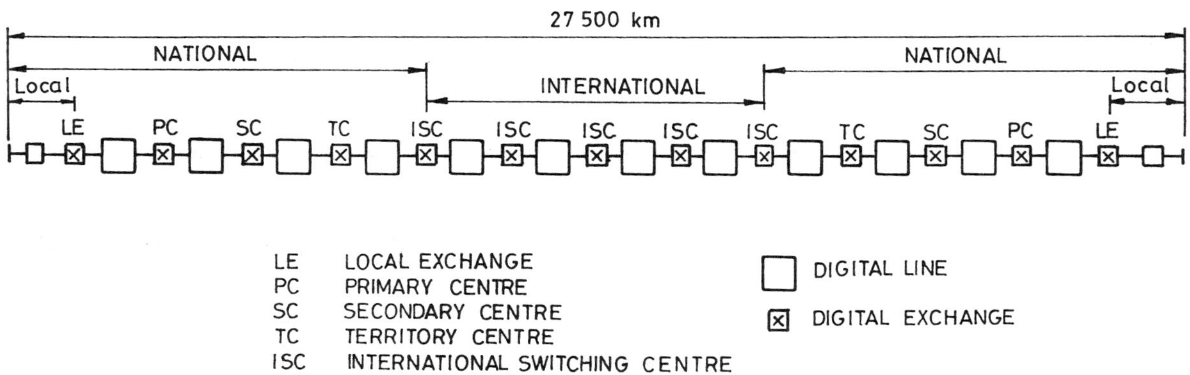

Figure 8.7 Standard Digital Hypothetical Reference Connection (longest length)

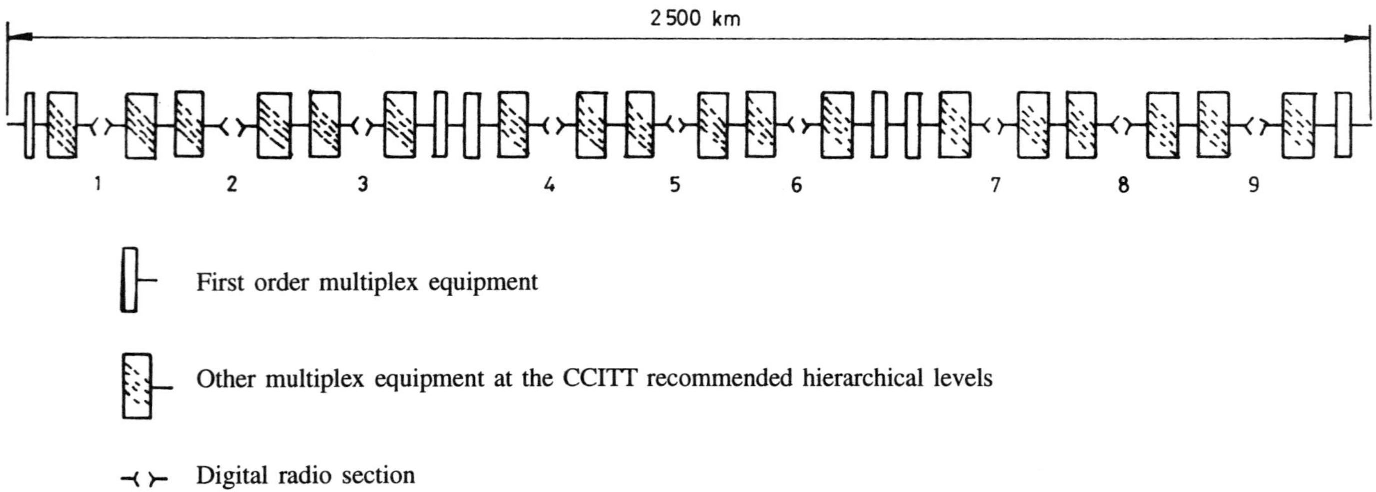

Figure 8.8 HRDL for Radio-Relay Systems With a Capacity Above the Second Hierarchical Level

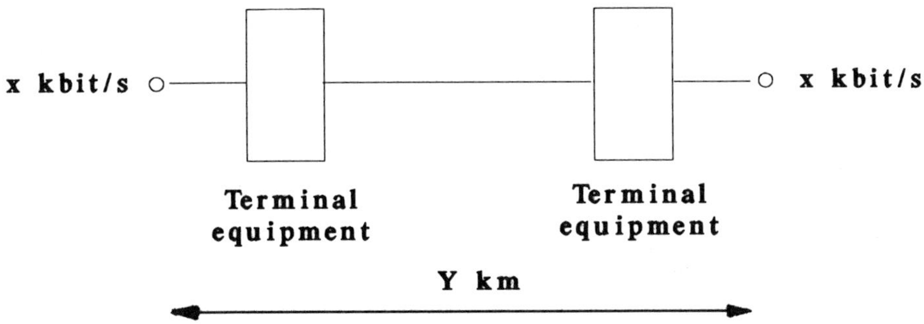

Figure 8.9 Hypothetical Reference Digital Section (HRDS)

Special attention should be paid to interconnections to other networks. Such interfaces should preferably be at 4-wire. However, 2-wire connections to the public network must be taken into consideration.

The attenuation from the transmitting 2-wire termination in a digital network, through the switched digital 4-wire sections to the receiving 2-wire termination (digital connections between subscribers with existing 2-wire subscriber lines on digital exchanges) is given in CCITT Rec. G. 121 Annex E for PTT networks in different countries.

On the basis of that information , it is proposed to define for WIS

$$T + R = 7 \text{ dB}$$

where T is the attenuation on the transmit side and R is the attenuation on the receive side. It is to be noted that different values may be appropriate for connections in the evolving mixed analogue/digital network.

The characteristics of a typical telephone set used in WIS may be such that:

$$SCRE + RCRE = 7 \text{ dB}$$

where SCRE is the Corrected Reference Equivalent (CRE) for the telephone sending side and RCRE is the CRE for the receving side (for definitions, see CCITT Rec. G. 111).

Based on the above assumptions, the overall CRE will be:

$$OCRE = T + R + SCRE + RCRE = 14 \text{ dB}$$

This figure is within the range recommended in CCITT Rec. G. 111, although a little on the high side with respect to the optimum. The calculations must be revised on the basis of rating for actual telephones to be used in WIS.

b) Transmission Loss Variation in Time

The CRE values given above do not cover variations in time of the loss of various parts of the system. Rec. G. 151, para 3 gives the objectives recommended by the CCITT for transmission loss variations on international circuits and national extension circuits as compared with the nominal values.

c) Distortion

Distortion has two elements: Attenuation distortion and group delay distortion. General guidelines are given in CCITT Rec. G. 113, 114, 132, 133.

d) Delay and Echo

CCITT Rec. G.131 gives the guidelines for permissible delay in relation to stability and echo.

For WIS, assuming that the longest HRX path length is 3400 km and number of A/D converters, (one at each access circuit) is two, and the number of digital nodal switches is 10; than the mean one-way delay should not exceed

$$0.004 \times 3400 + 2 \times 0.3 - 10 \times 0.45 = 18.7 \text{ ms.}$$

From Figure 2 of CCITT Rec. G. 131 it can be taken that for a maximum one-way delay of 18,7 ms (approximately), the echo problem can be described as follows:

i) 1% of the users will experience an objectional echo if the "corrected reference equivalent" (CRE) of the echo path is:

$$\text{CRE/E} = 34 \text{ dB}$$

ii) 10% of the users will experience an objectional echo if:

$$\text{CRE/E} = 28 \text{ dB}$$

An example of calculation of CRE/E is given in Annex A to Rec. G.131. A calculation with typical figures can be summarized as follows:

Sum of sending and receiving CRE

For the telephone set	:	7 dB
Talker side attenuation T+R	:	7 dB
Mean echo loss: E+T+R=11+7	:	18 dB
	+	32 dB

Where E is the four-wire to four-wire attenuation through the listener 2/4 wire hybrid.

It can be concluded that for the longest connections within WIS, objectionable echo may occur for more than 1% of the users. The problem is however not so serious that general use of echo cancellation is required. It is to be noted that this preformance is only expected in a degraded network and that for normal operations there would be no echo problems: for instance for a connection of 1500 km long involving 5 nodal switches, the mean one-way delay would be 8.85 ms giving a CRE/E = 28 dB for 1% and 22 dB for 10% of the users which are both less than the 32 dB calculated above.

For circuits terminating outside WIS, the problem is more obvious. The application of echo control devices in the connected network must be taken into account for IVSN, use of the standard 4-wire IVSN telephone set will of course prevent echo problems.

e) Noise

In a digital network, noise is dominated by the quantizing noise of the voice encoding prosess. This noise is well known for a pure PCM network (CCITT Rec. G. 113). For interconnection with other network, a survey of the noise situation should be produced.

8.6.2 Digital Transmission Model for WIS

The Hypothetical Reference Digital Link for WIS may be defined and the performance criteria determined accordingly. The maximum connection length in WIS is expected to be about 3400 km. under most severe damage conditions and about 1500 km. under normal conditions.

8.6.3 Performance Objectives for a Standard HRX

The performance objectives give the user an indication as to the expected error performance under real operating conditions and form the basis upon which performance standards are derived for transmission equipment and systems. Two important parameters of the transmission performance are the "availability" and "quality" for a HRX, and these objectives apply to,

- the percentage of seconds or minutes during which a certain threshold of error rate is reached,
- the percentage of error-free seconds.

8.6.3.1 Quality

The CCITT Rec. G. 821 states the quality performance of an international digital connection forming part of an integrated services digital network.

In CCITT Rec. G. 821 the objectives are given for a 64 kb/s channel traversing a 27.500 km HRX, on the basis of different performance classifications, namely the degraded minutes, severely errored seconds, and errored seconds (Table 8.2). The connection fails to satisfy the objective if any one of the requirements is not met.

Table 8.2. End-to-end error performance objectives for a 64 kb/s connection

Performance Classification	Objective
Degraded minutes	Fewer than 10% of one-minute intervals to have a bit error rate worse than 10E-6
Severely errored seconds	Fewer than 0.2% of one-second intervals to have a bit error rate worse than 10E-3
Errored seconds	Fewer than 8% of one-second intervals to have any errors (equivalent to 92% error-free second)

Three distinct quality classifications have been identified representative of practical digital transmission circuits. They are independent of the transmission systems used. These classifications are termed local grade, medium grade and high grade and their usage generally tends to be dependent on their location within the network.

With reference to Fig. 8.10 the allocation of the objectives for the three circuit classifications are given in Table 8.3 in accordance with CCITT Rec. G. 821.

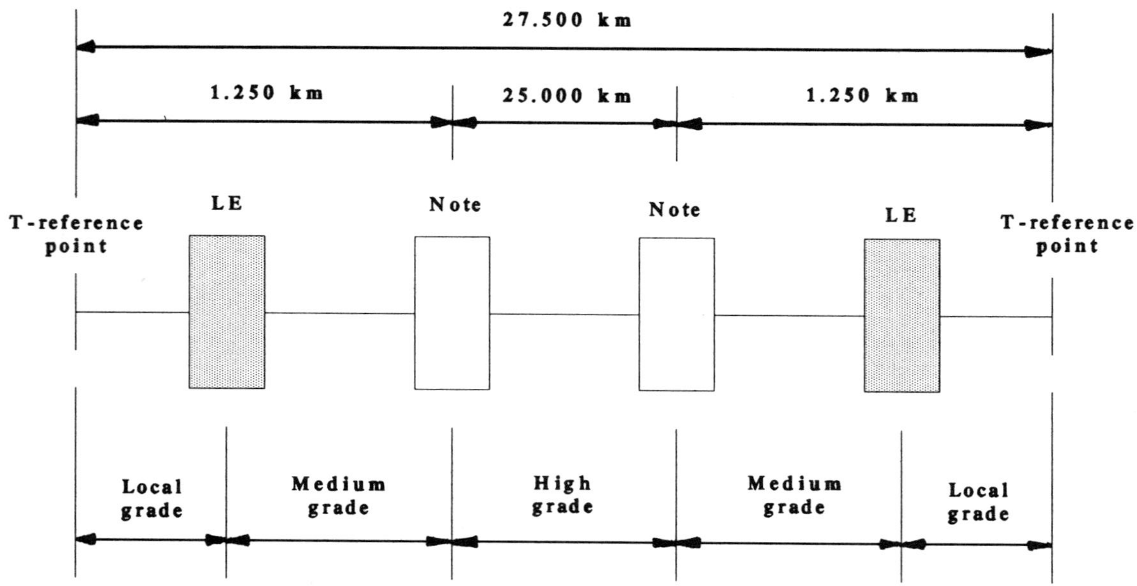

Fig. 8.10 Circuit quality demarcation of longest HRX

Table 8.3 Allocation of objectives with circuit classification

Circuit Classification	Allocation of degraded minutes and errored seconds	Allocation of severely errored seconds
Local Grade (2 ends)	15% block allowance to each end (Notes 1,4,5)	0.015% block allowance to each end (Note 5)
Medium Grade (2 ends)	15% block allowance to each end (Notes 2,4,5)	0.015% block allowance to each end (Note 5)
High Grade	40% (Notes 3,6,7)	0.04%

Note 1- The local grade apportionment is considered to be a block allowance, i.e. an allowance to that of the connection regardless of length.

Note 2- The medium grade apportionment is considered to be a block allowance, i.e. an allowance to that part of the connection regardless of length. The actual length covered by the medium grade part of the connection will vary considerably from one country to another. Transmission systems in this classification exhibit a variation in quality falling between the other classifications.

Note 3- The high grade apportionment is divided on the basis of length resulting in a conceptual per kilometre allocation which can be used to drive a block allowance for a defined network model (e.g. Hypothetical Reference Digital Link).

Note 4- The local grade and medium grade portions are permitted to cover up the first 1250 km of the circuit from the T-reference point extending into the network.

Note 5- Administrations may allocate the block allowances for the local and medium grade portions of the connection as necessary within the total allowance of 30% for any one end of the connection.

Note 6- Based on the understanding that satellite error performance is largely independent of distance, a block allowance of 20% of the permitted degraded minutes and errored second objectives is allocated to a single satellite HRDP employed in the high-grade portion of the HRX.

Note 7- If the high-grade portion of a connection includes a satellite system and the remaining distance included in this category exceeds 12,500 km or if the high-grade portion of a non-satellite connection exceeds 25,000 km, then the objectives of this Recommendation may be exceeded. The occurrence of such connections is thought to be relatively rare and studies are continuing in order to investigate this. The concept of satellite equivalent distance (the length of an equivalent terrestrial path) is useful in this respect and it has been noted that a value in the range 10,000 to 13,000 km might be expected.

Based on the fact that satellite error performance is largely independent of distance, a block allowance of 20% of the permitted degraded minutes and errored second objectives is allocated to a single satellite HRDL employed in the high grade portion of the HRX. For a satellite HRDL operating in the high grade portion there is a block allowance of 0.02% severely errored seconds.

8.6.3.2 Availability

A connection is considered unavailable when the signal is interrupted at least in one direction (i. e. alignment or timing lost) or when the bit error rate is worse than 10E-3 for 10 consecutive seconds. The 10 consecutive seconds condition is implied in order to seperate availability objective from severely errored seconds objective.

This distinction is necessary because the performance of a system is calculated from propagation parameters whereas unavailability is the result of factors such as equipment failure, propagation, interference, support facilities, facilities and human activity.

Availability objective for a hypothetical reference digital path of 2500 km is given as 99.7 % of the time in CCIR Rec. 557-1. This value is a provisional one and it is recognized that in practice, the objective may fall into the range of 99.5 to 99.9. The choice of a specific value in this range depends on the optimum allocation of outage time among the various causes which may not be the same when local conditions are taken into account.

Satellite link availability is given in CCIR Rec. 579-1 as follows:

- The unavailability of a hypothetical reference circuit or digital path in the fixed-satellite service due to equipment should provisionally be not more than 0.2% of a year,
- The unavailability due to propagation should not be more than

  0.2% of any month, and

  0.1% of any year.

8.6.4 Performance objectives for WIS

8.6.4.1 Quality of WIS Network

In the WIS network all the internodal transmission and main access links must be considered as high grade portion of the network. This is because, if some portions of the network is defined as medium or local grade with x % block allowance, there will be a possibility that a message passes through more than one "Medium Grade Section" in damaged conditions (when the routing is changed), and the overall performance objective will then not be met. Most PTT's also adopt the same approach and requires all R/L links to be installed as high grade. Therefore, the performance objective for any real digital link of length L is calculated using the general expression for HRX performance:

$$Degraded minutes < 0.1 \times \frac{0.4}{2500} L$$

$$Errored\ seconds < \frac{0.008 \times 0.4}{2500} L$$

The severely errored seconds, following the apportionment strategy given in section 3.3 of CCITT Rec. G. 821 can be calculated as follows:

$$Severely\ errored\ seconds \left[\frac{0.0004}{2500} + \frac{0.0005}{2500}\right] \times L$$

It should be noted that the above results are given in decimal form and not in percentage. Comparing these results with CCIR Rec. 634 (Error performance objectives for real digital radio-relay links forming part of a high-grade circuit within an integrated services digital network), it is seen that they are exactly the same.

For the WIS HRDS (3400 km) where the transmission media consist of radio link system (LOS), the CCIR Recommendation 594 would apply. The following requirements apply for each direction of a 64 kbit/s digital reference circuit. Fading, interference and other sources of degradation are all included in the figures below. Therefore, for the WIS HRDS of 3400 km, the Bit Error Rate (BER) shall not exceed the limits given below:

a) Degraded Minutes

BER shall not exceed 10^{-6} for more than

$$\frac{0.4 \times 10\ \%}{2500\ km} \times 3400\ km = 0.544\ \%$$

of time in any month, where the integration time for error counting is one minute.

b) Errored Seconds

Total number of seconds with error shall not exceed:

$$\frac{0.4 \times 8\ \%}{25000\ km} \times 3400\ km = 0.4352\ \%$$

of time in any month.

c) Severely Errored Seconds

BER shall not exceed 10^{-3} for more than

$$\frac{0.04\ \%\times 3400\ km}{25000\ km}+\frac{0.05\ \%\times 3400\ km}{2500\ km}=0.07344\ \%$$

of time in any month (integration time = one second).

8.6.4.2 Availability of WIS Network

The maximum HRDL length for WIS is assumed to be 3400 km. (without any tropo or satellite communication links). To determine the 3400 km. segment unavailability U, (U=1-A), a proportionality is established as follows:

3400 km segment unavailability (U)

$$U=\frac{3400\ km}{2500\ km}\times U\ (CCIR)$$

where U (CCIR) = 0.003, then

$$U=\frac{3400\ km}{2500\ km}\ (0.003)=0.00408$$

Hence, 3400 km segment availability (A), then

A = 1 - U,
A = 1 - 0.00408 = 0.9959,
A = 99.59%.

8.6.5 Design of a Digital Radio Link

The main propagation effect influencing the reception of a microwave signal and overall system reliability is fading which shall be considered below. The jamming threat has not been taken into consideration as a design parameter in the study at this stage (See Section 8.2.2).

8.6.5.1 Fading

Random variations in the refractive index of the atmosphere are the primary cause of fading on LOS radio links. It is generally agreed that, with adequate path clearance and in the absence of a single specular reflection on a path, deep fades are due to multipath propagation through the atmosphere (flat or deep fading). Amplitude and group delay distortions across the channel bandwidth can also occur because of the multipath nature of the transmission medium and this gives rise to selective fading.

a) Flat Fading

The analysis of LOS radio path has shown that the fading depth exceeded for a small percentage of time is a function of the following parameters:

- path length
- frequency
- climate
- nature of terrain
- transmission bandwidth
- modulation techniques used.

In narrow-band systems, for large fade depths, the average worst-month fading time, $P_a(W)$, can be approximated by the asymptotic equation (CCIR Rep.338)

$$P_a\,(W) = K\,Q\,(W/W_o)\,f^B\,d^C \qquad \text{(Eq. 8.6.1)}$$

where,

d : path length (km.)
f : frequency (GHz)
K : a factor for climatic conditions
Q : a factor for terrain conditions
W : received power (W)
W_o: received power in non fading conditions (W)

and the factor W/W_o can be expressed as

$$\frac{W}{W_o} = 10^{-M/10} \qquad \text{(Eq. 8.6.2)}$$

where M (dB) is the fade margin defined above.

Above equation can be considered valid for fade depths 15 dB or the value exceeded for 0.1% of the worst month, whichever is greater. The range of validity in path length is 15-100 km and in frequency range 2-37 GHz.

It is not possible yet to give a general set of rules for the parameters K, Q, B, and C in the above equations.

The parameter B lies in the range 1.0 - 1.5. For Northern and N.W. Europe and for America, B = 1.0. However for Japan it is 1.2 and for U.K. B = 0.85. The parameter C on the other hand is 3.0 for Northern Europe and America. Assuming B = 1.0, C = 3.0 for the country in our case the following formula can be derived to calculate the outage time, F_a (F), due to flat fading

$$P_a(F) = P_o 10^{-M/10} \qquad \text{(Eq. 8.6.3)}$$

where the parameter P_o "fading occurrence factor" is defined as

$$P_o = 0.3\ a\ b\ (f/4)\ (d/50)^3 \qquad \text{(Eq. 8.6.4)}$$

In this expression;

a = normalized climate coefficent (0.25 < a < 4)
a = 0.25 in dry or mountain climates
a = 1 for temperate climates
a = 4 damp climates or strong thermal inversion
b = normalized roughness coefficient (0.25 < b < 4)

For rolling terrain with a roughness r in the range of 5 to 100 m, the following formula can be used:

$$b = (r/15)^{-1/3} \qquad \text{(Eq. 8.6.5)}$$

b) Selective Fading

A notch in the hop transfer function occurs due to multipath reflections from atmospheric layers of different refractive index and the effect is called selective fading which causes signal distortion. Selective fading is considered only for digital radio links and not for analog LOS links because its effect on the latter can often be negligibly small.

Outage time due to selective fading, $P_s(F)$, is computed according to the following empricial formula:

$$P_s\,(F) = n\ 4.3\ AB\ (T/T_s)^2 \qquad \text{(Eq. 8.6.6)}$$

where,

n = The fraction of time during which fading is active.

$$n = (P_0/\ln 2)\ [1 + 10^{9/5}\ P^{0-6} + 10^{0-31a}\ P^2]^{-1/2} \qquad \text{(Eq. 8.6.7)}$$

A,B = Modulation dependent coefficients
T_s = Symbol time
T = Mean delay of the echo.

$$T = 0.7\ (d/50)^{1,5}\ nsec \qquad \text{(Eq. 8.6.8)}$$

The improvement given by baseband IF equalizers and by spectrum shaping is considered by changing the coefficients AB of the signature of the demodulator. This shows that selective fading value is related to the digital modulation technique used in the radio system.

Then the total outage time due to fading, P_T (F), can be found as the sum of the outage times due to flat fading and selective fading:

$$P_T(F) = P_a(F) + P_s(F) \quad \text{(Eq. 8.6.9)}$$

In order to reduce interruptions (unavailability of system) due to multipath fading, diversity techniques are employed as outlined below:

8.6.5.2 Diversity Improvement

Multipath fading may be made less by the use of diversity techniques such as frequency and space diversity. In radio systems protected with a stand-by bearer working at a different RF "frequency diversity" technique is employed. If two different antennas are used at the receiving site with vertical separation (the suggested vertical spacing is about 150-200 times the wavelength) then "space diversity" technique is used to enhance the performance of the system.

In most PTT systems, all radio-link systems operate with (1+1) frequency diversity technique, to mitigate against the effects of multipath fading. The links in the WIS radio system must also be protected against fading using frequency diversity. The use of space diversity must be examined for each link and used where necessary (taking into account fade level and frequency space availability) remembering that this technique would be more expensive than the frequency diversity technique.

Outage time due to fading in a diversity system is equal to the compound fading probability relevant to two (or more) RF diversity channel and can be determined from:

$$p' = p_T/m \quad \text{(Eq. 8.6.10)}$$

where,

p' = Probability of reaching BER = 10E-6 (or 10E-3) after diversity improvement.
p_T = Probability of reaching BER = 10E-6 (or 10E-3) in one diversity channel.
m = Uncorrelation coefficient, $m = n\,(1\text{-}K^2)$.
n = Fraction of the unit of the during which fading is active.
K = Correletion coefficient between channels.

The following empirical formulas are used to compute the correlation coefficient (K_f^2 and K_s^2):

For frequency diversity:

$$K_f^2 = e^{-2.5\, f_s/f_c} \quad \text{(Eq. 8.6.11)}$$

where,
f_s = Frequency spacing,
f_c = Carrier frequency.

For Space diversity:

$$K_s^2 = e^{-4\times 10^{-6}(s/\lambda)^2} \qquad \text{(Eq. 8.6.12)}$$

where,
s = Antenna spacing
λ = Wavelength

and

$$K^2 = K_s^2 K_f^2 \qquad \text{(Eq. 8.6.13)}$$

8.6.5.3 A Case Study

In this section a case study is given to demonstrate the application of the previous formulations.

a) System Model and Assumptions

In order to calculate the performance of digital radio links, it is required to make some assumptions as given below:

Tx Power (P)	26 dBm (400 mW)
Rx Noise Figure	5.5 dB
Rx Threshold Level (BER 10E-6)	-81 dBm
Rx Threshold Level (BER = 10E-3)	-84 dBm
Frequency (f)	7575 MHz
Frequency diversity	56 MHz
Telephone capacity	480 channels
Maximum allowed bandwidth	20 MHz
Climatic coefficient	1.5
Roughness coefficient	0.1
Radio hop length (d)	50 km
Space diversity	None
Antenna (A) diameter (D)	1.80 m
Antenna (B) diameter	1.80 m
Feeder (A) length	70 m
Feeder (B) length	35 m
Feeder (A) attenuation (0.047 dB/m)	3.29 dB
Feeder (B) attenuation (0.047 dB/m)	1.64 dB
Other attenuations	1 dB
Branching (A + B) attenuation	0.9 dB
Tolerance	1 dB

b) Outage Time Calculations

i) Antenna gain (Parabola)

$$G = 20\ \log f\,(GHz) + 20\ \log D\,(m) + 17.8$$

for a radome loss of 1.3 dB,

$$G = 39.2\ dB$$

ii) Free Space Loss (FSL):

$$FSL = 92.4 + 20\ \log f\,(GHz) + 20\ \log d\,(km)$$
$$FSL = 144\ dB$$

iii) Total Losses:

free space loss	: 144.0 dB
feeder att (A)	: 3.3 dB
feeder att (B)	: 1.6 dB
Branching (A + B) att	: 0.9 dB
Other attenuation	: 1.0 dB
Tolerance	: 1.0 dB
+	151.8 dB

iv) Total gains:

antenna gain A	: 39.2 dBm
antenna gain B	: 39.2 dBm
+	78.4 dBm

v) Received input level without fading:

The receiver input level in no fading condition is given by:

$$RL = P - Total\ losses + Total\ gains$$
$$RL = -\ 47.4\ dBm$$

vi) Fading occurance factor, P_o :

$$P_0 = 0.3\ ab(f/4)\ (d/50)^3$$

Taking a = 1.5, b = 0.1,

$$P_0 = 0.085$$

vii) Fade margin for flat Fading (BER = 10^{-6}):

It is the difference between the receiver input level and receiver threshold level for BER = 10^{-6}, and found to be 81 - 47.4 = 33.6 dB.

viii) Fade Margin for Flat Fading (BER = 10^{-3}):

Similary, the fade margin for flat fading for BER=10^{-3} is, 84-47.4=36.6 dB.

ix) Time Percentage Relative to Thresholds:

Probability of reaching the threshold level corresponding to a given BER, in a channel without any diversity protection:

$$P(F) = P_0 \ 10^{-M/10}$$

where M is the fade margin calculated in (vii) and (viii)

a) P(F) = 0.0037 for BER 10E-6
b) P(F) = 0.0018 for BER 10E-3

x) Outage time due to selective fading is computed using (Eq.8.6.6)

$$P_s(f) = 0.0016$$

The A, B constants which depend on the demodulator design and modulation technique used assuming % 0.0016 as the outage time due to selective fading, outage time for the hop is found to be:

$$P_h(f) = P_a(F) + P_s(F)$$
$$P_h(f) = 0.0018 + 0.0016 = 0.0034$$

xi) Before calculation of the diversity emprovement in tme percentage, the outage times for BER 10E-6 and 10E-3 must be checked against CCIR limits. If outage time (without diversity improvement) is below the CCIR limits,there is then no need to use diversity techniques.

According to hypothetical reference connection for max. length in WIS (National) of 3400 km and with an average hop length of 50 km the number of hop is 3400/50 = 68. Outage time for one hop is calculated as

$$P(f) = 0.0034 \, for \; BER \; 10E{-}3$$

Overall outage time $P_o(f) = (0.0034)\,(68) = 0.23$.

From Section 8.6.4.1(c),

$$P_{WIS} > 0.0734$$

Since $P_0(f) > P_{WIS}$, diversity must be used, in order to reduce outage time.

c) Improvement Factor Calculations

xii) Diversity improvement factor:

The diversity improvement factor is calculated by using the equations Eq. 8.6.10 - Eq. 8.6.13

$$P' = P_T^2/m$$

with $K_f = 0.981$ and $K_s = 1$ (no space diversity),

$$n = (P_0/ln2)[1 + 10^{9/5}\, p^{0.6} + 10^{0.318}\, p^2]^{-1/2} = 0.03$$

and

$$m = 0.00057$$

xiii) Total outage time with diversity improvement:

$P_h(f) = 0.0034\%$
p' : outage time with diversity improved,
p' : $P_h(f)^2/m = (0.000034)^2/0.00057 = 0.000002$
p' % : 0.0002

As it is shown, the outage time without diversity (0.0034%) reduces to 0.0002 % after the frequency diversity. If this result is applied to a HRDP of 3400 km in WIS under damaged conditions the total outage time way be calculated with 68 hops:

One hop outage time = 0.0002
Total outage time = 0.0002 x 68 = 0.00136 for BER 10E-3

As per to the previously obtained values, this limit is 0.0734 and the objective is achieved; i.e. the the total outage is less than CCIR limits.

8.7 Frequency Allocation and Planning

Of all the transmission parameters discussed so far, choosing a frequency band by virtue of which the communation services are to be carried out is utmost important and needs a detailed analysis. The design of a digital radio link system, therefore, must include a frequency allocation plan. This plan however, is subject to some criteria which must be taken into consideration.

The basic criteria affecting the determination of a frequency allocation plan are outlined below:

a) Transmission capacity,
b) Interference,
c) Tower heights,
d) Hop lengths,
e) Network size,
f) Diversity,
g) Anomalous propagation,
h) Cost.

The long-haul backbone radio-relay subsystem of WIS is based on digital links with a transmission rate of 34 Mbit/s. The transmission capacity is therefore limited to 480 telephone channels. It is also pre-determined that a (1 + 1) frequency diversity is adopted, and the height of the towers are to be as low as possible.

Another fact about the radio-relay system of WIS is the parallel operation and co-location of sites with that of the public system. Hence, the frequency plan being utilized by the public network operator should be taken into account with regards to interference and channel occupation.

RF interference is an important consideration which may occur from sources internal or external to the radio system. The interference sources in the area of each radio link, including their frequency, power and directivity are to be studied. The effect of RF interference on a radio system depends on the level of the interfering signal and whether the interference is in an adjacent channel or in a cochannel. A cochannel interferer has the some nominal radio frequency as that of the desired channel. Cochannel interference arises from multiple use of the same frequency without proper isolation between links. Adjacent channel interference results from the overlapping components of the transmitted spectrum in adjacent channels. Protection against this type of interference requires control of the transmitted spectrum, proper filtering within the receiver, and orthogonal polarization of adjacent channels.

Another factor to be taken into account in selecting the frequency band to be used is the probability of interception of the transmission by a potential adversary. This generally improves with increasing frequency.

The available RF bands and their present usage are given in Table 8.4 below:

Table 8.4 Present channel allocations

Frequency Range (MHz)	Comments
1400-1530	Used by PTT although reserved for military applications
1700-1903	" " " "
1900-2095	" " " "
2108-2300	" " " "
2491-2681	Available
3400-3600	" "
3600-4200	Used by PTT. This band together with 3400-3600 MHz band may however be shared.
4400-5000	This band is allocated for fixed and mobile military services.
5850-6405	Used by PTT's
6405-7125	Used by PTT's
6725-7050	At present, reserved for fixed satellite systems
7125-7475	Used by PTT's
7425-7725	Used by PTT's
7125-8200	Used by PTT's and SATCOM services
7260-8115	Used for military satellite systems.
8115-8400	Reserved for satellite services
8400-8500	Available
10700-11700	Available
11700-12500	Reserved for direct broadcasting satellite services
12750-13250	Reserved for PTT's
14050-15350	Available

The frequency bands below 10 GHz would, in general, be more suitable for the long-haul radio links because:

1) RF bands above 10 GHz are much more affected by rain attenuation. However, in a switched network this may not be too objectionable. These frequencies can, of course, be used for shorter paths (access links).

2) Extensive use of below 10 GHz bands ensures a wide variety of equipment suppliers and an up-to date technology.

Notwithstanding the above the maximum hop lengths for some of the above frequency bands have been calculated for the same transmission quality and availability as given in Section 8.6.4.1 and are given in Table 8.5 assuming:

- Transmitter output power	: 30	dBm
- Antenna diameter	: 3	m
- Receiver threshold level	:-78 to - 81dSm	
- Total Feeder Lnegth	: 10	m
- Barnching Loss	: 0.9	dB
- Tolerance and other losses	: 2	dB

Table 8.5 Max. hop lengths with respect to mose frequencies

Frequencies (GHz)	Max. Hop length (km)
2	110
4.5	100
7.4	90
8.3	86
11	76
13	70
15	60

The length and number of hops in a realistic WIS internodal network are given below:

Hop Length Between (km)	No of hops
0-10	5
11-20	24
21-30	28
31-40	48
41-50	35
51-60	39
61-70	32
71-80	15
81-90	7
91-100	2
101-110	3
111-120	1
Total	239

It can be seen from above that 179 hops are shorter than 60 km, for which any of the frequency bands shown in Table 8.6 may be used.

Table 8.5 and annex-A show that there is only a difference of about 3 dB in the Net Fade Margins between 2 and 15 GHz links. This difference, however, can always be compensated for with the system design by increasing slightly the antenna size or the transmitter output power.

The frequency bands listed in Table 8.6 should therefore be considered for use in WIS radio-relay subsystems:

Table 8.6 Candidate frequency bands for WIS

RF Band (MHz)	COMMENTS
2491-2681	Lower frequencies of this band are often used by PTT's and there is a limited application of this band because of its narrow bantwidth (200 MHz). Furthermore, this band is susceptible to relatively easy hostile interception.
3400-4200	This band should be shared with PTT's. Bandwidth is about 800 MHz.
4400-5000 (Upper 4 GHz band)	Increased bandwidth, used also for tactical systems. Hence interference eventually produced by tactical systems will have to be taken into account throughout the band.
7125-8200	This band can be used successfully for the internodal links but will have to be shared with PTT and SATCOM systems.
10700-11700	More suitable for short haul connections.
12750-13250	Coordinated use of this bandwith PTT is possible.
14050-15350	Suitable for short haul connections.

Radio frequencies to be used in WIS can be grouped in two or more frequency bands for reasons of availability and effectiveness. The most robust band should be assigned to long-haul links and the others should be used for access or short-haul links. This implies that two types of radio equipment will be used.

Upper 4 GHz and 7 GHz bands have been shown to be more suitable for long hops and should therefore be used generally for the long links of the WIS nodal network wherever possible. Consideration should be given to the use of equipment with frequency agility (e.g. equipment capable of being tuned in both, say, 4 and 7 GHz bands) for countering ECM. 10 GHz band and above should be considered for use in access links. The final decision on the choice of frequencies to be used for the hops of the internodal network must, of course, be a compromise between frequency availability and suitability taking into account factors such as path profile, climatic conditions, ECM and cost effectiveness.

8.8 PROCUREMENT ISSUES

The WIS Transmission Subsystem Concept is defined with the aim utilizing existing available equipment without any significant modifications required. this is driven mainly by the requirement to reduce cost and technological risk. In the preparation of detailed specifications and conduct of procurement processes for WTS these objectives must be kept in mind so that the desired cost optimization may be realized. Main elements of this strategy must therefore be:

a) Fully standardized equipment wherever possible,
b) Full compatibility between the different phases as well as within each phase,
c) Multivendor competitive procurement consistent with the Authority's technological objectives,
d) Carefull planning of alarm, maintenance and spare parts systems.

8.9 REFERENCES

[1] CCITT, Volume III, Redbook, Oct.1984.

[2] Ince A. N. et al "Digital Satellite Communication Systems and Technologies", Kluwer Academic Publishers, Boston, 1992.

[3] Ince A. N. et al "Digital Speech Processing", Kluwer Academic Publishers, Boston, 1992.

[4] Ince A. N. "Design, Testing and Operation of X-Band Satellite Communications Systems", IEEE Trans. on Communications, Vol. COM-22, No.9 September 1974.

[5] Ince A. N. "Design Studies for Reliable Long-Range Ground-to-Air Communication", IEEE Tran. Com. Technology, Vol.15, Oct.1967.

ANNEX-8A

CALCULATION OF THE MAXIMUM HOP LENGHTS

To calculate the maximum hop lengths for various frequency bands given in Section 8.7, the following assumptions are made for a sample digital link operating at 34 Mbit/s:

Transmitter output power	: 30	dBm
Antenna diameter	: 3	m
Receiver threshold level for BER = 10E-3	: - 81	dBm
Receiver threshold level for BER = 10E-6	: - 78	dBm
Total feeder length	: 10	m
Branching loss	: 0.9	dB
Torelance and other losses	: 2	dB

The hop length is adjusted for each frequency until the availability and qualitiy criteria given in Section 8.6.4 are satisfied for the system parameters given above. The hop lengths for which the system availability and quality are just satisfied are taken as the maximum hop length at that frequency. The frequency dependent losses are taken into account for each case and the rain attenuation is calculated in accordance with CCIR Rep. 338-5 for a rainfall intensity exceeded 0.1 % of time.

Details of these calculations are given in Table 8.A.1 for the two extreme frequencies of 2 and 15 GHz ad for a path length of 100 kms. This table shows that there is only about 3 dB difference between the two cases.

Considering that the rain attenuation assumed is unlikely to occur on more than one link at a time the difference calculated may mot be too significant for a fully switched network with adaptive routing feature.

Table 8A.1 System Calculations for 2 GHz and 15 GHz

Frequency	2 GHz	15 GHz	
1. Transmitter Output Power	30	30	dBm
2. Hop length	100	100	km
3. Total Antenna Gain	66.7	101.7	dB
4. Total Feeder Loss	0.28	1.6	dB
5. Torelance and other losses	2.9	2.9	dB
6. Rain Attenuation	0.07	21.3	dB
7. Free Space Loss	138.5	156	dB
8. Net Loss [(4+5+6+7)-3]	75	80.1	dB
9. Receiver input level [1-8]	- 45	- 50.1	dBm
10.Receiver Threshold (BER = 10E-3)	- 81	- 81	dBm
11.Receiver Threshold (BER = 10E-6)	- 8	- 78	dBm
12.Flat Fade Margin (BER = 10E-3) [9-10]	36	30.9	dB
13.Flat Fade Margin Redyction (BER = 10E-6) [9-10]	33	27.9	dB
14.Selective Fade Redyction (BER=10E-3)	3.5	1.1	dB
15.Selective Fade Redyction (BER=10E-6)	2	0.4	dB
16.Net Fade Margin (BER=10E-3) [12-14]	32.5	29.8	dB
17.Net Fade Margin (BER=10E-6) [13-15]	31	27.5	dB

CHAPTER 9

TERMINAL AREA SUBSYSTEM

The objectives of this chapter are to describe the requirements baseline that was established for the Terminal Area Subsystem (TAS) of WIS, to describe and define generically its elements and features, to discuss architectural options for WIS users towards recommend solutions. Recall that WIS will be a digital network, largely conforming to CCITT ISDN recommendations and following the ISO's Open System Interconnection Reference Model (ISO/RM).

It should not be surprising if a survey of present terminal area subsystems in most existing networks would reveal the following criticisms:

1) Inadequate Automatic Data Processing (ADP) support for information processing, for fusion and decision making and slow coordination.

2) Inadequate security (traffic flow, interception, spoofing, info protection), limited capability for key compartmentation, vulnerable key distribution.

3) Lack of survivability in that, facilities may not always be adequately protected against conventional threat, chemical/biological effects, EM, electronic warfare. Furthermore there is lack of diversity and redundancy, and there are few provisions for auto back-up and restoral in information handling systems.

4) Inadequate capacity and connectivity, limited availability, growth capability, lack of integration of major subsystems, limited communication service for mobile and itinerant users.

5) Limited interoperability between networks, limited interworking between devices, use of different standards nationally or for different functional areas.

Thus, there is an urgency in planning and coordinating the terminal area subsystems for increased interoperability, efficiency, security and survivability on the one hand, and for evolution of services toward integrated terminals, higher functionality and quality on the other hand.

The Terminal Area Subsystems take place in user locations. The user locations are typically command centers of strategic headquarters, which are the nucleus of Command, Control and Information Systems (CCIS). Information in the form of intelligence, reconnaissance, status of friendly and enemy forces, weather and logistics flow into them. Decisions concerning deployment of units, weapons utilization and additional intelligence collection requests flow from them. While voice communications, secure and non-secure, comprise a great portion of the incoming and outgoing flow of information, long-term strategy considerations are based upon authenticated formal, recurring and non-recurring messages.

9.1 TYPICAL USER AND EXISTING SYSTEMS

9.1.1 Typical HQs Structure

A typical HQs, from functional and administrative viewpoints, consists of:

- The Commander and his staff,
- Divisions (Personnel, Intelligence, Operations, Logistics, Communications),
- Sections within each Division,
- Branches within each Section,
- Message control center,
- Communications center (comcenter).

Each organizational unit within a HQs is established under specific terms of reference and has well defined responsibilities. It is these responsibilities that, when translated into command and control (C^2) functions and processes, establish types and means of communications.

9.1.2 CCIS Functions And Processes

Command, Control and Information System (CCIS) functions are those information processing activities of military nature which produce specific results or direct benefit to the commander or the HQs staff. The sets of CCIS functions contributing to the same type of operational goals constitute functional areas. The CCIS functions are directed to:

1) Status maintenance,
2) Situation assessment,
3) Planning and
4) Execution and coordination.

The following classes of functional areas could be considered for a typical HQs:

- Resources and operation monitoring,
- Situation monitoring,
- Situation and operation analysis,
- Decision making at operational control level,
- Operation control (alert status/execution),
- Plan generation and modification,
- Briefing preparation.

The interaction among these functional areas is shown in Fig 9.1. The processes which occur amongst these functional areas generate both intra-HQs and inter HQs information flow. These processes require secure, timely, accurate and reliable information on the status of friendly and enemy forces, resources and environment, and provide input to the commander's decision.

9.1.3 Message Handling Procedures

A detailed review of current message handling procedures is given in the sequel. This is intended on one hand to describe somewhat critically a typical example of what could be the present status. On the other hand it will be of guidance to establish the message requirements baseline and will be indicative of how WIS implementation should automise these manual routines.

The progress of a message from the originator to the recipient can be divided into three areas:

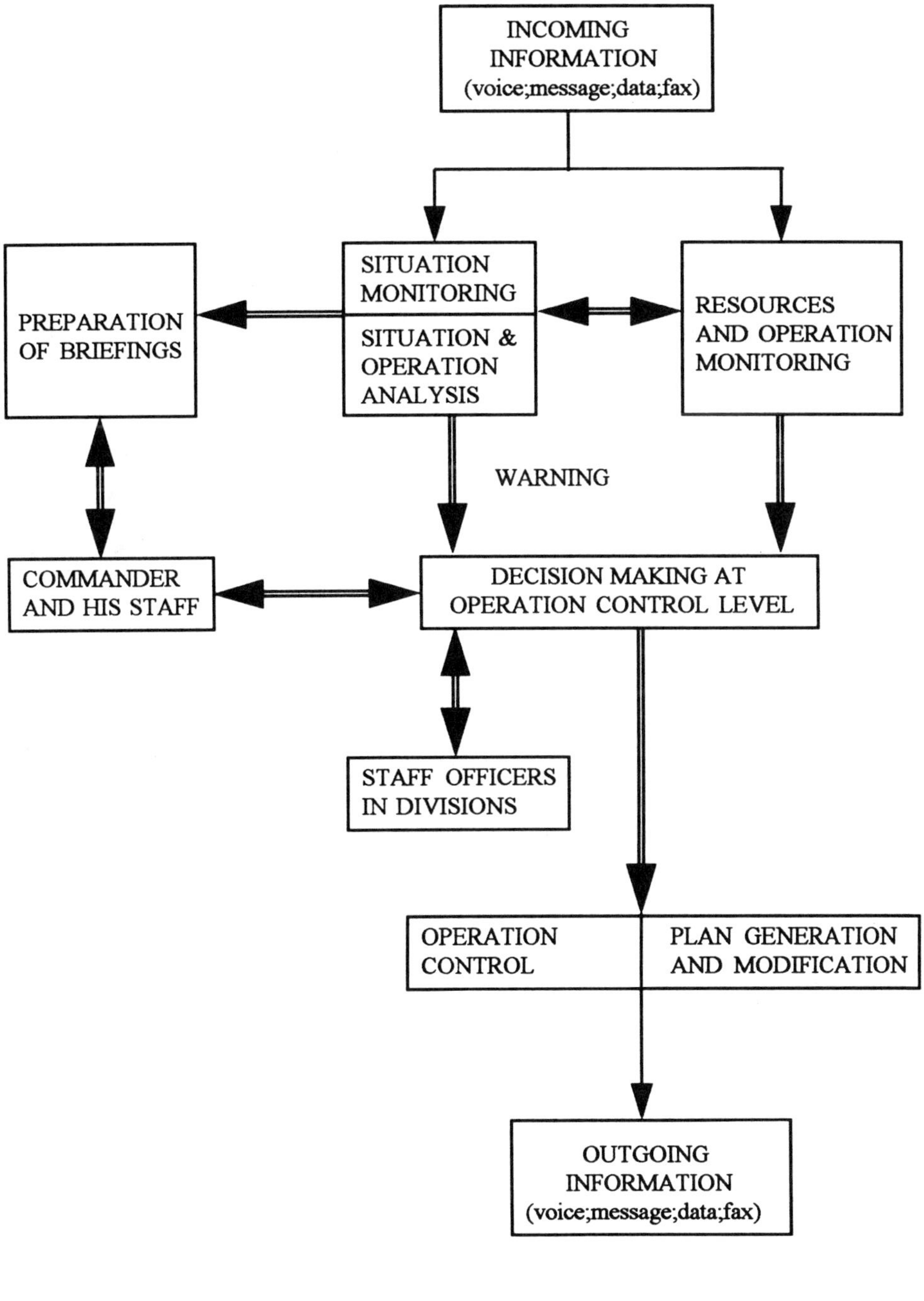

Figure 9.1 A Generic Model of a HQs CCIS Functions

- Message preparation at the originating HQs and entry into the communication systems,
- Message transmission, including relaying, between HQs,
- Message distribution at the receiving HQs.

9.1.3.1 Outgoing Message Handling

Outgoing message handling includes processes such as preparation, release, logging, filing and transmission. Each process may involve several office functions. These functions could be categorized as being clerical, monitoring and tracking or decision making. These outgoing message handling processes and allocation of their functions in the hierarchical structure of a headquarters could best be described in a matrix form. Fig 9.2 depicts the flow of outgoing messages through the HQs hierarchy. At present, all these functions are performed manually by the staff, and furthermore the HQs community is served by manually operated message control and communications centers. This, together with other attributes such as repetitive logging, filing and the use of messengers, result in introducing unnecessary delays in the intra-HQs message handling periods.

Message Preparation: Message preparation is the responsibility of the originating division/section/branch. It includes the functions of drafting, typing, proofread, coordination, approval for submission to the releasing authority, logging, and filing. Message preparation is initiated by the project officer who drafts and types it on approved forms. The draft message, after being proofread, is taken to the other divisions for coordination, if required, and to the coordinating officers for operational and administrative purposes. After the approval of the section chief and of the division head, the message is logged in the originating division, it is converted into established message format, i.e., the ACP 127 NATO Supp-3 format [1] prior to being released by the authority.

Releasing: It is usually the commander, or another officer designated by him, who has the authority to release messages. Once the message is signed by the releasing officer, it is delivered to the Message Registration and Control Center.

Message Registration and Internal Distribution: Message, with the proper releasing officer's signature, are assigned a unique serial number and a Date-Time-Group (DTG) and are logged in the Registration Unit. Usually the time of release becomes the Date-Time-Group (DTG) of the message. Based on the Subject Indicator Codes (SIC's) and any other special instructions given by the message originator, outgoing messages are allocated an internal distribution. The required number of copies are reproduced and delivered to the internal addressees and to the Message Control Unit.

Preparation of Messages for Transmission: It is the responsibility of the Message Control Unit to prepare the messages in a form suitable for electrical transmission. The activities include:

1) Assignment of Message Control Unit serial number (station serial number) together with Time of File,
2) Allocation of Routing Indicators,
3) Addition of those information elements required to compose ACP 127 NATO Supp-3 envelope,
4) Off-line encryption, if required,
5) Preparation and checking of paper tapes,
6) Logging.

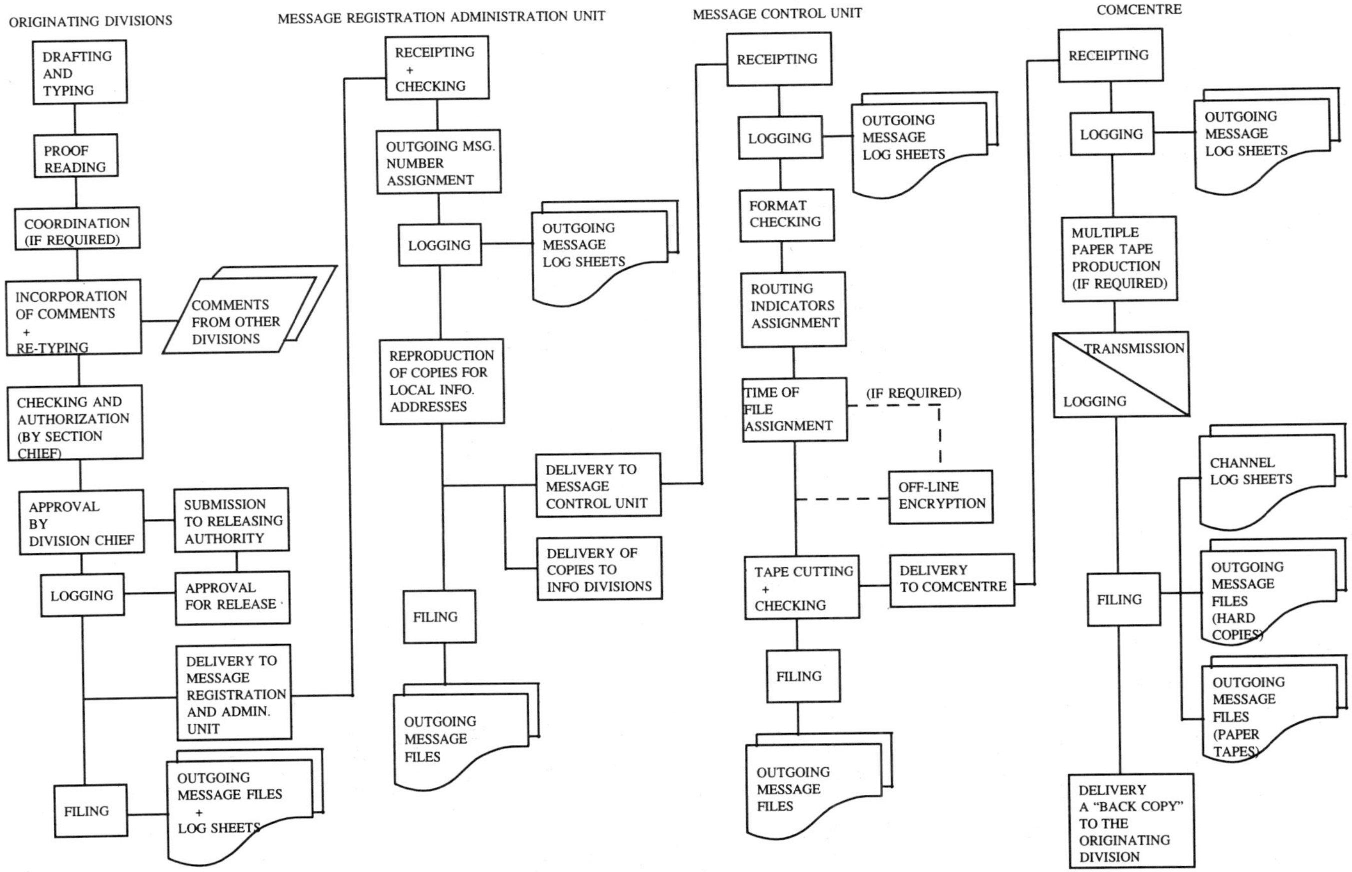

Figure 9.2 Current Outgoing Message Preparation Process

Transmission of Messages: The complete message in the paper tape form is delivered to the comcenter for transmission. The message is logged on the Channel Control Sheets and transmitted via the appropriate communications ports. The order of transmission is first determined by its precedence level and then within each precedence level on a first-in-fist-out basis. A copy of transmitted message is sent back to the originator (project officer) for information. All paper tapes are kept for a 24 hours period to meet retransmission requests.

Logging: Each message from its drafting "stage" to its "transmission" is logged in at different organizational units. For this purpose, in each organizational unit, daily log sheets are maintained. They also serve as a tool to trace and retrieve messages. hese log sheets are stored for historical purposes for a period of 30 days.

Filing: Hard copies of outgoing messages are filed at several organizational units within a HQs to facilitate retrieval of messages and to provide physical security until they are finally destroyed.

9.1.3.2 Incoming Message Handling

Messages arrive at the HQs either via courier services or via communications systems. Incoming message handling includes processes such as reception, allocation of internal distribution, delivery to internal addressees, receipting, logging and filing. Each process may involve several office functions; some of which are routine and clerical, and others may require some decision-making.

At present, all these functions are performed manually and the HQs community is served by manually operated message control and communications centers. This, together with other repetitive time-consuming office functions such as reproduction of copies, delivery, receipting, logging and filing, result in introducing unnecessary delays in delivering messages from the comcenter to the staff officers' desks. The flow of incoming messages within the HQs is depicted in Fig 9.3.

Reception: Upon arrival of a message from the communications systems in the comcenter, the operator performs the following functions:

1) Receipting (i.e., signing in acknowledgement of receipt), if required,
2) Validation of channel identifier and serial number,
3) Validation of address information,
4) Checking the readability,
5) Logging,
6) Filing.

Provided that the channel identifier and the channel serial number are correct, the operator then checks the associated ACP 127 NATO Supp-3 format lines to decide whether the message received is for the local HQs or it is to be relayed to another HQ. In cases where the message is garbled, the operator requests for a retransmission from the parent tape-relay center or from the transmitting comcenter. Messages, excluding those which are service messages, are then stamped with the time-of-receipt stamp, recorded on the channel log sheets and delivered to the Message Registration and Administration Unit.

Allocation of Internal Distribution: Incoming messages are reviewed and allocated internal distribution at several hierarchical levels within the HQs till they reach staff officers' desks. The functions pertaining to each level could be different. For example in the Message Registration and Administration Unit HQ-wide distribution information is taken care of

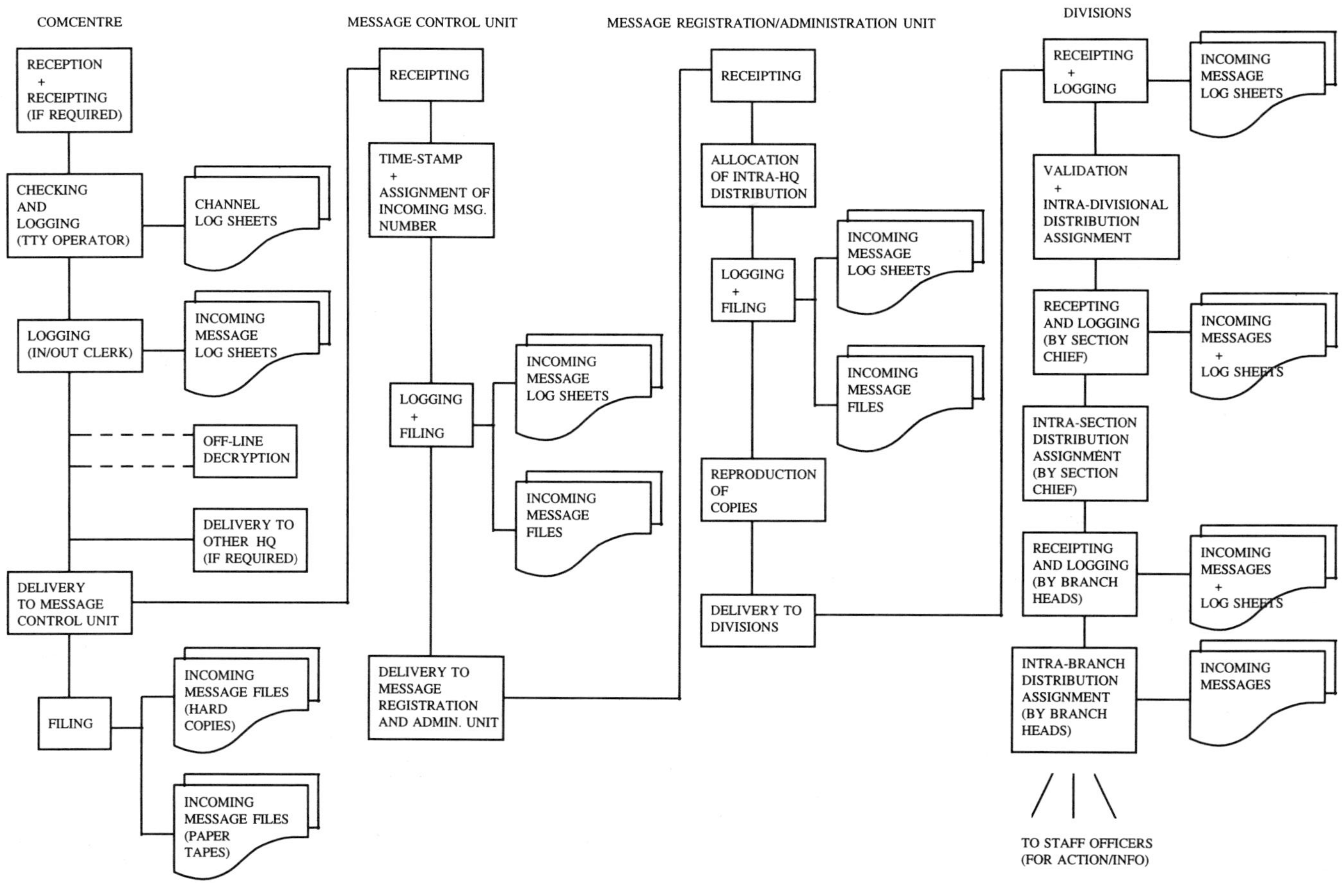

Figure 9.3 Current Incoming Message Handling Processes

according to the Standard Distribution List of each HQs. In cases where the introductory text includes special message distribution/internal handling instructions such as FOR COMMANDER'S EYES ONLY or PASS TO......or FOR....., it is the responsibility of message registration and administration unit to handle and deliver such messages as requested by the originator. On the other hand, in the Division, each Division Head, after reviewing the messages delivered to his Division, allocates an intradivision distribution. This information, recorded on the message, identifies the Section for ACTION and the Section(s) for INFO. Finally in the Sections, it is the responsibility of the Section Chiefs to allocate an intrasection distribution. This information is recorded on the message or on a distribution slip identifying the individual staff officers for ACTION or INFO.

Reproduction of Copies and Delivery: It is the responsibility of the Message Registration and Administration Unit to reproduce the number of copies required for intra-HQs distribution. In order to meet the message accountability and traceability requirements as dictated by national and allied security procedures, reproduction of copies are controlled strictly. Those messages with SECRET and above are stamped with the information showing very clearly the total number of copies produced and the copy number (e.g., COPY 2 OF 8 COPIES). The messages are delivered to the Divisions by messengers.

Receipting: Receipting is a process imposed by the security procedures to meet the message accountability requirements. Each time the message moves within the hierarchy some sort of receipting takes place, the detail of which is determined by the security classification of the message. Receipting information is stored for a period of 30 days.

Logging and Filing procedures are similar to these tasks in the outgoing message case.

Performance of Existing Message Systems

Sender-to-receiver delay is the performance measure of message handling systems most felt by the users. Delay statistics obtained in various exercises are given in Figure 9.4 and Table 9.1 [10]. As it can be seen, the performance of the existing Message Handling Systems falls far short of the delay objectives. The message handling system must therefore be improved through the use of computer based systems.

9.1.4 Information Handling Procedures

The staff officers have to maintain maps, charts, journal log sheets, data files (e.g., enemy forces order of battle, own forces order of battle, equipment inventory) in order to monitor own and enemy resources, prepare message texts, provide inputs to the commander's briefings, update operational plans etc. The activities listed above, when carried out manually, entail cumbersome, repetitive and time-consuming information handling processes.

9.1.5 Voice Communications

At present, voice services are supplied by local exchanges and PABXs of various types such as crossbar, electronic and digital ones. They basically provide intra-HQs voice services, and access to the PTT network and for some subscribers to the military analog voice network. Very limited networking facilities exist among these switches.

Voice connections between the HQs are established through the extension of a number of subscriber lines (i.e. long lines) from one HQs to another. PTT leased lines are used to provide "long lines" services. Most of the exchanges used at HQs offer basic telephone

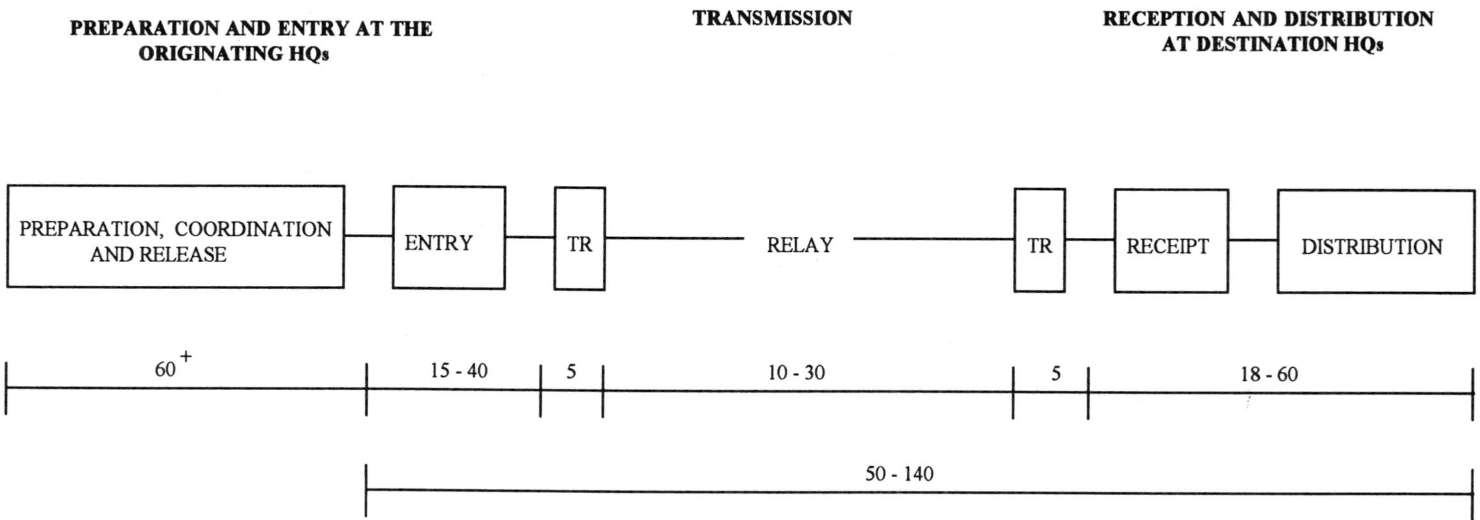

Figure 9.4 Handling of FLASH Messages (Median Times in Minutes for Various Component Processes)

Table 9.1 Average Delivery Delay of Telegraph Messages During Some Exercises. These Delay Figures do not Include Time Spent for Message Coordination and Preparation in The Staff Cells

(a) Average of total delivery delay of telegraph messages during recent military exercises.

PRECEDENCE	EXERCISE 1	EXERCISE 2	EXERCISE 3	EXERCISE 4	ACP-121 Objective
FLASH	139 mins	115 mins	120 mins	129 mins	10 mins
IMMEDIATE	297 "	171 "	260 "		30-60 mins
PRIORITY	480 "	285 "	not available	207 mins	1-6 hrs
ROUTINE	610 "	396 "			3 hrs-start of business next day

(b) Delay statistics for FLASH messages during a typical exercise.

HEADQUARTERS	TOTAL DELAY (MINUTES)		INTERNATIONAL DELIVERY* (MINUTES)	
	MEDIAN	25% **	MEDIAN	25% **
HQs # 1	120	240	55	73
HQs # 2	50	100	19	26
HQs # 3	45	75	13	23
HQs # 4	120	180	13	29
HQs # 5	135	300	12	18
HQs # 6	140	300	12	18

* The internal delivery times are measured from the receipt of the copy in the COMCENTER.

** Time taken for more than 25% of all messages of given precedence.

(c) Delay statistics for IMMEDIATE messages during a typical exercise.

HEADQUARTERS	TOTAL DELAY (MINUTES)		INTERNATIONAL DELIVERY* (MINUTES)	
	Median	25% **	Median	25% **
HQs # 1	270	500	75	100
HQs # 2	180	270	20	27
HQs # 3	150	300	40	50
HQs # 4	120	200	15	30
HQs # 5	240	> 500	15	23
HQs # 6	240	> 500	15	23

* The internal delivery times are measured from the receipt of the copy in the COMCENTER.

** Time taken for more than 25% of all messages of given precedence.

services only, with neither subscriber features nor supplementary services. Subscribers are equipped with analog telephone sets which are connected to the exchanges via two-wire loops using Dual Tone Multi-Frequency (DTMF) or dial pulse (DP) signaling. Only a rather limited number of secure voice equipment is available for communications between voice subscribers.

Performance of Existing Voice Communications Facilities

The shortcomings of present voice services can be summarized as follows:

a) Limited networking: Wide area connections are provided through permanent connections over the PTT leased lines. This approach is uneconomic, insecure and inflexible. These limited wide area facilities are offered to subscribers through operators, resulting in unduly long delays in connection setup processes.

b) Few subscriber features: Subscriber features and supplementary services in the context of WIS are instrumental in maintaining a high communication performance as well as meeting the differing operational requirements of users. Most of the existing switches and telephone sets can support only a very limited set of these supplementary services. These voice supplementary services are of civilian varieties. They offer therefore no military features such as preemption, precedence, secure communications etc.

c) Communications security: There is no systematic deployment of secure voice equipment and/or of bulk encryption.

9.1.6 Facsimile Communications

The analog facsimile machines used provide only a limited service mainly due to the facts that:

- There are no wide area switching facilities,
- Analog facsimile machines do not support end-to-end encryption, which is an essential requirement for message services,
- There exists no answer-back code which indicates the acknowledgement of the receipt of transmission and the destination address.

9.1.7 Video

Presently video information is not networked and is being displayed/distributed as follows:

a) For briefings and presentations CCTV or video tapes are used.
b) For alphanumeric information display, a conventional computer terminal connected via a V.24 interface with an asynchronous line running at 9.6 kbit/s is used.
c) For graphical/map information, rarely deployed intelligent terminals with bit-mapped graphics is used.

9.1.8 Communication Centers

Communication centers house all the message handling facilities which serve the HQs community. Messages are primarily exchanged between HQs via point-to-point telegraph lines, using mostly teleprinters operating at low speeds. The telegraph network between the major HQs are protected by on-line encryption equipment operating continuously for data flow security reasons.

The electronic message distribution units accept paper tapes as input and distributes them automatically to the external addressees determined by the originators. It is in the form of a star network, and is dedicated to low speed telegraph transmission. Messages with multiple addressees are transmitted either by rerunning the tapes as many times as required or by preparing the tapes in multiple copies. The teleprinters at both ends of the telegraph lines can receive or send messages. An automatic telegraph network is also used for message transmission among the transmission speed between the nodes in this network is 1200 Baud, and between the network and its users is 50 Bauds. All this communication is protected by data crypto equipment.

9.2 USERS AND THEIR REQUIREMENTS

The WIS user types and of their requirements baseline are discussed in this Section. The "WIS user" means "any authority" (i.e., HQs, organization, unit, base or station) who is entitled to use the communications services of the WIS" [see Chapter 2]. The "user location" is "a geographically defined area where the subscribers of one or more users are located sufficiently near each other that they can make use of the same local communications facilities. In some cases, one WIS user may need WIS services at several user locations, such as peace, war and alternative HQs.

Recall that WIS is a military integrated services digital network. The access network [Chapter 5], uses the nodal network for its basic interconnections and comprises access equipment of different types, access switches, PABXs, LANs, remote subscriber units or other remote access arrangements, and will be connected to the network either via line-of-sight (LOS) links or by means of fibre optic cables consistent with cost and survivability. The type of configuration will be determined by the operational requirements, user population, physical environment of the user location, types of traffic, traffic statistics and cost. Access arrangements for a WIS user locations are shown in Fig. 9.5.

The improvements/developments to be expected, with the implementation of WIS, in the user terminal area include:

- Improved quality of service, higher availability and a larger variety of services,
- More automation of information flow and management resulting in manpower savings and reducing intra-HQs delays,
- Integration of services,
- Easier procedures to recite or to reconfigure local facilities, and to accommodate new services (i.e., open-ended evolution),
- Limited set of standard interfaces to the network,
- More cost-effective operations,
- Faster and more flexible services acquisition and more effective secure communications and key distribution due to the employment of outband ssignaling

9.2.1 Users

WIS will recognize two groups of subscribers: Direct WIS Subscribers (DWSs) and Indirect WIS Subscribers (IWSs). DWS is "a WIS subscriber who is authorized precedence ROUTINE and above, hot lines, or other special WIS services and who is therefore required to be directly connected to an access switch" [Chapter 2]. IWS is "a WIS subscriber who is not authorized the services of a DWS and therefore can be connected indirectly (e.g. thorough a PABX) to an access switch".

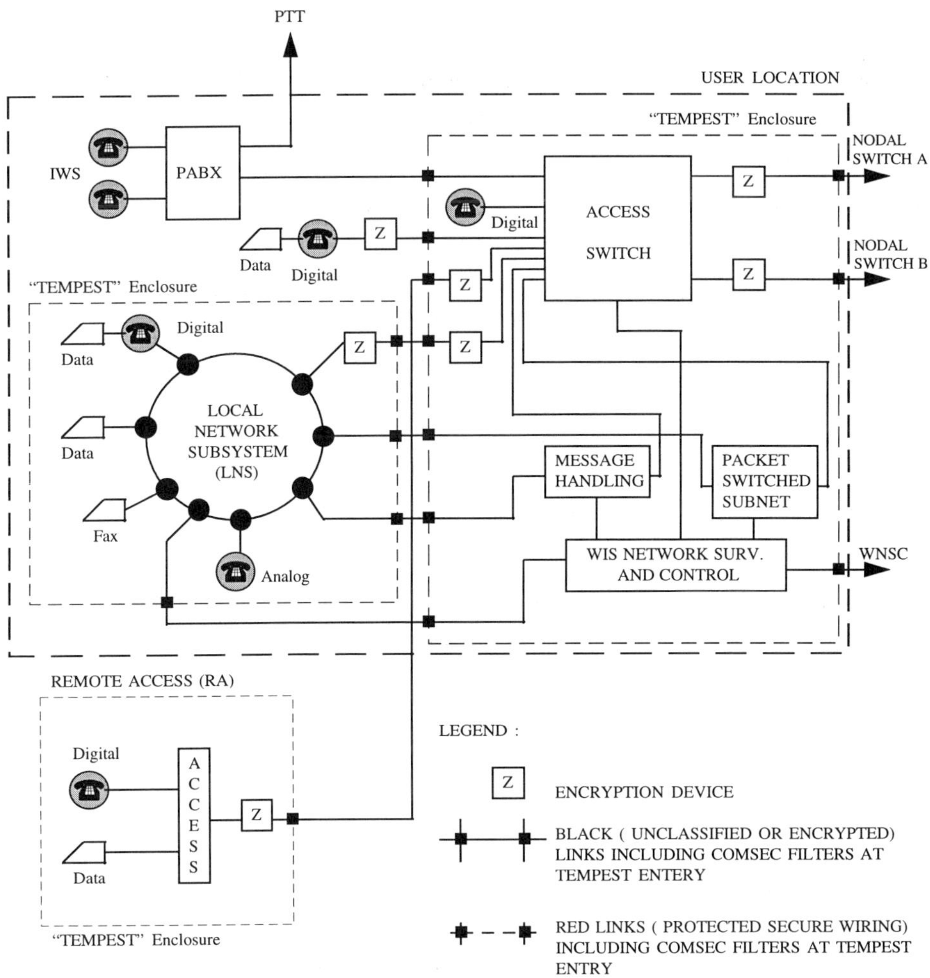

Figure 9.5 Generic Access Arrangements for a WIS User Location

WIS users are defined via criteria of eligibility. These users can be grouped under two broad categories: Military users and civil users. Military users include the static, itinerant and mobile users in each major group of land, navy and air forces. "Itinerant users" are those "who may operate from alternative static locations with different access points into the WIS" and mobile users are those "who have the facility to change their geographical locations frequently and freely while maintaining communications". Civil users include the main representatives of the government. WIS will meet communications requirements for all the operational, logistics and administrative functions of the military users. The civil users will be served by WIS for political and military consultations and for other administrative functions which may be specified by the Authority.

Itinerant (transportable) users (e.g. So-called "mobile-HQs") which may operate from alternative static locations will have access into WIS in two different ways: (a)directly via one or two access links into predetermined pickup points where users carrying WIS terminal devices will receive the full switched services of WIS, (b)via tactical networks which will itself be interconnected with WIS at a number of predesignated gateways which execute necessary conversions (e.g. rate, format, signaling). Mobile users are those who have the requirement to communicate while on the move such as ships and aircraft. They will have access into WIS always via a gateway where code, speed and other conversions will be carried out.

9.2.2 User Requirements

WIS subscribers at a user location need to communicate among each other locally (i.e., intra-HQs communication) and with other HQs (i.e., inter-HQs communication). The intra-HQs and inter-HQs communication is in the form of voice (plain/secure), messages, interactive and bulk data, facsimile, image and video communication, telemetry and teleaction activities. Terminal Area Subsystem (TAS) in WIS must provide the users the means for timely, secure and reliable communication.

The flow of information into, out of and within a WIS user location will be precedence controlled, with precedence levels as described in Chapter 2.

For telephone subscribers, precedence above ROUTINE will only be available to DWSs. In cases where IWSs require higher precedence, this will be provided by operator assistance and controlled by local procedures. Higher precedence voice will have, when necessary, rights of pre-emption over the lower precedence. Message and data traffic will be handled in order of precedence, on a first-in-first-out basis within each precedence level.

Types of services and their requirements have been detailed in Section 2.3.13. In the sequel information complementary to that Section and/or relevant to intra HQs practices are given.

9.2.2.1 Voice Requirements

WIS users require the ability to communicate in both plain and secure mode locally and with the subscribers in the other user locations.

a) Intercom requirements: Represent the need for both person-to-person dialogues and conferences involving staff from other Divisions within a HQs. The material for discussion will generally be of classified nature. The intercom requirements can be listed as conferencing between closed subscriber groups; paging of groups of subscribers or of all subscribers; broadcasting command briefings and announcements. The paged subscribers should be able to "talk-back".

b) Plain voice requirements: HQs staff require the ability to communicate in clear voice with non-WIS users (i.e. PTT subscribers) and with WIS subscribers who do not possess secure voice facilities (cf. 2.3.13.3.b).

- Direct indialling to subscribers, as a minimum to all DWSs, and with operator assistance when required,
- Abbreviated dialing for selected users,
- Call forwarding, transferring and follow-me capability,
- Preemption facilities for DWSs,
- Ability to connect with national PTT systems, NATO owned networks, CZCSs and other national strategic networks,
- Automatic hotline and off-hook services,
- Indicating and warning signals to inform that called number is not available, busy, or call is being preempted, etc.

c) Secure voice requirements: Those subscribers, particularly DWS's, that requires secure voice communications with other DWS's need to be provided with cryptographic protection without decryption at any intermediate point (i.e. end-to-end encryption), (cf. 2.3.13.3.d). There are two ways in which end-to-end secure voice communications can be provided:

- End-to-end encryption where encryption devices are collocated with the telephone sets,
- A pool of encryption device is shared by a group of subscribers who are connected to the pool via secure wiring.

d) Voice mail requirements: Voice messaging involves typically 10 to 70 seconds duration voice records. Besides communications with absentee parties, voice mail will find applications such as making voice reports to main office at times suitable to personnel remote from base, status reporting automatic reminder, broadcasting/narrowcasting messages for dissemination of general information.

9.2.2.2 Message Handling Requirements:

Messages handling services within the HQs, must be improved to meet the total message delivery times as stated in ACP 121 [2], and as detailed in 2.3.13.3.c.

9.2.2.3 Data Requirements

HQs staff require the ability to transport data locally between the local users and the host computer, and externally between the local users and the host computer at another user location, and between the host computers at different user locations. Data subscribers need to be provided with:

- Dial-up capability and automatic directory services,
- Ability to handle classified traffic,
- Capabilities to run applications, such as file transfers, remote job entry, data base inquiries etc.,
- Capabilities to be able to account and trace all generated or received "enquiry-response" data messages,
- Capabilities to exchange multimedia information.

9.2.2.4 Facsimile

Facsimile services will find many applications in WIS, especially since there exist forms of intelligence information that can only be transmitted through a facsimile. A requirement exists for the transmission of magazine quality imagery, primarily to support targeting efforts. Additionally, secure facsimile is crucial for rapidly transferring captured enemy documents to an intelligence center with the capability to translate, and exploit documents during a war. The availability of ISDN bandwidth and connectivity and the deployment of Group IV facsimile machines will improve the features and speed of this service considerably.

9.2.2.5 Video

Digital images and low bit rate digital video will play an integral role in the WIS information processing and communication activities. The requirements can be summarized as follows:

- Storage, archiving, and transmission of compressed digital images, (e.g., based on JPEG algorithms); there should also be provision for encrypting image file,
- Transmission of digital video streams (e.g., with MPEG or motion-JPEG algorithms),
- For both still images and video streams scalable compression algorithms should be preferred so that the these images can be decompressed and displayed on various terminals with different capabilities.

9.2.2.6 Multimedia Messaging

Multimedia is defined as the property of handling several types of representation media. In WIS users would need to use multimedia messaging, retrieval and conversational services. In these services the types of information representation media could be voice, graphics, text, fax, image and video. The user requirements can be listed as:

- Conversational services, i.e., bidirectional communication with real-time end-to-end information transfer from user to user or between user and host machine. Examples of conversational services are videotelephony, videoconference, autographic communication, interchange of processable or final form multimedia office document.
- Messaging services that offer user-to-user communication via storage units with store-and-forward, mailbox, message handling functions.
- Retrieval services to retrieve information stored.
- Human factors: Multiple use of terminal components for different functions (multifunctionality), and integrated use of several services within one terminal set higher demands on the users. Therefore more user friendliness in the form of user guidance for operational support of users is required such as icon-driven or monodriven screens, on-line help.

9.2.3 Traffic Characteristics

The following assumptions were made for the traffic characteristics in WIS.

9.2.3.1 Voice

Assume that the total inter-HQs voice and facsimile traffic in WIS has been estimated to be λ_{inter} Erlangs. Most of this traffic will be generated by those user locations which qualify as major HQs, since the others will be served by remote access units connected to the access switches. If this total traffic is partitioned equally among the N_{HQ} number of HQs, it results in λ_{inter}/N_{HQ} Erlang per HQs, which means that inter-HQs traffic is about 20 x λ_{inter}/N_{HQ} calls per hour, assuming 3 minutes call holding time. Finally assuming that of all voice calls, 30% constitute inter-HQs communication, then the overall call rate (both intra and inter) generated in a HQs becomes about 67 x λ_{inter}/N_{HQ} calls per hour. Thus, the intra-HQs traffic is about 47 x λ_{inter}/N_{HQ} calls/hour. More details are given in Chapter 4 on the traffic estimation used in WIS dimensioning.

9.2.3.2 Message

There exists very limited statistical data available on the message traffic, or on the percentage distribution by precedence/classification/special handling categories. However the following estimations and/or assumptions are considered to be portraying the incoming and outgoing message traffic characteristics for periods of tension and war.

a) Volume (sum of both incoming and outgoing): The message traffic among the WIS users has been estimated to be $\lambda_{message}$ Erlangs. If this traffic is carried over C bit/s channels and assuming an average message length of N_{char} characters, then one has:

   Volume of message data/sec = $\lambda_{message}$ C bit/s

   Messages handled per sec = $\lambda_{message}$ C/8 N_{char}

   Average Message traffic/HQ = Messages handled per sec/N_{HQ}

b) Message traffic distribution: By message category it can be assumed that 90% are operational messages, while 10% are service messages. On the other hand the precedence percentages are estimated to be 20% P1, 35% P2, 30% P3 and 15% P4. Finally by classification the following figures can be quoted: 10% C1, 30% C2, 20% C3, 20% C4 and 20% C5 (C1 being the highest and C5 being unclassified).

c) Message Lengths: Operational and service messages have an average length, respectively, of 1500 and 400 characters, both being exponentially distributed. The message header on average comprises 25% of the total length of a message.

With regard to the intra-HQs message traffic generated due with message preparation and distribution, the following assumptions are made:

- Each outgoing message is coordinated on average with three subscribers,
- Each message prepared is subjected to editing on average four times,
- Each incoming message is distributed on average to 10 subscribers.

9.2.3.3 Data

With a data traffic volume estimated as λ_{data} Erlangs in the WIS switched network, one can carry out computations similar to the message and voice cases. For example one can assume an average of 3.0 seconds of data call duration, and that most of the data traffic is generated by the major HQs. However since data traffic is typically burst in nature the average Erlang figures may be less significant in characterizing data traffic. On the other hand a much larger percentage of the data traffic will be of intra-HQs nature, in the form of LAN traffic, server accesses, distributed data processing etc.

9.3 TERMINAL AREA ARCHITECTURAL DESIGN

9.3.1 Overview

It is the "terminal area" where the WIS subscribers will most directly experience the changes and the improvements of modern technology. The changes will be in the types of terminals that can be supported, and the services that will be offered.

The TAS, as depicted in Fig.9.6, will generically comprise the following major elements:

- Terminals (see section 9.3.2),
- Automated message processing system,
- Data processing facilities and data bases,
- PABXs,
- LANs.

The implementation of ISDN principles in the TAS will be characterized by:

Services integration: A wide range of voice and data services will be provided using a limited set of multipurpose user-network interface arrangements. Each basic ISDN access provides two communication channels (2 B-channels) and thereby potentially becomes a multiservice access.

Out-of-band subscriber signaling: The ISDN D-channel is used for signaling between the subscriber and the network. The subscriber has thus a flexible access to all the call and service manipulation capabilities of the network. In particular, this allows for end-to-end encrypted communication over two wire subscriber loops with electronic key distribution on a call basis.

Single subscriber directory number: The subscriber will possess one subscriber number under which different terminals will be identifiable via ISDN protocols.

9.3.2 Terminals

The terminal types that will be supported by the TAS can be classified in a sense as monomedia and multimedia terminals. These ISDN terminals will also support authentication, and encryption features More explicitly they can be listed as:

- Voice terminals,
- Message and data terminals,
- Integrated voice and data terminals (IVDTs),
- Facsimile terminals,
- Video,
- Multimedia terminals.

9.3.2.1 Voice Terminals

The functional diagram of an ISDN telephone is shown in Fig. 9.7(a). Thus,s in addition to the classical telephone transducers, the ISDN telephone has keypad for various functions and an alphanumeric display. It would be desirable to have, e.g., an A4-size, low-flicker, flat liquid crystal display, with a resolution of 10E6 pixels. The screen will be used not only for screen-based directory, but also for word processing tasks and image display, browsing. One of the major application of the ISDN D-channelsignalingg facility in telephony, will be in the direct establishment and handling of secure communications. The

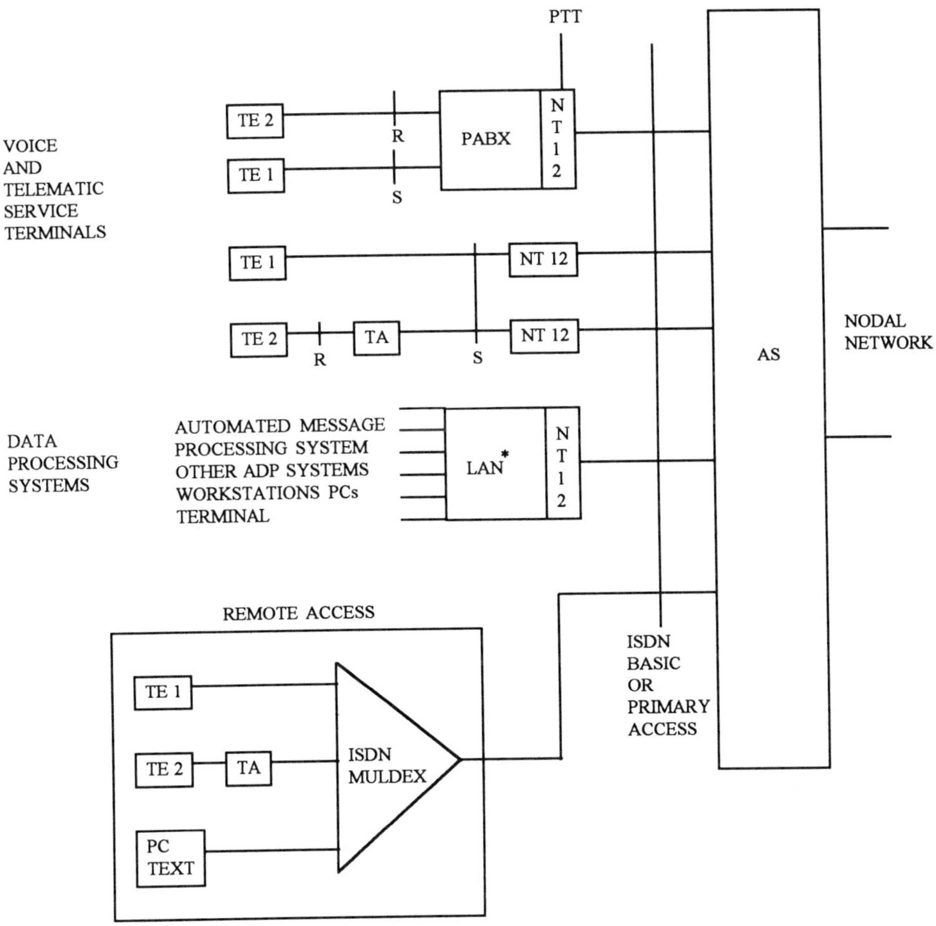

Legend :

TE 1 : ISDN Terminal
TE 2 : Non -ISDN Terminal
TA : Terminal adaptor
NT 12 : ISDN Network Terminations NT1+NT2
LAN : Local Area Network
AS : Access Switch

* LAN may well be considered as a subscriber to PABX

Figure 9.6 A Generic View of TAS Architecture

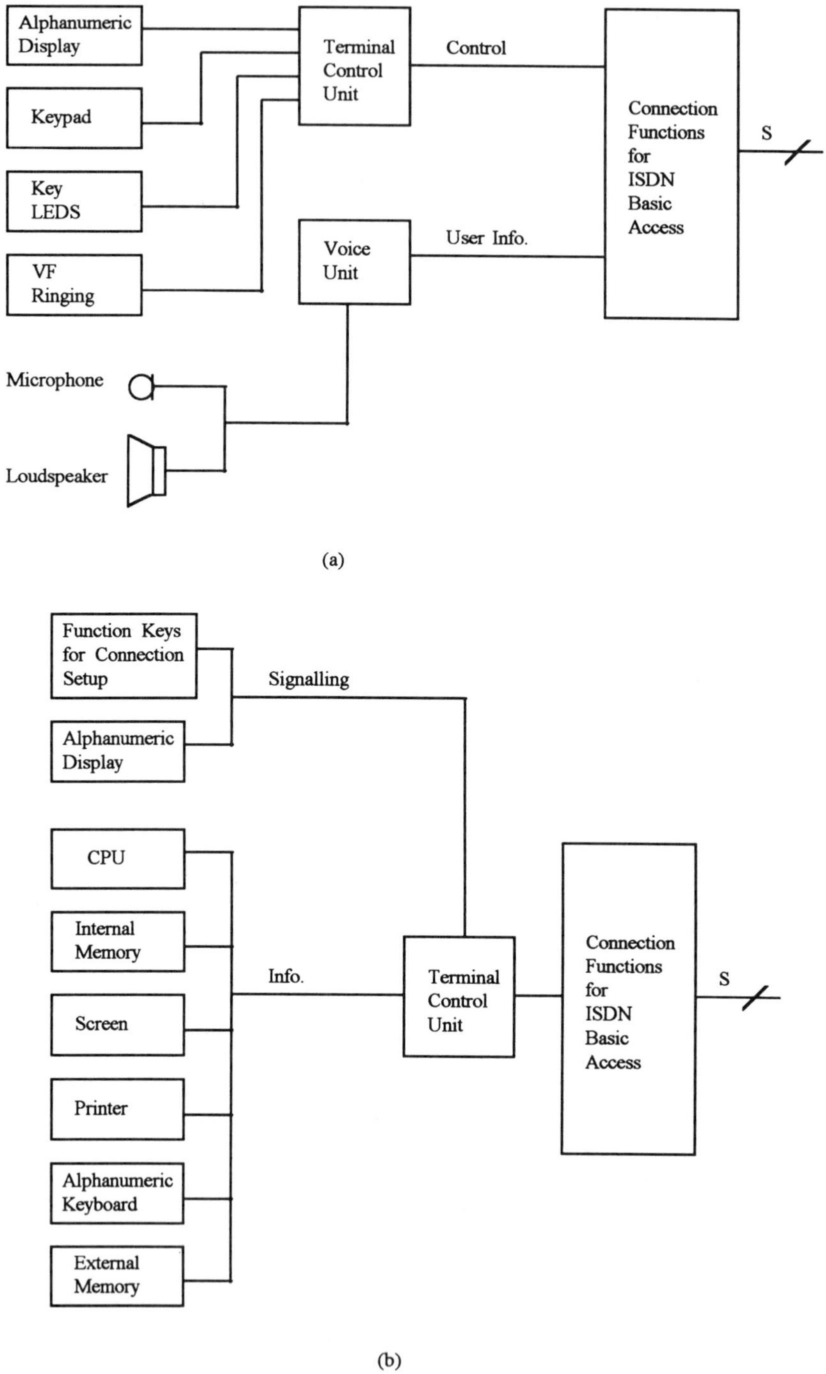

Figure 9.7 **(a) Functional Diagram of the ISDN Telephone Set, (b) Functional Diagram of an ISDN Terminal for Mixed Mode Operation (Textfax)**

ISDN sets will have programmable units, to negotiate/attain various supplementary services. Keys should be arranged according to function groups, such as group of name keys, group of call establishment keys (e.g., call forwarding), group of supplementary service keys (e.g., registration of incoming calls).

9.3.2.2 Message/Data Terminals

WIS subscribers will be provided with message terminals which will give them capabilities for message handling and access to central message processing facility. There is a declining trend for dedicated data terminals, e.g., telex, while the tendency is to provide and support all these services on personal computers (PC) and workstations (WS). In other words all the communication services, such as telex, facsimile, video, etc., will be handled on the PC or WS, either by means of specific cards, like fax-modem card on a PC, or via software and appropriate LAN gateways. In other words, the PCs/WSs used in the terminal area can be expanded through additional hardware and software to meet the functional and user interface requirements for various message/data service applications in WIS.

The criteria for selecting the message and data terminals (PC, WS, or else) are their capability in handling encryption, security issues such as electromagnetic compatibility, design of their man-machine interface, i.e., their user friendliness, memory type and capacity, processing speeds, display quality and resolution, printing features.

9.3.2.3 Multiservice Terminals

Multiservice terminals can have the advantages of creating a more ergonomic working environment, less wiring and possibly better cost/performance ratio. The user is provided with a single, compact desktop terminal capable of voice, data, graphics and text communications, and is linked over a single interface to the switched network. The integrated terminal occupies much less space than the corresponding set of single-service terminals, and change of service during a call or simultaneous communication in two or more services are made much easier. Note, however, that multiservice terminals should allow both individual use of single services (e.g., telephony only), or simultaneous use of several services (e.g., videophone + information retrieval + telewriting). Within each service, service specific operational functions should be easily handled. For example, for videophone, one could switch from partner's portrait to own portrait or to a document image or to the zoom facility; for videotext mixed mode, to switch from summary of contents options, to automatic page changing; for a recorder or video image database, picture search with fast forward or reverse etc. should all be feasible.

At WIS user locations, possible multiservice terminals could be as follows:

- Combined voice terminal (secure or nonsecure) and high speed facsimile terminal operating at 64 kbit/s,
- Combined voice terminal (secure or nonsecure) and narrow-band video transmitter/receiver (picture-phone),
- Combined voice/fax with telewriting tablet (autographic terminal),
- Combined voice/fax/data terminal with;
    * Image processing capability (i.e., compress and decompress JPEG and MPEG images, browsing capability over an image base, user friendly querying over image/video bases by topic and content),
    * Word processing capability,
    * Computing capability, e.g., to run application programs, like spreadsheets, etc.

A multiservice terminal will potentially consist of a monitor, a portrait camera (for videophone), a document camera, stereo loudspeakers and microphones (for hands-free telephone), telephone handset, writing tablet, monitor for word processing and image display purposes, a printer (to obtain hard copies), a facsimile, telex, or combined telefax unit, a personal computer, operational control unit in the form of a keypad. In addition there will be a memory or recorder to record part of a conversation, of a videophone sequence, or still images like documents, maps.

9.3.2.4 Facsimile Terminals

With the ISDN, present Group 2/3 analog facsimile machines will be replaced by more efficient Group 4 facsimile machines or to the ISDN adapted Group 3 machines. Due to the availability of 64 kbit/s transmission and full duplex capability, the transmission time for an A4 size page will be reduced from 30 sec to about 1 second.

The facsimile terminals in TAS must have the following features:

- Local copying,
- Incorporation of crypt options,
- Data and time display, and automatic inclusion in the communication journal,
- Abbreviated, delayed and repeated dialing,
- Storage capability,
- Identification of calling and called station directory number.

It is also desirable that the facsimile machines could be used for other applications, such as a textfax workstation, electronic mail system or archival system for storing scanned document and associated retrieval keys (e.g. date, keywords), as detailed above.

9.3.3 Automated Message Processing System (AMPS)

The type of improved message handling facilities envisaged for the WIS HQs will depend primarily on the size of the HQ. Two generic types are foreseen: the first is a multiuser system (AMPS) for large HQs and the second is a single-user message processing system suitable for small (e.g., less then 30 users) HQs. In these small HQs the messaging system can be built around a minicomputer with its peripheral devices and a message entry/delivery unit consisting of a VDU with a keyboard and a hard copy printer. Conformance both to ACP 127 NATO Supp-3 procedures [1] to the CCITT X.400 recommendations will be the goal.

9.3.3.1 AMPS Configuration

In the generic configuration of an AMPS in WIS, user locations comprises the following major elements:

- Input-output devices (VDU, keyboard, local processing, hard-copy printer...),
- Central processor and system control console,
- Storage facilities,
- Network interfaces.

The AMPS central processor will include two host computers, a watchdog processor and the associated peripheral hardware. The two host computers will be the exact duplicates of each other and will work in main and stand-by modes. The watchdog processor will monitor the host computer working in main, and in cases of failure, will bring the host computer in standby mode into the main mode. Each host computer will have its own dual-ported disk drive and magnetic tape drive. The associated peripheral hardware for each host computer will include a system console, and a magnetic tape/CD-ROM drive. The system console will consist of a VDU with a keyboard and a high speed printer for the system control and supervision functions.

The fundamental properties of the AMPS software will be its quality, integrity and maintability. The software must meet the WIS user message handling requirements together with the requirements for security, ease of maintenance and modification [see also Chapters 2, 5 and 12], reliability and accountability (e.g., adherence to AQAP-13 Software Standards). The software should consist of three subsystems: the system software, the applications software and the support software. The Man-Machine-Interface (MMI) on the AMPS will mostly be conducted through the use of menus and formatted screens.

For long time storage of messages (archiving), nonvolatile type of media will be used in the AMPS such as magnetic discs, diskettes and magnetic tapes. In the AMPS, the disk must be partitioned for messages of different classification. These secret partitions must not be easily accessible even by the people who maintain the system. The magnetic disks can be used as on-line storage media for the AMPS, with proper procedural security measures taken for classified messages. Diskettes and magnetic tapes are transportable storage media, and, if proper security measures are observed, they can be employed for off-line storage purposes.

Internal/external interfaces: The AMPS processor will support several types of message/data terminals at the WIS user locations, such as terminals with limited intelligence, PCs, workstations.

9.3.3.2 Functional Capabilities

The AMPS users in the HQs Divisions/Branches/Sections will be provided with capabilities to:

- Prepare and edit messages, ask for or submit comments for message coordination,
- Send messages to the releaser's position,
- Receive messages or retrieve messages from on-line and off-line storage media,
- Compile log records and statistical information,
- Obtain hard copies of messages.

The releasing authority (i.e., the commander) will be able to receive messages, reject or release the messages sent to him, and obtain hard copies of messages.

The system will translate the released messages into the defined format and transmit them in accordance with their security classification and precedence level. Also the system will allocate internal distribution to incoming messages and deliver them to the ACTION/INFO addressees. The system will provide the comcenter with the capabilities for supervisory control, message handling assistance, message servicing and engineering functions. The COMSEC/COMPUSEC policy and principles adopted for WIS [Chapter 12] will be fully implemented in the AMPS hardware and software design. The flow of messages in the HQs and the processes supported by the AMPS pertaining to outgoing messages are shown in Figures 9.8 and 9.9. The incoming message processes are quite similar.

9.3.4 Data Processing Facilities

Types of CCIS activities that would require automatic data processing support include:

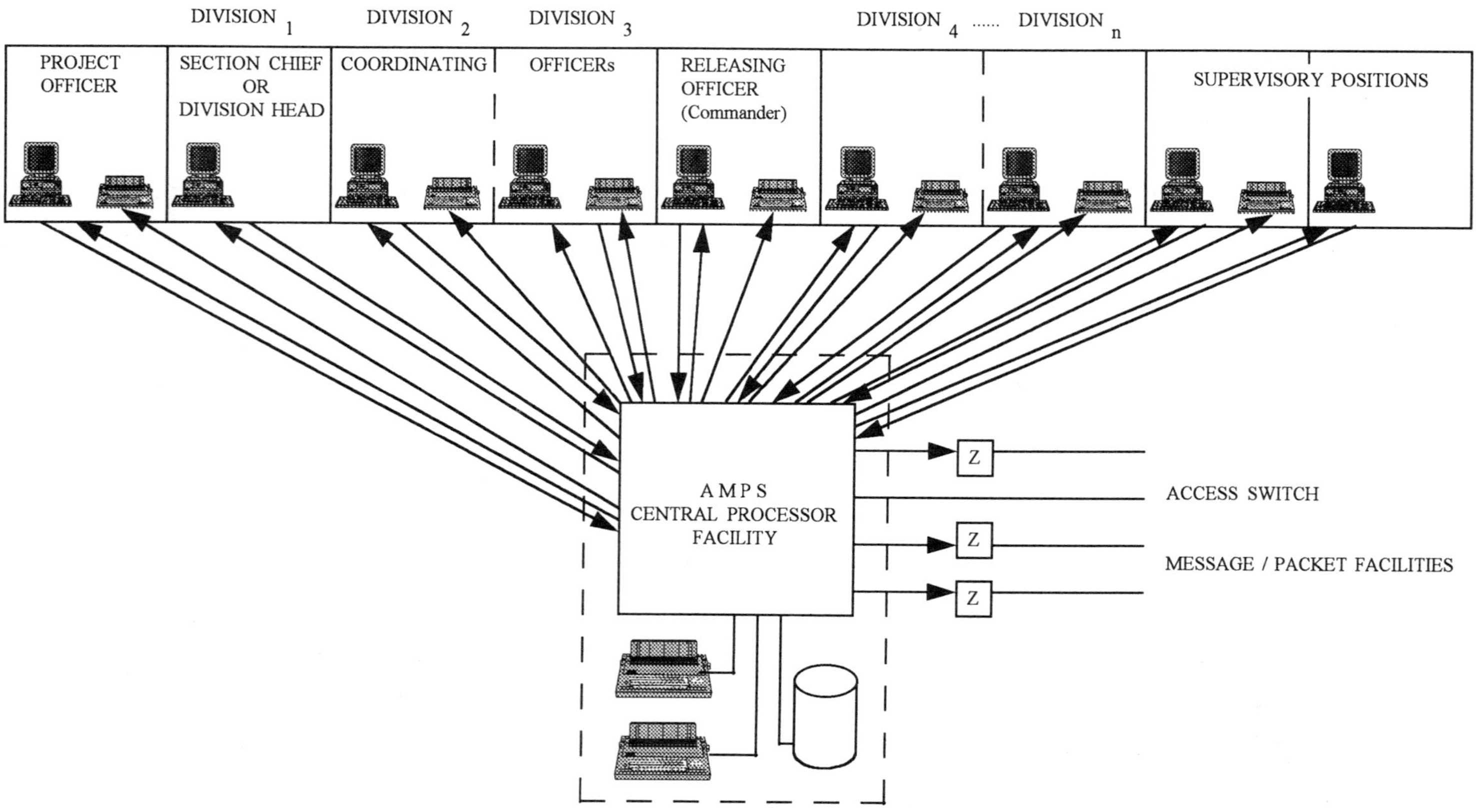

Figure 9.8 Flow of Outgoing Messages in the AMPS

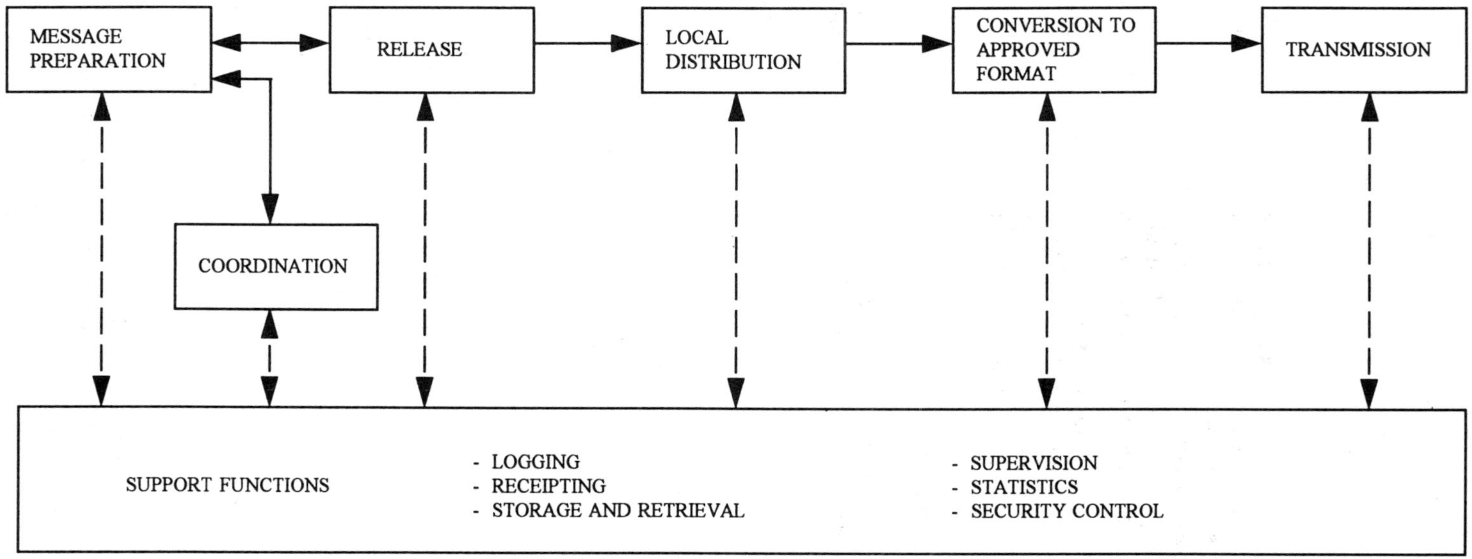

Figure 9.9 AMPS Outgoing Message Processes

- Information storage and retrieval (archiving),
- Statistical procedures,
- Simulations (e.g., war games),
- Data base access and updates,
- File transfer,
- Electronic mail,
- Handling of various types of data (e.g., radar track data, sensor/air picture data, meteorological data, weapons control data). Planning and tasking data,
- Software loads, data dumps,
- Logistics, materials movement and management,
- Collection and fusion of battlefield information,
- Image and video processing functions, computer vision applications.

These CCIS activities would be run on a variety of machines such as personal computers, mini and mainframe computers, workstations, data entry/output machines, and would be employing a variety of application programs. Both application program packets and data would reside on servers, which necessitates, among other reasons, local networking of the CCIS devices.

9.4 TAS ARCHITECTURES

Choice of a terminal area architecture will depend upon such factors as the population at WIS user locations, types and throughput of traffic, and physical environment of WIS user locations.

Alternative architectures conceived for Terminal Area Subsystems would in all probability be based on an ISDN PABX, which provides the switching functions and supports voice, message, and narrowband data terminals, or an ISDN PABX dedicated to the functions of voice and circuit-switched data terminals while LANs, e.g., FDDI variety interconnect packet-based wideband terminals. On the other hand fast packet switching can be considered as a unifying alternative, which supports both low-delay interactive services like voice, and low-loss data communication or wideband, e.g., video, data terminals. Various architecturesare discussed below.

PABX-only Architecture: The ISDN PABX will meet all the switching requirements for intra-HQs communications and will provide accesses to the WIS network, via the local access switch, and directly to PTT networks. It will cater for voice, message and data communication services as well as providing networking for the AMPS and data processor terminals in circuit switched mode. The switching fabric is based on 64 kbit/s TDM/PCM. The PABX will support a variety of supplementary services such as conferencing, intercom, announcements, group hunting, voice mail, abbreviated dialing etc. for voice subscribers and multiaddressing, abbreviated dialing, time and date stamp, CUGs, calling line/called line identification for data subscribers. The PABX will also be equipped with operator positions for various operator functions.

The advantages of using a PABX for intra-HQs connectivity in the TAS architecture include low connection cost, the fact that cable failures affect only a limited number of subscribers, easy implementation of supplementary services and security measures. In addition the central switch acts as a controller. The disadvantages of an ISDN PABX based TAS are that the bandwidth is limited to n x 64 kbit/s (multirate ISDN), which will constitute a handicap in delivering multimedia services necessitating bursty bandwidth assignments, especially involving video and images. The circuit switching provides only point-to-point connections, which precludes broadcast messages. Finally there may be too much complexity in one switch, and the cost increases rapidly with large switches.

The major components of a modern PBX are shown in Fig. 9.10. The switching network provides the actual circuit switched connections to connect the various ports together. These ports are to be interfaced with analog telephones, digital telephones, data terminals, digital trunks and other equipment. The intelligence of the switch resides in the processor.

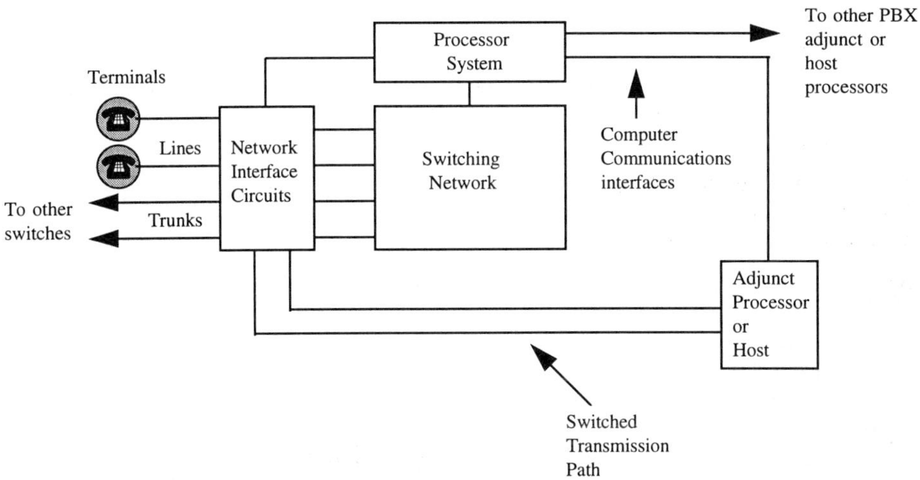

Figure 9.10 The Major Components of a Modern PBX [6]

The processor system consists of one or more processors and peripheral equipment that are used to run the program that controls the system, and to monitor and receive information from the interface circuits.

Interfaces: The PABX will support analog and ISDN basic access interfaces for the subscriber side and analog and digital line interfaces (tie line and ISDN basic access, respectively) and ISDN primary rate access (PRA) interface for the connections to external networks (i.e. WIS WAN and PTT networks). In order to provide switching services for data terminals, the PABX must have the appropriate interfaces (i.e., X.20, X.21b, X.21, and V-series interfaces) for circuit switched data terminals. Packet terminals (X.25 type) will be served at the physical layer (i.e., X.21 interface), while higher layer functions will be performed by the nodal packet switches. It must be able to perform protocol conversion functions (e.g., signaling conversion, rate conversion, format conversion). In this sense, the PABXs will accomplish both protocol dependent and protocol independent switching.

Data Communication Features: Data communication features are those features in a PBX designed to provide data connectivity between data end points and ensure the reliability and integrity of these connections. The main features are as follows:

- Protocol conversion,
- Gateways to convert circuit-switched data to packet switched data for connection to LANs and/or for connection to packet node in the WIS wide-area network,
- Multiplexing circuit-switched data endpoints.

Signaling: The PABX will support DP and DTMF subscriber loop signaling for analog telephones and ISDN signaling (CCITT Recs. I.430, I.440, and I.450) for digital telephones. The PABX will employ DC loop signaling and CCITT PABX No.7 signaling as defined in the CCITT Q.710 for analog and digital connections to external networks, respectively.

Wiring topology: Cabling configuration will be star, where every attached device will be connected to the PABX over twinax cables for indoors networking. For distances up to 1000 meters S-interface can be used with the network termination (NTI/NT2) functions in the PABX.

Security: All the terminal equipments both from hardware and software viewpoints, and installations, including wiring, will conform to the WIS COMSEC/COMPUSEC policies and principles.

Despite the above advantages, the "PBX-only" solution may come short, for example, in handling PC to host or to PC, work station to work station traffic, in dealing with fast file transfers, or in handling variable bit rate video traffic.

PABX + LAN Architecture: In this alternative architecture for TAS the PABX will meet all the switching requirements for voice and circuit switched data terminals for their intra-HQs communications, and will provide accesses to WIS, via the local access switch, and directly to PTT networks. The LAN, on the other hand, will support the intra-HQs message and data communication. For LAN implementation, among the choices of Carrier Sense Multiple Access/Collision Detection (CSMA/CD), token bus, token ring and Fibre Distributed Data Interface (FDDI), the token ring LAN implemented on a star-wired ring configuration is proposed at low speeds, and for higher data requirements (e.g., video) FDDI would be a solution.

Since both are token-based media access schemes, the factors leading to such a LAN choice are worth discussing:

- Hardware simplicity: The simplicity of the point-to-point connections between the wire centers on a ring offers the advantage that the interconnect complexity is moved from the analog hardware domain (as in CSMA/CD) to the digital logic domain of the station itself. Furthermore, the point-to-point connection is inherently the least complicated and most reliable type of connection, with the most noise immunity. Moreover, the continuously available clock signals simplify the receiver design, improving its reliability.
- Reliability: It is attained with suitable physical partitioning and dual attachment capability. In the event of failures, this configuration allows for immediate substitution of redundant standby components and the eventual isolation, repair and replacement of failed components. Another aspect of ring reliability is that the point-to-point nature of the connections provided between the stations. As they repeat and check the data, stations are able to monitor the frame error rates in the connection between them and their upstream neighbor.
- Optical Fiber: The point-to-point connection characteristic of a ring can easily accommodate the use of optical fiber. This is desired for reasons including high data bandwidth, security, immunity to electromagnetic interference, reduced weight and size.
- Ease of Configuration: Stations or links can be added and deleted without detrimental effects on the ring traffic. A ring inherently imposes no restrictive logical limit on the length of ring links, the number of stations or the total extent of the network that can be accommodated. However the token circulation time may become impractically long if too many stations in serial are accommodated.

- Performance: Token rings provide time-bounded access delay, even under impressed loads approaching the full capacity of the ring. Rings are insensitive to load distribution and the performance achieved is not degraded significantly by the presence of inactive stations.
- Full-duplex capability: Such applications as voice and video involve the transmission of "isochronous" data (i.e. data which comes in fixed amounts and at fixed time intervals as clocked by some external clock). While in a nonring topology, isochronous data must be treated in a half-duplex mode, a ring can offer two point-to-point connections, resulting effectively in a full duplex operation.

Fast Packet Switching Based Architecture: In one such alternative architecture, all the devices - data and voice - are connected to one or more LAN. The bandwidth of the LAN and the access protocol are chosen in such a way as to cater for both low-delay (near real time) loss-tolerant services interactive services like voice and video, and lossless but delay tolerant services like data communication. To this effect IEEE 802.9 Multimedia Interface standards can be adopted. In IEEE 802.9 the total bandwidth is either 4 or 20 Mbit/s, which can be divided into:

- The ISDN channel structure, i.e., two 64 kbit/s B-channels and one 16 kbit/s D-channel,
- IEEE 802-series LAN channels,
- Channels for circuit switched traffic.

The multimedia access is effected via two UTP (unshielded twisted wire pair) which simplifies the workstation interface by reducing the number of cables. The LANs connection to the access switch will take place in a gateway.

The LAN-based alternative can, however, provide a restricted number of isochronous 64 kbit/s channels and the software effort to furnish the PABX services for voice would be impractical. In this respect the wideband switching provided by the ATM becomes a very attractive alternative.

In conclusion it can be stated that the TAS architecture in WIS will be based on ISDN PABX, while LAN(s) provides the local interconnectivity for terminals characterized by bursty wideband traffic. ATM PABX, on the other hand, as ATM evolves into a mature technology, will present an attractive and viable alternative for TAS requirements.

9.5 REFERENCES

[9.1] "Message Relay Procedures", Allied Communications Publication: ACP 127 NATO SUPP-3, Jan. 1983 (NATO Unclassified).

[9.2] "Communications Instructions General", Allied Communications Publication: ACP 121 (NATO Unclassified).

[9.3] CCITT Rec. X.400, Message Handling Systems.

[9.4] CCITT Rec. T.411 and T.412: Office Document Architecture and Interchange Format.

[9.5] "Installation of Electrical Equipment for the Processing of Classified Information", AMSG-719D, 1987 (NATO Con.).

[9.6] S. Bush, C. Parsons, Private Branch Exchange Systems and Applications, McGraw Hill, 1994.

[9.7] R.J. Horrocks, R.W.A. Scarr, Future Trends in Telecommunications, Wiley, 1993.

[9.8] D. Deutsch, "Electronic Mail Systems", in Digital Communications, Ed. T.C. Bartee, Howard Sams Co., 1986.

[9.9] D.N. Hatfield, "Transmission Media" in Data Communications, Networks and Systems, Ed. T.C. Bartee, Howard Sams Co., 1985.

[9.10] A. N. Ince et al., Survey and Comparison of Message-Handling Aids for ACE Headquarters, SHAPE Tech. Center, TM-449, April 1975.

[9.11] IEEE 802.3 CSMA/CD Access Method and Physical Layer Specifications.

[9.12] IEEE 802.4 Token-Passing Bus Access Method and Physical Layer Specifications.

[9.13] IEEE 802.5 Token-Passing Ring Access Method and Physical Layer Specifications.

[9.14] IEEE 802.6 Fiber Data Distributed Interface.

APPENDIX 9A

CCITT X.400 MESSAGE HANDLING SERVICES

CCITT has proposed an extensive set of rules in the X.400 series Recommendations [3,8] for message handling services in private or public data networks.

Functional model of the MHS: MHS functions are carried out by two specific entities: user agent (UA) and message transfer agent (MTA). UAs constitute the interface between operators and the MHS, and they are the terminals which act as sources and sinks for messages. Each UA has a keyboard, screen and/or printer, editing facilities, and storage for holding received messages and copies of transmitted messages. MTAs implement all functions required for message switching nodes which include routing, forwarding, copying, holding and accounting.

The X.400 configuration is shown in Fig.9A.1. A service provider must have at least one MTA to serve all his customers. The User Agent (UA) typically contained in a PC or an intelligent terminal, is attached to a parent MTA. However both the MTA and UA can be collocated and resided in the user's terminal, so that users can directly communicate with each other without needing a third party service provider.

If a UA is operative, it will receive messages sent to it. If it is not operative, the MHS,or one or more MTAs in the MHS, will hold messages for that UA, until it becomes operative. An operative UA must always communicate with its own MTA in MHS.

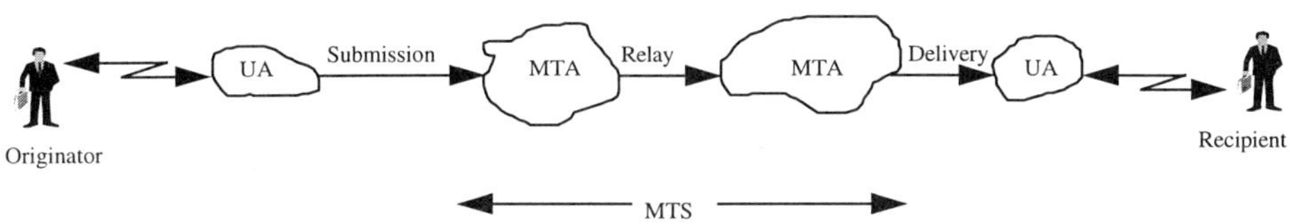

UA : User Agent
MTA : Message Transfer Agent
MTS : Message Transfer System

Figure 9A.1 A Functional View of MHS Components

MHS is divided into Management Domains (MD) and each MD is responsible for serving all the subscribers associated with it, their connections, their passwords and entries in the name server, etc. In WIS each user location will function as a MD as illustrated in Fig 9A.2.

Protocols: Protocol definitions, required between each object pairs shown in Fig 9A.3, have been developed based on the ISO reference model for Open System Interconnection (OSI). The process of message transfer between two MDs is governed by the three protocols developed for the application layers:

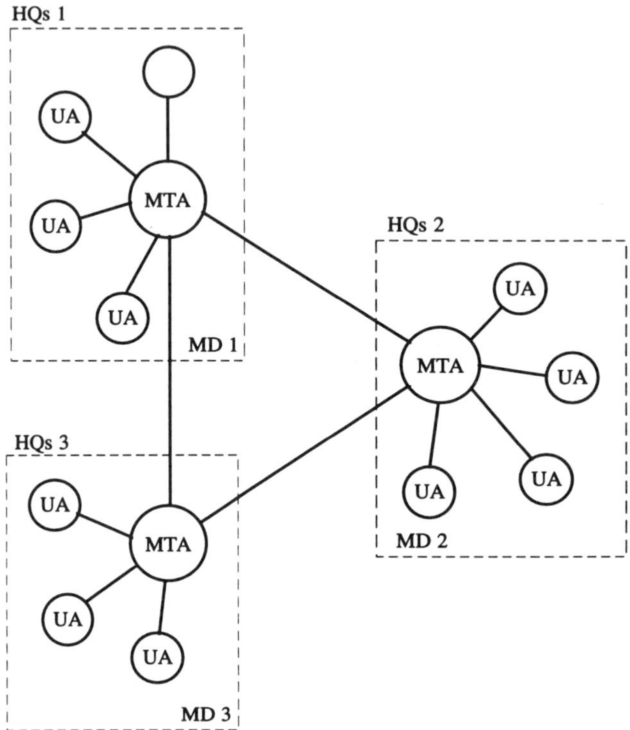

Figure 9A.2 Management Domains in the WIS MHS

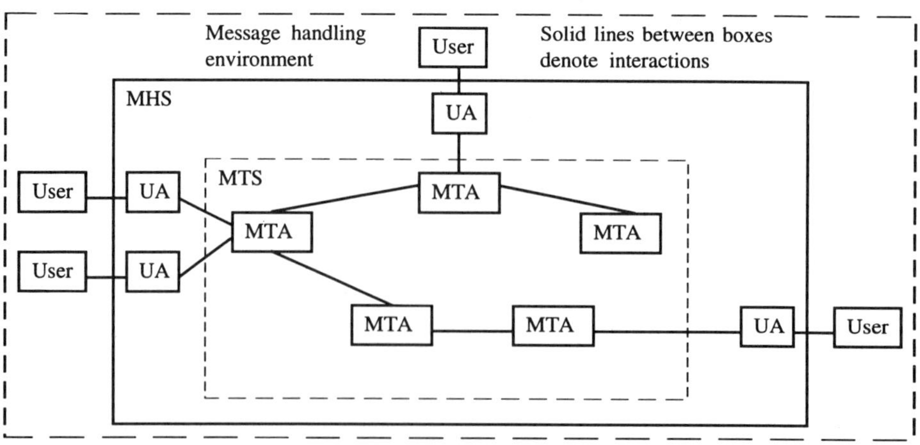

Figure 9A.3 X.400 Functional Model

- MTAs exchange messages with each other using the Message Transfer Protocol (P1).
- MTAs exchange messages with the associated UAs using the Submission and Delivery Protocol (P3).
- The control information in the "heading" and the mixing of various media in the "body" of the message exchanged between the UAs is specified by the End-to-End Protocol (P2).

Addressing: A major feature of the X.400 MHS is that, it is intended to permit addressing in a manner similar to that of the postal services. Name-servers within the network translate the "addresses" to "network addresses" for those UAs which are permanently connected and the network addresses of their parent MTAs. The Reliable Transfer Server (RTS) (i.e., the transport layer protocol) will route a message from its originating MTA to its destination MTAs. The destination MTA will use the RTS to route the message to a permanently connected UA, or hold it until a destination UA "logs in" and identify itself to the MTA.

In principle, several subscribers could share a UA, if sorting and distribution of messages could be done in a satisfactory manner. However, in WIS, multi subscriber configurations can be employed for closed subscriber groups only. This limitation is imposed mainly by security reasons and need-to-know principle that must be applied in dissemination of classified messages. In WIS, each UA must be assigned to one authorized subscriber or a closed group of subscribers.

Services: Every MHS offers various "services elements" to its users. Service elements include individual facilities which may always apply, or may be invoked when required, or which are optional. The Message Transfer Service (MTS) basically enables UAs to access and be accessed by the MTS in order to exchange messages. The service elements available in Message Transfer Service are as follows:

Basic Services:

- Access management,
- Content type identification,
- Converted identification,
- Delivery time stamp indication,
- Message identification,
- Non-delivery notification,
- Original encoded information types identification,
- Registered encoded information types,
- Submission time stamp identification,

Submission and Delivery:

- Alternate recipient allowed,
- Deferred delivery,
- Deferred delivery cancellation,
- Delivery notification,
- Disclosure of other recipients,
- Grade of delivery selection,
- Multi-destination delivery,
- Return of contents.

Conversion:

- Conversion prohibition,
- Explicit conversion,
- Implicit Conversion,

Query, Status, Inform:

- Probe,
- Alternate recipient assignment,
- Hold for delivery.

Some of the important service elements, which may find useful applications in WIS, are explained below:

- Access management controls the access via the use of password, or similar means, permits UA and MTA to identify and validate the other. It provides a capability for the UA (i.e., subscriber terminal) to authenticate its originator or recipient (i.e., subscriber) name and maintain access security.
- Message identification can be used for later reference to the message, in particular for delivery confirmation.
- Non-delivery notification can be used by a destination MTA (i.e., AMPS at another WIS user location) to advise the sending UA that the destination UA (i.e., action/info officer) has not received a message.
- Original encoded information types indication: A UA can register with the MTS what types of coded information it can receive.
- Alternate recipient allowed: An originator of a message may specify that., if the selected destination UA is unavailable, then an alternative recipient UA may get the message.
- Deferred delivery: The MTS holds the message until a specified time expires before delivering it.
- Delivery notification: An originating UA requests that an explicit notification be returned to the originating UA when a submitted message has been successfully delivered to a recipient UA.
- Disclosure of other recipient: If a message is delivered to multiple recipients, each recipient can be informed as to the identities of the others.
- Multi-Destination delivery: An originator sends a message to multiple recipients.
- Probe: An originating UA can check the availability of a recipient UA before sending a message.
- Hold for delivery: A recipient UA may request that the MTA hold incoming messages for it, and not deliver them until authorized.

<u>Interpersonal Messaging (IPM):</u> The IPM service builds on the MTS and is provided by means of the end-to-end protocol (P2). It meets the following requirements:

- To provide services for communicating memoranda (i.e., intra-and inter-HQs organizational unit document),
- To provide framework for the message body,
- To provide conversions that will permit the exchange of messages among UAs and terminals supporting different content types.

This application-independent service provides a subscriber with services to assist him in communicating with other subscribers locally or in other MDs. IPM services provide the functions necessary for preparation and editing of messages, submission and delivery interaction with the MTS, functions necessary to present messages to its addressees and provides functions to cooperate with other UAs in order to help its user in dealing with messages. IPM services can provide editing capabilities to assist users in preparing and editing messages. A local database, to help the user find previously received and filed messages, can be provided by the IPM services as well. These kinds of services are known as local services, so they do not require coordination or cooperation with other users.

APPENDIX 9B

SERVICES PROVIDED BY PABXs

Classical PABXs handle primarily voice traffic with a limited set of functions and features. Recent developments have expanded the range of functionalities and features of PABXs significantly. At present, they are evolving into functionally integrated voice and data switches.

Voice Teleservices: WIS voice teleservices comprise the following features:

- Local calls,
- External calls (Direct in Dialing/Direct Out Dialing),
- Operator facilities such as break-in, hard to reach traffic,
- Transfer of calls during external conversation,
- Automatic ring-back,
- Abbreviated dialling,
- Automatic call forwarding or call back to operator in case of no answer,
- Follow me,
- Group hunting,
- Local closed user groups,
- Preemption for incoming higher precedence calls,
- Nonsecure conferencing,
- Announcements,
- Call waiting,
- Paging,
- Voice mail.

Data Teleservices: Data Teleservices can be effected on a PABX by providing the appropriate interfaces or line access modules to interconnect telematic terminals and other data equipment such as word processing systems, personal computers, intelligent/dumb terminals, mass document storage systems, mainframes, office micros and LANs. Subscriber equipment may belong to any of the following categories:

- Circuit-oriented data terminations (e.g., CCITT Rec. V. 24/28, X.20, X.20bis, X.21),
- Packet-oriented data terminations (CCITT X.25),
- Wide-band terminations.

Data teleservice features comprise the following:

- Local and external calls,
- Both protocol-independent connections, where the interworking relies on the user equipment, and protocol-dependent connections, where speed, signaling, format conversions are supported by PABXs.
- Abbreviated dialing,
- Local closed user group,
- Store and forward functions for text type messages (i.e., internal mail).

APPENDIX 9C

LOCAL AREA NETWORKS

LANs will serve as a general purpose subscriber loop with distributed switching ability, and with gateways to an access switch, to a packet switch or to a message switch. These gateways will serve for LAN-WAN interconnections. Extensions and interconnections of LANs through bridges and routers will also be considered.

In WIS, a variety of data processing equipment such as mini and microcomputers, mainframes and their peripherals (printers, tape units) and various terminals will be served by LANs. Recent developments in voice packetization indicate that the LAN medium can handle efficiently voice and data in an integrated manner by using simplified protocols. In this respect it is possible to view a LAN as a general purpose subscriber loop, resulting in a much simplified wiring plan for voice and data subscribers.

9C.1 LAN Topologies

Topology, in the context of TAS, refers to the HQs layout and cabling strategy adapted in connecting the subscribers and resources to a LAN. There are three basic forms of topologies (Fig 9A.4):

- Star,
- Ring,
- Bus (and its variations, e.g., tree).

a) Star Topology: As depicted in Fig 9A.4(a), in the star topology the wiring to each subscribers winds around a hub. The hub makes possible monitoring, repairs, crossover manipulations, and it possesses the property that a line break has a minimal effect, in that one user loses service while the rest of the system continues working. The disadvantage is that the wiring loops twice as long as in a noncentralized scheme.

b) Ring topology: Ring structure Fig 9A.4(b) are characterized by each subscriber being connected to two adjacent subscribers, forming a complete loop. Each subscriber attaches to the local wiring net at a repeater which device is capable of receiving data on one link and transmitting it, bit by bit, on the other link with no buffering at the repeater. The links are unidirectional, and messages are passed in one direction around the ring until they reach their destination subscriber. The ring topology is susceptible to cable breaks, and in this case the whole system would stop working.

c) Bus/Tree topology: In bus topology Fig 9A.4(c), all subscribers are connected to a central bus which permits transactions in both directions in a broadcast mode. The tree topology is a variation of the bus topology as in Fig 9A.4(d). The bus/tree topology is able to handle a wide range of subscribers in terms of the number of devices, data rates and traffic types. However, from reliability viewpoint, it is not very reliable, since a break in the cable can disable a large part of or all of the subscribers.

The main considerations in the choice of a LAN topology for TAS are:

a) Diagnostics and troubleshooting: It should be easy to locate the fault domain in the cabling net and to remove the faulty segment for repair or maintenance without cutting off all the subscribers.

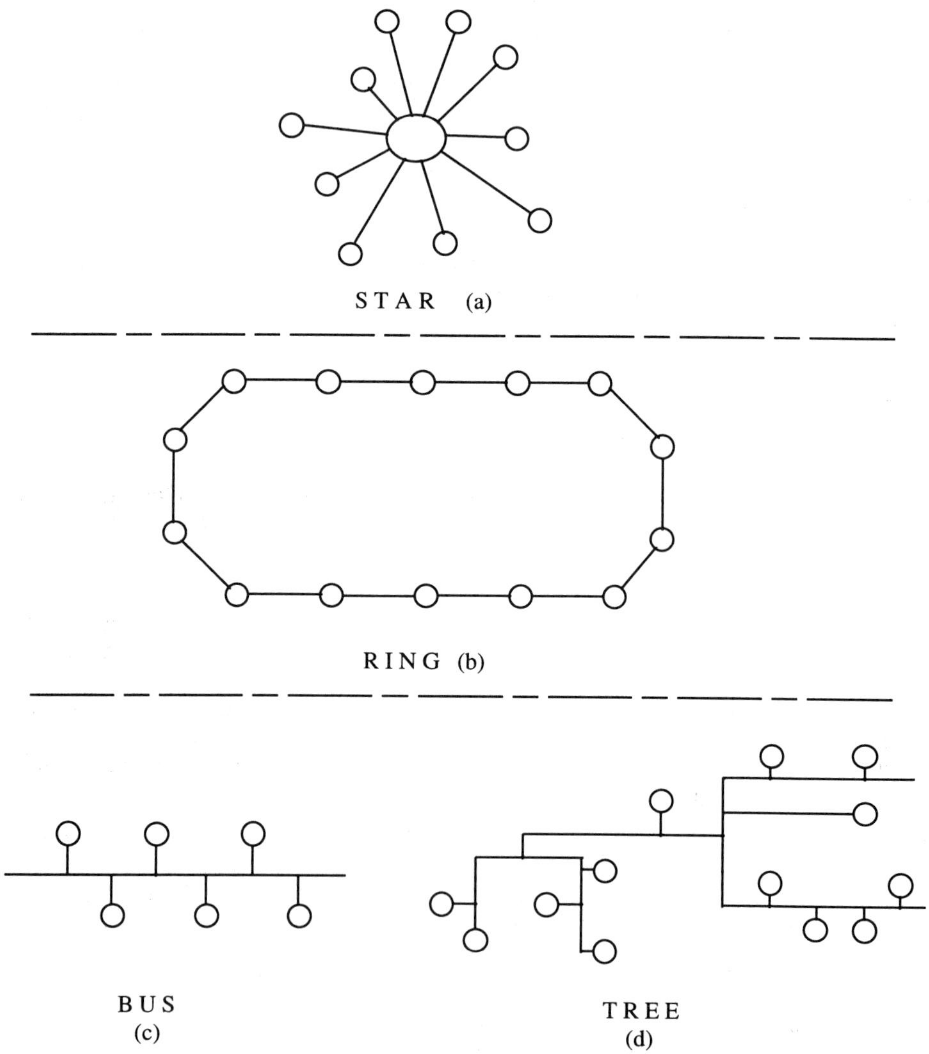

Figure 9A.4 Basic Forms of LAN Topology

b) Expansion/growth: It should be ensured that subscribers can easily be added, relocated or removed, and that the cabling net (i.e., transmission media) itself can easily be expended to accommodate future requirements.

c) Cost effectiveness: Maximum use should be made of existing cable layouts.

Table 9A.1 gives a comparative summary of the advantages and disadvantages of the topologies. Accordingly, it can be argued that the most effective topology, both operationally and coastwise for TAS LAN is star-wired "ring" topology.

9C.2 Transmission media

The choice of transmission medium for LANs in the terminal area subsystems in WIS involve a number of factors such as security, capacity, reliability, topology. These factors are:

- Handling characteristics and resistance to physical abuse,
- Physical description and transmission characteristics,
- Susceptibility to problems associated with radio-frequency interference, electromagnetic pulse and compromising emanations (RFI/EMP/TEMPEST), ease of tapping,
- Suitability for use in bus systems or in sequential systems (slotted ring, token ring).

A comparative summary of the four possible transmission media is given in Table 9A.2. It should also be noted that the choice of transmission media and topology for a terminal area not independent. For the bus topologies (baseband, token bus and broadband bus), coaxial cables are appropriate for indoors cabling. However, their outdoors use is not appropriate due to security requirements (see Appendix 9.D). Fibre optic cables stand out to be the best candidate for wiring at WIS user locations, especially where LANs will be installed and outdoors wiring is required, otherwise the use of braided twinax cable would suffice.

9C.3 Medium Access Technologies

The main medium access methods are:

- Carrier sense multiple access/collision detection (CSMA/CD),
- Token bus,
- Token ring,
- Fiber distributed Data Interface (FDDI).

Their characteristics in terms of bandwidth, transmission media, topology, protocols used, data rates are discussed below.

CSMA/CD LAN (Ethernet):

CSMA/CD LAN is a baseband LAN with the following characteristics:

- Cable: 50 Ohm coaxial cable (rather than 75 Ohm commonly used for CATV) is used to minimize reflection and loading problems.

- Topology and Protocol: Bus topology is being used. Tree wiring and star wiring alternatives exist. Maximum allowable distance between stations on a single segment is 500 meters.

Table 9A.1 Comparison of the Topologies for TAS Installations

TOPOLOGY FEATURE	STAR	RING	BUS/TREE
CONNECTION METHOD	Subscribers are radially connected to the central node.	Subscribers are physically connected to two adjacent subscribers to form a closed loop.	Subscribers are connected to a common backbone and are directly connected to each other.
SUBSCRIBER INTERFACE	Data is switched at the central node and other subscribers see only traffic intended for them.	Transmission is from left to right and interface units handle some traffic not intended for them.	Each interface unit sees all the LAN traffic and picks up only those packets addressed to it.
EFFECT OF CHANGES	Configuration changes occur at the central node with no interruption in LAN traffic.	Ring must be broken to add or delete a subscriber.	Connection of a new subscriber often requires breaking the bus.
IMPACT OF ERRORS	None, since defective subscribers are not switched by the central node.	Defective subscribers must be switched out of the ring and Bypassed.	None, even if a defective subscriber is not switched out of the bus.
TRANSMISSION MEDIA (See also App. 9E)	Coaxial cable, fiber optic or twisted pair.	Coaxial cable, fiber optic or twisted pair.	Coaxial cable is common.
COMMON ACCESS METHOD	Carrier sense multiple access with collision detection (CSMA/CD), switching.	Token passing.	CSMA/CD, token passing.

Table 9A.2 Comparative Evaluation of Different Forms of Transmission Media for TAS Installations

TRANSMISSION MEDIA	ADVANTAGES	DISADVANTAGES
TWISTED PAIR	Mature,well-understood technology. Simple to install. Quick, easy installation. Least expensive cable medium. Same wiring used for telephones. Very flexible.	Crosstalk between adjacent wires may cause errors. Subject to external electromagnetic interference. Emanations can be intercepted. Exterior cable must be protected against lightning. Limited maximum bandwidth.
BASEBAND COAXIAL	Low maintenance cost. Simple to install and tap. Wider bandwidth than twisted pair. Resists interference. Carries signals longer distance than twisted pair.	Expensive to install. Less flexible (in some cases thick coaxial cannot be used). Lower noise immunity than broadband coaxial. Bandwidth can carry only 40 percent load. Limited distance and topology. Conduit required for hostile environments.
BROADBAND COAXIAL	Supports voice and data applications. Tolerates 100 percent bandwidth loading. Good immunity to noise and radiation. More flexible topology (branching tree). Large geographical area coverage. Uses off-the-shelf industry standard CATV components. Rugged equipment needs to conduit.	High maintenance costs. More difficult to install and tap than baseband. RF modems required at each user station.
FIBER OPTIC CABLE	Extremely wide bandwidth. Difficult to tap. Very durable (glass does not deteriorate). Immune to outside electromagnetic or radio frequency interference. Supports many channels: voice, data, video. Small size and light weight. Low signal loss.	Relatively more expensive. Requires skilled installation and maintenance personnel. Poor flexibility. Fiber optic can't be bent sharply. Difficult to attach directly to devices on a LAN. Taps not perfected. Limited to high-traffic, point-to-point connections (backbones).

- Data rates and stations: 1 Mbit/s and 10 Mbit/s are the standard data rates. Up to 100 stations with 1000 bit average frame sizes can realistically be supported.
- CSMA/CD Protocol: The access protocol is CSMA/CD, which is, though a simple protocol, is prone to instabilities when the media utilization rises. Typically the performance of CSMA/CD LAN degrades sharply at utilizations in excess of 35% [11]. Collision occurs when two stations sense idle medium and try to transmit approximately at the same time. The collision probability is dependent upon the number of stations, the traffic rate, the length of the LAN, and the collision window given by the one-way propagation delay. Stations do not regenerate or amplify the signal. Only a source station transmits on the cable, all other stations listen and read the signals.

The disadvantages include the fact that collision detection is based on signal strength, and considerable fine tuning is necessary to install a network; the fault domain is difficult to identify; and no back-up path is available with standard topology. Furthermore no Medium Access Control (MAC)-level priority mechanism exists. The bridging is complex and limited in capacity. Single messages (e.g., corresponding to speech activity file transfer, full screen, image) require several frames with no deterministic way to predict response times.

The advantages are simple implementation, simple protocol, and good performance at very low data rates.

Token bus LAN [12]:

- Token bus LAN is a broadband bus built on a 75-Ohm coaxial cable.
- Protocol: Token-passing bus protocol, where physically broadcast media are used while logically sequential access is effected. Thus, physically a bus, but logically a ring, topology is realized.
- Priorities and Access Classes: Stations can be declared to be nontaken holding (i.e., receive-only stations). Token holding stations monitor the traffic load and adjust their timers, including the token holding time. Eight access priorities and four access classes are available. The purpose of priority mechanism is to allocate bandwidth.

The advantages include the facts that there is guaranteed bandwidth, as a station receiving the token is granted control of the common bus, for a specified amount of time. On the other hand, limitation on the token holding time guarantees that the token circulates fast enough. Therefore, the traffic monitoring is based on token-rotation time. Furthermore prioritized traffic can be handled, and protocol provides automatic recovery/management. The main disadvantages are the increased protocol overhead in each station, the difficulty in reconfiguring, fault identification and repairing.

Token ring LAN [13]:

- Token ring LAN uses baseband transmission at 4 or 16 Mbit/s on shielded twisted pair and implemented on a star-wired ring topology.
- Deterministic access protocol:
  * Controlled access via circulating token,
  * Single token, multiple frame protocol,
  * Sequential media, sequential access.
- Priority: Eight access priorities are available.

The advantages include high bandwidth utilization and easily extendibility via bridges. Furthermore all stations can detect hard errors like broken connections, failing terminals (reconfiguration to bypass errors) and soft errors (counter thresholds exceeded). LAN traffic is actively monitored by Active Monitor which also plays a role in the automatic recovery. The disadvantages are inefficiency in transferring long items of data (e.g., voice samples) on a large ring, and continual transmission at a high priority by stations that could prevent the access of lower priority transmissions.

Fibre Distributed Data Interface (FDDI) [14]:

- FDDI is an optical fibre based LAN which uses optical media with two counterrotating rings. Bridging to copper based LANs is also permitted.
- Nominal data rate: 100 Mbit/s.
- Protocol: Token passing protocol, closely identified with Token-ring LANs.
- Default timer values established allow up to 500 stations with total of 100 kilometers of fiber optic cable.

The advantages of FDDI include possibility for various configurations; provision for redundancy and recoverability; therefore, high availability. FDDI also meets requirements of real-time applications (e.g., voice, image, composite documents). Frame sizes from 30 to 100 Kilobytes can be accommodated. The disadvantages include the complexity of the protocols and the fact that FDDI attachments, transmitters and receivers are more costly.

Comparison Of The LAN Options:

1) CSMA/CD protocols (IEEE 802.3) is a poor solution for high traffic volumes, or to realize backbone LANs and high availability LANs. Their growth potential is rather restricted. Baseband LANs need specialized cable and eventually expensive bridges and repeaters.

2) Token passing bus (IEEE 802.4) LANs are best suited to meet the requirements for broadband multichannel transmission and time-dependent applications, providing guaranteed bandwidth in time-critical situations. They are also capable of coping with bursty traffic pattern. Token-passing bus does not really suit for backbone LAN (due to overhead) or high availability LAN (cost) implementation and it is not advisable for installation in a rapidly growing environment.

3) Token-passing ring (IEEE 802.5) LANs are best suited as general purpose LANs attaching workstations and systems. They allow very flexible and virtually unlimited growth. Token ring LANs are well suited for multisegment bridged LANs and hierarchical LAN configurations. Their bridging capabilities meet very well the requirements of high availability LANs. This type of LAN does not meet the broadband LAN requirements and the protocol is not sufficiently efficient in an ultra-high speed LAN environment. Token-passing rings may be wired using a variety of standard twisted pair cables and optical fibre cable sections. They are most closely aligned with FDDI.

4) FDDI LANs meet the requirements of very high speed LANs and are therefore best suited for high capacity backbones, highly utilized services, image and engineering applications as well as high bandwidth applications (video, integrated voice/data).

APPENDIX 9D

TAS SECURITY ISSUES

The appropriate communications and computer security measures for the TAS in secure areas will be combined with the overall security concept applicable to the WIS user location [Chapter 12].

Typical measures will include physically separate and controlled wiring, subject to regular physical inspection. Optical fibers are proposed to be used for intra-HQs wiring due to their superior EMC performance and, hence, resistance to crosstalk and interception. However, twinax cables can also be used within the secure enclosures. The optical fibre cabling arrangements should be used to interconnect separate TEMPEST protected terminals. At those WIS user locations, such as airbases, where the HQs may span several building in an area of a few square kilometers, the links between the subscribers or clusters of subscribers may need to be protected. In these cases, it will be the responsibility of the WIS user to apply end-to-end encryption.

The minimum requirements of secure cabling in HQs are as follows:

- All cables should traverse open areas and inspection of cables should be easy,
- It should be a simple task to trace cables from subscriber location to access switch and/or PABX premises,
- All new cabling should be authorized and installation be carried out under strict supervision,
- Cables should be splice free, whenever this is possible,
- It should be difficult to wiretap cables, with suitable warning in case of attempted tampering.

At present, all WIS equipment, including those in the terminal area will satisfy MIL-STD 411 requirements for EMC and TEMPEST requirements to limit compromising emanations. TAS elements will also comply with the TEMPEST requirements as described in Chapter 13.

The security and integrity of all messages/data and software stored or utilized by data and message processing facilities in the terminal area will be the responsibility of the end users.

Software Security: Security in the context of TAS is very similar to computer security (COMPUSEC). There are three modes of operation defined for computer systems which are also applicable in case of WIS TAS [Chapter 12]. They are summarized below.

- Dedicated systems: The computer system stores and processes information in a specific area of interest, serving a closed user group cleared to the highest classification level of information handled.

- Multi-level systems: In this mode, the system stores and processes information of mixed classification and it serves a user community having different levels of security clearance and need-to-know.

- System High: System operates in a similar way to multilevel, but main difference being that information handled is regarded as of highest classification, despite its actual classification, and all users are cleared to the highest level of information processed.

Multi-level system software poses the greatest difficulty in a testing, evaluation and verification. It is also well understood that formal certification, that will provide a high level of confidence in software, is only possible for relatively small systems operating in dedicated mode. WIS TAS will offer integrated services to a common-user local network, and therefore it has to be a multilevel system.

The software security problem in a multilevel, distributed system such as WIS TAS, becomes a manageable problem if and only if distributed systems can be divided into several dedicated systems (e.g. CCIS, AMPS, PABX). Then, the software for each system can be verified individually, assuming that interactions between dedicated systems are accountable. These arguments also increase the importance of using end-to-end encryption in WIS, because it is the only way of achieving COMSEC at the highest level of assurance in multilevel systems, where trusted computer security standards are still missing.

Development of Secure Software: Recognizing the fact that verification and certification of programs are only possible for software developed under close supervision and via observing strict rules, it is recommended that any program development for WIS TAS should comply with the following rules:

- High level languages such as ADA, PASCAL, C, CHILL or any other suitable language shall be used.
- Software shall be extremely modular, well documented and accountable. AQAP-13 should be used as the development and documentation control standard.
- All input/output parameters of software modules shall be clearly defined, all input/output procedures shall be traceable.
- Special built-in software and/or firmware modules shall guarantee software integrity by detecting and putting up alarms, in case of tampering with software.
- Software test/verification routines should be designed during software development.
- Special software monitoring modules should be developed that will constantly monitor all commands issued by all subscriber terminals and will compare them with fixed directories containing lists of authorized commands for that particular terminal or class of terminals. These modules shall discover any suspicious activity by detecting the terminals from which attempts are made to execute nonauthorized commands, and shall trigger suitable alarms.
- Similar modules shall be used to guard the system against attacks based on cracking password systems by trial and error approaches.

Secure Software Maintenance: Software maintenance has two objectives: Assurance of users that no nonauthorized changes in software are permitted; and guaranteeing that all software changes are made properly, being in accordance with rules set for secure and accountable software development. All software changes shall be fully tested; verified and then certified by an independent team of technically qualified experts with required level of security clearances. The management of the changes must be formalized in a reliable computer-based configuration control system.

Periodic and spontaneous checks should be made on the software running on the system components to verify that there are no nonauthorized modifications in the software. Very critical software modules, tampering of which may cause irreparable damage to communication system security, may be put on silicon to eliminate modification by ordinary software tools.

APPENDIX 9E

BUILDING WIRING

Cables are assemblages of conductors, screens, strength members and insulating material that carry one or more circuits. A circuit itself consists of two or more conductors and provides the connection between terminals and nodes. The main considerations in selecting a cable type for WIS user locations are:

- Loop resistance, i.e., the resistance in ohms for a specified length of cable,
- Attenuation, the loss in dB/km for passband frequencies of the signal,
- Characteristic impedance, while it is normally 600 ohms for telephony, for digital services using higher frequencies it is a function of cable dimensions and the cable must also be properly terminated. Typical high frequency impedances are 50-80 ohms for coax and 100-150 ohms for twisted pair,
- Crosstalk can be expressed as a loss in dB between circuits for a specified length of cable and a frequency range,
- Handling characteristics and resistance to physical abuse,
- Susceptibility to problems associated with radio-frequency interference, electromagnetic pulse and compromising emanations (RFI/EMP/TEMPEST), ease of tapping,
- Suitability for the intended application, i.e., plain telephony, token ring, multiservice terminals etc.

9E.1 Wiring Categories

The predominant type of the physical medium to be used in WIS user locations will be cable varieties. However indoors radio communications such as wireless PBXs and LANs are not entirely precluded for a limited number of users. The cable varieties can be categorized as follows:

a) Unscreened Twisted Pair (UTP): Twisted pair cabling is cheap, light, easy to handle, reasonably resistant to physical abuse, and by far the most common transmission medium for both analog and digital signals, for telephony and LANs The number of pairs within a sheath and high frequency performance are the considerations for selection. It is very susceptible to RFI/EMP/TEMPEST problems.

b) Screened Twisted Pair (STP): In installations with unpredictable noise source, screened cable is used instead, which consists of pairs of twisted conductors with shielding braid, with single outer screen, or with both inner and outer screens around the wires. Screening is intended to reduce crosstalk and external radiation (EMC). Earthing of such a shield may create problems of its own, particularly in an environment where EMP is to be expected. Protective devices such as spark gaps, surge arrestors, filters and voltage limiting semiconductors would still need to be installed at bulkhead connector panels to reduce the residual effects of EMP/RFI to safe levels. However, on the positive side, it is relatively difficult to tap into STP cabling without altering its transmission characteristics so significantly as to allow detection. Attenuation and waveform distortion are both significant problems with long lengths of twisted pair cable (STP or UTP). Sequential systems, as in the Token Ring, on the other hand, regenerate the signal at every repeater as a matter of course. Thus 10 Mbits/s speeds typically found in such systems can be readily supported.

c) Flat Copper Cable: This type of cable will provide the final connection between a wall socket and a terminal apparatus. Since it causes high levels of radiation and crosstalk, it does not run more than three meters. TEMPEST considerations preclude its use in WIS locations.

d) Baseband Coaxial Cable: This is cheaper grade of Community Antenna Television (CATV) coaxial cable, with a braided sheath. It has a diameter of 1 cm and is reasonably flexible. It should, however, be handled carefully, when used in CSMA/CD systems as reflections are easily induced by bending the cable beyond its minimum radius of 25 cm. Repeated winding alone would be sufficient to cause degradation. However, cable characteristics are far less important for sequential systems and transmission over 500 m cables would be possible. This cable is a 50 ohm cable and is exclusively used for baseband digital transmission. A form of Manchester encoding is employed, and data rates of up to 10 Mbit/s can easily be achieved. The coaxial cable used for baseband generates a low, but still significant, level of emanations by comparison with unshielded twisted cabling and is also less susceptible to the effects of RFI and EMP. It can easily be tapped without noticeable disturbance to the system.

e) Broadband Coaxial Cable: The coaxial cable used for broadband transmission has a greater diameter, typically 2 cm, and incorporates an other shield of solid extruded aluminum rather than braided wire. It is extremely difficult to bend and is therefore totally impractical for use in indoors cabling. This cable is a 75 ohm cable and can be used for both analog and digital transmission. Various modulation schemes, including FSK and PSK, are usable for digital data transmission. The efficiency of the modems employed determines the bandwidth needed to support a given data rate. With current technology, a data rate up to 50 Mbits/s is achievable. This cable exhibits lower radiation levels than baseband coaxial, partly because the shield is solid and partly because transmission is usually modulated.

f) Optical Fibre: Optical fibers are exceptionally small, flexible and light, but their fragility requires that considerable strengthening material be added to cables especially intended for field use. Typical, for a 4 core cable, the diameter would be between 0.5-1.0 cm. Optical fibers themselves are totally immune to the effects of RFI/EMP, but if metallic strength members are employed then the problem of providing protection against such threats remains. Optical fibers are emanation free, therefore TEMPEST considerations do not arise. Tapping into optical fibers without interrupting transmission is not possible. Optical fibers have very low attenuation values; high silica glass offers very high speed-distance products reaching 100 Mbit/s/km. Plastic fibers are limited to distances of 1 km. Typical attenuation values of commercial glass fibre are 2-6 dB/km at 0.8 μm wavelength and 0.62 dB/km at 1.3 μm wavelength.

The other components of the cable assemblages are connectors, terminating blocks, boxes and distribution frames, patching apparatus, line terminations, transformers, signal regenerators, line drivers, overvoltage protection devices, and for fibers, electrooptic and optoelectronic converters.

9E.2 Wiring Performance Requirements

The wiring system performance requirements are as follows:

- End-to-end transmission and noise performance of a cable circuit within a building (crosstalk effects excluded),
- Crosstalk effects between circuits within a cable sheath,
- Earthing,

- Electromagnetic compatibility (EMC) issues,
- Earthing,
- Resistance to tampering and physical abuse.

Transmission Performance: The building wiring plays a small role in the apportionment of end-to-end impairments in the international calls as determined by CCITT.

- Analog voice band terminals (speech, fax, voice-band data) employ normally unscreened twisted-pair wiring. Such a loop, one km. in extent, would encounter an attenuation of 2 dB. Notice that the connection between the wall socket and the terminal apparatus, being unbalanced, may cause, during signaling, significant impulsive noise that can interfere with data circuits.

- ISDN terminals have the relevant wiring attached to the S or T interface as in CCITT Rec. I.430. In this recommendation the attenuation and delay figures are given for various wiring topologies, such as point to point or multipoint.

- V.10 Interchange Circuits (CCITT) are unbalanced and twisted pair wires, for which the length should not exceed 1 km for data rates below 1 kbit/s.

- V.11 Interchange Circuits (CCITT) are normally balanced, twisted pair wires, for which the length depends upon the data rate above 10 kbit/s, while for rates below the length should be less than 1 km.

- V.28 Interchange Circuits have a suggested length limit of 15 m on unbalanced twisted pairs.

- V.36/37 Interchange Circuits have suggested limits of 60 m, the former being recommended for data rates up to 72 kbit/s, the latter for rates above 72 kbit/s.

- G.703 Physical and Electrical Characteristics of Hierarchical Digital Interfaces have implicitly a limitation on wiring length. A maximum attenuation of 3 dB at 64 kbit/s with twisted pairs of 120 ohms characteristic impedance, and 6 dB maximum at 2048 kbit/s with 75 ohms coaxial cable.

- IEEE 802.3 (CSMA/CD) Ethernet LAN has many wiring options some of which are:

  1) 10 Mbit/s, 50 ohms coaxial cable with up to five 100 m segments,
  2) StarLan option using a 1 Mbit/s twisted pair with up to 5 segments,
  3) A 3.6 km. segment carrying data at 10 Mbit/s modulated onto a carrier at a frequency in the CATV band (this option is of interest only if a CATV wiring already exists).

- IEEE 802.4 (Token Bus) LAN has as baseband transmission medium the 75 ohms cable with short stubs consisting of 37 to 51 ohm drop cables

- IEEE 802.5 (Token Ring) LAN employs screened twisted pair wiring, where the overall cable attenuation end-to-end should not exceed 26 dB at 4 MHz.

- Fiber Distributed Data Interface (FDDI) has a maximum loss of 2.5 dB/km within attainable link lengths of 2 km.

Crosstalk is the result of near-field coupling between transmission media. Far-field coupling, which may be correlated with near-field coupling, is of concern in the context of EMC. Obviously in a cable sheath there will be multiplicity of disturbing circuits (crosstalk generating) for every disturbed circuit. The factors involved in crosstalk assessment are the crosstalk properties of the cable, the spectra of signals involved in crosstalk, and the noise requirements or margins for the disturbed circuit.

From the point of view of crosstalk attenuation the balanced and unbalanced state of the disturbed and disturbing circuits is of relevance. In fact, one can have:

1) Both circuits unbalanced,
2) One circuit balanced and the other not, which improves the crosstalk attenuation by 20-25 dB,
3) Both circuits balanced, which improves the crosstalk attenuation further by another 40 dB.

The crosstalk increases with the length of the cable, and especially the Near End Crosstalk (NEXT) behavior is dominated by cable attenuation. For short lengths and low frequencies, voltage or current addition of crosstalk terms applies and for long lengths and high frequencies crosstalk terms can be considered to be uncorrelated, and hence power sum law applies. It may be surprising that if coaxial cables are laid side-by-side, at lower frequencies (e.g., below 100 kHz) the crosstalk can be worse than for twisted pairs, because they are unbalanced and there is mutual inductance between them. At frequencies above 5 MHz the crosstalk performance of coaxial cables becomes superior to twisted pairs.

Crosstalk introduced into a disturbed circuit is treated as noise on that circuit, and hence crosstalk limits are determined by the noise limits acceptable for that circuit. The crosstalk disturbance depends upon the spectrum of the disturbing signal. The procedures for assessment of crosstalk depend also on whether the disturbing circuit is digital and the disturbed circuit is analog, or the disturbing circuit is analog and the disturbed circuit is digital, or finally whether both circuits are digital.

Earthing: Earthing arrangements can be protective earths for safety (including equipment protection) and functional earths for prevention of coupling between different earthing paths, which has also a bearing on the EMC. The two functional earthing techniques are:

- Earthing of screens,
- Common bonding networks.

Earthing of Screens: When cables extend appreciable distances and are screened either with an inactive screen (e.g., a screen around a balanced pair) or with an active screen (e.g., as in a coaxial cable) earthing of screens must be considered. Foremost it is necessary to prevent earth currents associated with the mains from flowing through the cable screens, especially due to mains harmonics extending into the baseband signal spectrum. Normally the screen is earthed at one end only. For example in a connection from a PBX to a terminal equipment the earth would be at the PBX. From the point of view of EMC it is preferable to earth screens or to connect them to the signal common return at both ends, especially above 30 MHz.

Common Bonding Networks: It is common practice in buildings containing large amounts of equipment for each work area to have its equipment earthen locally to an earth point which is joined to the building earth by a separate connection resulting in a star configuration. Reducing thus the earth impedance is advantageous from the point of view of both reducing coupling between apparatus and improving the EMC situation.

CHAPTER 10

INTEROPERABILITY

10.1 OBJECTIVES

A network like WIS will have to interoperate with a number of other networks, in order to meet various operational requirements, to optimise flexibility and survivability, and to obtain cost effective rationalisation. For this purpose, WIS should be capable of establishing interconnections between various networks and/or interchanging transmission capacities or access facilities. Any interconnection should be established in a secure and automatic mode and they should not compromise the security and internal homogeneity of the network.

For reasons of minimum complexity and maximum maintainability, flexibility and robustness it is desired that WIS be as internally "clean" as possible. This desire leads to the concept of gateway solutions, and to the necessity of wide scale deployment of gateway units implemented separately from WIS switches. The superiority of the solution with gateways implemented as separate units as compared with embedding inter working functions and integrating them with the switches can be listed as follows:

- Portability,
- Flexibility and expandability to changing network architectures and services,
- WIS remains as a homogenous network,
- Incompatible precedence and signalling mechanisms can be bridged,
- Considerable path diversity and alternate routing capability can be achieved by the use of multiple gateways.

The complexity of gateways will depend on the differences and incompatibilities in the signalling, numbering, routing systems and in other characteristics of the two networks. As a consequence, generic solutions do not apply, and therefore the gateway functions and responsibility for the interface must be clearly defined and delineated on a case by case basis to cover such aspects as specific boundaries of responsibility, funding responsibility, standards, maintenance, operating procedures and control, provision and funding of circuits and modification of existing facilities.

It should also be realised that there are a number of major constraining factors that will make the achievement of ideal interoperability a long term goal. Typical of these constraining factors are the transitional nature of networks, funding limitations and lack of mature and stable standards.

10.2 GATEWAY FUNCTIONS AND FEATURES

The WIS network as well as the other networks with which WIS will interoperate, will be marked by a series of transition periods. Furthermore, the differences in timing of replacement schedules, varying requirements and a continuous technological evolution will mean that new systems will be appearing and old systems will be phased out during the implementation phases and operational lifetime of WIS. This reality will necessitate the deployment of flexible interoperation units i.e. "gateway units", between WIS and these adjacent networks. The gateway concept provides a basis for the switched interconnection between WIS and the other networks on a call by call basis and can be used to support a variety of options. In this sense gateways represent an open-ended evolutionary solution.

Major inter working functions to be implemented in a gateway are listed below:

- Signalling conversion, i.e. supervisory and control signalling and possible user to user signalling,
- Speed, code and format conversions,
- Multiplexing hierarchy conversion,
- Maintenance and management procedures,
- Mapping and conversion of supplementary service sequences, e.g. precedence and pre-emption,
- Crypto key interchange for end to end encryption applications,
- Directory services, e.g. to enable the telephone numbers for itinerant or absent users to be located through an enquiry point or directory number translation,
- Flow control, e.g. through class-marking at the gateway.

In addition for data communications:

- Error control,
- Flow control,
- Address conversion (e.g. for logical devices).

Other considerations in planning gateway functions are as follows:

- Operational and security considerations may impose certain restrictions on the inter working functions at the gateways. For example it may not be desirable to have interoperability in the network management, data collection, monitoring and maintenance areas. The gateway units should therefore be able to restrict data for functions like these from traversing between the networks.

- An attractive potential solution would be that each system responsible for communication with another system should terminate his connection with gateway units converting from his internal standards into a commonly agreed set of standards and profiles. Thus, for the case of M different networks rather than having to design and maintain M times (M - 1) different gateway conversions it would suffice to support only M +1 different interfaces, and M conversions to this one agreed general interface.

- These gateway standards would reflect the services and functions in the network on either side of the interface. The real world of existing and future networks contains a plethora of such services both in number as well as in implemented functionality. When the number of networks exceeds a few, then the common interface defined as the highest common denominator would become, at best, the possibility to exchange basic calls. No standard for a general interface with comprehensive functionality exists, therefore, and the gateways described later are basically defined as bilateral definitions, or definitions between classes of networks (e.g. ISDN to ISDN or tactical to tactical etc.).

- In order to maintain the generic software and hardware of the switches as homogenous and maintainable as possible, the interface should be implemented as a separate unit or as a subsystem in the switch. Some of the interface functions may, however, be integrated with the main switch software, to utilise the processing power and database of the central switch.
- Gateway units should be implemented with such features as simplicity and modularity, while reflecting the service features and functions of both networks. They should also be able to translate new functions and services as they evolve in time.

10.3 CLASSES OF INTEROPERABILITY

Interoperability between adjacent networks can be discussed in three primary classes:

1) Inter working in which a user in one network can communicate with a user in the adjacent network using their regular terminals and service procedures. An example of this type of interconnection is when a WIS user can originate a call to a user of an interconnected tactical network.
2) Inter working in which a network is used to pass traffic between two other networks using the intermediate network as a transit media. An example of this is when two subscribers connected to two different tactical networks communicate using WIS as a transit network.
3) Inter working in which two networks rationalise their transmission resources by interchanging capacity on a dedicated circuit or group link basis. This interchange of capacity will typically be at the 2 Mbit/s level.

Class 1 offers the benefit of true inter-connectivity at the user level. In the public telephone networks this is today a service we take for granted, and the interface level that is standardised is an analogue 3,1 kHz telephone channel. For data communications this is not so much an obvious commodity, and the examples of the most widespread interoperability is to be found in the Internet, based on the now "de facto" standard protocols TCP/IP. In ISDN networks this type of interoperability is a goal behind all efforts in standardising interfaces and protocols, and the agreed levels of standard protocols and profiles are increasing with respect to the basic bearer services, and some of the simpler teleservices. In the longer term it is expected that an increasing number of teleservices and also important supplementary services, like for example conference calls etc. will become widely adopted. Between inherently different networks, like tactical networks, conventional telephone networks, dedicated data networks (e.g. packet switched networks) and ISDN networks, the complexity of this class of interoperability will remain a major technical and "political" challenge.

Class 2 offers the advantages of cost-effective exploitation of resources, while the user to user protocols can be negotiated without the involvement of the intermediate network. It is therefore possible to achieve this type of interconnection without bridging difficult differences between networks, and as a special example the achievement of security compartmentation can more easily be achieved since the intermediate network does not need access to the content of the user traffic.

It can be argued that the class 3 interoperability is merely a transmission interchange, and as such will not even involve the switches in the interconnected networks. The only necessary interface standards are related to transmission parameters like in CCITT Rec. G.703, and agreement on timing and network control data. In conventional networks as seen today including ISDN, this may be the whole story, but in the future we will see broad band networks using ATM (Asynchronous Transfer Mode) switching techniques where this form of interoperability will become a common service.

10.4 CONCEPTS FOR WIS INTEROPERABILITY

During the design of the WIS network it was identified that the network was required to provide degrees of interoperability (classes 1, 2 or 3) with a number of networks.

10.4.1 Interoperability With Similar ISDN Networks

Being a secure and survivable ISDN-based network WIS was anticipated to interoperate at all three classes with adjacent, subordinate and superior ISDN networks. These were identified as:

- The public ISDN network, operated by commercial or government owned agencies.
- A superior multi-national defence ISDN network operated by an alliance in which the WIS Authority is a member.
- ISDN-type private networks serving subordinate user formation as sub-communities under the WIS user organisation.

For all these cases the following observations were found to apply:

- At class 3, the interoperability was readily achievable by standard digital transmission interfaces at 2 Mbit/s. The challenges identified were mostly in the procedural and financial arena, where both parties were assumed to offer transmission capacity to each other, while the timing at the interfaces would be assumed to be transparently carried through the providing network, that would also offer a simple form of status reporting regarding the operational status of the link.
- At class 2, the interoperability is requiring agreement on the above, plus the agreement of signalling protocols supporting the basic bearer services offered by the intermediate network. Since no termination of calls would take place in the host network, all call set-up related protocols would be restricted to the negotiation of a bit-transparent non-restricted 64 kbit/s teleservice. The gateway signalling system could be either CCITT SS No 7 or Q-sig, depending on the status of the adjacent network. All other tele-service and supplementary services would be the responsibility of the originating and terminating network, as would the higher level services of the terminals utilising the channel.
- At class 1, however, the complete interoperability profile would need to be negotiated, and the extent of the interoperability features would be depending on the level of common services identified for the two networks. The common services may, however, be differently implemented, giving increasing complexity in the signalling conversion functions of the gateway.

The developments in public and private ISDN networks will undoubtedly create a powerful thrust in availability of standardised services and functions that will benefit a network like WIS. The basic bearer services will allow terminals to interoperate, and interoperable terminals will with and without the use of terminal adapters provide end user functionality's for voice and non-voice services. The networks will interoperate over two types of signalling interfaces:

a) CCITT Signalling System No 7: This interface will develop into the standard interface between national commercial PTT networks. The international PTT gateway will be the interface that a network like WIS can use to interconnect to such a network. The interesting challenge to the PTT is the delineation of the network control on their network borders, as well as the settlement of network charges for calls crossing the network borders. Since WIS is a network for defence communication, being non-commercial by nature and secure and self-contained by nature, these issues will not further reduce the freedom of access across these gateways. On the other hand will the combination of the issues listed above limit the range of problem areas to the standardisation of inter-network procedures for multi-level precedence and pre-emption and the control of the use of cascaded satellite hops, which must be strictly limited.

b) ETSI Q-sig/ ISDN signalling: Originally designed to be the interface between a PABX and the public ISDN network, this interface is now significantly developed in the direction of being a balanced network interface between private switches (PABX's) designed to operate in private corporate networks. Since these applications are similar to the WIS original application, and since the same user driven approach applies to these networks, it is envisaged that many of the gateways that will be developed for WIS will be based on Q-Sig. The bilateral agreements required to solve security issues will easier be resolved in this profile, and a number of supplementary services are already becoming standardised in this interface.

10.4.2 Interoperability With Analogue Networks

In developing inter working schemes between WIS and existing analogue networks, the following limitations of these networks must be observed:

- These networks were all designed as closed networks with limited scope for inter-networking.
- These network are typically implemented with limited capability of network extension.
- The digit handling capacity in these networks are often limited, both in inter-network numbering as well as for sub-network addressing.
- The switches in these network can seldom support any introduction of new call processing or signalling functions.

Since the limiting factors in inter working with analogue networks will be on the side of the analogue network, the obvious first observation is to use the WIS implementation as an incentive to replace the analogue networks with digital alternatives. WIS itself can in some cases be extended to handle for example the PABX requirement of a user site. The public networks on the other hand is rapidly being upgraded to ISDN standards, leaving WIS only in interim periods to need to inter work with the old analogue networks.

For the classes of interoperability listed above the following comments apply:

1) For class 1, the only inter-operating bearer service is a 3,1 kHz analogue channel. This channel can be used for non-secure voice, in which today's standard international telephone service will be the dominant tele service. The channel can

further be used for modem type data traffic up to about 2,4 kbit/s (depending on channel quality up to 9,6 kbit/s). This data service can then support data links, as well as narrow-band secure voice, provided that the user define and specify the terminal interoperability (the networks will not take any responsibility for this).

2) For class 2, WIS can offer transit services up to 64 kbit/s to allow analogue networks to interoperate, as the case is for the tactical networks. WIS will in this case only represent a transmission medium for the interconnected networks, and no signalling conversion is required for this purpose.

3) For class 3, as for class 2, WIS can provide transmission bearer service, in this case with a transmission capacity of 2 Mbit/s. For both these classes will no relevance exist for the opposite direction, in that there will be no need for the analogue networks to offer any capacity to transport WIS traffic.

10.4.3 Interoperability With Dedicated Data Networks

WIS is expected to need to interconnect to two different types of data networks:

a) Connection oriented packet switched wide area networks (WAN's) based on X.25/X.75 data protocols.
b) Connection less local area networks (and possibly WAN's) with networking interfaces based on TCP/IP.

The dominant factors in setting the standards for these interoperabilities are:

1) WIS as is ISDN is a connection oriented network where wide area packet switched data will be carried through an integrated network of packet handlers collocated with the ISDN circuit switches.

2) The end user application of switched data will predominantly be user applications installed on servers and workstations locally internetted using Local Area Networks, using TCP/IP software protocols.

The natural and pragmatic choice for this inter working is therefore to install internet type routers at the gateways to the LAN's, and let the router to router traffic subscribe to the WIS-internal data services (circuit switched B-channels or packet switched service through the packet handlers).

As a special case of this inter working will be the handling of the message traffic from WIS to and from message networks that it will be interconnected with. The WIS message traffic will internally be routed between internal message switches, specified according to the CCITT X.400 principles. The main concern in this context is the security aspect of interconnecting various message systems that have different standards and security principles. Some of these problems have to be solved by establishing manual review and release gateway stations between the systems, even in some cases with real "air-gap" to secure against non-authorised illicit channels.

10.4.4 Interoperability With Tactical Networks

The following considerations are to be taken into consideration in planning inter working functions between WIS and tactical area communication networks:

- The tactical networks are subordinate networks to WIS.
- The tactical networks will follow the standards in STANAG 4206-4214, STANAG 4249 and EUROCOM D/1.

- The interoperability between the tactical networks and other networks will be arranged through WIS. Operational requirements and geographical considerations will necessitate the deployment of several tactical networks inside the area covered by WIS. Consequently WIS will act as the wide area network (WAN) between tactical networks, and for any other network that these network need to interoperate with.

Two types of interfaces are needed for the WIS to tactical network inter-connections:

a) Digital Single-channel Interface (DSI) (16 or 32 kbit/s channels from the tactical network, i.e. switched channels),
b) Trunk or trunk group Multi Channel Interface (MCI) (n-fold 256/512 kbit/s trunks transported in bulk through WIS, i.e. non-switched).

These interfaces are illustrated in Figure 10.1, and are described in some more detail below:

10.4.4.1 Digital Single-Channel Interface (DSI)

The inter working between strategic and tactical networks on a call by call basis is crucial as a wartime operational requirement. A certain percentage of these calls will be secured by end-to-end encryption.

The main functions of this interface are:

- The PCM coding of voice signals in WIS (CCITT Rec. G. 711) is converted to EUROCOM CVSD voice coding (EUROCOM D/1) and vice versa.
- Data traffic will in the tactical network be packaged inside the standard traffic channel of 16 kbit/s. This applies to circuit switched data at rates up to 9,6 kbit/s according to CCITT X.1 user rates, as well as unformatted 16 kbit/s primarily intended for digital facsimile. For these traffic types the gateway will convert the signals into standard WIS user rates, via WIS terminal adapter functions or PAD functions for packet formatted traffic.
- Message traffic has yet to be standardised inside tactical networks, although the use of a stripped down version of CCITT Rec. X. 400 is foreseen in some networks. When this service become standardised, it will be natural also to include the conversion to standard WIS X.400 protocols at this interface.
- Signalling messages according to the tactical network signalling system (EUROCOM, TRI-TAC or STANAG 4200-series) is converted to the WIS common channels signalling system (CCITT No 7 or Q-Sig.). A common set of supplementary services and the corresponding signalling sequences need in detail to be scheduled for this conversion. The outcome will be that some services in either network will be unavailable across the interface, due to incompatibilities between the networks.
- Due to the signalling and data conversion taking place at this interface, all user data and signalling inf. will have to appear at the interface in non-encrypted form. The consequence of this is that both networks will have to have trust in the gateway should any classified information be needed to be transferred. As the final service must therefore be mentioned the situation that the interface transfers voice or data in encrypted form. In the case of voice this has to be either 2,4 kbit/s NBSV, or 16 kbit/s encrypted CVSD coded voice. In the latter case as well as with the general case of 16 kbit/s encrypted data will the main issue then be to find terminal equipment in WIS that are compatible with these formats.

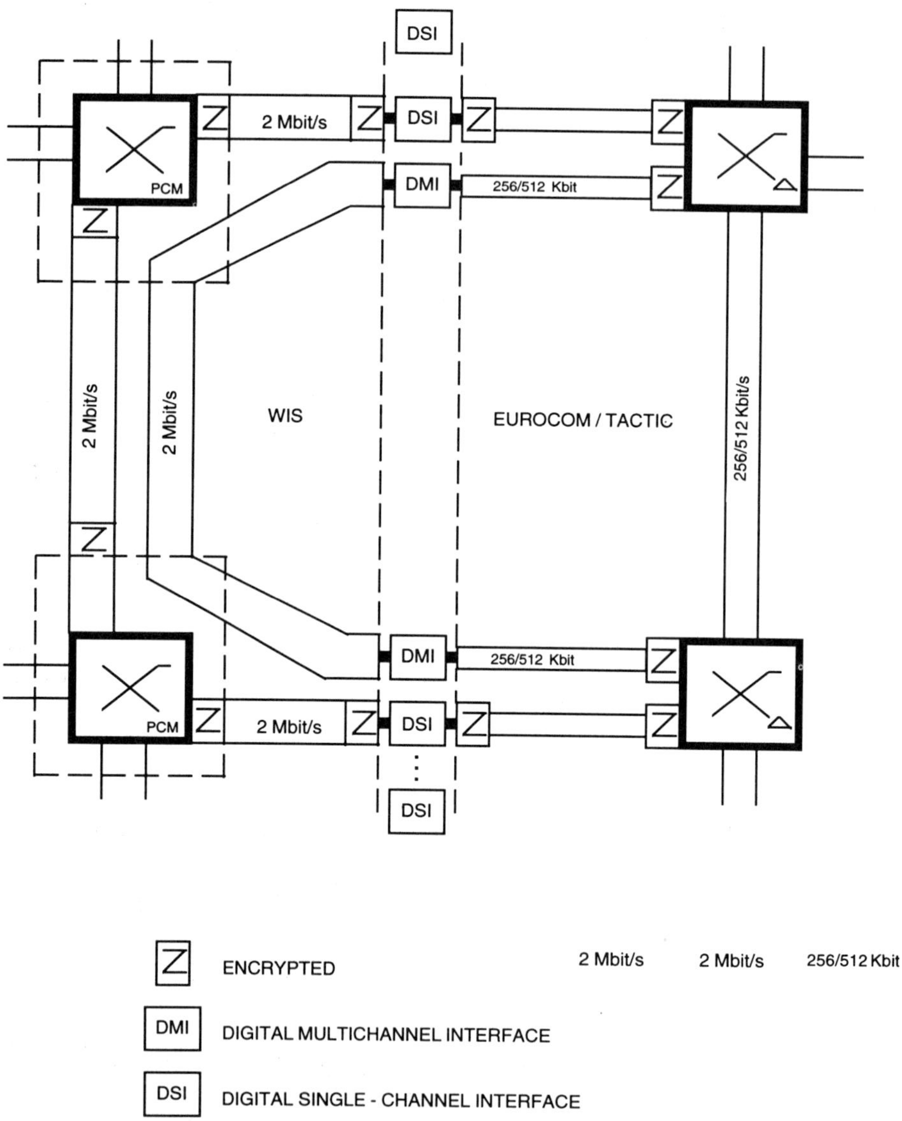

Figure 10.1 Switched and Non-Switched Interconnection Between EUROCOM Type Tactical Networks

The interface will in summary then support the following modes of traffic:

- Plain Voice,
- Narrow band and medium band (16 kbit/s) secure voice,
- Data (including facsimile and telegraphy),
- Video (slow-scan).

The transmission rates on the tactical side will be 16 kbit/s (or 32 kbit/s), while on the WIS side they will be 64 kbit/s. Rates below 16 kbit/s can exist in the form of rate adapted streams that will be converted to standard X.1 rates at the interface.

Depending on the type of traffic, the main options for rate conversion can then be summarised as shown below:

1) Transcoding, where the conversion between PCM and CVSD is provided using cascaded processes of D/A and A/D encoding. This option is only suitable for plain voice or data converted by voice channel modems, but the interoperability achieved becomes class 1.

2) Multiplexing, where the tactical channels are converted by multiplexing 2 or 4 channels into one 64 kbit/s channels in WIS. In this case the WIS network is used as a transit network, and the interoperability achieved is only class 2. The multiplexing is done by interleaving the subrate streams within each B-channel octet in accordance with CCITT Rec. I.460.

3) Rate Adaptation, where the lower rate is entered into the B-channel rate of WIS by means of bit replication an/or bit stuffing. The rate adaptation can be made according to CCITT Rec. I.460 and I. 461. The rate adaptation preserves the bit transparency requirement, and could be applied also for class 1 interoperability, if as a special terminal adapter case in WIS a tactical CVSD terminal could be hosted as a "visitor" terminal. Otherwise the method will only support class 2.

4) Multi-sampling, where the 16 or 32 kbit/s bit-stream is PCM converted according to CCITT Rec. G.711. It is conjectured that for 16 kbit/s secure voice application, the resulting loss of synchronisation from jitter may be intolerably high.

On the basis of these techniques, the variants of the Single Channel Interface (SCI) for plain voice interconnection, may be as shown in figure 10.2. Here it can be seen that the interconnection of plain voice terminals can be obtained using the following functions:

- To interconnect a CVSD tactical voice terminal with a digital or analogue WIS terminal the gateway performs trans-coding (TC) between STANAG 4209 and CCITT Rec. G.711.

- To interconnect a CVSD tactical voice terminal in the tactical network with a tactical terminal "visiting" in WIS, rate adaptation (RA) is performed from 16/32 to 64 kbit/s. The tactical terminal in WIS must have a special type of terminal adapter reconstructing the 16/32 kbit/s bit stream.

- The gateway unit can also perform multiplexing (MUX) of 16/32 kbit/s bit streams into a 64 kbit/s WIS bearer circuit. The terminal adapter function in this case need to be a similar subrate multiplexer.

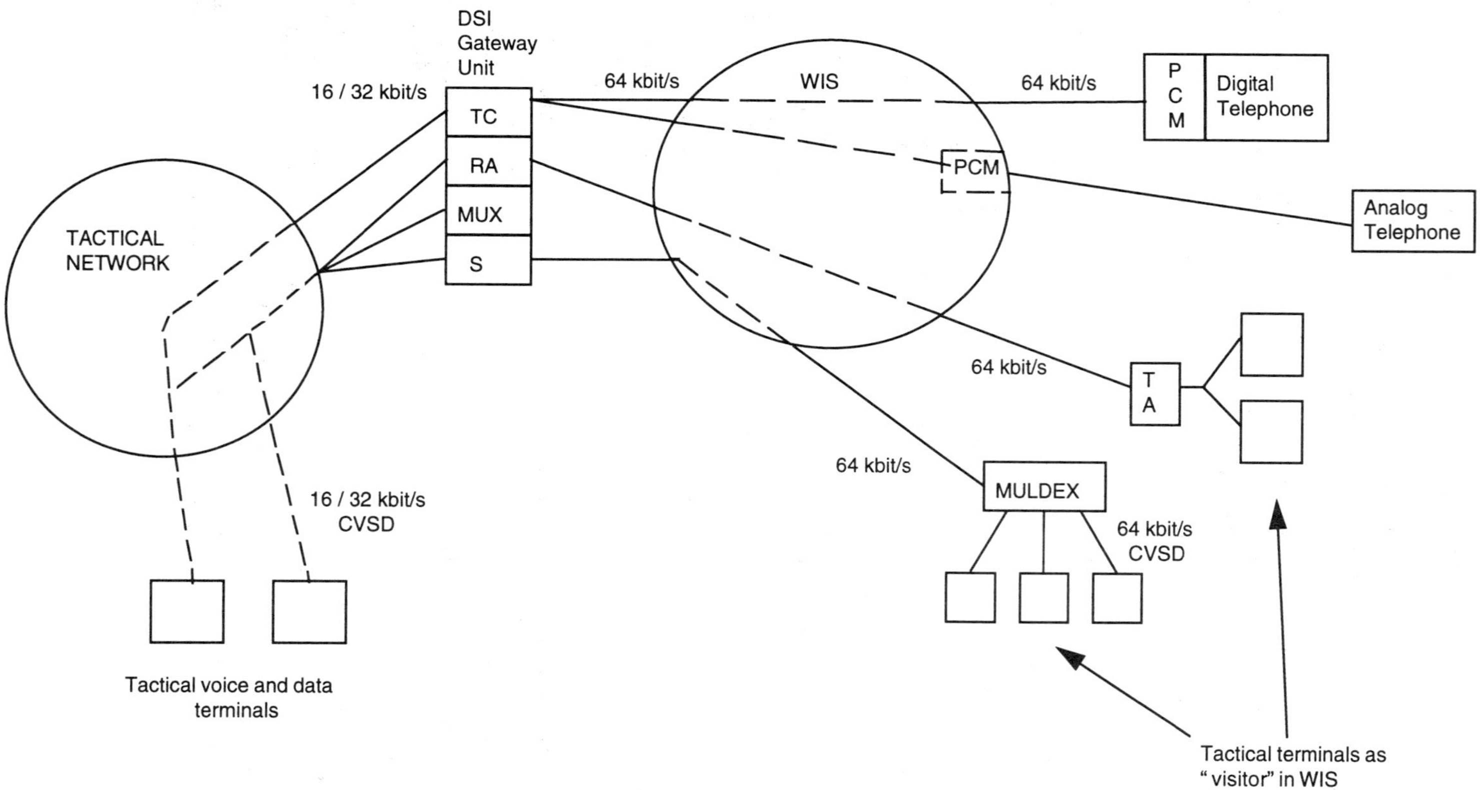

Figure 10.2 Interconnection of Terminals Between WIS and Tactical Networks (S:Signalling Conversion)

10.4.4.2 Digital Multi-Channel Interface (DMI)

In this case, as can be seen from figure 10.1, the WIS is merely a transmission carrier for the tactical network, which presents a multiplexed (and encrypted) signal at the interface point. It is shown in the figure that the 256 or 512 kbit/s signal is packed in a 2.048 Mbit/s PRA-group in WIS and patched through on a dedicated group basis. This represents the technically easiest solution, but it can also be envisaged that an ISDN bearer services of n times 64 kbit/s (n = 4 or 8) could be used in this case giving higher flexibility and less waste of capacity in WIS.

CHAPTER 11

NETWORK SURVEILLANCE AND CONTROL

11.1 REQUIREMENTS AND OBJECTIVES

The study of the network surveillance and control system for WIS was carried out in close harmony with the established requirements of the Authority as presented in Chapter 2. The development was firmly based on the so-called WIS concept as described in Chapter 4.2, and it was accordingly decided to provide:

- surveillance,
- control and maintenance,
- service and resource provisioning,

that will support

- planning activities

for all WIS network elements with a view to maintaining maximum communications performance under changing traffic conditions, natural and man-made stresses, disturbances and equipments disruptions.

Control considerations for any defence communications system operating in a common user multiple interest configuration must take into account not only the facilities making up the system, but also the users of the system and the environment in which the system operates.

The WIS Network Surveillance and Control Subsystem, hereafter called WNSC, will therefore be required to:

a) ensure the required communication availability for network users during times of peace, tension, crisis and war,

b) maintain the network during the same conditions at a stipulated level of communications performance, i.e. grade of service, quality of service,

c) provide and restore service to users in their appropriate priorities under all operating conditions,

d) support centralised and decentralised modes of operation,

e) have a level of survivability at least equal to that of the subsystems which it is controlling,

f) have associated with it security measures designed to assure a maximum of protection against all anticipated active and passive enemy actions directed at deteriorating network performance or integrity, and

g) have an organisation with adequate manning, sufficient control and test equipments and other facilities to meet the above requirements.

The WNSC subsystem must support in a time efficient manner the following control functions:

a) The status and performance monitoring of all WIS elements, i.e. switches, links and terminal equipments in order to identify where and when faults and degradations are occurring, or are likely to occur in the network, ideally before these affect overall system performance and service provided.

b) Maintenance of the required service and performance through timely fault detection, location, isolation, traffic and configuration management and repair as appropriate. In particular, the capability to restore service to designated critical subscribers within minutes of a failure of a network element.

c) Maintenance of current data on the users and the status of resources and assets which may be used to support and restore service under stress condition.

d) Collection of selected data on traffic flow, equipment utilisation and failure rates, user to user and link availabilities, grade of service, and message delivery times which will be used for configuration and traffic management. This data is also to be used in support of longer term system planning and management functions.

e) Data base management and the audit of hardware and software resources.

f) Effective planning and direction of logistics and maintenance activities on the network

g) The electronic distribution of encryption keys if required.

h) Prevention of unauthorised access and interference to system processors, data and programs stored on the processors and their peripherals.

i) Co-ordination with other and interconnected networks control organisations at appropriate levels, and furnishing the agreed reports and status information to these networks.

The WNSC design must be based on ISO OSI-RM, with the following features:

a) It must consist of computer based control systems (operations systems) supporting various network operations and maintenance functions. These systems must have modular hardware and software architectures so as to facilitate evolutionary development and expansion without major redesign.

b) The WNSC subsystem resources (processors, terminals etc.) must be collocated with the communications equipments at hardened locations so as to have a level of physical survivability at least equal to that of the other subsystems that comprise the network. Survivability must be further improved by equipment redundancy, function duplication and database distribution.

c) The system and terminals comprising the WNSC subsystem must be easy to operate, to maintain and be highly reliable and must have an adequate level of software and hardware documentation.

d) The WNSC subsystem design must where appropriate conform with TEMPEST regulations, COMSEC criteria and to ISO, CCITT and CCIR Recommendations, where applicable.

11.2 SYSTEM CONCEPT

11.2.1 Introduction

This section develops the WNSC concept, including the general principles and objectives for operations, maintenance and management of the WIS network and planning, designing, operating and maintaining the WNSC subsystem itself, to meet the requirements stated in the above sections. In section 11.2.2 below, the system and operational requirements are elaborated. Computerised operations, multi-vendor environment and related questions are discussed.

There is a current trend towards the use of commercial and civilian standards as the basis for the design of military communications systems. ISDN standards and the architecture implanted on Open Systems Interconnection Reference Model (OSI/RM) of ISO, have already been accepted as the basis for NATO's digital communication plans. WIS was also designed and implemented in conformity of these standards. As a consequence, network management systems for WIS is also structured in consonance with the OSI reference model.

The WNSC subsystem is furthermore designed and implemented within the framework of the CCITT "Telecommunications Management Network" (TMN) concepts and architecture, with enhancements to meet the military requirements. TMN concepts, with particular reference to WNSC, are discussed in section 11.2.3 below.

The WNSC concept is developed on the basis of TMN with the following additional characteristics:

- A hierarchical distributed control system, for survivable, flexible network control and autonomous operations in local and regional levels.
- Integrated operations system, for effective and cost-efficient network control.
- A surveillance-based control system, for responsive, speedy and efficient network control.

These specific issues, together with the other military characteristics like survivability and security are discussed in section 11.2.4 below.

11.2.2 Operational and System Requirements and Factors

The WNSC subsystem, as other subsystems, should be designed and implemented such that it will meet the operational requirements stated and detailed in chapter 2. Here the following requirements apply:

a) Survivability,
b) Communications Security,
c) Readiness,
d) Interoperability,
e) Standardisation,
f) Reliability,
g) Availability, Flexibility and Responsiveness.

Telecommunications networks are becoming more intelligent and sophisticated as enhanced software technologies are introduced into equipments and facilities. The WIS network elements (NE) will all have intrinsic intelligent features for their operations and maintenance. These features include: performance monitoring capability, remote threshold setting, self diagnostic testing, remote data manipulation, and enhanced communications with humans, and with the computer based control systems (Operations Systems (OS)). In this context, an OS is a conceptual, computer based system which assists telecommunications network administrators in performing various network surveillance and control functions, i.e. application functions. It can take the form of a single intelligent terminal or a multiprocessor system. Operations Systems will be implemented at 3 hierarchical control levels, master or network level (NCC), regional control level (RCC) and local or node level (TCC). They will automate network operations and management activities, providing dynamic and responsive control of the WIS network.

WIS network equipments and facilities may be supplied by several different vendors, or future service requirements may necessitate products by a supplier different from the original ones. This will present problems as regards uniformity of operations. Thus adopting international standards, open system architecture and standard interfaces for the WNSC subsystem, will tend to reduce the complexity of these problems. The availability of such established standards to a sufficient level of detail, will be a essential prerequisite to achieve a cost-effective multi-vendor environment. This is not obvious to happen, and close monitoring is required to assess this question.

Military communications networks are moving away from providing the traditional and non-integrated plain telephone, telegraph and manual message services, to more sophisticated and computer aided and integrated voice, data and picture services. Many of these services which will be introduced in WIS will require more stringent performance criteria than the traditional ones, and will be introduced as functional additions to the initially implemented bearer services. The WNSC design should be based on OSI-ISO reference model to accommodate the control requirements for such future network enhancements.

WIS will be a digital network largely following CCITT IDN/ISDN standards and recommendations. The key characteristic of the IDN network is that digital technology allows for a very efficient integration of transmission and switching resources. This gives significant advantages to a more efficient network control organisation, if these integration features are utilised. WNSC will have similar characteristics with the network control systems developed for IDN/ISDN networks.

WIS operational requirements demand fast response to:

- Equipment failures,
- network damage, and
- operational changes.

The WNSC subsystem will be a computer assisted system, automating the process of detecting and verifying troubles, protecting/ restoring service, isolating troubles and returning systems to service. Further this computer assisted system will collect performance and traffic measurement data from the network elements and process and analyse these data for the dimensioning, planning, operation and management of the WIS network.

Military communications networks are increasing in complexity due to the diversity of equipments being connected and the variety of services provided. At the same time, the volume of management information from network elements is increasing. Despite the availability of computerised support tools, effective network management requires a high degree of operator alertness, coupled with a great deal of expertise which will be difficult to acquire and maintain. Thus, knowledge-based systems are required to help operators manage their network more effectively and cope with the information overload inherently present in these demanding environments.

Initially, in the WNSC, filtering, correlation and fusing of management data will be done by trained and experienced operators, assisted by computerised support tool and suitably established written routines and procedures. As experience is gained through this operation, knowledge based tools will be developed and deployed in the WNSC subsystem.

11.2.3 Telecommunications Management Network (TMN)

The basic idea behind a TMN is to provide an organised network structure to achieve the interconnection of various types of Operations Systems (OS) and the telecommunication equipment using an agreed architecture with standardised protocols and interfaces (CCITT Rec. M. 2x)

The TMN is a "Computer Integrated" control network with a generic interface, common to all Network Elements (NE) and the Operations Systems (OS). The generic interface specifications include the definition of the protocol and languages used, and the detailed enumeration of the message specifications.

A TMN can vary in size from a very simple connection between an OS and a single piece of telecommunication equipment, to a large network interconnecting many different types of OSs and telecommunication equipment. It may provide a whole series of management functions, and offer communications both between the OSs and the various parts of the telecommunication equipments and associated support equipments.

Figure 11.1 shows the general relationship between a TMN and a telecommunications network which it manages. It is worth noting that a TMN is conceptually a separate network that interfaces a telecommunications network at several different points to receive information from it to control its operation. A TMN can be implemented by overlaying it physically on the existing telecommunications network that it supports. The communication links required by the TMN can be separately established, or can use some of the communication services already existing:

- Internodal common channel signalling network,
- Access or D-channel protocol interfaces,
- Packet switched data network,
- Circuit switched data network,
- Dedicated circuits,
- Or combinations of the above.

As seen in Figure 11.1, an integrated control system is envisaged for the switches, and transmission elements of WIS. The TMN-type WNSC will provide:

- Integrated operations,
- Dynamic network control, using surveillance-based operations,
- Flexible and autonomous operations,
- Enhanced survivability and
- Efficient use of resources

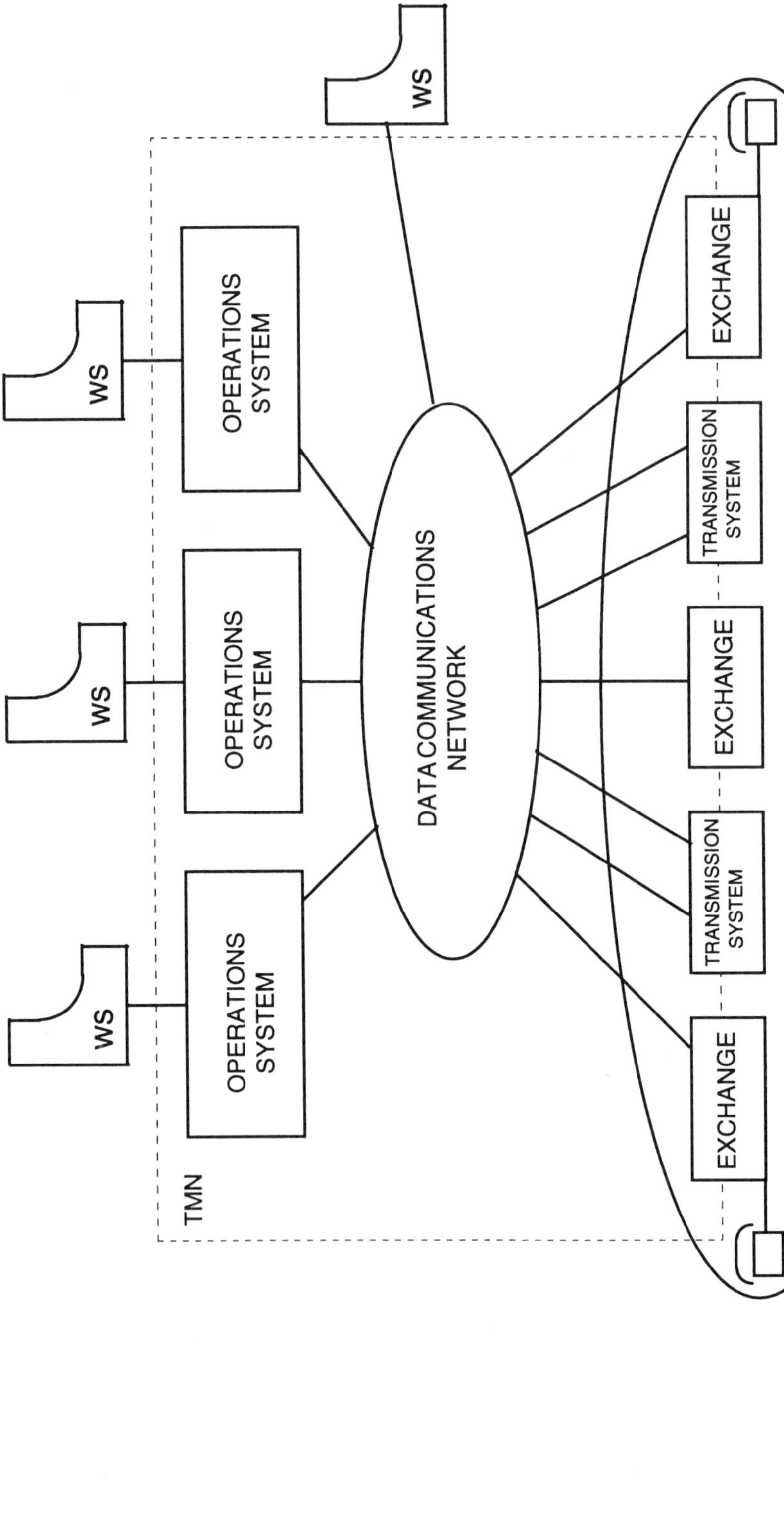

Figure 11.1 General Relationship of a TMN to a Telecommunication Network

11.2.4 TMN Functional Architecture

A TMN functionally provides the means to transport and process information related to the management of telecommunications networks. As shown in figure 11.2, it comprises operations systems function blocks (OSF's), mediation function blocks (MF's) and data communications function blocks (DCF's). The function blocks provide the TMN general functions which enable a TMN to perform the TMN application functions. A TMN is also connected to network element function blocks (NEF's) and workstation function blocks (WSF's). As shown in figure 11.2, are all identical reference points (q-q, f-f, x-x) connected through the facility connection external to the TMN.

As defined in CCITT Rec. M.2x, are the following function blocks relevant to a WIS TMN implementation:

a) Operations System Function Block (OSF)
The OSF processes information related to telecommunications management to support and/or control the realisation of various telecommunications management functions.

b) Mediation Function Block (MF)
The MF acts on information passing between NEF's and OSF's to achieve smooth and efficient communication of primitive functions and processes involving decision making and data storage.

c) Data Communications Function Block (DCF)
The DCF provides the means for data communications to transport information related to telecommunications management between blocks.

d) Network Element Function Block (NEF)
The NEF is a functional Block which communicates with a TMN for the purpose of being monitored and/or controlled.

e) Work Station Function Block (WSF)
The WSF provides means for communications between function blocks (OSF, MF, DCF, NEF) and the network control operators.

The following reference points define conceptual points of information exchange between non-overlapping function blocks. A reference point becomes an interface when the connected function blocks are embodied in separate pieces of equipment:

a) Q-Reference Points
The q-reference points connect the function blocks for NEF to MF (q1), MF to MF (q2) or MF to OSF (q3), either directly or via the DCF.

b) F-Reference Points
This reference point connect function blocks OSF, MF, NEF and DCF to the WSF.

c) G-Reference Points
This reference points connect the WSF and the user.

d) X-Reference Points
The X-reference point connects a TMN to other management type networks including other TMN's.

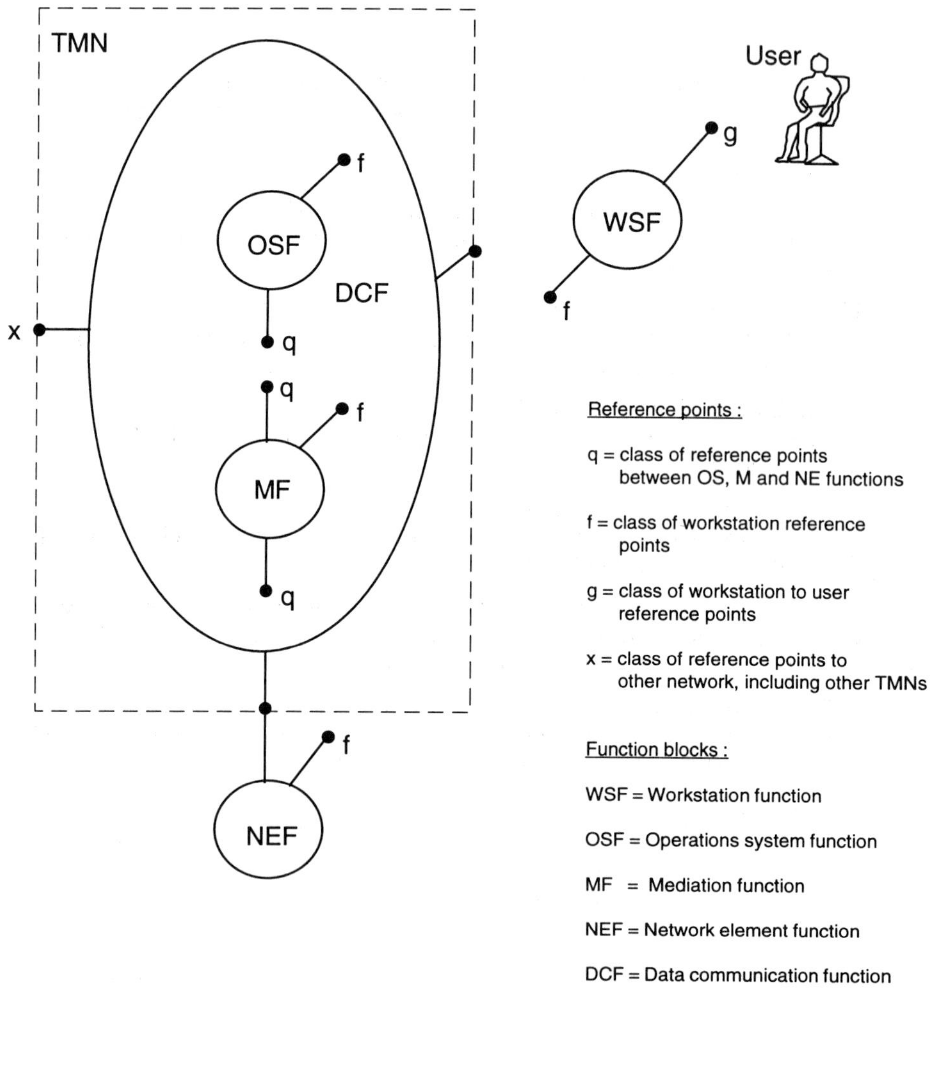

Figure 11.2 A generalized Functional Architecture for a TMN

11.2.5 TMN Physical Architecture

Figure 11.3 shows a generalised physical architecture for the TMN, the WIS specific NE models and the physical architecture of the WIS- specific TMN are developed in the later section 11.4.

TMN functions can be implemented in a variety of physical configurations. The functional modules can be listed as:

- Operations Systems (OS) performing OSF
- Mediation Devices (MD) performing MF
- Data Communications Network (DCN) performing DCF at OS level
- Local Communications Network (LCN) performing DCF at NE level
- Network Elements (NE) performing NEF
- Workstations (WS) performing WSF

Local Communications Network (LCN) is a communication network within a TMN which supports the DCF normally at reference points q1 and q2.

Network Element (NE) is comprised of telecommunications equipment (or groups/ parts of telecommunications equipment) and support equipment that performs NEF's, and has one or more standard q-type interfaces.

11.2.6 Definition of Standard Interfaces

Standard Interfaces, corresponding to the reference points are defined below. The Q-interfaces are applied at the q-reference points, and have three members, providing flexibility of implementation, within the class of Q-interfaces:

1) Interface Q1 is intended to connect NE's containing no MF to MD's or to NE's containing MF via a LCN.
2) Interface Q2 is intended to connect MD's to MD's, NE's containing MF to MD's or to other NE's containing MF via a LCN.
3) Interface Q3 is intended to connect MD's, NE's containing MF and OS's to OS's via a DCN.

The WIS specific TMN interfaces are described below in section 11.4, and are related to the WNSC system architecture.

11.2.7 TMN Protocol Families

The Q interfaces as present on the DCN and the LCN determine protocol families PQ (DCN) and PQ (LCN).

PQ (DCN) is a family of protocol suites for use with the DCN applied to the Q3 interface

PQ (LCN) is a family of protocol suites for use with the LCN applied to the Q1 and Q2 interfaces.

Protocol alternatives for Q2 and Q3 interfaces for the WNSC are given below in section 11.4.

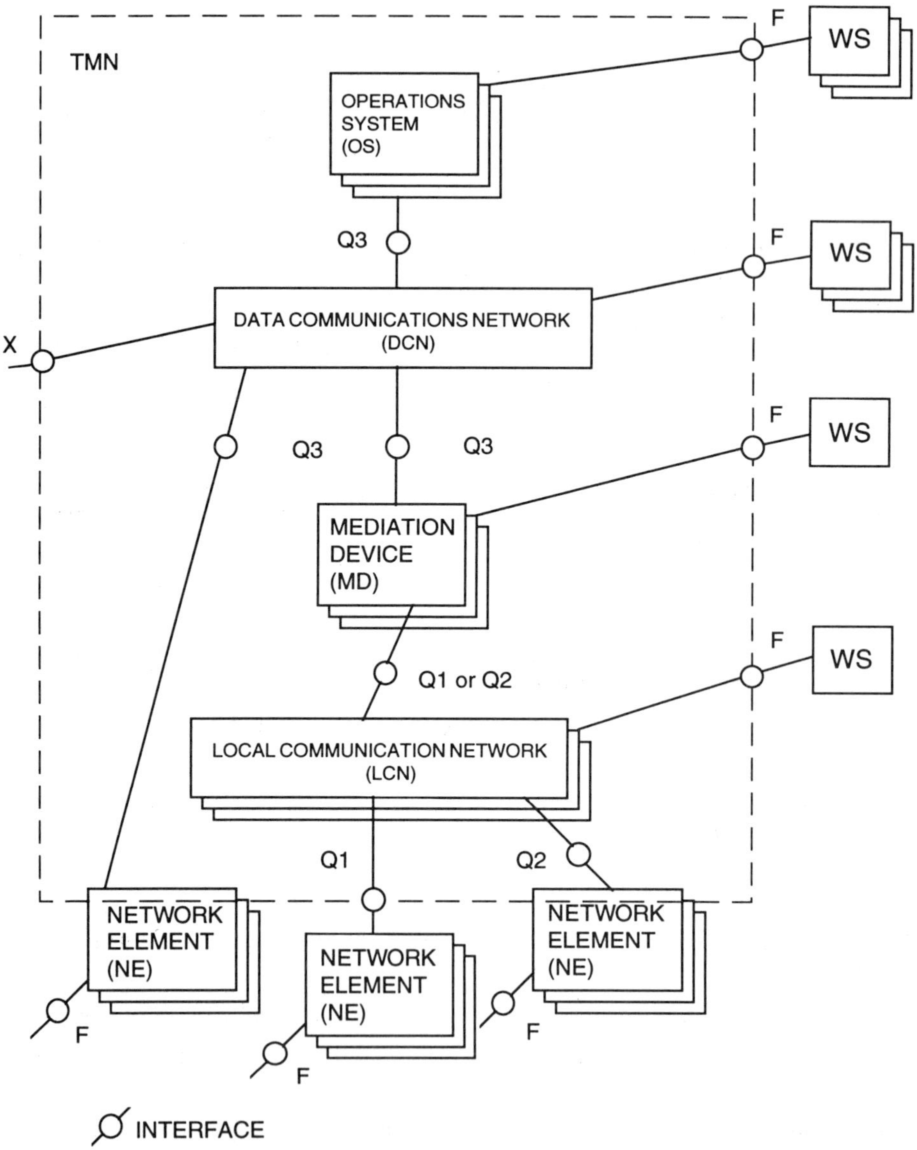

Figure 11.3 A Generalized Physical Architecture for a Telecommunications Management Network (TMN)

11.2.8 Concept Development Issues

11.2.8.1 Introduction

Having taken the CCITT TMN concept as a basis for developing the WNSC concept, it is appropriate to discuss here in which areas different or additional solutions must be prescribed based on the special application of the WIS network.

It is important to justify and explain why a general concept like TMN can not be directly adopted to a network application like WIS. The reasons for this are found in a description of the differences between the type of network that the TMN concept is developed for (typically a public ISDN network), and the characteristics of the WIS network.

The TMN has been conceived in the environment of the advent of ISDN public networks. The areas of similarities with WIS therefore obviously include:

- Digital Transmission and Switching (IDN)
- Integrate digital subscriber access with multi-functional teleservices
- Communication protocols defined and supported for interconnection of multi-vendor type "open systems"
- Combination of local communications requirements and wide area common user systems
- A basic set of modern and effective supplementary services adding value for the user to the communications services he is offered by the network.
- The basic requirement for efficient control of the network resources to optimise their use with respect to network performance

WIS will, however, face a set of different operating environments that needs to be included in the WIS concept as well as in the concepts of the individual network elements:

- The failure mechanisms in the network will include both normal equipment failure as well as man-made enemy inflicted hostile actions, that will require more drastic maintenance actions.
- The traffic volume may vary significantly and quickly during times of tension, requiring immediate capabilities for rearranging resources and restricting non-essential traffic.
- A civilian network, while in an error occurrence context in a relatively stable state will focus on revenue from traffic volume. A military network on the other hand must face intended damage inflicted and must then by all possible means be able to satisfy a relatively low but essential traffic requirement.
- The error occurrences will not be distributed over time in the same way as in a civilian network, and protection must be provided against overloading central resources. This will lead to much more reliance on automatic control actions and decentralised decision making.
- The network operation must not for any of its vital missions depend on any centralised resources, since an enemy destruction or capture of such a resource would effectively stop the network operation.
- A civilian network does not normally assume any threat against its control functions to be presented by unauthorised human intervention. For WIS this threat is real and of significant probability, and this must lead to extensive precautions and measures to avoid all types of harmful actions caused by unauthorised access to control elements.

11.2.8.2 Hierarchical Control Structure

Centralised control of a network improves system efficiency, enhances restoral and reconstitution capability, and reduces operation and maintenance cost. More specifically it reduces manning cost. Decentralisation on the other hand improves survivability, without which even limitless capacity is meaningless in wartime conditions. The WNSC concept is therefore based on a hierarchical structure supporting both centralised and decentralised control operations.

In WIS in peace time, the network wide quality of service and resource management activities will be low and time delays in implementing these activities and activating network control actions will not be critical. It is therefore possible to co-ordinate any operation throughout the WNSC echelons with the centralised mode of operation. However, under stress conditions, service and resource management activities and network control measures have to be implemented very quickly and sometimes without the time to wait for centralised co-ordination. Further as a consequence of network fragmentation the network control may also be fragmented and some of the WNSC resources may not be available to support a centralised operation. Thus, network operations have to be functioning autonomously at local and regional levels.

Decentralised network control requires distributed control cells, distributed control functions and distribution of the databases to locations close to the relevant users and their operational needs. WNSC TMN will be designed so that the application functions and information are available (i.e. distributed) to all control levels.

In peace time, i.e., centralised mode, the OS's at the regional level (RCC), primarily provide network surveillance and control functions for the assigned region to initiate or perform various operations. The OS's will also process the surveillance data to form an input in contribution to the overall network performance to the MCC. The OS's at the network level (NCC) will collect the pre-processed surveillance data from regional control centres to analyse these data to establish an overall network performance view. The NCC OS's will also perform service and resource management functions. Network configuration and traffic management decisions will be taken at the NCC level for the decisions involving several regions and at an RCC for the actions specific to a region

Since WIS is designed to function primarily during war-time conditions, special attention must be given to the principles of distribution of control responsibility for any given network element.

As a basis, the control authority must reside with the local TCC. If access is available to another authenticated TCC or an RCC, the TCC will allow its control to be transferred. The same applies to the RCC that will transfer its control upon request to the NCC or to another RCC that is correctly authenticated.

In the presence of severe stress, damage or suspected compromise, the lower level centres will automatically take back the control and will have to be equipped with all means locally to perform all of its required WNSC functions.

To ensure this, more equipments, functions and databases will have to be located closer to the Network Elements (NE) in WIS, compared to a conventional TMN, where more centralised OS functions communicate with NE's using distributed Mediation Devices (MD). With this approach the hierarchically distributed control structure is designed to support a dynamic combination of centralised and decentralised control.

11.2.8.3 Integrated Operations

In the TMN based WNSC subsystem, a set of operations systems will be implemented at the NCC, RCC and TCC levels. These OS's will form a pool of resources interconnected via the data communication network (DCN) (figure 11.1), OS's, NE's, workstations (WS's) are thus forming an integrated system.

Control functions related to WIS switching and transmission subsystems will be performed using the same resources. Any additional control functions relating to other WIS subsystems, for example COMSEC, will also be carried out in this integrated system.

Each OS may be built to support different functions of the overall job of operating the WIS network. However, a WIS operator may need to access several OS's to perform a specific job. This, in turn, will increase the load on the system and will require the operator's knowledge of each operations system involved. In integrated operations, either all control functions will be implemented in a single OS or the OS's will be interconnected amongst themselves. The complete OS cluster even though implemented separately will appear to the operator as a single system.

This common user interface will contribute significantly to the increase in the quality of operations by allowing for an increased span of control and a reduction in operator error rates.

If a single OS is used to implement all the functions, then each function will be realised as a separate functional building block in the OS. This modular approach will allow new functional modules (applications) to be added without modification to the overall structure of the systems. Likewise, support within existing functional modules for the new network elements could be added with no impact to the existing supported network elements.

11.2.8.4 Network Surveillance and Alarms

WIS network surveillance and maintenance operations will be based on automatic detection of anomalies. It is a surveillance based approach that capitalises on the fact that WIS network elements can have self monitoring capabilities, can generate alarms when a unit or a component fails or when measurable values in a unit exceed their defined thresholds, and can issue warning messages when performance degrades. Monitoring can aid in preventing system outages with early detection of system degradation. When degradation is detected, system service can be scheduled (protective mode) as opposed to reacting to a hard failure. In addition, system performance can be guaranteed or confirmed to a given level of quality. This strategy enhances the effectiveness of the WNSC.

WIS NE's will be interfaced to an OS and alarms may be collected either automatically or via polling. Once alarm messages are received by the OS located at the related control centre, they can be analysed and diagnosed to determine the nature of the trouble, the location, the potential impact on the network, and whether service can be protected/ restored, further repairs can be initiated and perhaps completed even before the users detect the problems.

Surveillance based maintenance strategy will consist of three basic steps:

- The NEs will collect surveillance data on itself and the facilities and circuits connected to it.
- The NEs will send this data to the OSs in the generic OS/NE interface format

- The OS's will collect data from all NEs and using them to initiate control actions and direct the activities of the maintenance force in an effective manner.

The NE should be able to verify and isolate troubles using automated error analysis and diagnostics procedures. However, when these fail to identify the source of the trouble, the OS must be able to obtain additional detailed information for further analysis.

Network surveillance will be performed at network, regional and local levels. An OS at the NCC will have access to all of the NE's and can provide a network wide view. The OS implemented in an RCC may only access to the NE's in its region and those in the adjacent region for which the RCC may take over responsibility. Further the OS implemented at a TCC will cover the nodal switch and the connected transmission equipments and access switches.

11.2.8.5 Survivable and Secure Network Control

Survivability and security are the two most important military requirements to the WIS network, in general, essentially giving similar focus for the WNSC, which is required to be as survivable and secure as the network itself.

Survivability

It will be appropriate to address survivability in terms of:

- physical survivability and
- operational survivability.

For physical survivability, the WNSC control centres will be located at the sites already hardened to defined requirements, so the physical survivability of the WNSC subsystem will automatically equal that of the network itself. The operational survivability on the other hand, requires a high degree of system and organisational flexibility and redundancy and replication of system hardware and software resources in a site and at least in one other site.

The WNSC subsystem must provide a high degree of flexibility to ensure a survivable network control in various operational conditions. Accordingly, a network control architecture based on the hierarchically distributes system of control cells will be designed and developed for the WIS network. The distribution of control resources will provide the flexibility, and at the same time, hierarchical structure will ensure the integrity of network control operations.

A distributed control system with distributed databases will be very effective in maintaining local and regional communications, when portions of the network are isolated due to damage. It will increase flexibility in controlling the network fragments by alleviating the burden of routine actions from the higher control levels.

The WNSC subsystem will be designed with the following considerations and objectives for the operational survivability:

- The operation of the network will not depend on the continuous functioning of the complete WNSC subsystem. The network will continue to operate in the absence of continuous access to resources in the WNSC subsystem
- Each OS and related databases with a specific network management function will be duplicated on site and at least in one other site at the same and next higher level.
- Each NE or MD will be associated with at least two OS's located at two different control centres for a specific function. These OS's will be called primary and secondary OS's.

- In order to minimise the impact of damage to a higher control centre, or to its communications with subordinates, each control function will be delegated to the lowest hierarchical level.
- Every control centre can take-over the functions of any of its sub-ordinate centres and also every control centre can take-over the functions of its superior control centre at a reduced level (i.e. emergency operations).

Accordingly:

i) There will be an Alternate Network Control Centre (ANCC) located at one of the Regional Control Centres (RCC)
ii) Every RCC will have a designated alternate at another RCC
iii) Every TCC will be able to replace another TCC or act as an RCC in its region when required.

- Two levels of network operations will be defined namely: (1) normal operation and (2) emergency operation. In emergency, a limited number of basic network management functions will be executed. These functions may be implemented in a separate OS at selected control centres.

In the WNSC most of the operational decisions will be based on the information contained in databases. Thus the maintainability and survivability of the databases are very important. For survivability, information will be kept at the lowest hierarchical level where the knowledge is required and responsibility for database updates is most effectively concentrated. Survivability considerations further indicate that each database and database segment be replicated at the same and next higher control levels. On the other hand, the multiple locations and multiple copies of data items in a distributed database can mean distributed chaos if transactions and database update schemes are not carefully implemented and monitored. Thus appropriate transaction procedures and update schemes must be designed and developed to maintain the consistency and integrity of the databases.

Security:

The WNSC subsystem will consist of a number of multi-processor systems sited at WNSC control centres each with the ability to operate locally on a locally held database or process management data from the connected network elements and data from other control centres. Each centre will be sited in a WIS facility although provisions may have to be made for some remote terminals. All communication lines will be protected.

The WNSC will be defined to operate in a "SYSTEM HIGH" mode of operation from a COMPUSEC point of view, containing all necessary mandatory access control functions, and protected to the security level of the most sensitive information in the system.

The security of WNSC hardware and software will meet the security standards stipulated by the Authority, and will be ensured by identifying users, restricting access to sites, equipment, performance data, specifications etc.

The software security is considered to be a more difficult area, and require additional measures such as:

- Each piece of software will be carefully screened and evaluated before acceptance and installation into the system
- Input/ output ports will be tested to ensure that no unauthorised changes to programs from local or remote stations are possible.
- Tests will be conducted to ensure that the system can not be overloaded or dead-locked artificially

- Security-relevant events will be recorded.
- Systems and applications programmers will be security cleared.
- Programs, data and back-ups will be treated as highly classified material in all aspects (issue, up-dating, destruction etc.)

11.3 SYSTEM FUNCTIONS

11.3.1 General

The functions to be performed by the WNSC subsystem can be partitioned in several ays:

a) According to system process, or

b) According to the control level at which it is performed

The first partitioning highlights the surveillance and control functions relating to a particular activity or subsystem of the WIS, while the second partitioning defines and itemises functions at each level of the control structure both at equipment and network levels.

The system functions are logically partitioned as follows:

- Network Surveillance
- Network Control and Maintenance
- Resource and Service Provisioning

11.3.2 Network Surveillance

11.3.2.1 General

The NE surveillance data are of three types:

1) event data
2) status data, and
3) performance and traffic measurements data

Figure 11.4 shows the relationship of these three types. As shown here, event data are generated whenever an NE detects (or does) something that affects the NE performance. In WIS event data will be collected from the NEs at the time of occurrence or via polling at a rate of order of seconds as applicable.

Status data indicate the service state of an NE or one of its subparts. If an NE makes a measurement concerning an event and stores it for a fixed period of time, it would become performance data. A performance datum may be as simple as a counter incremented at the occurrence of a particular event, or may be a complex calculation based on a series of events. The status and performance data will be collected from the network elements according to a schedule or whenever requested by the OS. However, when a monitored performance parameter exceeds the pre-set threshold value, it will constitute an event which will be reported automatically. Based on the surveillance data the WNSC subsystem determines the nature and severity of failures and degradations in the network and initiates maintenance activities and network control actions.

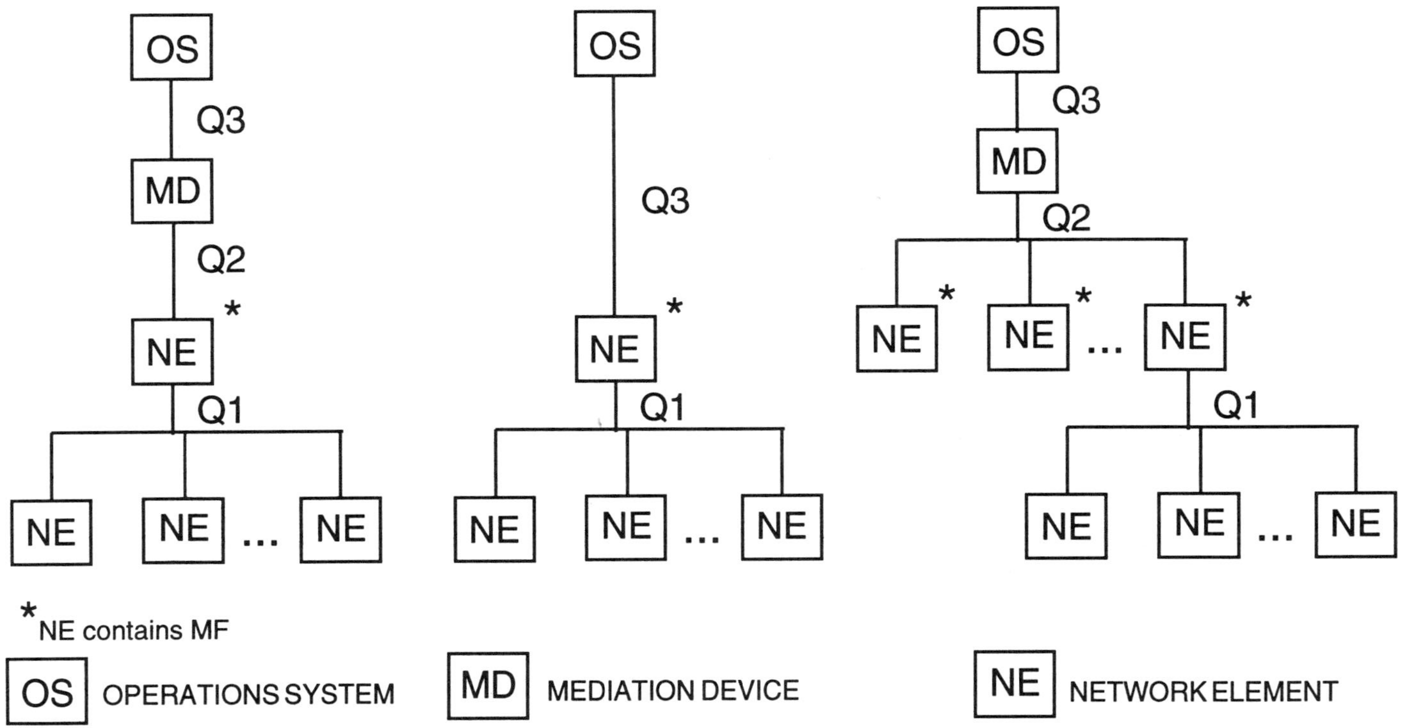

Figure 11.4 Examples of Cascaded Network Elements (Physical Configurations)

11.3.2.2 Alarm Reporting

Event and status data are analysed and classified as alarms in NE's and OS's. A NE/MD is responsible for gathering surveillance data on itself and the transmission facilities and circuits connected to it. The NE/MD may also be responsible for gathering surveillance data from connected NE's. The NE/MD may transmit event and status data as raw data to an OS which processes them to general alarms or the NE/MD may pre-process the event and status data to generate alarm messages to form an input to the OS, depending on the intelligence of the NE/MD.

Three alarm categories are defined in WIS. These are "URGENT", "NON-URGENT and "INFORMATION". Alarm indications will be visually provided on the equipment, collected at convenient alarm panels on site and at appropriate remote control centres. Alarm messages will be displayed , printed and stored as required on site and/or at remote control centres and must contain at least the following information:

- description of failure
- location of failed item or information which can be useful in locating failed items
- possible consequences of failures and the automatic actions executed by the NE.

The WNSC subsystem will have means to request, schedule and condition alarm reports at the NE's. The WNSC subsystem will also be capable to:

- route alarms to designated locations
- allow/ inhibit alarms either local audible/ visual alarms or remote alarms, and
- initiate alarm reset at an NE.

11.3.2.3 Performance Monitoring

Performance monitoring includes functions to evaluate and report upon the behaviour of network elements and the effectiveness of the network or the network element. Its role is to gather status and statistical data for the purpose of monitoring and correcting the behaviour and effectiveness of the network or network element and to aid the maintenance and network control functions.

Performance data for switches is derived from bids, seizures, answer signals, clears and the time of their occurrence. The data can be used to measure:

- circuit group performance,
- destination code performance,
- switch performance,
- common channel signalling performance.

The actual parameters to be used for circuit group and destination code performance will depend on such factors as application of switch i.e. access or nodal switch and adaptive routing scheme implemented for WIS.

11.3.2.4 Traffic Measurements

The WNSC subsystem collects traffic measurements data from WIS switches. These data will provide the database from which the dimensioning, planning, operation and management of the network will be carried out.

Information gathered from these measurements can be used for:

- identifying traffic patterns and distributions on a route and destination basis
- determining the amount of traffic in the switch and the network
- monitoring the continuity of service and the grade of service

The above data and information are gathered with the purpose of supporting the following fundamental activities:

a) dimensioning, planning and administration of the switch and surrounding network,

b) performance monitoring of the switch and surrounding network,

c) network management,

d) operation and maintenance of the switch and surrounding network,

e) forecasting,

f) dimensioning, planning and administration of the common channel signalling network,

g) performance monitoring of the common channel signalling network.

The generation, collection and output of raw data will be achieved by continuous as well as periodic and non-periodic measurements carried out in the switch.

The traffic data analysis may be performed by the switch or by an operations system depending on the following:

- total amount of data,
- need for analysis of data from the network,
- where knowledge is acquired.

11.3.3 Network Control and Maintenance

11.3.3.1 Introduction

WIS NEs provide surveillance and traffic data about their operational conditions to the network surveillance operations systems (OSs). The OSs, which log these data, perform routine analysis and present results to initiate network control actions and maintenance activities. Some control actions are initiated by switches in response to pre-set levels being exceeded in the switch.

The network control and maintenance activities refer to all actions and procedures required to maintain the individual NEs and the overall network performance at a defined level of quality of service and to restore service to all users or at least to the designated critical users under all conditions.

These activities may be categorised as follows:

- traffic management,
- configuration management (provisioning not included),
- maintenance.

Controls will usually be activated or deactivated in steps that are intended to avoid surge effects in the network. In WIS, network control actions and maintenance activities will usually be executed at the lowest echelon of control architecture in order to minimise the response time to network degradations and to enhance the survivability of the network control, but all the actions will be controlled and co-ordinated by the higher control levels to maintain the network integrity.

11.3.3.2 Traffic Management

Real time traffic management or automatic and/or dynamic network management actions are initiated by the switches automatically. These actions, which are preassigned, can respond automatically to conditions detected internally by the switch, or to status signals received from other switches. Dynamic network management controls will not thus be considered within the WNSC framework, they will be assumed to be part of the switching subsystem functions e.g. routing. On the other hand "near real time" traffic controls will be functions of the WNSC and will take the form of operating the network at reduced load and/or restricting the traffic on a destination code basis and rerouting of traffic from circuit groups experiencing congestion to the parts of the network which are lightly loaded with traffic.

Traffic controls automatically activated by switches will depend on the WIS routing scheme, see chapter 7. However they will include:

- code blocking control,
- cancellation of alternative routing,
- call gapping,
- restriction of direct routing,
- skip route,
- temporary alternative routing,
- circuit directionalisation,
- circuit turn-down, busying,
- recorded announcements.

On the other hand, traffic controls initiated by the WNSC will include:

- Trunk group blocking; to stop traffic from entering certain links for various reasons
- Access level management
- Priority reassignment
- "Minimise" procedure (e.g. blocking all traffic from lower access levels or priority levels)
- Initiate and bring in service reserve or reallocated resources (trunks and common equipment)

Traffic controls specified above could be activated by manual or automatic means. When automatic activation is provided, however, an ability for manual override must also be provided.

11.3.3.3 Configuration Management

Configuration management provides functions related to circuit connectivity, reconstitution, reconfiguration of sub parts or components of individual NE's or the other resources of the network to optimise the use of the network and above all to ensure continuity of critical communications by taking necessary actions or implementing contingency restoral plans.

Management functions will be supported by:

- Service and network data bases: providing up to date information related to operational conditions of the network resources and the services provided,

- Contingency plans: covering reconfigurations, rerouting, service modifications plans to be implemented under various threat and damage scenarios.

Service and network database segments will be created and maintained at the location where the knowledge is required, and that the content, form and degree of detail of these data shall match their local use. The network data base must contain information on:

- hardware of all transmission systems and switches,
- inter-switch trunks,
- access lines and point-to-point circuits,
- spares related to any of the above.

To achieve shorter response times in damaged network conditions, contingency plans will be developed for various threat and damage scenarios. These plans will be partitioned to control locations and the details of plans will match their local use and levels of skills of manpower available in the location.

11.3.3.4 Hardware Maintenance

As stated before, a surveillance-based maintenance strategy will be employed in WIS. Failures will be detected by the continuous supervision of network elements, preferably before a user detects the failure. This strategy allows the controlled maintenance principle of the CCITT Rec. M. 20 to be applied to the network.

On the occurrence of a failure in an NE, failure information will be transmitted to an OS which logs and processes this data. The result and possible actions will be displayed or printed at control centres which are responsible to maintain this particular NE and the selected fault information will also be delivered to the higher control centre.

The network operator who receives this information performs the network control function. During execution of this function the operator can access the OS or the NE to have historical data on the fault then to request tests from the NE. The goal of the control function is to quickly determine the type of action, if any, is needed to protect services, and then to identify the cause of symptoms reported such that the maintenance process can be initiated. It is, however, important to note that control actions which can be executed almost instantly must be applied to protect services before initiating any maintenance process which may require longer time.

A maintenance process is initiated by a WNSC OS upon a detection of a failure in the network and covers the following phases:

- system/ service protections,
- failure location,
- failure correction,
- verification, and,
- restoration.

System/ service protection:

When a failure has been detected in an NE, the following functions must be performed by the WNSC itself and by the NE involved with the failure:

WNSC:

- Receives necessary information from the NE and processes this, initiates network control actions such as reconfiguration, traffic rerouting, protection switching etc.

- Modifies operational condition of the faulty item(s) in the NE, e.g. putting an item "out of service" or "in testing". The switch NE's will perform these functions themselves.

NE:

If a fault occurs either in maintenance entities without automatic change-over capabilities or in those with automatic change-over capabilities but no standby available, the following actions should be executed in the NE (CCITT M.30):

- Initiate "URGENT" maintenance alarm indication to designated OS's
- Transmit an alarm indication (AIS) in the direction affected (down-stream)
- Initiate a service alarm indication at the appropriate entities, e.g. at the primary multiplex or digital switch interface so that as a consequence the affected circuits may be removed from service.

If a fault occurs in a maintenance entity having automatic change-over capability with a standby available, then the NE should automatically change over to standby and initiate a "NON-URGENT" alarm to its designated OS's.

Failure Localisation:

When the alarm message from an NE is insufficient for failure localisation, an OS in the WNSC performs actions to diagnose the failure by various means of loop-backs and diagnostics programs in co-operation with the affected NE or its neighbours.

Failure Correction:

Failure correction normally requires changes on repair items of an NE. Maintenance strategies should be developed, identifying the level of corrective maintenance (i.e. repair), card replacement, unit replacement etc., allocating spare parts and components at each control level to meet the overall maintenance objectives with a minimum number of visits.

Verification:

After the failure has been corrected, various tests will be initiated by the OS to assure that the NE is operating properly.

Restoration:

After the failure has been corrected and verified, WNSC or the NE puts the corrected unit(s) into service.

11.3.3.5 Software Maintenance

The development in the software production technology have found an important field of application in telecommunications. It is therefore expected that WIS software will be developed and implemented in accordance with "state of the art" practices in modular design using current high level languages such as C, CHILL or ADA. It is also expected that the switch software development, i.e. upgrades, modifications etc. will be the responsibility of the supplier, because of the high cost of software development and the need for specially trained engineers. The Authority must, however, have a strong Software Maintenance and Development Centre (SMDC), in order to assess the quality and security of the software, to define software requirements and to be able to maintain system software in the absence of the manufacturer's support.

The quality of system software will be checked continuously by WIS operational units. The software quality control includes the implementation, co-ordination and supervision of the system software, the acceptance of upgrades, modifications, the reporting of software quality status (e.g. restarts, crashes etc.), and the initiation of compensating measures (e.g. the returning to previous software load in case of high defect rates).

WNSC subsystem will handle audits, modifications, additions, deletions in databases of the entire WIS system and those of the WNSC subsystem itself. The databases and their maintenance roles and functions are described in detail in section 11.3.5.

11.3.4 Provisioning

Provisioning refers to the activities related to providing necessary resources to the network, and establishing service, testing, modifying, discontinuing service; and maintaining accurate records for:

- access circuits and user services,
- inter-exchange circuits, and
- network resources.

There are two categories of activities in provisioning: (a) service provisoning and (B) resource provisioning.

Service provisioning operations include interfacing to the user, service order control, obtaining circuit and facility assignments, and handing off the service orders for installations.

Resource provisioning operations involve translating the forecasted traffic demands and the new service demands to equipment and facility demands, planning network growth, creating contingency plans, performing equipment economic studies, generating equipment work orders, managing trunk assignments, controlling installations, inventory updating and acceptance testing.

In peace time both types of provisioning will primarily be the responsibility of the network operating authority which processes validated service requests and the traffic forecasts into service and work orders. These orders are executed through "on-line provisioning". On-line provisioning refers to a process where resources and services are provisioned from a computer terminal interactively and without manual intervention in the field. This is made possible by the intelligent network elements which are able to store configuration and user data and allow local and remote data manipulation by the operators.

During stress conditions, the rate of change in the network and the services will become very high, as users change locations, and damage occurs and is repaired. Further the network and the resources of the WNSC subsystem itself may also be damaged. As a result of this, the centralised mode of operation may not be possible. Thus the provisioning decisions will be taken and implemented at the regional and local WNSC control units in accordance with the requirements of the local commanders. However, in order to maintain the network integrity, comprehensive contingency plans will be developed in support of the decentralised (distributed) provisioning operations during stress conditions.

Provisioning relies on large databases containing information about the network facilities and equipment as well as data on the users. The network and service data bases will be organised in such a way that they will support both centralised and decentralised modes of operations.

11.3.5 Data Base Management

System functions will be supported by various databases in the WNSC subsystem. Management of this distributed database must be both distributed and centralised. A hierarchy of control will therefore be imposed with network-wide functions being managed by the central operating organisation and control of other database activities being distributed in a hierarchical fashion parallel to the levelled structure of the overall WNSC organisation.

The principal data bases are:

- Network database,
- User database,
- System database, and
- Statistical database.

11.3.5.1 Network Data Base

This database contains the WIS transmission network topology. The transmission network has two profiles, i.e. the transmission link level and the transmission line level. For transmission efficiency, the circuits are bundled to form a transmission link, which terminates at a transmission multiplexer. Links can be multiplexed to higher order links for more efficient transmission. A transmission link goes through a chain of transmission lines which are terminated at the transmission terminal equipment. These profiles make the transmission network structure complicated because the links of different destinations are carried by the same transmission line for economy and flexibility of the network. A flexible network provides easy configuration and/or expansion of a transmission network to meet the requirements of a multi-service digital network. A network data base which contains the information on the network profiles is necessary to operate the network in a global view. The network data base which contains the information on the network profiles is necessary to operate the network in a global view. The network data base will be kept at the network control level and duplicated at the alternate NCC. Further, segments of this data base will be also kept at the related regional centres.

The network database will comprise the following:

- Transmission link data base,
- Transmission line and equipment data base.

Transmission link data base will have information on the connections of links with other links, and names and kinds of transmission systems supporting it. Finally the transmission line and equipment data base will have equipment number, connection with other equipment.

Procedures must be established to maintain and manage the network data base and its distributed segments in synchronisation.

11.3.5.2 User Data Base

This data base contains information on WIS users. These information include:

- Directory number,
- Type of service,
- Class of service assignments,
- Service restriction,

- Line assignments,
- Relevant dates etc.

A user data base will be kept at the NCC and duplicated at the alternate NCC. Further relevant segments of this data base will be kept at RCC's and TCCs.

11.3.5.3 System Data Base

This data base keeps the switch and equipment system programs. AT least one previous version of the program for an equipment or switch will also be kept to enable a return to previous load under a fault condition. System data bases will be kept at a control centre associated with the equipment. There will also be a back up in the next higher control centre.

11.3.5.4 Statistical Data Base

The following information will be kept in a statistical data base:

- Traffic measurements,
- Performance data,
- Grade of service indexes,
- Maintenance activity data etc.

Statistical data bases will be kept at the NCC and RCCs. The content of these data bases will be delivered to the operating administration at prescribed intervals.

11.3.6 Communications Security

WNSC will be designed so that it contributes to the fulfilment of communication security requirements of WIS. It will also have its own security measures to prevent

- Unauthorised access to control information,
- Unauthorised alteration of control information,
- Insertion of control information,
- Removal of control information.

The crypto control elements are not conceived to be directly part of the WNSC subsystem. The control and operation of the crypto devices may be carried out designated crypto cleared WNSC personnel using the resources of WNSC subsystem.

The WNSC data communications network (DCN) may be used to distribute the crypto keys. An application function can be defined for this purpose. The possibility for further integrating WNSC and the COMSEC Control System was decided as an issue requiring further study, particularly in areas of compartmentation, software security and protocol optimisation.

In line with the security principles/policies, communication security of the WNSC itself will be realised by implementing the following measures:

a) Encryption: WNSC connections will be encrypted to protect control information during transmission

b) Authentication: Network control communication between control centres and between personnel and control centres will be authenticated.

c) Physical protection and electromagnetic shielding: Access to the control centres must be controlled and control information in storage and transmission must be protected against compromising electromagnetic radiation.

d) Access restriction to terminals and equipment: Access to the switching system processors and that of the WNSC will be restricted to authorised agents through the implementation of passwords and card-key locks on terminal devices. Database areas will have a different security level, thus requiring access by different and/or multiple passwords. In addition, the commands will be categorised into functional areas and different levels of privileges which will also require the use of different passwords. Some commands may even require approval in the process of execution. Another security measure to be adopted is keeping records of all, or some of the database modifications and network control operations performed by the users of WIS and the WNSC personnel. This log will consist of:

   - the time of access,
   - the identity of operator(s) and terminals(s),
   - the operation executed.

11.3.7 Man-Machine Interface

11.3.7.1 General

Man-machine Interface (MMI) represents the users' perception of WNSC. Facilities provided to an operator to fulfil his given tasks forms the functional part of the WNSC MMI. These functions are provided by the terminal equipment which includes display devices, data/command entry devices and hard copy devices. The MMI functions of the WNSC will conform to the CCITT Z. series recommendations.

At all WNSC centres MMI will have the following characteristics, as a matter of course:

- Handling databases, storage, retrieval based on queries, updates
- Handling text
- Providing graphical output, i.e. graphs, charts, symbols etc.
- Providing tabular output
- Providing reports

Normally, a user will be trained for the tasks he is assigned. MMI must, however, take care of the situations where the user may be:

- attempting un-authorised operations
- inexperienced in some aspects of the operation
- obliged to share some facilities with other users
- obliged to work with degraded MMI equipment
- asked to carry out other tasks
- interrupted during a task
- away from the station etc.

The MMI must be able to provide effective interaction under such situations.

11.3.7.2 Access Control

The first task of MMI is controlling access to the system. At the WNSC centres, files containing operator profiles will be kept. Based on these profiles, standard operation procedures will be incorporated into MMI in order to:

- Recognise user attempt,
- Identify user,
- Monitor user activity,
- Recheck user identity for privileged access and commands,
- Record user activities,
- Perform log-off procedure.

11.3.7.3 Man-Machine Service

Services provided by the MMI must be easy to use, they must be tolerant to user errors, adaptable to user training and experience. For an (authorised) user, MMI will provide the following services:

- Presenting menus, prompts, questions
- Validate and act upon user response
- Accept commands
- Provide "Help" facilities at every stage of the operation
- Store sequences of operations, re-run if required
- Accept system reports, alert user
- Accept user reports, prepare and format messages
- Accept and distribute mail
- Perform conferencing
- Perform operator training

It is required that the basic MMI functions will be identical for all types of WNSC operator consoles. The use of advanced workstations are assumed at all types of centres. On these, additional MMI functions will be given with the aim to increase operator efficiency. These additional MMI-functions will be built on top of the basic functions described above.

11.4 SYSTEM DESIGN

11.4.1 Introduction

In this section, a proposal for a system design for WNSC is developed with "Telecommunications Management Network " (TMN) principles and architectures of CCITT taken as a basis. An integrated control architecture is developed for WIS transmission and various switching subsystems with network elements architectures and interfaces. This management network may be divided into two levels, reflecting the different characteristics of the functional levels of the WNSC. The highest level serves the inter working between the Operations Systems (OS) located at the TCC, RCC and NCC. The lowest level serves the inter working between Network Elements (NE) and their OS's. In some cases NE's and OS's are collocated, e.g. for switch nodes and in other cases they are not collocated, e.g. for transmission repeaters.

The functions associated with a TMN are split into (a) TMN general functions and (b) TMN applications functions.

The TMN general functions are the basis for the TMN application functions. These functions include information retrieval, transport and storage, security and user terminal support. The TMN application functions are those required for the operations, administration and maintenance of a telecommunications network.

The WNSC/TMN will support the application functions which were presented in section 11.3. In planning and designing WNSC/TMN, the following requirements will be taken into consideration:

- Survivable network control
- Flexible autonomous operations
- Fast and efficient operations
- Network elements as defined during network design
- Other WIS subsystem considerations
- Multivendor environment
- Modularity, flexibility and expandability
- Maintainability
- Technical standardisation
- Security
- Cost

The WNSC/TMN as shown in Figure 11.1 will consist of:

- Operations Systems
- Workstations and intelligent and dumb terminals
- Data Communications Network, and
- Network Elements and Mediation Devices

The following section describes implementation aspects of these elements.

11.4.2 Operations System

In the WNSC subsystem, a three level hierarchical OS architecture will be used to distribute the network control functions as shown in figure 11.5 and to provide flexible and survivable network control. These three levels being NCC, RCC and TCC. The OS's at these centres will be systems of similar specifications with increasing capability from TCC up to NCC.

The application functions can be implemented in a single OS or in a set of OS's. In the WNSC, an integrated approach will be adopted: either each application function will be implemented in an OS and all the OS's will be configured in a network, e.g. in a LAN and with some OS's dedicated for only TMN general functions or application functions will be implemented in a single OS where each application function forms a functional building block independently from the other functions. In both cases, the system will be called an "Integrated Operations System". The integrated operations system is discussed further below.

The OS's will support the same software, hardware and MMI. Only difference will be the size, processing power and content of the databases supporting the operation. This vertical and horizontal standardisation will provide major benefits, as regards:

- Procurement,
- Maintenance and logistics support,
- Technician and user training,
- Mutual back-up of technicians for survivability.

An integrated operations system can be implemented on a single processor with multiple functions or on a cluster of processors. Below these two approaches are discussed.

Using a single processor has several disadvantages among them the following can be mentioned:

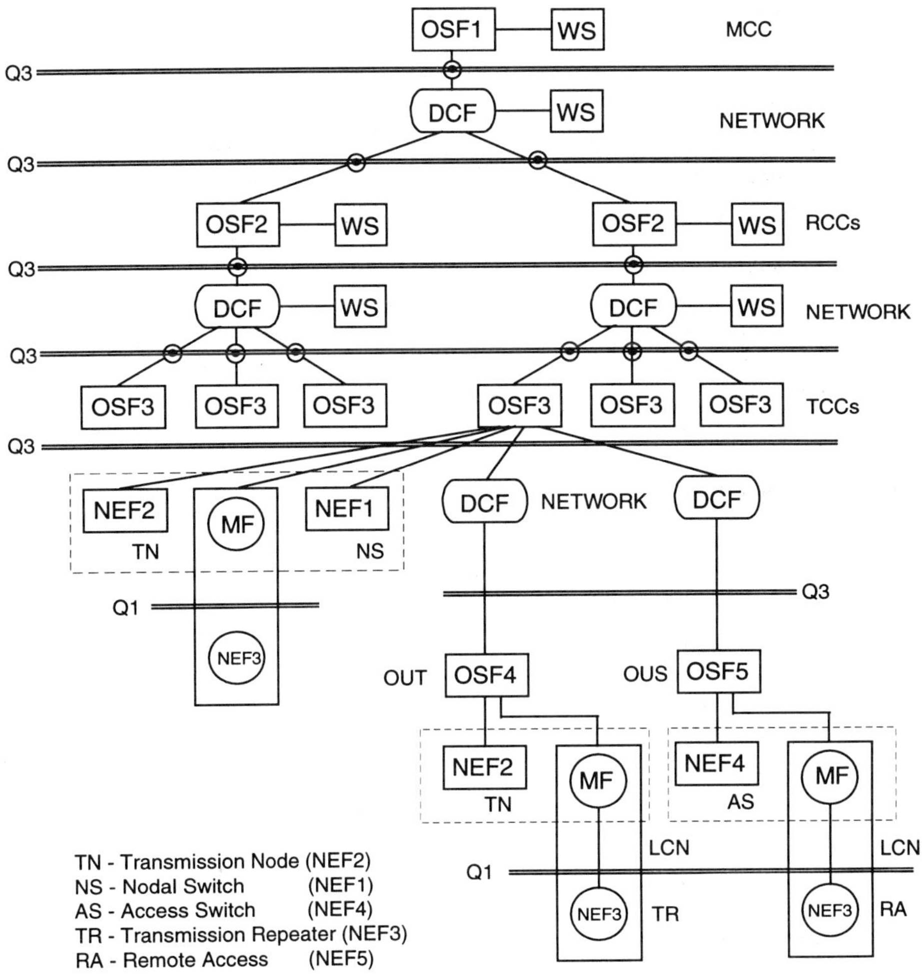

TN - Transmission Node (NEF2)
NS - Nodal Switch (NEF1)
AS - Access Switch (NEF4)
TR - Transmission Repeater (NEF3)
RA - Remote Access (NEF5)

OSF1 - NCC
OSF2 - RCC
OSF3 - TCC
OSF4 - OUT (Transmission)
OSF5 - OUS (Switching)

Figure 11.5 The TNSC / TMN Architecture

1) Processes for separate application functions will be claiming the resources of the same processor.
2) Upgrading to a larger processor, in order to increase the system capacity would also bring software portability problems
3) New operational features can be realised by updating the software, however these updates may again involve system downtime
4) With a single processor, outages will affect all applications, as no priority can be designated among applications in the event of an outage

With a multiprocessor architecture, each processor is assigned a specific application function. The architecture is then used in the following ways:

1) According to the requirements for additional application functions, a processor and its appropriate software can be added to the system
2) Additional capacity growth can be achieved by replicating the processors
3) Reliability can be achieved by adding spare processors
4) It is possible to assign one processor to perform common functions for the others, e.g. MMI. From the user's viewpoint, this can help mask operational differences among the various functional applications in the system and makes the system integrated.

In conclusion, the WIS OS's will be multiprocessor systems with a flexible architecture.

11.4.3 Workstations

Workstation access points are shown in Figure 11.5. The workstations will be used at the NCC, the RCC's and the TCC's. At other control locations (OU's), intelligent terminals will be used instead. A Workstation can be connected to an OS or directly to the DCN.

Operators obtain information about the status and performance of the network and perform network control functions via a workstation. WIS workstations will be personal computers equipped with multicolour displays to be used to display geographical and logical maps of the network. Users can chose to display the entire network or zoom in on the user selected regions or individual NE's. Basic connectivity and status information about switches and transmission equipments, access and internodal links are displayed. This enables operators to monitor the status of the entire or part of the network at a single glance.

The WNSC workstations will have a built-in menu-driven command interface, which provides basic operations and management functions in a pictorial manner. If a command function requires additional information, the workstation prompts the operator using windows and "fill in the blank" forms. Extensive error checking is performed on operator entered data.

The workstation also allows the operator to create multiple display window to perform simultaneous operations, help panels, menus and forms which are employed to provide the operator with a tutorial like environment. This reduces the operator training time, and increases operator accuracy and efficiency.

11.4.4 Data Communications Network

The WNSC Data Communications Network (DCN) will consist of a backbone network and some Local Communications Networks (LCN's). The backbone network will provide inter-connections between the OS's and major network elements, i.e. nodal and access switches and transmission nodes as shown in figure 11.5. An LCN will connect minor network elements in a certain area. Examples are transmission repeaters. The LCN will then be connected to the backbone network via a mediation device or a major network element incorporating the mediation function.

The OS's and WS's at a control centre may also be interconnected amongst themselves and then to the backbone network via a LAN. The LCN's and the backbone network design must ensure that their implementation provides appropriate degrees of availability and network delay. Proposed interfaces and protocols for the DCN connections must be such that the TMN provides speedy and efficient service.

The function attributes that should be taken into account when planning the design of the DCN are:

- Grade of service or delay, which should be in the order of < 1 sec
- Reliability or accuracy with should be of a level not resulting in service affecting errors
- Availability which should be high, > 99.95 %
- Quantity, which would be medium, in the order of 256 octets per transaction
- Frequency, which would be of medium level, i.e. around 1 transaction per minute per OS
- Priority, which should be high supporting the availability requirement

It is expected that the data communication channels and the interfaces will support the required transfer time requirements. Thus only the data collection procedures and speeds and operating systems efficiency of the OS's have to be considered.

The backbone network for the WNSC may be implemented using the capacity of the packet switched an circuit switched data networks in WIS or its common channel signalling network. In addition, the WNSC elements (e.g. OS's, NE's and WS's) may be connected via dedicated communication channels to the backbone network. These various options are shown in figure 11.6. Note that it may be feasible to employ combinations of these alternatives depending on the WIS functions available at the specific sites. The DCN option selected for the WIS should meet the requirements listed above, and the common channel signalling network is the preferred option.

WIS access and nodal switches NE's will have direct connection to the backbone network as shown in figure 11.5. Transmission elements NE's on the other hand will access the backbone network via a Local Communications Network (LCN). The LCN will consist of Embedded Operations Channels (EOC's), or order wires, and the local processors associated with the transmission equipments. The EOC's are dedicated channels in digital signals formats which transport network operation and maintenance traffic. The LCN will be designed such that it provides at least two data channels between an NE and its related MD and its standby MD.

11.4.5 Network Elements

A Network Element (NE) is the grouping of telecommunications equipment that communicate operations and administration messages with WNSC/TMN over one or more standard interfaces to be monitored and controlled.

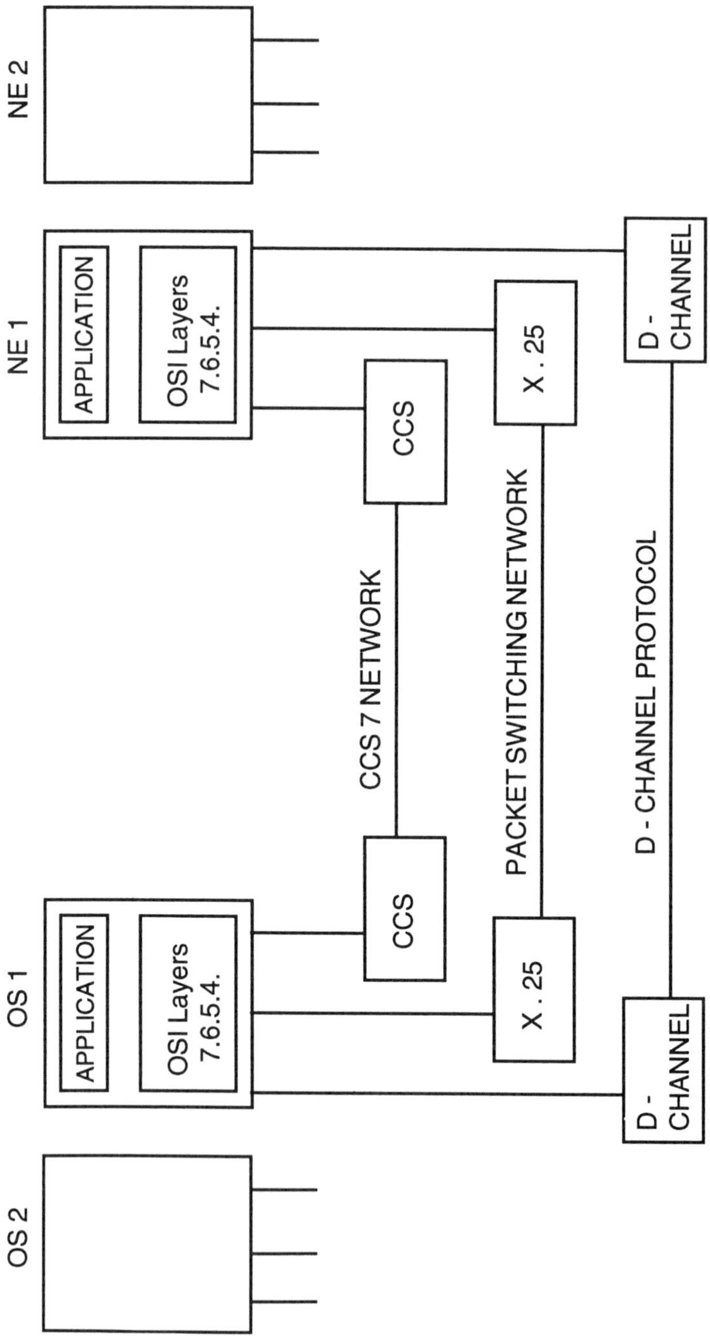

Figure 11.6 WNSC Networking Architecture for WIS Switched Subsystem

The WIS network will include many different types of NE's with different complexity and functional characteristics. An NE may contain the following function blocks:

1) The Maintenance Entity Function block (MEF) is involved in the telecommunications process. Typical MEF's are switching and transmission. A Maintenance Entity (ME) contains one or more MEF's.
2) The Support Entity Function block (SEF) is not directly involved in the telecommunications process. Typical SEF's are failure localisation, protection switching, bulk encryption etc. A Support Entity (SE) can contain one or more SEF's.
3) The Q-Adapter Function block (QAF) is used to connect to TMN those ME's and SE's which do not provide standard TMN interfaces. QAF's are typically interface converters. A Q-Adapter (QA) can contain one or more QAF's and may also contain Mediation Functions (MF).
4) The Mediation Function (MF) provides communications with the TMN. Its function includes communication control, protocol conversion etc.

The NE's with internal mediation functions will always interface with DCN network with standard Q-3 interface. Examples of this are WIS switch elements. However, WIS transmission elements may have non-standard mediation functions implemented in stand-alone devices or combined with other unrelated functions. A number of transmission NE's (e.g. repeater sites), will be connected to an MF in a transmission node via a local communications network as shown in figure 11.5.

11.4.6 Interfaces and Protocols

The TMN interfaces and protocols have not yet been finalised. International study groups of both CCITT and CCIR are working on the recommendations for these. Further, a group of international suppliers have formed a forum to contribute to the standardisation of TMN. It is expected that the recommendations for TMN will be finalised in the near future.

Activities also ongoing in ISO encompasses the extension of the OSI Reference Model to include management and the specification of protocols to support the communication of the management information between open systems. The protocol currently under development, known as the Common Management Information Protocol (CMIP) is intended to be a general purpose management protocol that is suitable for the management of both OSI resources and the real resources used to provide communication services

The protocols for the Q1, Q2 and Q3 interfaces are defined and standardised by CCITT. The Q1 protocol is of limited interest in WIS, since this may be solved with vendor specific solutions. While products are now available supporting a multi-vendor Q3 environment, the Q2 functionality will still for some time have product specific variants.

CHAPTER 12

COMMUNICATIONS AND COMPUTER SECURITY

WIS is being planned and developed to meet the communications requirements of the Authority. WIS will be a digital network, largely following CCITT, IDN/ISDN standards and recommendations. The study of the WIS COMSEC/COMPUSEC subsystem, as developed and presented in this chapter was carried out in the light of the requirements summarized in Chapter 2.

This chapter addresses issues related to identification of threats to communications security in WIS, describing general principles of COMSEC/COMPUSEC policy, system and software security, physical and electronic security for network installations, encryption for protection of information, authentication, key management, and COMSEC related organizational issues.

The COMSEC subsystem as defined and studied, is not limited to conventional transmission network security. The COMSEC approach for WIS must also include areas like installations and communication at user sites, in order to provide full interoperability and consistent COMSEC postures between local and wide-area transfer of information.

The performance of other subsystems has in some cases an impact on the COMSEC subsystem. This is in particular true for the encryption devices to be used either on end-to-end connection or as bulk encryption devices on WIS transmission links. The requirements imposed on the transmission system for these systems to operate properly have been included in the requirements to the WIS transmission system and are therefore not treated in this chapter.

12.1 THREATS TO COMMUNICATIONS SECURITY

12.1.1 Introduction

For the network, the complete assessment of the COMSEC threat would require an extensive investigation of the capability of the opponent with respect to all active and passive measures at his disposal. It is not the intention to perform this assessment fully here but some possible counter measures will be discussed in this section to allow the COMSEC concept to be developed for use in the network.

In this chapter, the term "opponent" is used to describe any adversary who is trying:

a) To gain access to sensitive information by electronic or any other means,
b) To deny the normal exchange of communication between WIS users, or
c) To alter or insert dummy or false information of any kind into the system.

It is usual practice to classify threats in two broad groups as active and passive threats, depending on the method used to obtain information or to deny correct information. Mainly passive threats and their countermeasures were the subject of the study reported here. ECM type active threats such as jamming are not considered since these can be taken as measures similar to physical damage in the context of denial of communication. This area is therefore treated more appropriately in the chapter on WIS Survivability Design (Chapter 13).

Whether it is an active or a passive threat, it is to threaten:

1) Confidentiality,
2) Integrity,
3) Availability

of the communication network. Therefore, it is the minimum requirement that the COMSEC system shall provide sufficient protection against any opponent activating measures in these three areas.

In this sense, the following will describe how the system will be affected by possible threats against the network.

1) Confidentiality: WIS will be trusted to receive, process and transmit information of a high level of sensitivity. Through the provision of conference facilities and message handling, the network will also copy such information to multiple destinations. Interception of this information would be of great value to an opponent and it is expected that efforts will be made to intercept and interpret transmissions and emissions.

2) Integrity: The threat against the integrity of WIS communications will be effective in three different areas:

   i) Message Integrity:

   Information transmitted over the network may include highly critical signals, such as data for the direction of operational units. The accuracy of this data will be essential for the end users. It is possible that an opponent may seek to modify or corrupt this traffic, thus threaten message integrity.

   ii) Authentication:

   Information over the network will directly initiate significant actions by the operational units. The identities of the sending and receiving parties must then be assured. It is possible that an opponent might seek to connect himself into the system and masquerade as a bona fide user to subvert the system and misdirect or confuse other users, by transmitting misleading information and commands.

   iii) System Integrity:

   The network will have the ability to route and copy sensitive information. It must do this in a reliable and trustworthy fashion. Accidental or malicious errors in control systems and their software could disrupt valid communications channels or create invalid ones. They could affect link availability, damage confidentiality by creating illicit channels (e.g. setting up unauthorized conference calls), and subvert authentication (e.g. allowing one entity to masquerade another). Corruption of system behavior and its integrity would then provide an opponent with a powerful means of attack.

3) Availability: The network will provide the principal communications medium for the co-ordination of all the Authority's organizational elements. It is considered probable that an opponent would seek to render the network unavailable. Survivability of WIS against physical attack, ECM and EMP is beyond the scope of this chapter. It is worth noting, however, that in addition to this, an opponent might seek to make WIS unavailable, by seizing or corrupting the software and processes of the network's surveillance and control system or its common channel signalling system. This threat is therefore considered here since it represents a requirement to the security system of the network.

In the following sections possible types of threats shall be identified and discussed to emphasize the measures to be taken during COMSEC system planning.

12.1.2 Active Threats

The type of hostile activity which requires extensive electronic measures either to:

- Jam a transmission,
- Transmit a fake message (spoofing),

or

- Inject traffic during a transmission,
- Seizure of control channels to disrupt traffic flow,
- Seizure of control channels to divert traffic by creating illicit addresses,

are called active threats.

Counter measures against active threats are basically of two different types.

1) Efforts made to make the network insensitive to loss of certain transmission links or nodes or resources of any kind. This is typically included in the network survivability design and is treated in chapter 13.

2) Efforts made to identify and reject unauthorized attempt to take control over the network functions, to perform opponent directed disruption of service or diversion of traffic to unwanted places.

12.1.3 Passive Threats

The following are major passive threats to communication which enable the opponent access to sensitive information without affecting integrity of system or transmission channels.

In contrast to active threats where the hostile activity usually can be detected, passive threats are always covert or silent activities that can only be detected under special circumstances:

- Human intelligence, usually abbreviated as HUMINT, obtains intelligence from human sources. Protection against HUMINT is provided by logical and physical access restrictions.

- Communication intelligence, COMINT, derives information from electromagnetic signals radiated intentionally over the regular transmission media. Unless suitable protections in the form of encryption or protection are used, it can be assumed that all types of communication channels are vulnerable to COMINT.

- Electronic intelligence, where the abbreviation is ELINT, extracts sensitive information from electromagnetic signals emitted by sources which are not communication transmitters. Protection against ELINT is mainly affected by means of proper shielding and physical protection measures.
- Crypto-analysis of intercepted messages is another threat which requires special attention. The chances of success will be greatest if the opponent has access both to the plain text as well as its encrypted form, and uses "the known plain text attack technique". It is obvious that this technique requires a significant level of opponent capabilities in correlation techniques and computer resources. It is quite clear that the requirements for protection of this type lies in the areas of personnel security and cryptological quality.

The success of crypto-analysis depends on resources available to the opponent, and because of secrecy involved, it is usually very difficult to judge the capability of an adversary in this field.

12.1.4 Threat to Computer Security

The network is developed to meet the present and future communication requirements of the Authority. There is a steady and growing trend in the area of defence communications towards the increased use of computerized C3I-systems. In the future these systems will be developed and implemented in the WIS environment having attributes like:

- Distributed intelligence,
- Complex and multi-function software systems,
- Multi-user environment with time shared resources,
- Local and wide area networking features.

System security is a vital and growing requirement for such systems, and it represents a difficult problem due to the underlying contradiction; the objectives of system security and networking C3I systems as described above are fundamentally different. The C3I system is characterized by general, flexible access, while security seeks to impose limited access rules with rigidly controlled conditions.

The threats to information system security in WIS therefore requires special attention in this study. The composition of the threats are typically based on those described in the previous sections (i.e., active and passive threats to communications media), but additional threats arise from the nature of the vulnerability of the C3I systems using the network for their connectivity. Furthermore it can easily be seen that WIS itself will have a very similar characteristics, being itself a system of distributed intelligence, mainly software based communication system.

In this respect, the most critical threat to the system security will be those that are inflicted by human access, since they will cause the most damage through:

- Non-authorized access to or manipulation of sensitive information being processed by the system,
- Harmful manipulation of computer software (programs or data) at any stage of the software cycle.

Hence, these are threats that can destroy the system confidentiality, system integrity and availability in systems like WIS by:

- Creating unintended links enabling access to sensitive information,
- Creating unintended links which make spoofing possible,
- Disrupting valid communication links.

These active threats can materialize in large computer based communication systems by planting software bugs or by modifying software in order to abuse services provided by the system. Examples of such abuses are:

- Establishing non-authorized conference calls,
- Enabling non-authorized call diversions

in computer controlled switches.

Such threats must be considered as real and quite probable if the system is not properly protected against software tampering.

For the Message Preparation and Distribution Systems (MPDS's), the threat may take to form of:

- Non-authorized multi-destination message distribution,
- Altering messages before delivery.

12.2 OBJECTIVES AND REQUIREMENTS

12.2.1 Objectives

In the light of the threats presented in the previous section, the top level COMSEC/COMPUSEC system objectives for WIS may be expressed as follows:

1) There should be no deliberate or inadvertent access to inappropriate information by unauthorized persons as a result of WIS communications and data processing services.
2) There should be no deliberate or inadvertent unauthorized functions performed by WIS communications and data processing services.

The first statement covers the risks of loss of confidentiality (illicit read access of user data) and loss of message integrity (illicit write access to user data). The second statement covers the need for system integrity and availability (illicit functions). Both statements imply strong authentication processes.

Since absolute security can never be achieved in practice, a third statement may be added:

3) Means should exist to detect, report and act upon security breaches and attempted security breaches.

Finally, two pragmatic statements are included:

4) The above policy objectives shall all be met without impairing the communications usefulness of WIS or affecting its availability.
5) The above policy objectives shall also be met by adopting suitable design approaches to arrive at cost effective solutions, without compromising security.

12.2.2 Requirements

In principle, all the sensitive communication originated by the WIS users, irrespective of whether it is internal traffic or external traffic, shall be protected.

All of the direct WIS subscribers (DWS), and approximately 25% of all indirect WIS subscriber (IWS), will require the possibility to send and receive classified information through the network. The majority of the traffic thus generated will have classification at low to medium level, while a small set of the users and a very limited volume of traffic will have the highest level of security classification.

All access links and all internodal links shall be protected. All transit switches, access switches, subsystems and all terminal equipment shall meet full or partial EMC and TEMPEST requirements.

Careful planning must be performed to avoid excessive and non-balanced protection measures, keeping the cost for the overall WIS implementation within reasonable boundaries.

COMSEC system design shall support computer security (COMPUSEC) and software security performance in computer based subsystems such as switches, CCIS systems etc.

Physical security will be provided in WIS installations through access restrictions and access controls of various types. Finally, COMSEC devices used in WIS will assume that the WIS transmission standards (quality, availability, slip rate etc.) as given in chapter 7 are fulfilled.

12.3 POLICY ISSUES

12.3.1 Introduction

This section describes how the overall COMSEC objectives described in the previous section relate to the different types of WIS areas and interfaces.

WIS is defined as the integrated strategic communications network for the Authority. WIS and the related COMSEC measures discussed in this chapter will not be limited to the wide area network components, but will also cover the user locations and equipments for local area communications.

The totality of all precautions to be taken shall constitute the COMSEC subsystem. The measures to be provided involve physical security which prevents direct access to sensitive information and electronic security aimed at preventing an adversary to alter or extract useful information from communications.

The policy is to attain a COMSEC posture which enables the evolution toward total communication security for all traffic. It is a clear requirement that the evolutionary implementation of WIS COMSEC elements shall build the COMSEC subsystem gradually to its final configuration with a minimum of interim measures that are obsolete in the final configuration.

It is possible to categorize all installations being either in secure or insecure areas.

1) Secure Areas: The term "secure areas" is used for areas accessible only by trusted and specially cleared personnel with "need to know". Such areas typically are protected by strong physical access controls. These areas will have a security perimeter, and entry and exits are recorded and audited. Strict supervision of activity exists within the perimeter. Secure areas may be further compartmented into a number of separate security zones.

The security perimeter may be physically extensive (e.g. an entire building), or it may be very small (e.g. a room, or indeed a single piece of equipment). Within this definition, the term is applicable to transportable as well as fixed installations.

2) Insecure Areas: All other areas are termed "insecure areas" since there is no strong control over physical access to them.

The policy statements applicable specifically to each of these areas are as stated in the following sections.

12.3.2 COMSEC Policy in Secure Areas

The first top-level policy objective implies that need-to-know principles will be enforced even within secure areas.

When the secure area is under the control of the end user, the appropriate communications and computer security measures must be combined with the overall security concept of the user site. Typical measures for such installations may be physically separate and controlled wiring subjected to regular physical inspection. Optical fibres should preferably be used due to their superior radiation performance and, hence, resistance to cross-talk and interception.

These cabling arrangements are then used to interconnect internally separate TEMPEST enclosures and TEMPEST protected terminals. The end-user will be responsible for the security of the computer installations and for all aspects related to the integration of WIS with other secure subsystems located at the same place (PTT installations and other electronic equipment for non-operational use). Secure areas under WIS control will consist of all WIS installations where the user traffic is found in non-encrypted form. These are the switching nodes, and all WIS Network Surveillance and Control installations.

All user traffic, except traffic with special classification (end-to-end encrypted), will be handled in non-encrypted form within the switching nodes or control systems. These will be manned with specially cleared personnel, and no specific additional mechanism will be required to enforce mandatory or need-to-know rules. Specific policies on end-to-end encryption are considered further in the related sections.

All sensitive information for which WIS is itself the end-user (i.e. sensitive data other than end-user traffic) will be treated following the same principles as WIS user traffic.

The first and second top-level policy objectives presented in section 12.2.1, dictate that end users will be responsible for the security and integrity of all data and software stored or utilized by data processing functions in the terminal area. In particular it will be their responsibility to ensure that computer equipment meets all requirements on COMPUSEC policy, accountability and assurance that may be specified by the relevant security authorities.

The installations at the user sites can be split into the terminal equipment and cabling part which is discussed above, additional to the Access Switch installation. It is anticipated that this will be housed in one or more separate secure areas, containing various functions like, Access Circuit Switch, Access Packet Switch, Packet Assembly Disassembly (PAD) equipments, Message Preparation and Distribution System, Local Area network Components (other than the terminals themselves) and the equipments for the WIS Network Surveillance and Control System (WNSC). For these elements it may be assumed that the WIS COMSEC Authority will be responsible for the security arrangements, like in the cases of separate WIS installations.

All data processing functions for which WIS is itself the end-user, such as switch control processors, will be treated following exactly the same principles as in the paragraph above. In particular, WIS control software will meet defined specific policy, accountability, and assurance levels, such as to fulfil the requirements of all four top-level statements presented in section 12.2.1.

12.3.3 COMSEC Policy in Non-Secure Areas

Since there is no control of access in insecure areas, the first top-level policy statement implies that all sensitive information intentionally transmitted into or through insecure areas must be encrypted to assure confidentiality. Since WIS will support concentrated traffic with multiplexed bearers, these must be bulk encrypted when leaving the secure areas.

It also follows that the bulk encryption should be continuously active, so that no implicit information is disclosed through lack of traffic flow security. For the same reason the bulk encryption must moreover include all signalling and routing signals.

The bulk encryption process should further assure message integrity, preventing undetected malicious modification of signals.

It is noted that all the criteria specified in this chapter would normally be understood to apply to, and be provided by, standard available "off the shelf" bulk encryption equipment.

The second top-level policy objective implies that the bulk encryption equipment will detect error and alarm conditions, adopt "safe" modes in the event of failures, and will promptly signal all such conditions to system management.

The fourth top level policy objective requires that the bulk encryption equipment should not affect the bit-error rate, slip rate, or availability of WIS links.

The WIS network, including any bulk encryption process, must consequently be able to support end-to-end encryption methods.

It is the end-user's decision and responsibility to use end-to-end encryption to protect highly sensitive information which crosses boundaries between secure and insecure areas.

The fifth top level policy objective requires that application of end-to-end encryption in WIS shall be limited by constraints on overall system cost as long as the application does not violate the fourth top level policy objective.

12.3.4. COMSEC Policy for Interfaces

The top level policy statements imply that there should be no un-intentional transmission or reception of sensitive signals across these interfaces between WIS and other systems.

The security perimeter of insecure areas should therefore include protection against unintentional emissions and transmissions. The area may be screened as a whole, or else each individual equipment supporting sensitive information should conform to appropriate TEMPEST, EMC and RED/BLACK separation specifications as designated by the relevant authorities.

12.4 PHYSICAL AND ELECTRICAL SECURITY

12.4.1 Access Restrictions in WIS

Service integration, in the form of a military ISDN, and Communication Security are two conflicting requirements for WIS or for any other military environment. They are conflicting requirements because the nature of a modern C3I system with ISDN type features is specially designed to be open and accessible to a variety of users at different locations, shall have distributed intelligence and will necessarily contain a set of complex software programs. All these characteristics tend to make the task of system security very difficult.

Furthermore, the user friendly services such as call diversion and conferencing can be very dangerous in case of deliberate manipulation of hardware or software. If a wire -tapper can successfully initiate participation in conference calls, for example, then he will gain access to sensitive information without attracting any attention at all.

In other words, new and improved services and facilities presented must not be abused by an opponent located anywhere in the network.

Hence, physical protection of WIS installations against would be wire tappers, is extremely important from a communication security point of view.

In WIS, all switches will be treated as Red Areas with proper marking at all entrances, and precautions such as:

- Only those who are authorized can enter premises,
- Well kept records of persons entering the premises shall be maintained,
- All maintenance shall be done under strict supervision,
- Intrusion alarms shall be used at all possible entry points. Area protection, whenever possible, must be the preferred way of limiting and controlling access,
- All unauthorized entries should immediately be reported to Network Control and COMSEC Control Centres.

All the areas that contain end user equipment such as terminals, CCIS systems, ADP systems etc. must be properly protected and access limiting as well as access control be provided to the extent possible.

12.4.2 Secure Cabling Inside Headquarters

Service integration in any system eventually leads to greatly simplified wiring through the combination of several different circuits within the system boundaries.

In applications such as WIS, where users accessing common communication facilities of a user site are normally confined to a well defined area, the cost of wiring replacement may not be a prohibitive factor.

In those cases that re-wiring is feasible, carefully planned and implemented re-wiring shall lead to more secure cabling, that is difficult to tap and monitor.

The minimum requirements of secure cabling at a user site are as follows:

- All cables should traverse open areas and inspection of cables should be easy,
- It should be a simple task to trace cables from subscriber location to switch premises.
- All new cabling should be authorized and installation be carried out under strict supervision.
- Cables should be splice free, whenever this is possible.
- It should be difficult to wire-tap cables.

There are four possible cable types that can be used for cabling at a user site:

- Twisted pairs,
- Twisted pairs in multiple pair cables,
- Coaxial or twin-axial cables,
- Fibre optic cables.

In the first three cases, all the requirements can not be met unless special precautions such as using metal conduits, are taken. Among these solutions, coaxial or twin-axial cables are expensive and difficult to splice, multiple pair cables are also expensive and not practical for subscribers distributed over large areas. Twisted pair cable is inexpensive and easy to install but very vulnerable to wire-tapping.

Fibre optic cables satisfy all the requirements for secure cabling, including cost factor. When compared to copper cables in metallic conduits, fibre prices become competitive.

The wide bandwidth of fibre cable makes it attractive to multiplex subscriber lines at suitable locations to reduce number of fibre cables entering switch enclosures at big user sites.

The same large bandwidth also makes fibre cables attractive for use as carriers in LAN-type networks, which will coexist for local information system functions.

If properly constructed fibre cables with metal cladding are used, it will be physically difficult to tap the cable since special tooling and equipment are required and only experts can do such sophisticated tapping.

Therefore, it is recommended to use fibre optic cables to connect terminal equipment to the local switch or the local host computer, in accordance with established installation requirements. New or existing copper cables may be used in those applications where fibre optic cables are not suitable, provided that all installation work is performed according to the same requirements.

12.4.3 Signal Emanations from Cables and Links

It can be assumed that there are three boundaries across which sensitive information transfer can take place in the form of unwanted electrical signal emanations at a user site. The levels of these signals, unless suppressed beyond levels defined by the Authority, may lead to detection by sensitive electronic equipment, thus permitting successful electronic intelligence.

These three boundaries are:

- Interface between subscriber lines and nodal or access switches,
- Interface between network and switches,
- Interface between public network lines and access switches (PABX's).

These interfaces must be designed in such a way that information about any signal other than the intended one should not be transmitted through the boundary.

If the signals are digital, properly filtered and critical signal components properly masked, then the use of fibre optical cables for information transfer across the three boundaries stated above, will have the benefits that unwanted signal emanations (intelligible crosstalk) will, probably, be completely eliminated and expensive line filters are eliminated.

Therefore, unwanted signal emanations will be reduced to a great extent if fibre optic cables are used across the three afore-mentioned boundaries.

Across the switch and PTT line boundary, the outgoing lines from the PBX or access switch to the PTT network constitute the very critical connections and they are the most vulnerable part of the network from COMSEC point of view.

It is therefore strongly recommended to have a properly constructed fibre optic connection across this boundary which is terminated by a special adapter circuit located just outside the switch premises.

12.4.4 WIS TEMPEST Specifications

WIS equipment is expected to satisfy requirements for electromagnetic compatibility (EMC) and TEMPEST, to limit compromising emanations. Testing for these requirements are done according to established standards. However, testing for TEMPEST requires special test equipment, test area and properly trained personnel. For example, in NATO there is a special department devoted especially to carry out TEMPEST testing and inspections. All TEMPEST testing and installation of equipment shall be done in accordance with established standards approved by the Authority.

In principle, equipment compliance with TEMPEST requirements are obtained through application of proper design techniques during the design and manufacturing phases. The additional effort required in design, manufacturing and testing considerably increases the cost of equipment itself as well as its installation. Therefore, within the network, it is important to conduct a "Risk of Compromise Assessment Study" before deciding on a global determination of TEMPEST requirements to the networks equipments and installations.

The rationale of assessing risk of compromise is to avoid application of the same rigid standards to minor installations that process small volumes of low classified data as well as to installations in locations that are already sufficiently protected by other means or have a low risk of attack.

During such an assessment study, the following set of criteria will contribute to arrive at an accurate assessment:

- Physical Security,
- COMINT threat,
- ELINT threat,
- Computer and Software security requirements,
- Type and classification of information handled,
- Geographical location of the site.

Depending on the results of assessment, the degree of TEMPEST protection required shall be determined. It is clear that high risk installations should benefit from full measures while low risk installations could be given minimum amount of approved equipment (mainly terminal equipment).

Emission level of TEMPEST approved equipment and installations depend on several critical factors which might show variation with time. Emission level should be checked regularly and Radiation Surveys should be conducted at least once a year. Special attention must be given to repairing TEMPEST approved equipment because improper repair procedures normally result in leakage from equipment, violating TEMPEST requirement.

Therefore, a special organization needs to be established to conduct regular Radiation Surveys, install new TEMPEST approved equipment, test newly repaired TEMPEST approved equipment and oversee such repairs. It is also desirable if all maintenance work is undertaken by this organisation.

12.5 END-TO-END PROTECTION

12.5.1 Speech Encryption

The availability of non-encrypted 64 kbps speech traffic should not be affected by WIS transmission in normal conditions due to the high transparency of transmission media. Speech is very robust and would still be easily intelligible even in the worst-case conditions of 1 in 10 to the minus 3 error rate or slips every 20 minutes. Indeed the slip rate could be much higher (even to the G.822 threshold) since only 5% of slips cause audible crackles on telephony links.

Narrow-band speech would be much more seriously affected by the above worst-case conditions and almost 100% of slips would be audible; hence these conditions would be much closer to the intelligibility limit.

The network will support digital speech circuits with self-synchronizing (CTAK) end-to-end encryption. Users will however experience the elevated error rate characteristic of this mode, potentially tripling outage time as discussed in section 7.2. Slip events will also be much more intrusive, since each event will cause not a single byte error but a burst of around ten garbled bytes as the equipment self-re-synchronizes. Subjectively this means that all slips will be both audible and intrusive.

Intelligibility would still be retained even if the slip rate rose to the G.822 threshold level, although the user would experience elevated background noise and crackles every two minutes.

The network will also support stream ciphered speech (KAK mode) as long as the network slip rate remains within the specification. Each slip event in the network will disrupt the cryptographic synchronization and destroy the usefulness of the channel until re initiation has occurred. In 64 kbps speech there is no helpful frame alignment word to assist automatic detection and so the re-initiation may have to be manual and may possibly require link reconnection. This inconvenience is likely to be less acceptable than the elevated noise effect of using the self synchronizing (CTAK) mode. It may be supportable down to the threshold of one slip every 20 minutes but certainly not to the level of one slip every 2 minutes. Users employing stream ciphered speech over the network should consequently be aware of the risk to such communications if the exceptional slip conditions of Section 12.5.5 could arise.

These problems will not affect stream enciphering equipment employing continuous forced re-synchronization. In this case a few percent of the channel bandwidth is "stolen" to carry continuous synchronization update information. Slips event will then only temporarily disrupt the link and re-synchronization. will occur automatically. Users will experience crackles during the re-synchronization interval but manual intervention will not be necessary.

12.5.2 Data Encryption

WIS will be carrying a number of different types of data traffic. In the case of Circuit-Switched data, e.g., for bulk data transfer, digital facsimile or show seam video, identical solutions exist as for original voice. Depending on the nature of the data traffic the described error mechanisms will require different end-to-end error recovery protocols.

In the case of packet switched data (e.g., interactive remote terminal or LAN-to-LAN message traffic), the principles of pocket data encryption with application of the Trusted Communication Sub layer (TCS) on top of the network layer of X.25 will provide adequate end-to-end protection.

For ISDN terminal multi-mode traffic, it is expected that devices will be available providing end-to-end protection of the B-channels (as for circuit-switched data) and packet data encryptions for user data in the D-channels.

12.6 BULK ENCRYPTION

Bulk encryption equipment shall operate on each link to be protected on a bit-by-bit basis, i.e. it will neither delete nor add any bits in the stream of digits. Once encrypted, channels of the link can not be accessed at any intermediate point of the network such as relay stations or repeaters. If access is necessary, then the link has to be decrypted before such access can be achieved.

Most links in WIS will be continually bulk encrypted to ensure confidentiality and traffic flow security. This encryption should be transparent to all forms of user data and should not affect the network availability, error or slip characteristics.

Suitable bulk cryptos for G.703, 2048 kbit/s traffic are available and will be employed in WIS. Encryption at this level will permit compartmentation of network-internal traffic and through traffic from other sources.

Where 2048 kbit/s are further multiplexed to higher rates, new framing and supervisory information will be superimposed on the subordinate streams. Where bulk cryptos possess internal clocks, these clocks will be designed to accommodate variations in the incoming rates of the signals they handle, and will have a total excursion range of no more than 2048 kbit/s +/- 50 ppm, to conform with the accuracy requirements of G.703. It will also be verified that the accuracy of the encryption clocks in combination with the exchange clocks is also within the requirements of G.703.

For example, the clock timing of slave exchange equipment will vary from the master end due to cyclic effects such as wander, when the physical length of the transmission path varies with diurnal or annual climatic conditions. Where the signal is derived from higher order multiplexers, the received clock phase may also shift slightly during the free-wheel times of frame alignment and justification bit reception.

Where the resulting accuracy of the slave equipment is +/- X ppm, it is required that the maximum excursion of a self-clocked bulk encryption should be no more than +/- Y ppm, such that x+y is less or equal to 50 ppm.

The bulk encryption process will not affect the net error rate of the communication links. Use of equipment in the CTAK self-synchronizing mode is not favoured since each error in the transmission path would be expanded into a long burst of errors.

A typical error extension of about two orders of magnitude would mean that the availability of WIS links would be dominated by their 10 to the minus 5 performance instead of the 10 to the minus 3 level. For normal radio relay equipments this would correspond to a loss of about 3 dB in fade margin. Microwave links typically degrade at about one decade per 6.5 dB of fade if they have antenna space diversity, otherwise at one decade per 10 dB of fade. Consequently the use of CTAK mode would either compel the use of 3 dB higher EIRP's, or result in a tripling of effective link outage.

A stream-cipher mode (KAK) is the preferred mode instead, since it causes no error extension. Stream cipher mode is, however, not self-synchronizing and is only acceptable so long as loss of cryptographic synchronization is unlikely to occur. This is because the re synchronization protocol is typically lengthy and may require the link to be broken down for that period. Loss of synchronization will occur whenever there is a slip event between the enciphering and deciphering engines.

Therefore slips due to plesiochronous operation of the switch nodes must be absorbed within the switches themselves and not in the transmission links. Unless this condition is met, each slip could easily result in prolonged data loss and/or loss of connections.

Consequently the timing of receive bulk encryption devices will be derived from that of the incoming data stream or will fully compensate for any variations in it, so that the long-term clock rates of the transmit and receive encryption equipments on each link will be identical. There should then be no problem since the link bulk encryption devices will not normally suffer slip events. In the absence of slips, cryptographic synchronization should only be lost if there is a spontaneous equipment failure or interruption of the transmission path. These events should be rare.

Stream cipher equipment will nevertheless be capable of rapid recovery from any loss of synchronization, since all data intelligence is lost during periods of de-synchronization. Re-synchronization protocols should therefore be brief (not exceeding tens or low hundreds of milliseconds) and these protocols must be automatic and promptly initiated. Loss of G.732 framing word from the decrypted 2048 kbit/s plain text may be used to initiate the process, and encryption equipment for network-internal use should either be capable of detecting this condition or else able to respond to an externally supplied detection circuit. Detection would normally require about 6 frames, or less than one millisecond.

Circuits routed over the links should not require reconnection each time a stream cipher re-synchronization occurs. The bulk encryption equipment should conduct any re synchronization process sufficiently quickly, and while presenting suitable interface conditions to the switching node functions, it should ensure that circuits are held throughout the process.

Because of its superior error characteristics, stream cipher equipment is also preferred for channels not carrying the G.732 structure. Synchronization recovery processes must then be initiated by other alarm conditions.

Manual initiation of re-synchronization process should also be available in the event that automatic process are defective or inapplicable.

The encryption equipment will generate local alarm in the event where encryption is not applied to the transmitted signal. Under failure conditions, the equipment will apply the normal (non-encrypted) AIS alarm signal to the link. AIS will be detectable at receiving cryptographic equipment and the appropriate condition will be passed to the switch receiving equipment.

All bulk encryption equipment will conform to the appropriate standards for EMC TEMPEST and RED/BLACK separation and will be approved by the relevant authority.

All bulk encryption equipment will be located within the physically secure and electrically protected environments of the switching nodes. They will furthermore be located in a controlled area within that environment.

Physical access to all bulk encryption equipment will be restricted to suitably cleared and authorized personnel, and all such access will be recorded and audited. All maintenance will be strictly supervised. All unauthorized events will be immediately reported.

Key management of the bulk encryption system will be hierarchical. Data will be protected using session keys, which may be electronically communicated under key encryption keys.

Equipment shall be able to receive keys and change keys during normal operation. Simple partitioning will be achieved by using different keys for each bulk encryption pair. The relevant master keys for each pair will be generated at and distributed from Key Management Centres. The Key Management Centres will be controlled by appropriate authorized bodies.

12.7 SOFTWARE SECURITY

12.7.1 Introduction

As mentioned earlier, this section handles communication security problems in a wider sense, including user area communication and other computer based systems used at the user site. These systems are CCIS/MIS systems, computerized message preparation and distribution systems, and other data processing systems.

Each of these systems should have a link to the network for connection to other user sites. Furthermore, local connections of terminals to central processors will also be made over circuits provided by WIS. Therefore, computer security, COMPUSEC is not a problem specific to security of computer based nodal switches and access switches but is a part of a larger challenge which is the security of the entire network, including all data processing systems, CCIS systems and their software.

It should be noted here that although utilization of computers in switches and in information systems are totally different, their common feature is the dependence of trustworthy system operation on software. Hence, it should be the preferred approach to treat issues in COMPUSEC without making any distinction between switches and other computer systems.

In the following sections, general principles of secure software development and maintenance will be developed.

12.7.2 Secure Software Certification and Verification

Software security in the context of COMSEC is very similar to computer security, COMPUSEC. The trend towards using more complex computer and communication systems creates a new problem, that is, system users should have a high level of assurance that the system will perform its functions in the way they are intended, without any unpredictable side-effects.

Furnishing any level of assurance requires controls on software security which are truly internal measures and does not involve external measures such as physical security. This is a new field in software engineering and it is necessary to determine guidelines required for effective verification and certification of software.

It is important to note that software security comes from design, not from extensive testing during a certification program. Testing of a software package offers limited level of assurance because fully exercising a package is not possible. Even in moderately large systems, the number of possible paths in software is so enormous that testing each for some possible flaw or loopholes becomes impossible.

It was an overriding objective for the design of WIS to use commercially developed hardware to reduce overall system cost. This implies that a small part of software shall be customized and major part of it would be previously developed software. Hence, it is apparent that a large transit switch or even an access switch with fully implemented ISDN features shall have software which is extremely difficult to certify.

Distributed processing, which is employed in designs for more reliable and flexible operation, and modularity form the basis of modern switch architecture. This constitutes the primary factor which immensely complicates software verification.

It is the conclusion of preliminary investigations that this issue requires special attention and guidelines should be determined after extensive collaboration with agencies specialized in this field.

There are three modes of operation defined for computer systems which are also applicable in this case. They are summarized below:

- Dedicated Systems: The computer system stores and processes information in a specific area of interest, serving a closed user group cleared to the highest level of information handled.
- Multi-level Secure Systems: In this mode the system stores and processes information of mixed classification and serving personnel having different levels of security clearance and need-to-know.
- System High: System operates in a similar way to multi-level, but main difference being that information handled is regarded as of highest classification despite its actual classification and all users are cleared to the highest level of information processed.

Multi-level system software poses the greatest difficulty in testing, evaluation and verification. It is also well understood that formal certification that will provide a high level of confidence in software, is only possible for relatively small systems operating in dedicated mode

WIS is a communication system offering integrated services to a common user network and therefore it has to be a multi-level system.

The software security problem in a multi-level distributed system such as WIS, becomes a manageable problem if and only if distributed systems can be divided into several dedicated systems. These dedicated systems, which may be a CCIS, MPDS, PBX etc. have to be used in a star configuration. Then, software of each system can be verified individually assuming that interactions between the dedicated systems are accountable.

These arguments also increase the importance of using end-to-end encryption in WIS because it is the only way of achieving COMSEC at the highest level of assurance in multi-level systems where trusted computer security standards are still missing or unachievable.

12.7.3 Development of Secure Software

Recognizing the fact that verification and certification of programs are only possible for software developed under close supervision and observing strict rules, it is recommended that any program development for WIS should comply to the following rules:

- High level languages such as ADA, PASCAL, C, CHILL or any other suitable language shall be used. The selection of a language will clearly be influenced by the availability of application software. In a similar manner in the selection of an operating system (such as UNIX) consideration will have to be given to security in addition to availability. Thus the selection of a language and an operating system will have to be a compromise between security and availability

- Software shall be extremely modular, well documented, and accountable. The established software quality standards shall be used as development and documentation control standards (ISO 9000/AQAP13).
- All input/output parameters of software modules shall be clearly defined, all input/output procedures shall be traceable.
- Special built-in software and/or firmware shall guarantee software integrity by detecting and putting up alarms, in case of tampering with software.
- Software test/verification routine should be designed during software development.
- Special software monitoring modules should be developed that will constantly monitor all commands issued by all terminals and compare them with fixed directories containing lists of authorized commands for that for that particular terminal or class of terminals. These modules shall discover any suspicious activity by detecting terminals which attempts to execute non-authorized commands and trigger suitable alarms.
- Similar modules shall be used to guard systems against attacks based on cracking password systems by trial and error approaches.

12.7.4 Secure Software Maintenance

Software maintenance has two objectives which are first, to assure users that no non-authorized changes in software are permitted; second, to guarantee that all software changes are made properly, being in accordance with rules set for secure and accountable software development.

All software changes shall be fully tested, verified and then certified by an independent team of technically qualified experts with required level of security clearances. The management of the changes must be formalized in a reliable computer based configuration control system.

Periodic and spontaneous checks should be made on software running on system components to verify that there are no non-authorized modification in software.

Very critical software modules whose tampering may cause irreparable damage to communication system security may be put on Silicon to eliminate modification by ordinary software tools.

12.8. COMSEC SYSTEM ORGANIZATIONAL ISSUES

12.8.1 Responsibilities

It is the end-users responsibility to decide where to use and when to use end-to-end encryption equipment. It will, on the other hand, be the responsibility of the network operator to provide secure transmission of the user's sensitive information through bulk-encryption, but the network will not guarantee unconditional security of transmitted information.

The COMSEC Control Centre (CCC) will have the primary responsibility for the proper operation of all bulk-encryption equipment used in the system, and to notify the users by issuing warning in case of bulk encryption equipment failure or any other security compromising situation.

Distribution of all crypto key variables will also be the responsibility of the CCC.

It will be the Network Operators responsibility to provide physical security by access limiting and control. Similar security measures should be implemented at the user sites as well. All security measures should be evaluated and controlled by the CCC.

Although it will be the Authority's responsibility to distribute, maintain and account crypto equipment, it should be a separate and independent organization's responsibility to approve suitability of crypto equipment, algorithms and key management techniques to be used in WIS. This function may be carried out by a technical centre appointed by the Authority.

A software development centre should be responsible with all aspects of problems associated with computer security. This centre should, to the extent possible, test and verify all software used in WIS and oversee the new software developed.

A radiation survey group should assume the responsibility of risk assessment, determining requirements and planning use of TEMPEST approved equipment. The same group should also carry out or supervise installation and maintenance of TEMPEST equipment as well as monitoring the efficiency of TEMPEST related measures.

12.8.2 Organizations required for COMSEC Implementation

12.8.2.1 COMSEC Control Centre (CCC)

Since all access links and internodal links will be encrypted, this means that the number of bulk encryption devices in a network like WIS may be us much as 1500. If additionally the number of required end-to-end encryption devices is considered, it is obvious that a high-capacity and independent COMSEC Control Centre (CCC) becomes a necessity.

The function of the CCC will mainly be to monitor the status of the bulk encryption devices in the network. CCC will also generate and distribute the key variables for all encryption devices, using the Data Communication Network (DCN) of the WNSC subsystem (see chapter 11). It was recommended in WIS to install six CCC's in WIS distributed to and collocated with the NCC's and RCC's of the WNSC.

12.8.2.2 COMSEC System Operation and Maintenance

A single organization was envisaged in WIS to have the full responsibility for planning, equipment selection and procurement, installation, maintenance, key generation, key distribution, crypto material distribution and accounting, inspection of crypto facilities and personnel training. The same organization would also be responsible for evaluating transmission security as well as communications security of existing systems. There were no such competent organization or agency in existence at the time of the WIS network design. It was therefore recommended to establish such an agency under the command of the Authority, as a part of the WIS implementation.

12.8.2.3 Radiation Survey

The task of installing, testing and maintaining the TEMPEST approved equipment requires specially trained personnel, well established procedures, special test equipment and special laboratories.

Therefore, a dedicated group should be established to carry out the above tasks for WIS. This group should be independent and should report to the Authority for all of its activities.

The proper implementation of a full scale COMSEC system which uses TEMPEST approved equipment in a cost-effective manner, but in no way that compromises security, requires a very careful assessment of security risks.

Hence, it is recommended that a special organization is established to conduct regular "risk of compromise" assessment studies. Depending on the results of such assessments, the degree of TEMPEST protection for new installations will be determined. These studies will also determine the efficiency of installed TEMPEST equipment and, if necessary, the improvements in existing protective measures.

This second group should also be established as an independent group, being directly responsible to the Authority.

12.9 BIBLIOGRAPHY

[1] "Information Processing Systems - OSI Reference Model", International Standards Organization, Public No. 7498, October 1984.

[2] "L. D. Faurer and R. H. Courtney, "Computer Security, the Defense Department, and the Private Sector - A3-Part Dialogue about Fundamental Objectives and Needs", Computer Security Journal, Summer 1984.

[3] "DoD Trusted Computer Systems Evaluation Criteria", United States Department of Defense, Publication No. 5200.28, December 1985.

[4] "Guidance for Applying the Department of Defense Trusted Computer Systems Evaluation Criteria in Specific Environments", United States Department of Defense Computer Security Center, Publication No. CSC-STD-004-85, June 1985.

[5] "DoD Trusted Network Interpretations", National Computer Security Center, Publication No. NCSC-TG-005, Version 1, July 1987.

[6] "DoD Password Management Guideline", United States Department of Defense Computer Security Center, Publication No. CSC-STD-002-85, April 1985.

[7] B. C. Karp, L. K. Barker and L. D. Nelson, "The Secure Data Network System", AT&T Technical Journal, Vol. 67, No. 3, May/June 1988, pp. 19-27.

[8] Proceedings of the 1987 IEEE Symposium on Security and Privacy, ISBN 08186-0771-8, Oakland, California, April 27-29, 1987.

[9] Proceedings of the 10th National Computer Security Conference, September 21-24, 1987.

CHAPTER 13

SURVIVABILITY DESIGN

13.1 DEFINITION

In general terms, survivability can be defined as the system's capability to be continuously responsive to the needs of the users in spite of system damage caused by hostile attacks or other causes. A more refined definition of survivability is "the percentage of stations required both to survive physical attack and to remain in electrical connection with the largest single group of remaining stations". This definition takes into account the likely effect of enemy attack not only upon the system, but also upon the forces and commands using the system, and considers the effectiveness of the surviving network in terms of its usefulness to the majority of its users.

In order to achieve survivability thefollowing design requirements must be satisfied:

a) Diverse and richly interconnected transmission media must be used, to provide diversity and redundancy of routing required for a survivable system.

b) The system must not contain critical elements or features, the destruction or failure of any one of which would cause serious disruption of overall performance.

c) Key elements such as switching centres and control centres must be provided in sufficient quantities and with adequate dispersal, in order to minimise system degradation in the event of damage.

d) The location of switcing centres and critical communications equipment, must be selected to avoid potential target areas, where possible.

e) Where communications facilities must be located in vulnerable areas, some hardening must be provided.

This chapter addresses the issues related to the identification of threats to network survivability in WIS, describing the threats attempting to keep the network from satisfying its vital traffic and connectivity requirements. The definition and study of the survivability of the WIS must be based on the assessment of this threat, and must identify a certain set of survivability enhancement measures that aims to bring the survivability to a certain level compliant with the operational requirements. The most difficult part of this clearly being to establish a realistic threat scenario, flexible enough to analyse consequences of design decisions in the light of different types of possible hostile attacks that may be anticipated.

An important element of the study is to make a reasonable estimate of the cost elements specifically related to survivability, in order to put a "price-tag" on the quantified survivability improvements. To enable a reasonable evaluation of the "quantity" of survivability achieved for this price, a comprehensive method for analysing the traffic performance of a degraded network has been developed. This method gives the possibility to analyse the connectivity, traffic handling volume and grade of service for the various levels of priority traffic and for defined damage conditions.

WIS will have two seperate entities that for survivability considerations should be treated separately. These are the network itself with all its installations and the users of the network. The objective of the survivability design is to specify measures for the network which would give it a degree of survivability more or less equal to the survivability of the users it serves.

The justification for this "balanced design" is that the network should not represent a significantly more attractive target than the operational users and headquarters served by the network. Since the survivability of the users may be taken as given, with a known hardness and operational vulnerability, the survivability design is reduced to finding a set of measures for the WIS that in total provides grossly the same survivability as that defined for the users.

Basically, because of the uncertainties involved with assessing the intentions of a potential enemy, this task is complicated and must therefore be treated statistically taking into account the plurality of the communications functions and services that the network must provide and the assumptions that have to be made regarding:

- Threat definitions
- A certain minimum but vital level of connectivity and capacity required by the users.
- The possibility of overall and specific network performance analysis when network is damaged.

A number of survivability features are defined and specified as basic elements of individual subsystem of WIS. These are described in various other section of the book, and includes the following features:

- A dynamic, adaptive and flexible routing scheme that ensures optimal connectivity capabilities during periods of heavy degradation and significant and dynamic changes in network topology
- A strong multi-level precedence and pre-emption scheme with ruthless pre-emption and effective load reduction load capability for the switched network.
- Nodal switches with high efficiency with respect to blocking, through-connect time and signalling, and most importantly, the over-load protection mechanisms must be very strong and dependable. The software system of the nodal switches must be highly verified and dependable, in order to prevent adversary effects to be accomplished by remote control or following an enemy capture of a functioning installation.
- A survivable network surveillance and control system capable of both centralised and decentralised operation for local restoration and performance in phases of network partitioning and damage (i.ei any region and level must have all functions required, and not depend on any central resource for any vital control required inside its own physical area)
- A survivable network synchronisation system that ensures required performance during periods of dynamic network degradation.

- A COMSEC system that ensures the network against hostile attacks in the form of eavesdropping, active ECM, spoofing, intercepts and non-authorised network control actions.

13.2 THREAT EVALUATION

The political situation in the world has in recent years shown significantly more dynamic changes than we have been used to over the previous 45 years. The design and implementation of a system like WIS is aiming at a very long operational life time in the order of maybe 25 to 40 years. The protection of the system must therefore be as flexible possible allow for the best possible adaptability to changes in the threat that the system will be facing.

Although the task to make threat assessments entails sensitive and difficult predictions regarding potential opponents interest in denying your users access to communications services, it is strongly recommended to evaluate your system design with respect to its survivability. It may wise to establish more than one threat scenario, and the assessment of likelihood of attack to the system must be closely discussed with the Authority to make sure that reasonable threat levels are used. The survivability design could then be made in a modular way such that certain precautions may be designed but their implementation may be deferred until a later stage. It is also useful to be able to know in quantitative terms how much a certain measure would be able to increase the networks survivability.

The threat evaluation should be based firstly on a set of probable attack methods that an aggressor would be able to apply in order to destroy or disturb the network. These could typically be:

- Sabotage attack with:
  - special forces
  - local terrorists and saboteurs
  - regular forces
- Air attack with different types of aircraft and conventional weapons
- Jamming from ship-based, land based or air-borne jammers, including land-based jammers delivered by saboteurs
- High-altitude EMP burst
- Nuclear weapons

With the attack methods identified for the system a discussion must be made to define the likelihood and weight that should be placed on each threat in the survivability design of the system.

It will typically be found that attacks from sabotage teams are the most important threat to a ststem like WIS. During periods of tension and crisis, it can be expected than an opponent would find it feasible to stage attacks of this kind using trained special teams or local saboteurs. The vulnerability of WIS installations to these kinds of attacks must be analysed in each case, but it can be difficult to judge which level of protection that can completely protect the system. The best solution will therefore be to keep this threat in mind when making detailed design e.g. of building security, fencing, antenna locations, access control etc.

Air attacks as a part of a wartime scenario is a highly threat to communications installations. The key protection elements here are minimum physical sizes, burial of critical installations, and some physical hardening of above ground installation.

The threat represented by jammers and interceptors is considered as a likely event and must be dealt with in the survivability design. The most probable method to be employed for EW attacks would be jammers delivered by air or placed by saboteurs, in addition to airborne jammer platforms like helicopters or light fixed wing aeroplanes.

In the past it was assumed that the EMP threat a high altitude nuclear burst would be a major threat to communications installations. Numerous studies of this subject has been made in the past and suitable, albeit cumbersome, protection methods have been designed. We have chosen to include some material on this threat in the book, although the likelihood of such threats may be significantly reduced for most applications today.

The application of nuclear weapons in any foreseable crisis scenario seems today fortunately to become more and more an unlikely situation. It would in any way not be that such weapons would be used against communications installations.

The survivability design will now be continuing after a conclusion defining the level of threat that are assumed from the different attack methods as listed above, and for WIS this resulted in a definition of a general HEMP threat together with certain types of sabotage attacks as the most likely aggressor methods. This was followed by different types of air attacks with certain defined types of weapons. Of the prioritised attack scenarios, the decision was made to include the protection of the most likely ones in the basic design, while the remaining would be treated as probabilities of damage.

13.3 REQUIREMENT ANALYSIS

13.3.1 Quantitative

The level of survivability to be achieved by the WIS in the event of the threats as described above, is defined by the Authority and stated in section 2.3.4 of Chapter 2. It would be required to take certain central requirement statements to be the quantitative basis for the analysis of these requirements. In the case of WIS this become the following two statements:

1) WIS is to be no less survivable than the command and control elements that it support.
2) Under all but the most extreme conditions shall WIS provide communications for the vital traffic between primary users and the main command amd control elements.

The levels of survivability of the existing user sites that are subscribers to the network, will have to be determined in co-operation with the Authority and analysed with respect to the same types of threats as for the communications installations. The level of vital traffic could be determined as the equivalent to the volume and traffic interest of the highest procedence level identified by the Authority. In the case of WIS this was chosen as the voice, telegraph and data traffic on FLASH level added with a 30% overload margin.

The total traffic load between the primary users of the network was for WIS estimated to be about 2000 E, of which 15% was assumed to be of FLASH priority. The traffic load the network will have to carry even under stress and damage condition would therefore have to be more than the initial traffic volume at the that level totalling 300 E. It is required that this traffic shall not experience a blocking probability of more than 1%.

In order to be able to access the network performance under stress, it is necessary to assign a value to the probability of user sites becoming disconnected under damage conditions. It is assumed that this should be less than 15%.

13.3.2 Qualitative

In qualitative terms, these requirements are taken to mean that the survivability design should be such that for a given weifht of effort, would the assumed opponent expect to achieve greater and more lasting disruption to our command and control system by destroying the terminals or our commanders, than by attacking the WIS itself. Additionally, if by miscalculation, accident or as a target of opportunity a WIS installation is attacked, then it must be able to absorb the resulting damage without losing the minimum emergency level of connectivity. The perspective offered by this qualitative view of the requirement has been taken to simplify two aspects of the design study.

Firstly, the losses which the enemy could expect to face in delivering an attack to the command centres would be as high losses in attracking a WIS element which serve those command centres in the same geographical area. Therefore with the exception of maritime and airborne mobile users, to whom this generalisation would not apply, the design to achieve comparable user and communication survivability's has been made solely on the basis of the weight of attacks finally delivered each target.

Secondly, the definition of a minimum vital traffic level has been taken to mean that this level can be used as the ultimate measure of success of any survivability design feature.

13.3.3 Enhancements

The requirements statements approved by the Authority and described in Chapter 2.3.4 stipulates a number of survivability measures to make the WIS network inherently resilient to the anticipated threat. These are for exaple dual homing for all important users, the use of a grid network, independent transmission routes, rapid restoration capabilities, jamming protection, avoidance siting, EMP protection and physical hardening. Each of these measures has been incorporated in the design in full where the requirement has been specific, and as appropriate where needed to meet the overall survivability requirements.

13.4 METHODOLOGY

13.4.1 Minimum-Cost-Network

The first step of the methodology was a preliminary survivability analysis which was conducted on the basis of the first approximations of the threat and the vulnerability. The estimates were all made in order-of magnitude terms, and were used to indicate the survivability features which should be taken into account in the minimum cost network design. Two major features were identified at this stage. The need for dual access link connectivity for the primary users and hardening of the switch structures to be able to survive all projected conventional attack levels. The rationale to support these deductions as developed for the next step of the methodology is described in chapter 13.4.2 below.

In addition to the contribution which this first step made to the minimum-cost network design, it also influenced the network configuration by indicating general principles which should be adopted to maximise survivability wherever alternative design options were available at comparable cost. For example, in the choice of node locations remote from likely target areas, the use of diverse transmission media, the selection of internodal routes to avoid potential targets. These ad hoc methods, together with the meshed grid configuration conferred a high but un-qualified measure of survivability to the minimum-cost network. The following steps describe the methods used to meet the specified levels of survivability stated in the requirements.

13.4.2 Direct Convertional Attack

The second step of the methodology is aimed at determining the effects in quantitative terms of a direct attack against the network, compared to the effects of a similar attack against the users of the network. The complication at this stage is that the threat and attack methods will be different in the two cases. It is therefore difficult to establish a credible scenario for this comparison, i.e. it will be difficult to find a single weapon that is likely to be used against both users and the WIS.

This difficulty has been overcome in the study reported here by establishing a scenario of damage in terms of progressive elimination of WIS installations, without studying the exact combination of weapon type and weight of attack to achieve the stipulated damage. A number of such damage scenarios with up to ten progressive steps have been analysed, with and without the parallel destruction of user sites giving rise to reduction in offered traffic volume.

The user sites were defined as belonging to one out of three categories depending on the hardness of their location. While the majority of the network users were assumed to be housed in general office buildings without any sizeable hardening measures above regular construction practices the two other types are assumed to be progressively stronger in hardening. The strongest site is assumed to be an underground defence headquarters with full range of protection measures while the medium level would be basement installations or hardened above-ground constructions.

With these three categories identified, the vulnerability of each of them in terms of threat weapon, attack form and weight were stipulated. The normalisation were made towards a single attack type used as a fictitious study parameter of one "attack sortie" and the classification in our study gave the following result:

Table 13.1 User Site Survivability

Category	Attack Sorties	No of sites
A	200	2
B	75	50
C	4-25	200

For the purpose of our study, this evaluation of user area vulnerability was used as a measure for the balancing of the survivability of the network to support these users. It was naturally necessary to gain the approval of the Authority for this evaluation.

The emphasis given to this aspect of the methodology reflects the importance of the primary survivability requirement. By establishing categories by which each individual command and control element served by the network is rated, the same weight of attack can be assumed in the attack of the communications facilities, albeit with different attack methods. While an air attack may be most likely on a headquarters, sabotage teams and jamming may be more relevant threats to the communications installations which may, however, be made ineffective by measures such as hardening, fencing and patrolling, thus forcing the enemy to resort to air attacks.

As far as jamming is concerned and due to the low output power of LOS radio systems, relatively low power jammers can be quite effective as long as they can be used in the vicinity of a repeater or terminal station. This threat is therefore similar in scale to that of a sabotage team. The relative ease with which the enemy could deploy such teams for electronic or physical attack is primarily related to the degree of surprise which could be achieved. Covert sleeper groups could be expected to achieve initial success but once the local defences have been alerted the enemy would not expect to succeed without the use of well-trined special groups operating in a wide area.

The vulnerability estimates for all types of WIS installation are summarised in the table below. The measure used is that for a typical fighter bomber attack. The categories are then compared in quantitative terms with the user location types as listed above. As additional data it is estimated which user site category they would be close to for the threats of jamming and sabotage.

Table 13.2 Communication Site Survivability

Element	Air Attacks	Jamming	Sabotage
Buried cable installation	100 (A)	A	B
Buried switch node	50 (B)	B/C	B
Buried transmission node	30 (B)	B/C	B
Buried repeater	30 (B)	B/C	B
Conventional repeater	5 (C)	C	C

13.5 SURVIVABILITY

13.5.1 Conventional War

The complete design criteria adopted to provide matching survivability between all users and the WIS in the face of the range of threats which may have to be faced in periods of tension and conventional war can therefore be summarised as follows:

a) Build all transmission repeater station in buried constructions where possible. If burial is impossible or extremely costly, then make the buildings as small as possible and construct them in reinforced concrete with wall-thickness of about 18 cm to enable splinter protection.

b) Make all antenna constructions as low as possible, and preferably on the ground/low visibility constructions. If towers are required build them as hollow reinforced concrete masts with wave guides protected inside the construction.

c) Build all transmission nodes as buried constructions with lowest possible visibility.

d) Make all commercial power access as buried cable with the step-down transformer either buried or remote.

e) Build all Nodal Switch buildings as buried constructions. For especially important nodes consider additional hardening measures to ensure comparable level of hardness with the main headquarters they serve.

f) Dimension the selected transmission links to carry the minimum level of vital traffic under the worst total threat conditions.

g) Provide protection to buried or semi-buried structures.

h) Provide all stations with perimeter fencing at least 50 meters in radius, with intruder alarms and proper mechanical protection.

The level of survivability achieved by a design based on the above criteria represents the baseline from which further survivability measures need to be specified to counter additional threats likely to be present in limited and general nuclear war.

13.5.2 Electro-Magnetic Pulse (EMP)

Electromagnetic pulse resulting from a nuclear weapon detonation poses a serious threat to communications equipment. For low altitude bursts damage caused by thermal and over pressure effects will be dominant, destroying both the communications and the user facilities. At high altitudes, the situation is different. Over a very large surface area a very high time-varying electromagnetic field (HEMP) is experienced. The maximum peak value of this electric field on the earth surface can be as high as 50 kilovolt/meter, while most electronic systems are able to survive 5 to 8 volts/meter.

EMP ground coverage is determined on a line-of-sight basis from the point of burst and extends to a 1000 km radius for a burst altitude of 100 km. Therefore, a 1 MT burst at this altitude will potentially destroy all unprotected electronic installations inside an area of the earth surface of 1000 by 1000 km.

The response to the EMP threat is foreseen in the requirement chapter of this book, and hence included in the design of the WIS. Protective shielding is therefore required for all major installations. Design provision has, however, only been made to protect those elements that are defined as essential. These are:

- the nodes
- access and internodal links
- access switches for primary users, where the protection is assumed provided by the user installation itself.

Annex 13 A gives design information for implementing an effective EMP Protection System (EMPPS) for WIS, consistent with WIS survivability design and military requirements (Chapter 2). The design information provided covers the threat description, a detailed coupling analysis, general guidelines for EMP hardening and rules/processes for verification of the EMPPS.

13.6 PERFORMANCE ASSESSMENT

The performance of the network as designed to meet the conventional threats has been assessed using computer simulation techniques, on the basis of ten separate attack patterns derived from the threat scenario defined by the Authority. By using the cumulative effects of each attack it is possible to contain within worst case boundaries the wide range of the unpredictability associated with the attackers freedom of choice. Moreover the traffic load is allowed to increase following these attacks, thus enabling an indication of the design safety margin to be given with reference to the actual minimum communication requirement. The chosen ten attack patterns are selected on a nearly random basis. The method allows further work to concentrate on establishing patterns using other threat scenarios. The study can then be extended to analyse the network in more detail to identify potential local or regional enhancements.

Three numerical values have been used to measure the network's overall performance:

1) Number of internodal links remaining, this indicates the connectivity of the network,
2) Total handled traffic
3) Grade of service for vital traffic under damage and overload (100%) conditions. It is assumed that the vital traffic can be defined as the 15% allocated to highest priority during dimensioning, plus 20% of the next highest priority traffic giving 20% of the total traffic load.

The results obtained from the various simulation runs are plotted in figures 13.1 and 13.2.

It is seen from figure 13.1 that under the assumed damage conditions a margin above the designed minimum vital traffic capacity of about 400 Erlang will remain. This corresponds to a low connectivity level of 13 links and total handled traffic of more than 500 Erlangs under the highest damage scenario.

The accumulated damage represented by these 10 accumulated damage patterns are extreme and it should be noted that no restoration has been assumed. (i.e. the analysed damage represent the net growth of network damage). After stage 10,a total of 33 installations are assumed to have been destroyed, corresponding to a total of 1000 sorties, with reference to the assumptions in table 13.1.

In figure 13.2 the blocking probability for the four different priority levels are shown. As can be seen, the network starts showing signs of fragmentation from stage 6 onwards. Up to that point the network is capable of providing non-blocking connectivity for all P1 traffic. Level 7 and upwards causes fragmentation of the network and that the priority system stops working since the users that still have connectivity will have less and less blocking, while the users with P1 traffic requirements will have higher average blocking probability due to some users of this category not having the necessary access the network.

It is to be noted that the simulations leading to figure 13.1 and 13.2 assume at most 12% loss of traffic due to either loss of user sites or their acces links and up to more than 60% of the internodal links. These figures therefore represent the worst-case threat scenarios for the traffic-carrying capability of the network.

The performance of the access network can, on the other hand, be seen as the success in keeping users connected to the network. As shown in figure 13.1, a total of 215 headquarters remain connected, representing a reduction of about 12%, which is better than the required 15%.

A high-altitude nuclear burst particularly if aimed WIS elements would reduce network capacity primarily through simultaneous interruptions or cause permanent damage if the network is unprotected. However, assuming that all links in the internodal networks and access network, all nodes and all primary user access switches are fully EMP protected then the system could survive with full capacity provided there is no other damage to the network.

The speed with which the network is reduced to its minimum capacity in convertional war is determined solely by the amount of effort committed to the network by the enemy. The effort needed to attack all transmission sites is within range of a high-potency aggressor, but would require the commitment of more than 2700 fighter-bomber or equivalent attack sorties. Therefore if the enemy is credited the capacity of 100 sorties per day to attack on the network, after ten days the network could be reduced to the minimum configuration analysed. If the nodes remain as the most invulnerable element to convertional attack, and the same weight or effort was then devoted to the links, one or two such links could be cut per day. The ability of the network to maintain the required level of emergency service would then depend on the relatve rate of loss of primary users and links. However, the restoration facilities of the internodal network through redundancy, repair and using mobile facilities are likely in most regions to reduce the rate of loss sufficiently to keep pace with the requirements.

There was at this stage no defined threat on Electronic Warfare (EW). This threat on the other hand may test the minimum network capacity more quickly than direct attack in a localised area since trunk capacities would be reduced if LOS links were jammed using jammers with appropriate EIRP near the station. Special protective measures will have to be taken for vulnerable LOS links. These would best take the form of reserve troop security guards equipped with direction finding equipment to locate jamming sources since the sabotage threat would also be covered, and the capability to react to similar attacks on military LOS systems would also be available. Alternatively or in addition nulling antennas and/or frequency-agile radio equipment may be used to eliminate the effects of possible jamming.

13.7 SURVIVABILITY ENHANCEMENTS

13.7.1 General

Survivability enhancements for increasing the remaining network capacity, after a given level of direct attack or for reducing the rate at which the network capacity degrades under attack should be determined using the design criteria specified for the full network. Alternatively these enhancements may be considered as an improvement to the basic design safety margin and can therefore also be regarded as optional extras with the choice between alternatives on the basis of cost and effectiveness. The following setions describe some alternative options which may be used to enhance the survivability of the network.

13.7.2 Connectivity

The addition of more internodal links, particularly node-skipping links, to the network and of access links may offer a cost-effective solution for survivability enhancement in many locations. The necessary links/groups may be rented from the PTT. In future years, satellite links obtained from national or international SATCOM systems may be used to provide node-skipping internodal links and access links. This may improve the network significantly against network fragmentation. The nodal switches in WIS were dimensioned so that, if necessary, they may accommodate the additional links mentioned above.

13.7.3 Transportable Facilities

Transportable switching and transmission facilities are recommended as basic survivability measure for restoring damaged WIS elements. These facilities may also be used in some location and instances such as exercises as a peacetime temporary measure. Transportable transmission resources held in reserve may also represent a cost-effective alternative to the provision of additional fixed transmission links since they would be available for use over a wide area. For exaple a helicopter transportable relay facility would offer considerable flexibility to circuit restoration and may provide greater cost benefits than other alternatives.

13.7.4 Electronic Counter Counter Measures (ECCM)

The field of Electronic Counter Counter Measures (ECCM) may at any time offer, through technological breakthrough, a cheaper and more effective method of protecting a transmission system from EW attack than by direction finding and destroying local jamming sources.

ECCM techniques such as antenna nulling and frequency agility exist today to protect LOS system against EW, but at a cost. These have not been included in the WIS at this stage. It is believed that some of these features can be added at a later stage with only a limited additional cost.

13.7.5 Network Extension

The network itself contains some 100 access switches all of which are connected to two nodes. The modification of at least the larger switches to allow a certain level of transit switching capability would effectively extend the internodal network providing:

- Additional flexibility in routing
- Additional re-configuration capability for the network control system

probably at relatively little cost for the additional survivability benefit gained.

13.7.6 Hardening

The use of ground level (hidden), camouflaged and hardened antennas whose survivability matches the survivability of the associated hardened radio station building should be investigated. At such sites spare antennas should be stored.

Each of the options outlined above appears to be worthy of further study. Moreover, as the network itself is further developed, opportunities are likely to present themselves for enhancing survivability in particular areas. Therefore a continuous survivability activity should be maintained throughout the network implementation phases:

- To follow up the developments in the threat and
- to identify cost-effective improvements

while maintaining the survivability balance within and between each network region.

13.8 SURVIVABILITY COST

The cost elements that are included in the WIS cost estimates and direcly relating to the survivability requirements are:

- Added physical hardness to all WIS structures
- EMP protection of all WIS structures
- Provision of a second access link for all major user locations
- The purchase of transportable facilities

The cost for the internodal links are not judged to be related specifically to survivability, since the design was based on traffic requirements rather than survivability. Similarly the costs related to the Network Survivability and Control System (NSC) is purely justified on the ground of peacetime management and control requirements. It is to be noted, however, that the NSC will play an important role in maintaining the survivability performance in wartime. It can be argued that the requirements for routing, flexibility and multi-level precedence and pre-emption are survivability features likely to make the switches more expensive. These costs are not identified in this context since they are judged as generally oriented to military requirements on a wider range than survivability.

On this basis the identified cost estimates for survivability can be summarised in percentage of the overall cost estimate as follows:

- Civil works enhancements (ex EMPP) : 2.3%
- EMP Protection : 2.4%
- Access links : 3.7%
- Transportable facilities : 1.2%

- Total Survivability Cost : 9.6%

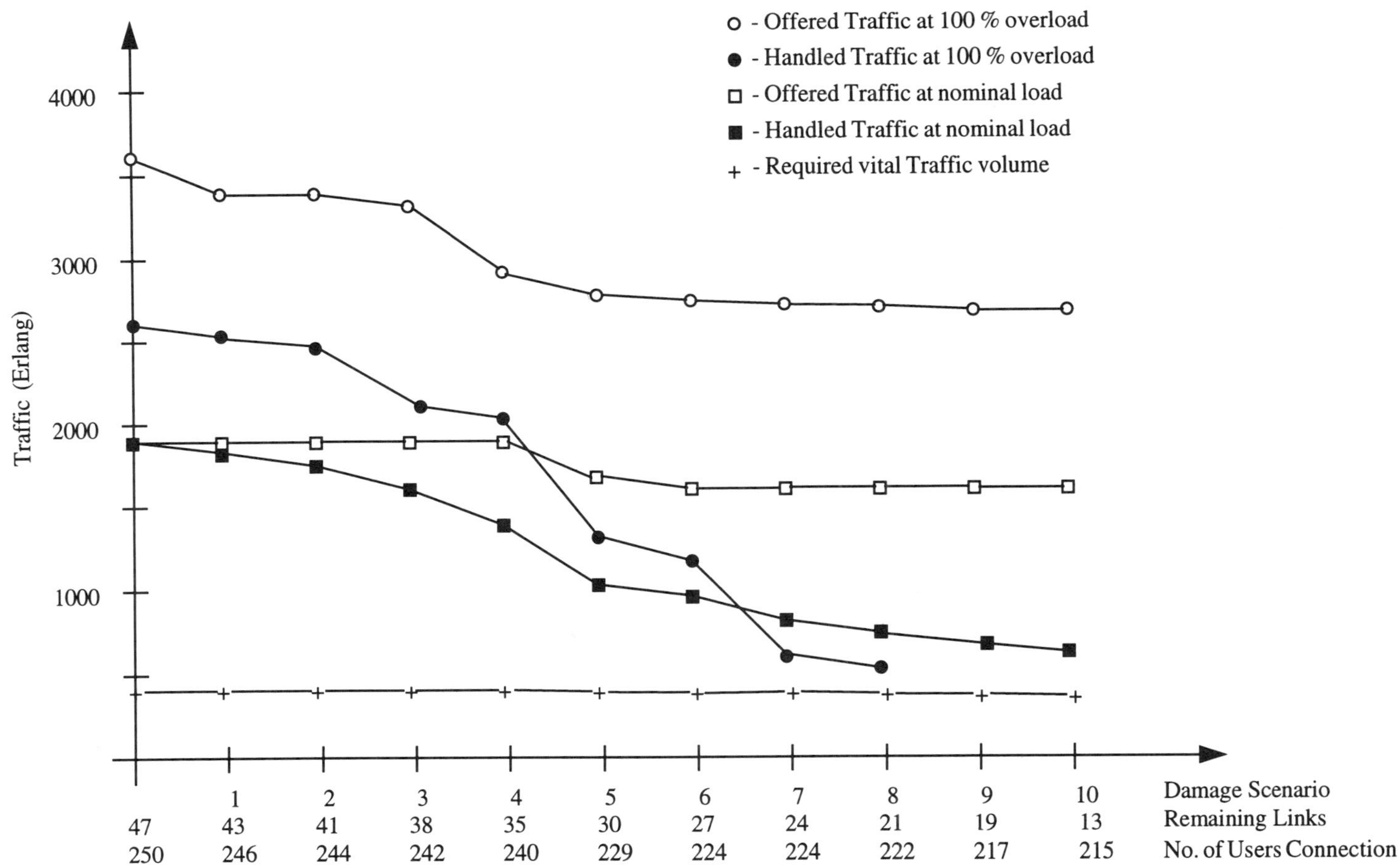

Figure 13.1 Network Traffic Capacity Performance

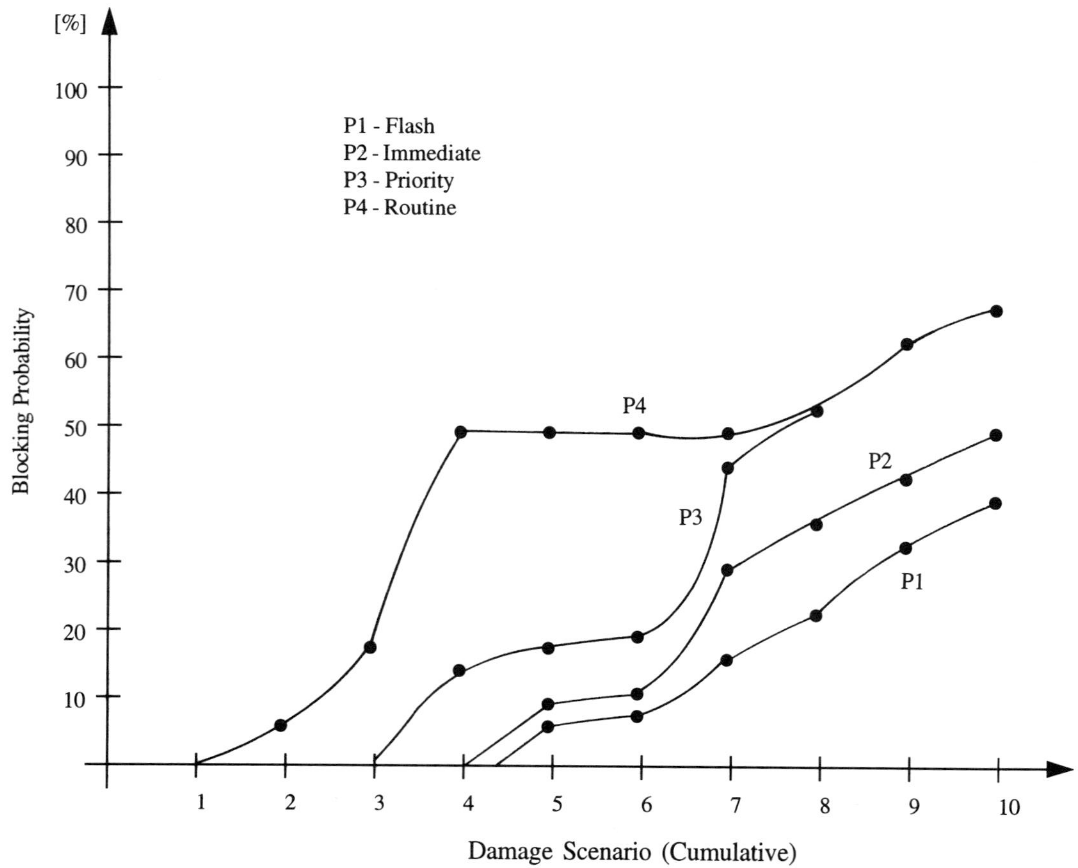

Figure 13.2 Network GOS Performance

ANNEX TO CHAPTER 13

EMP PROTECTION

13A.1 INTRODUCTION

13A1.1 Objective

A nuclear burst generates an intense electromagnetic pulse (EMP), which can reach a peak field strength of tens of kilovolts per meter within a few nanoseconds. The EMP produced by a high altitude detonation (HEMP) is the most significant in that such a detonation covers a large area on the surface and the EMP generating mechanisms become more effective due to the nature of the atmosphere. Under these conditions, all unprotected systems exposed to the HEMP may be subjected to direct or through transients induced on:

- Power lines,
- Communication cables,
- Grounding conductors,
- Antennas,
- Other metals attached to the system.

The induced transients cause detrimental damage or upset in the electronic systems.

As foreseen in Chapter 2 all of the major WIS installations are to be protected against EMP threat to a level as specified by the Authority. Protective shielding to provide an attenuation of about 100 db (design goal) is required for all user sites, all nodes and all interconnecting access and internodal links if the vital traffic requirements for WIS is to be met [Chapter 13].

The purpose of this Chapter is to provide a generic document for implementing an effective EMP Protection System (EMPPS) for WIS, consistent with WIS survivability design [Chapter 13] and military requirements [Chapter 2].

13A1.2 Structure of the Chapter

The chapter contains the threat description, a detailed coupling analysis, general guidelines for EMP hardening and rules/processes for verification of the EMPPS.

In Section 13A.2, the HEMP environment is overviewed and the threat is defined which will lead the design of the WIS EMPPS. A composite EMP field characteristic is obtained from various documents that give the threat definition.

In Section 13A.3, it is intended to give a detailed coupling analysis by overviewing different approaches for the early-time, intermediate-time and late-time HEMP fields. The effects of each of these time regimes for WIS installations are given taking into account all the elements of the EMPPS, individually.

Section 13A.4, gives the main rules and guidelines for an effective hardening design for WIS against the threat defined in Section 13A.2.

In Section 13A.5, the verification/testing requirements and procedures for WIS EMPPS are described. Also, a hardness maintenance and surveillance approach is included.

13A1.3 Basic Elements of the EMPPS

The EMP Protection System includes the following components:

- Shielded enclosures,
- Doors and entry panels,
- Terminal protection devices (TPD), and
- Grounding hardware as shown in Fig. 13A.1.

The Electromagnetic Pulse Protection System (EMPPS) should be properly designed so as to maintain the shielding integrity for each of the above components.

The EMPPS shall protect equipment against both direct and induced effects of the HEMP environment. Direct effects are those associated with the generation of electrical transients directly by the radiated HEMP fields, and the induced effects are the transients present on external conductors that penetrate the barrier. The shielding hardware provide for the mitigation of direct effects, while the terminal protection hardware protect against induced effects. The grounding and bonding hardware are included to enhance the overall shielding integrity of the EMPPS and to assure realization of the potential benefits provided by the terminal protection hardware.

The deleterious direct effects may be controlled using homogenous materials, either ferrous or non-ferrous, for sufficient field attenuation to include at least 20 dB design safety margin. Deleterious penetration currents via fortuitous and intentional conductors are controlled using voltage/current limiting devices.

The required shielding attenuation depends upon the susceptability of equipment. When the susceptability is known, from test data or from specifications, lesser values of attenuation may be acceptable. The safety margin is determined from the susceptability level of the equipment and the excepted EMP transient levels.

In cases where equipment susceptability is unknown, the attenuation provided by shielding should be at least 80 dB at all frequencies from 100 kHz to 100 MHz [1]. These attenuation standards apply at all locations within the shielded volume that are located 1 m. or more from the shield, expect for corners.

When higher fields exist near walls or in corners, the shield may still be acceptable provided sensitive critical equipment and unshielded cables are excluded from the high field regions.

The WIS baseline communication network consists of the major elements as user facilities (e.g., headquarters), transmission facilities, switching systems and power systems. Although all such elements are of potential concern, this Chapter mainly considers the EMPP requirements for a typical digital radio relay station and a nodal transmission/switching facility. However, the study is intended to be sufficiently generic so that the results are applicable to all WIS installations. Typical nodal switch and radio relay sites are given in Fig. 13A.2 and Fig. 13A.3, respectively.

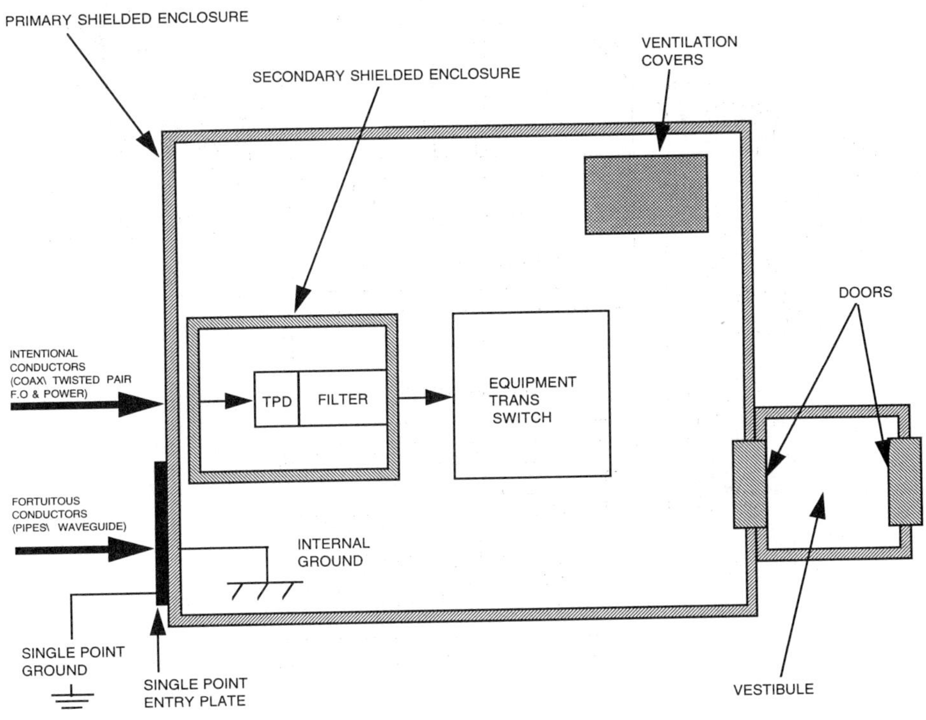

Figure 13A.1 Major Components of EMPPS

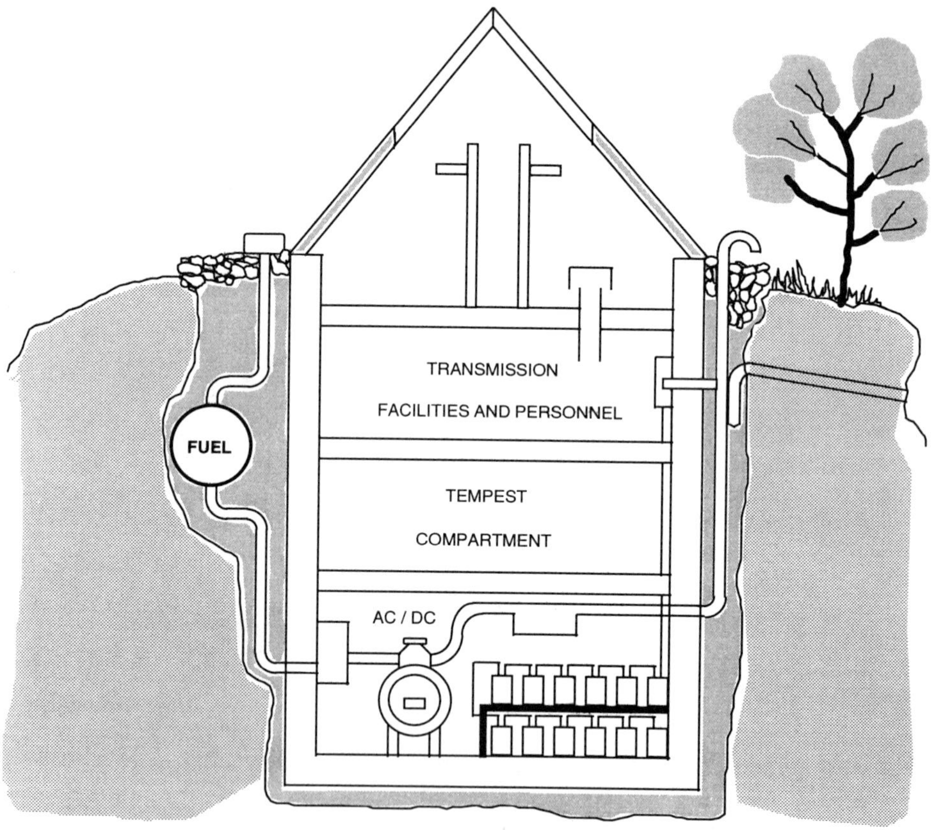

Figure 13A.2 Nodal Switch and Transmission Site

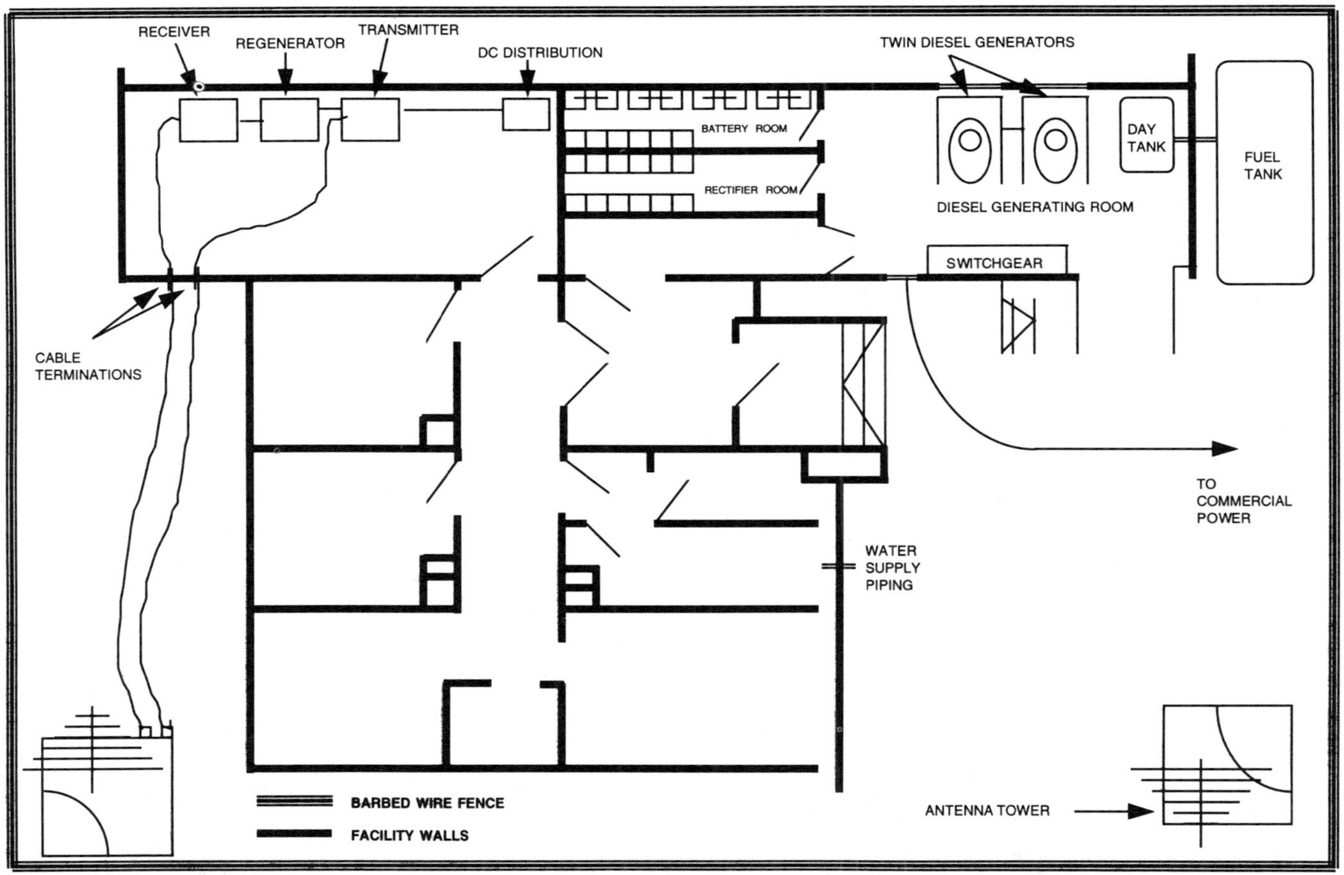

Figure 13A.3 Radio Relay Station

The basic elements in the WIS EMPPS are outlined below.

13A1.3.1 System External Interfaces

Each WIS facility does not have a full complement of transmission and switching subsystems, and therefore, external interfaces are site dependent. The WIS baseline facility contains a full complement of external interfaces. The external EMPPS interfaces include the hostile environment, intentional and fortuitous conductors and the earth electrode subsystem.

Typical conductor interfaces consist of cables, pipes and other fortuitous conductors that carry HEMP induced transients to the WIS system. If required, non-critical equipment shall be hardened to prevent secondary damage to mission critical equipment within the EMPPS. Such equipment includes tower heights, perimeter lighting, leads to wall receptables, fence control and alarm circuits.

The WIS facility interfaces may consist of the following penetrations:

1. The coaxial buried existing or new facilities
2. The waveguide for the microwave system
3. The fiber-optic bundles for either existing or new facilities
4. The landline interface consisting of twisted-pair bundles
5. The powerline interface consisting of the following conductors:

    a. Three phase conductors (A,B,C)
    b. One neutral conductor

Three are two ventilation opening in the EMPPS for air inlet/exhaust. An entrance is provided for personnel ingress/egress. Fortuitous conductors, such as water and sewage pipes do not penetrate the EMPPS primary shielded barrier, but are routed within the WIS facility. These conductors shall be treated to prevent re-radiation effects, and inadvertent arcing within the facility.

13A1.3.2. System internal interfaces

The EMPPS internal interfaces include interconnections of telecommunications equipment, power distribution and facility support. Internal interfaces between the WIS equipment shall be protected by the EMPPS at the global level. The primary shielded enclosure provides a benign environment for internal equipment and cable interfaces.

13A1.3.3. Dimensional Requirements

The shielded volume dimensions shall be capable of accomodating the planned WIS susceptible equipment. The EMPPS shall allow adequate aisle space for ingress/egress and walk-around space internal and external to the shielded volume. The shielded volume shall be designed to allow easy access for immediate inspection, testing and maintenance.

13A1.3.4. Survivability/Vulnerability

The EMPPS shall be designed to be as survivable, endurable, supportable, sustainable and reliable as the operational system being protected. The requirements and conditions specified.

13A1.3.5. Environmental Characteristics

The EMPPS shall be designed to survive and operate in, or be protected against the ambient environmental conditions. The environmentel conditions are either natural (wind, rain, temperature, etc.) or hostile. The EMPPS shall maintain shielding effectiveness and transient suppression overall environment extremes.

13A.2 HEMP THREAT DESCRIPTION

13A2.1 Overview of HEMP environment

In the determination of necessary and sufficient EMPP measures, in their implementation, and in the verification of having achieved the desired level of protection, the ideal situation would be to have a set of threat HEMP environments that describe the most stressing conditions likely to be encountered. The EMPP measures for the WIS must be developed and evaluted using waveforms available within the literature available. This section discusses identification of the waveforms that may be used for WIS.

The use of these waveforms will support the identification of appropriate EMP protective measures for WIS. This assertion is based upon the use of realistically conservative threat waveforms and the imposition of significant design margins for the various EMPP measures discussed in this Section.

As noted in several presentations at the 1987 SHAPE EMPP Technical Symposium [2], and in many documents before and since, the HEMP environment is described in three different time intervals, with each interval having its own characteristic waveform and underlying signal-generation mechanism or mechanisms. Each of these three intervals (early-time, intermediate-time, and late-time) is individually discussed below. The early-time waveform is given the most attention because of its greater field strength and because of the large number of candidate waveforms that have been suggested for it. The HEMP environment discussions conclude with the presentation of a composite waveform for use in further analysis in this report and in subsequent design, implementation, and verification of EMPP measures for WIS.

13A2.2 Early-Time HEMP fields

The prompt gammas from a high-altitude nuclear burst create large Compton-electron currents due to interaction with the atoms and molecules of the upper atmosphere. The early coherent motion of these electrons about the earth's magnetic field lines generates an intense radiated electromagnetic field with a rise time of a few nanoseconds (ns) and a pulse width of 1 ms or less. This field comprises what is called the early-time portion of the high-altitude electromagnetic pulse (HEMP). This signal reaches the earth's surface as a near-plane wave so that E(t) [in V/m] and H(t) [in A/m] are related by the expression H(t) = E(t)/Zo with Zo = 377 ohms.

Any observer on the earth's surface within sight of the burst point will experience the early-time field, except for observers whose line of sight to the burst lies along the earth's magnetic field lines. At other points on the earth within sight of the burst, the field generally exceeds one-half the maximum field strength (which, in the northern hemisphere, is experienced in an area centered a distance approximately HOB tanθ south of the point underneath the burst point, where θ is the local magnetic dip angle and HOB is the burst altitude.) In spite of variations in the field strength over the earth's surface, it is often assumed in systems studies that the earlytime field is uniform over the illuminated area. Except for systems that extend over hundreds of kilometers, this is a fully justified approach since one does not have control over the burst locations of EMP-producing detonations. For systems that extend over hundreds of kilometers, consideration of the spatial variation in the field intensity could lead to somewhat reduced loads; but it is not overly conservative to treat such systems as if all elements within sight of the burst are subjected to maximum threat-level early-time HEMP fields.

The radiated field is polarized with its electric vector perpendicular both to the direction of propogation (assumed nominally to be along the line of sight from the observed to the burst point) and the earth's magnetic field at the observer's location. The electric vector may be resolved into its horizontal and vertical polarization components where the vertical component lies in the plane of incidence and the horizontal component is at right angles to this plane with both components being normal to the propagation direction. This means that, for observers magnetically north and south of the burst point, the field has only a horizontal-polarization component. Thus, the maximum horizontal-polarization component of the incident field is equal to the maximum incident field strength. For observers on the magnetic east-west line, the electric vector lies anywhere between zero and (90-0) degrees off the horizontal, depending upon the elevation angle. Thus, to first order, the maximum vertical-polarization component of the earlytime HEMP field is given by cosθ time the maximum field strength. In the United States, a typical value for θ is 67° so that the maximum vertical component is approximately 40 % of the maximum field strength. In Turkey for instance, a more typical value for θ is 55°, so that the maximum vertical-polarization component is almost 50 % stronger than in the U.S. Since, as will be shown later, the vertical component couples strongly with both vertical and horizontal cables, transients induced on conductors by the threat-level fields can be higher in Turkey than in the U.S. or in northern Europe.

The field strength is usually expressed in analytic form to simplify subsequent coupling analysis. Two such forms that have historically been favored are:

$$E(t) = E_0 \, [e^{-\beta(t-t_0)} - e^{\alpha(t-t_0)}] \, U(t-t_0)$$

where

$$U(t) = 0. \; for \; t < 0., \quad U(t) = 1. \; for \; t > 0.$$

and

$$E(t) = \frac{E_0 e^{\alpha(t-t_0)}}{1 \div e^{(\alpha+\beta)(t-t_0)}}$$

The first of these forms is known as the Difference of Exponentials (DEXP) form; the second, as the Quotient of Exponentials (QEXP) form. The following lists several values of the parameters α, β, E_o, and t_0 that appear in the open literature:

13A2.2.1 Old DEXP Form

An unclassified DEXP fit was published over ten years ago [3] which has been widely used in EMP analysis and is still in use for some studies. The parameters for this fit are:

$$E_0 = 5.25 \times 10^4 \quad V/m$$

$$\alpha = 4.76 \times 10^8 \quad s^{-1}$$

$$\beta = 4.0 \times 10^6 \quad s^{-1}$$

$$t_0 = 0$$

This fit reaches a peak value of 5.0 x 10^4V/m at a time of 10.2 ns.

13A2.2.2 New DEXT Form From 1987 SHAPE Symposium

At the 1987 SHAPE EMPP Technical Syposium, a new DEXP was presented that approximates an early-time HEMP waveform consistent with recent calculational results [4]. The fit respesents a worst-case approximation in the sense that it is an upper bound to calculated results in the frequency domain and comes close to representing the time-domain envelope of those results. The parameters for this fit are:

$$E_0 = 6.34 \times 10^4 \quad V/m$$

$$\alpha = 1.2 \times 10^9 \quad s^{-1}$$

$$\beta = 4.3 \times 10^7 \quad s^{-1}$$

$$t_0 = 0$$

This DEXP form reaches a maximum of 5.4 x 10^4 V/m at a time of 2.9 ns.

13A2.2.3 New QEXP Form From 1987 SHAPE Sysposium

The same reference presents an alternative QEXP fit to recent calculational results. As noted in [4], i t avoids the "artifical" f^{-2} behavior of the frequency-domain transform of the DEXP form and, therefore, provides a more reasonable description of the initial rise of the early-time HEMP waveform. The parameters for this fit are:

$$E_0 = 6.4 \times 10^4 \quad V/m$$

$$\alpha = 2.0 \times 10^9 \quad s^{-1}$$

$$\beta = 4.3 \times 10^7 \quad s^{-1}$$

$$t_0 = 3.12 \times 10^{-9} \quad s$$

This QEXP fit reaches a peak value of 5.4×10^4 V/m at a time of 5 ns.

13A2.2.4 New DEXP Form From NATO CCPC EMP Expert Group

In [5], a DEXP fit is given with the following parameters:

$$E_0 = 6.4 \times 10^4 \quad V/m$$

$$\alpha = 4.0 \times 10^9 \quad s^{-1}$$

$$\beta = 3.0 \times 10^7 \quad s^{-1}$$

$$t_0 = 0$$

The fit reaches a peak value of 6.15×10^4 V/m at a time of 1.23 ns.

13A2.2.5 Comparison of Different Waveforms

A comparison of the waveforms for the various analytic fits found in literature is given in Fig. 13A.4 for the time-domain waveforms and in Fig. 13A.5 for the absolute value of the corresponding frequency-domain Fourier transforms, which for the DEXP and QEXP forms are, respectively:

$$E(\omega) = E_0 e^{-j\omega_0} \left[\frac{1}{\beta + j\omega} - \frac{1}{\alpha + j\omega} \right]$$

and

$$E(\omega) = \frac{2\pi j E_0 e^{-j\omega_0}}{(\alpha + \beta)[e^{\pi(\omega + j\alpha)/(\alpha + \beta)} - e^{-\pi(\omega + j\alpha)/(\alpha + \beta)}]}$$

It is concluded that, in the frequeny domain, the new waveforms suggest a possible decrease in the energy content of the incident wave at low rrequencies and strongly hint at an increase in energy content at frequencies above 10 to 20 MHz, relative to the Old DEXP fit. Consistent with the use of a relistically conservative approach to the specification of a design/evaluation waveform for the WIS EMPP Program, it is therefore

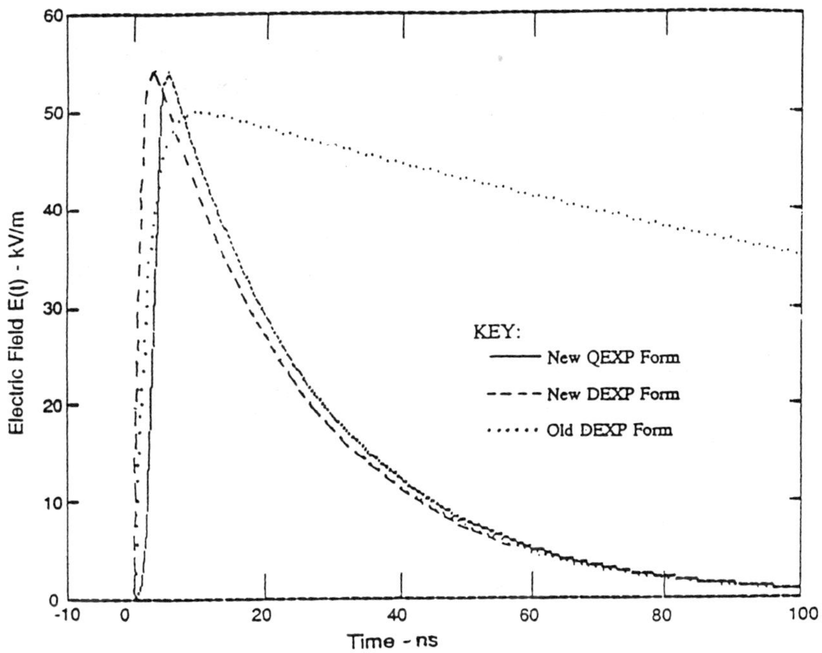

Figure 13A.4 Comparison of Various Early-Time HEMP Fields

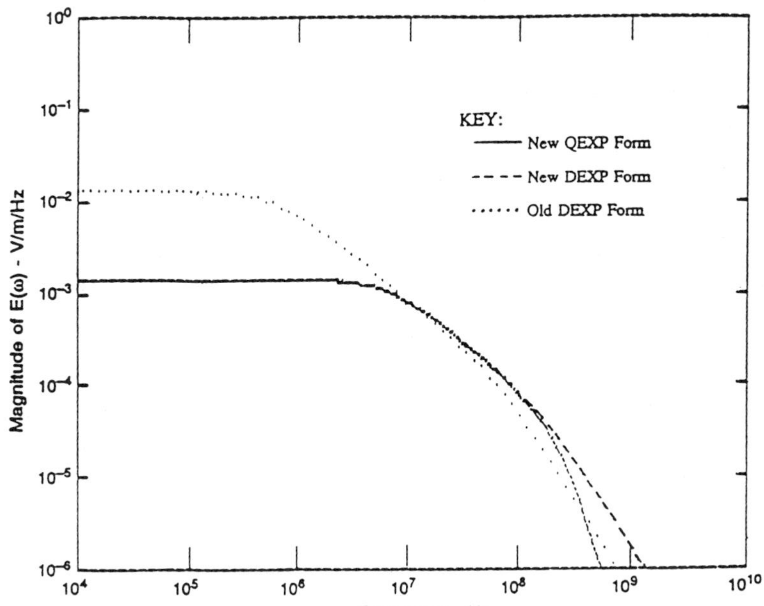

Figure 13A.5 Comparison of Frequency-Domain Early-Time HEMP Fields: Previous DEXP Form and the New Forms Presented at the1987 SHAPE EMPP Technical Symposium

concluded that one should assume an energy content at low frequencies equivalent to the Old DEXP form but should assume an enhanced high-frequency component.

In view of the above observations and discussions, the following recommendations are made with regard to the appropriate early-time HEMP waveform (s) to use in the design, implementation, and verification of EMPP measures in the WIS.

i) The transients on cables and other structures induced by the early-time HEMP wave should be calculated using the Old DEXP form and the New DEXP form presented at the 1987 SHAPE EMPP Technical Symposium [4]. The EMPP measures must be sufficient to protect equipment against either resulting set of transients.

ii) For analysis purposes, the incident early-time HEMP field is a plane wave that illuminates the entire earth's surface within view of the burst point with a field whose time dependence is as previously described. The magnetic field intensity is related to that of the electric field by the relation H(t)=E(t)/377, with E(t) in V/m and H(t) in A/m.

iii) The transients on exposed conductors should be considered as being generated by incident fields with any time-domain and frequency-domain intensities "up to and including" the threat fields described above.

iv) The maximum horizontal-polarization component of the incident field is given by E(t). The incident field may have any linear polarization vector subject to these constraints.

13A2.3 Intermadiate-Time HEMP Fields

The intermediate-time HEMP fields are those in the time interval between 1 μs and about 1s. These signals are due to four mechanisms [6]. First, there is the extension of the early-time HEMP field into the time period after 1 μs due to the continued low-level emission of gammas from the explosion and the generation of a radiated signal by the geomagnetic turning of Compton electrons produced by these gammas (the same mechanism responsible for the early-time signal). Second, there is the radiated signal due to the charge separation produced by multiple-scattered Compton electrons that generated the early-time signal. Third, there is a contribution from the geomagnetic turning of these multiple-scattered Compton electrons. And fourth, there is a contribution from the gammas generated within the upper atmosphere by inelastic interactions of weapon neutrons with the atmosphere.

From the discussions carried out in [6], the following attributes of this intermediate-time wareform can be assigned.

- The contributions from the first three mechanisms contribute to an intermediate-time waveform that reaches a peak of about 100 V/m and extends for a period of about 100μs. The peak field strength is noted to be about 0.002 of the early-time HEMP field evaluated at 10 ns and where a decay time of 100 to 1000μs is described as a plane wave portion of the intermediate-time HEMP field is described as a plane wave with the same polarization characteristics as the early-time field.

- Beyond 100μs, the field is due primarily to charge-separation effects induced by the Comptons from the gammas produced by neutron inelastic interactions in the upper atmosphere. The amplitude of this waveform is lower than for the first part of the intermediate-time signal, but none of the unclassified references makes explicit mention of the maximum field strength to assign to this part of the signal. However, an admittedly qualitative composite HEMP waveform shown in [4] and [6] permits to conclude that this latter portion of the intermediate-time HEMP signal has a peak value of about 10 V/m out to about 10 ms, at which time the field rapidly decreases to a backgound level of 0.001 V/m until the onset of the late-time HEMP signal. This part of the intermediate-time field is not a plane wave; rather, the magnetic and electric fields are such that the magnetic field in A/m is about 0.1 times the incident electric field in V/m. The polarization of this part of the signal is also different from the early-time HEMP field and the initial part of the intermediate-time field: the electric field is primarily vertical, with the horizontal component of the electric field being only about 10% as large [6]. Thus, the horizontal magnetic field in A/m is numerically about equal to the horizontal electric field in V/m. In addition this latter portion of the intermediate-time field is thought to exhibit significant spatial dependence, so that the maximum field strengths will not be experienced over the full length of extended systems. Unfortunately, the spatial dependence of these fields has not been very well described in the available literature.

Consistent with the above discussion, the following waveform is chosen to describe the maximum field strength for the intermediate-time HEMP signal:

$$E(t) = [90e^{-10^4 t} + 10e^{-10^2 t} + 0.001e^{-0.1 t}] \qquad V/m$$

This forms a peak value of 100 V/m in the neighborhood of 100μs, and drops further to a low-level background value after 10 ms, consistent with the behavior noted above. The maximum horizontal-polarization component of the field is equal to 100% of the first term of E(t) but only 10% of the latter two terms. The maximum vertical-polarization component is equal to 57% of the first term in E(t) consistent with the early-time HEMP signal plus 100% of the latter two terms in E(t).

In using this characterization of the intermediate-time HEMP signal, one should consider that the first term in E(t) describes a field that illuminates the entire earth's surface in view of the burst point with uniform intensity, consistent with the situation for the early-time HEMP field. The latter two terms, however, describe fields that exhibit some as-yet-unspecified spatial dependence so that extended systems need not experience the full intensity of the fields described by those terms. The results of [6] have been used to provide an interim characterization that can be used until the spatial dependence of these fields is further developed: the fields described in the latter two terms for E(t) can be assumed to be uniform on the earth's surface over a circular region of radius equal to three times the height of burst. Outside this region on the earth's surface, these fields can be assumed to drop off with an e-folding distance equal to the height of burst. This characterization of the spatial dependence of the latter two terms in E(t) for the intermediate-time HEMP fields means that full threat-level fields are to be used to determine transients induced on systems whose dimensions are less than six times the height of burst but it permits a reduction in the transients otherwise predicted on larger systems. This approach avoids the concerns of [6] in over-predicting coupling effects for extended systems for the latter portions of the intermediate-time HEMP fields.

As for the early-time HEMP fields, the EMPP measures for WIS must withstand transients induced by intermediate-time HEMP fields with any time-domain and frequency-domain intensities "up to and including" the maximum threat-level fields described above.

The time-domain and frequency-domain characterictics of the intermediate-time HEMP fields described by the expression for E(t) given above are shown in Fig. 13A.6 and Fig. 13A.7, respectively.

13A2.4 Late-Time HEMP Fields

In the time region beyond about 1 s, the HEMP fields are due to two different plasma effects and are sometimes called the magnetohydrodynamic (MHD) EMP fields. The earlier of the two plasma effects is associated with the ionospheric blast wave produced by the high-altitude detonation that displaces the earth's magnetic field lines, thereby producing a perturbation that is observed in the time interval between about 1 s and 10 s after burst. The second plasma effect is associated with the motion of the bomb debris and heated air from the detonation across the earth's field lines as this hot mass moves in response to what is called the atmospheric heave phenomenon. This latter process takes place in the time interval from about 10 s to 200 s after detonation. The peak fields for these MHD effects are of the order of tens of volts per kilometer according to [5]. Consistent with this characterization of the late-time (MHD) HEMP fields, the following expression is used to describe the maximum intensity for these fields:

$$E(t)=[k_1 U(t-t_{01})(e^{-\beta_1(t-t_{01})}-e^{-\alpha_1(t-t_{01})})+k_2 U(t-t_{02})(e^{-\beta_2(t-t_{02})}-e^{-\alpha_2(t-t_{02})})]$$

with

$$k_1 = 0.27 \quad V/m$$
$$\alpha_1 = 1.0 \quad s^{-1}$$
$$\beta_1 = 0.6 \quad s^{-1}$$
$$t_{01} = 1.5 \quad s$$
$$k_2 = 0.27 \quad V/m$$
$$\alpha_2 = 10 \quad s^{-1}$$
$$\beta_2 = 0.60 \quad s^{-1}$$
$$t_{02} = 15. \quad s$$
$$U(t) = 0. \; for \; t < 0.$$
$$U(t) = 1. \; for \; t > 0.$$

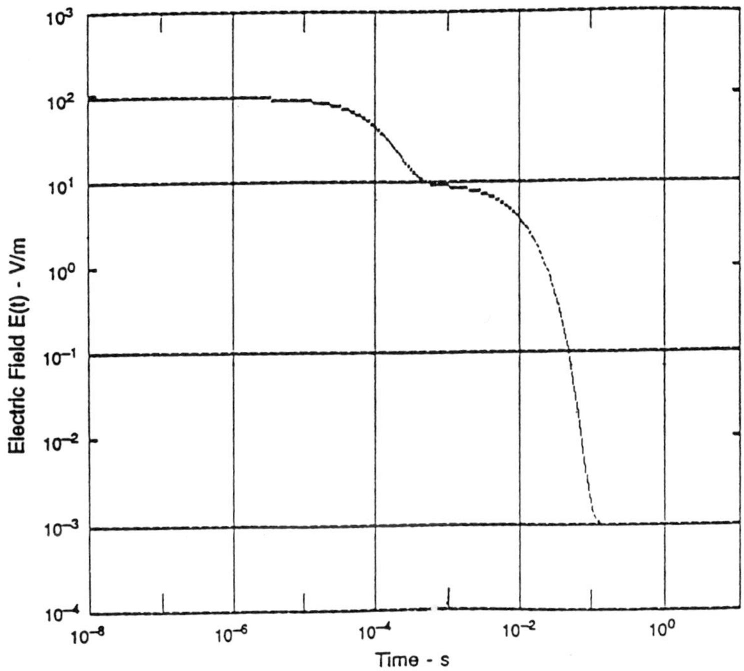

Figure 13A.6 Electric Field Intensity For Intermediate-Time HEMP

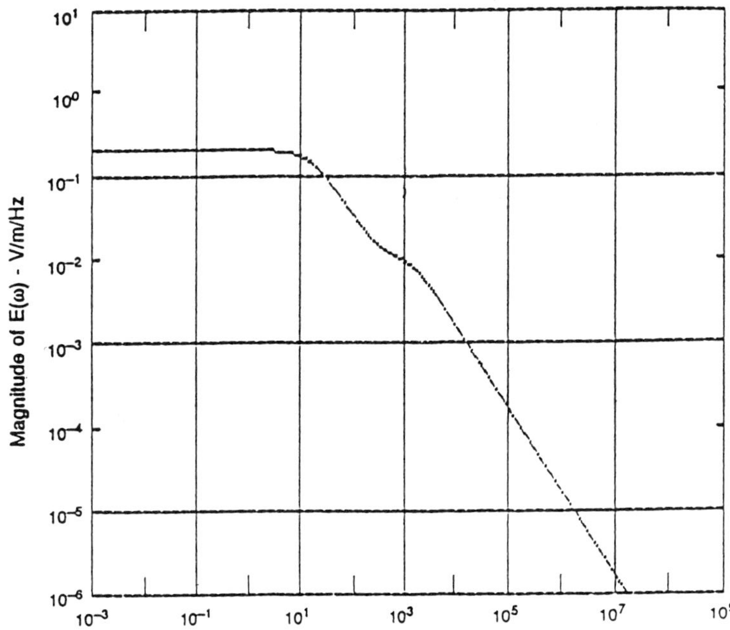

Figure 13A.7 Frequency-Domain Electric Field Intensity For Intermediate-Time HEMP

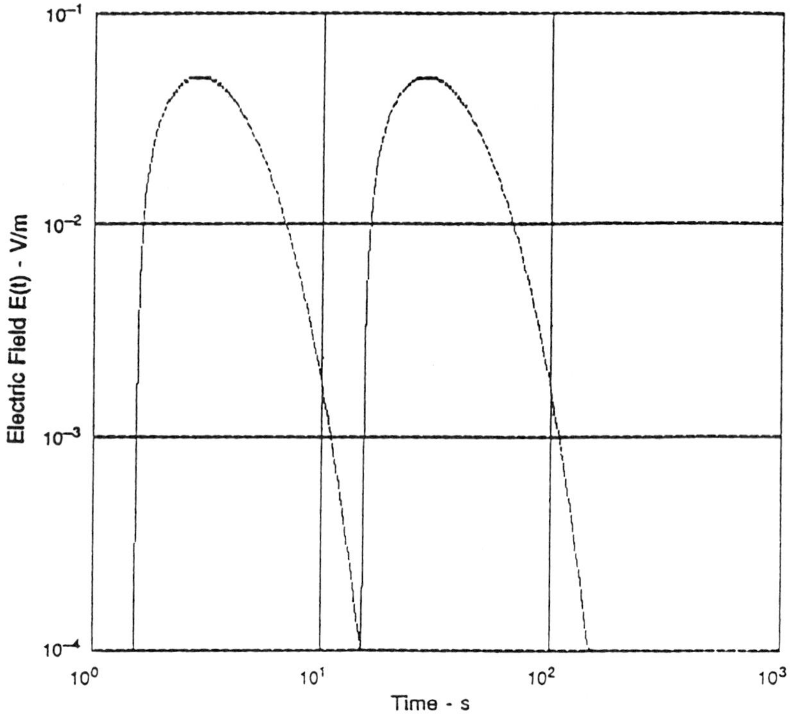

Figure 13A.8 Electric Field Intensity For Late-Time HEMP

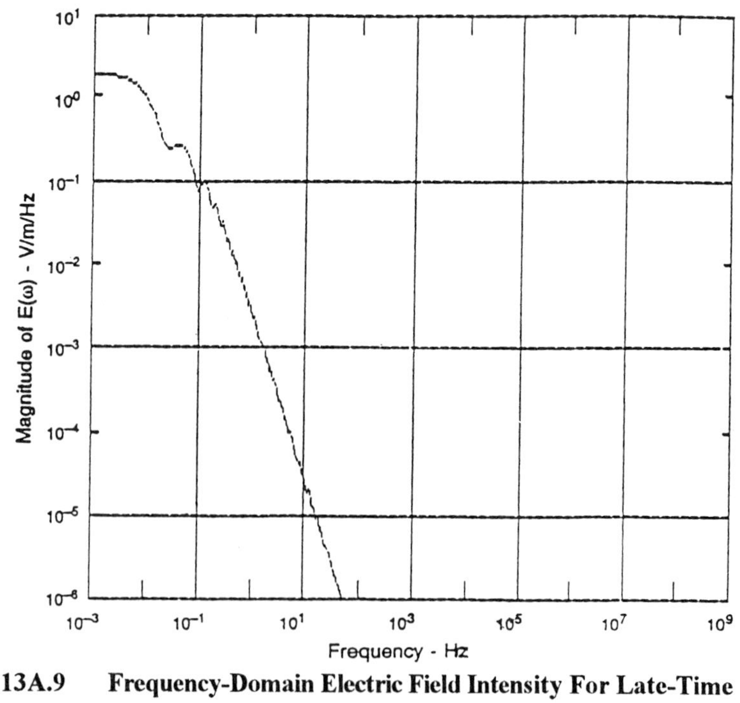

Figure 13A.9 Frequency-Domain Electric Field Intensity For Late-Time HEMP

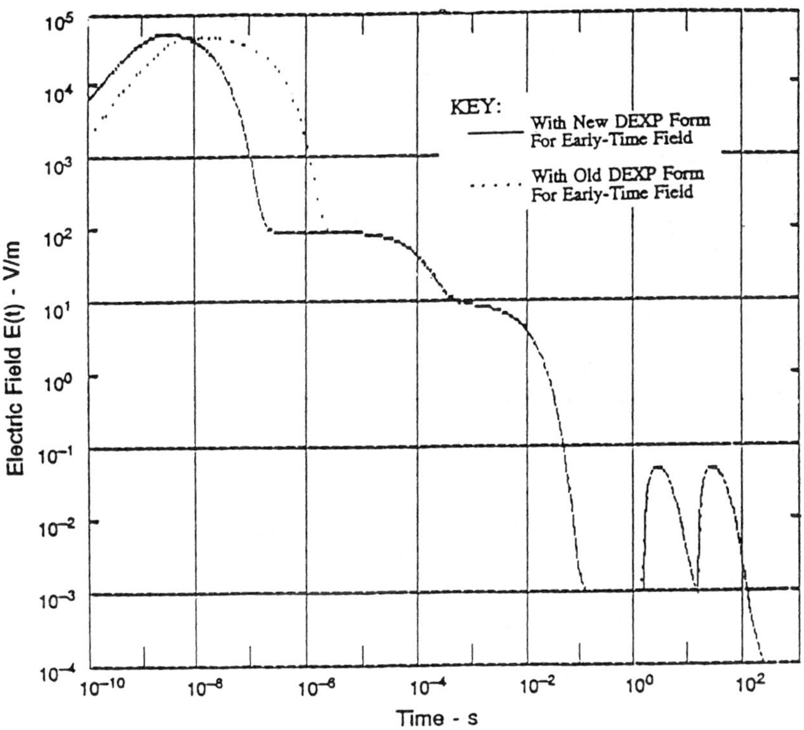

Figure 13A.10 Composite HEMP Total Electric Field Intensity

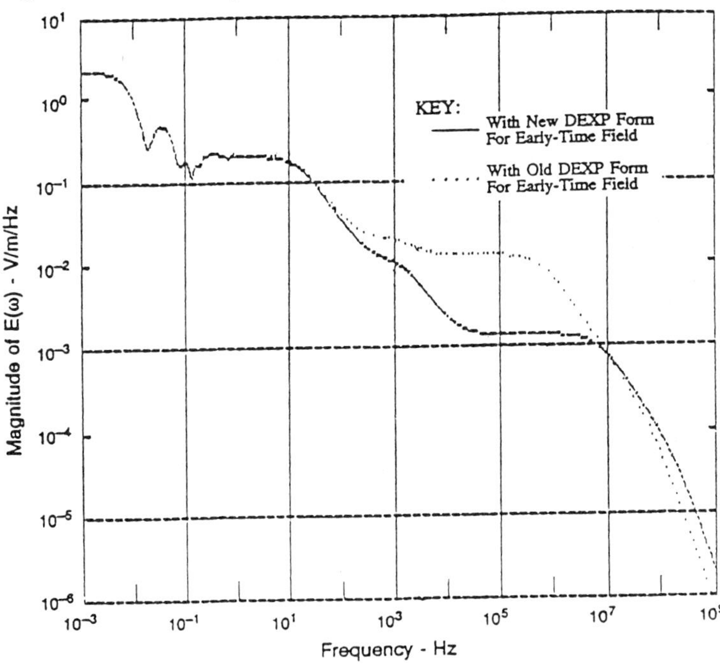

Figure 13A.11 Frequency-Domain Composite HEMP Total Electric Field Intensity

The above form provides two distinct low-intensity, low-frequency signals (see Fig. 13A.8) consistent with the MHD waveform characterization in the open literature and with the composite HEMP waveform presented in [4] and [6]. It should be viewed as providing the maximum expected field strength for the late-time HEMP signal. Loading of systems even over the horizon can occur for portions of the MHD signal. Thus, one should assume that an entire system, regardless of its spatial extent, will be exposed to the maximum late-time HEMP fields. The frequency-domain plot corresponding to the above time-domain late-time HEMP signal is shown in Fig. 13A.9. Peak MHD-field values of 50 V/km are shown at 2.8 and 27.8 seconds, consistent with the analytic form given above.

It should be noted that the late-time fields described above differ from the other fields defined in this Chapter in one important aspect: they represent the resultant fields after the effects of ground conductivity have been taken into account. Thus, these can be used directly to determine the effects of the late-time fields on long conductors (the only items on which significant transients will be induced by the late-time fields) whereas the effects of the other fields on all cables and structures must be determined by subsequently including the effects of ground reflections on the incident HEMP environment.

The polarization of the resultant late-time HEMP fields is dependent upon the burst-viewer-magnetic field geometry, but a worst-case orientation of this field is appropriate to assume for designing the EMPP measures for the late-time field: that is, one should assume that the electric field is always aligned along the longitudinal axis of long conductors.

The effect of the MHD field just described can be rather accurately modeled by assuming that the field imposes a slowly varying gradient [given by E(t)] on the nominally constant ground potential. The effects of the MHD portion of the HEMP environment are, therefore, experienced only by conductors that are grounded at spatially separated points.

13A2.5 Composite HEMP Waveforms

Using the recommendations made above for the various portions of the HEMP waveform composite waveforms can be given for the entire span of the HEMP environment. The result is a set of two composite signals, with the differences representing the use of the two different early-time waveforms, where survival must be guaranteed upon exposure to either earlytime signal. The time-domain plots of the resulting composite HEMP waveform are shown in Fig. 13A.10; the frequency-domain plots, in Fig. 13A.11. These plots describe the maximum total field intensity to be used in the WIS analysis. Fig. 13A.12 through Fig. 13A.15 show similar information separately for the horizontal-and vertical-polarization components. The full intensity of the MHD portion of the HEMP signal is used for both polarization directions, consistent with the assertion that an appropriate worst-case analysis will assume that this late-time electric field is always aligned along the conductor. In using the curves of Fig. 13A.10 through Fig. 13A.15 the following ground rules are to be observed:

i) WIS equipment and installations shall be protected against the transients generated by incident HEMP fields "up to and including" the waveforms shown.

ii) Except for the spatial dependence previously introduced for the latter part of the intermediate-time HEMP fields, one shall assume that the fields shown will illuminate any system within sight of the burst point.

iii) In determining the transients induced on systems, the effects of ground reflections shall be added to all portions of the incident HEMP wave except for those associated with late-time MHD portion of the wave.

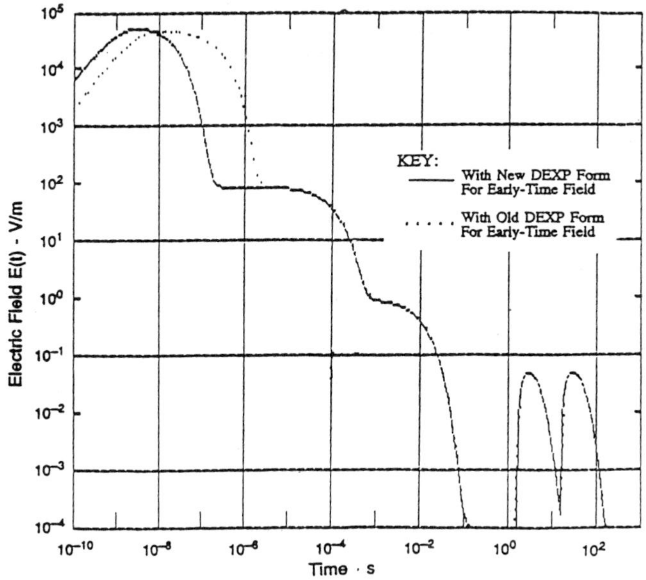

Figure 13A.12 Composite HEMP Electric Field Intensity, Horizontal Polarization

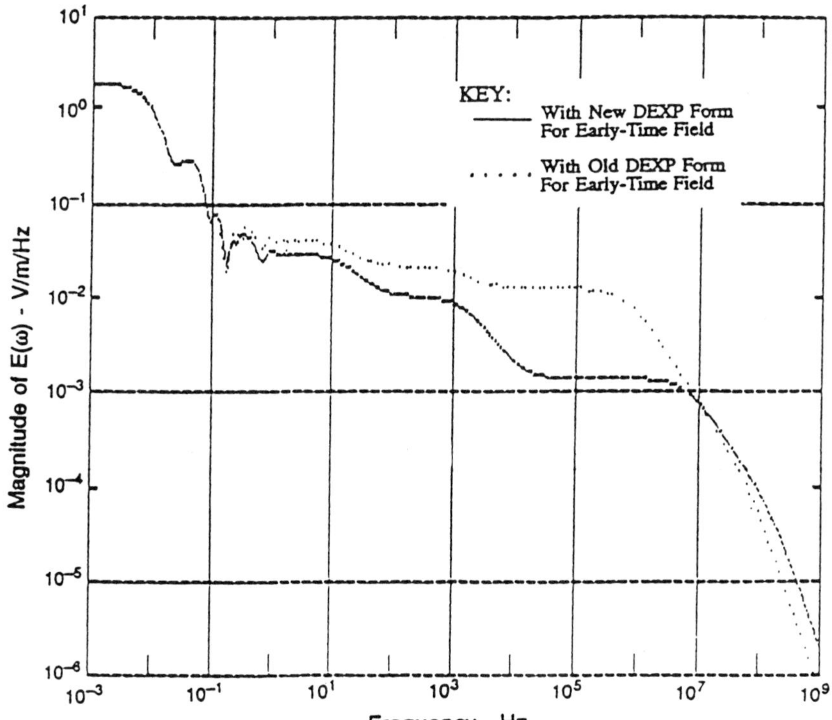

Figure 13A.13 Frequency-Domain Composite HEMP Electric Field Intensity, Horizontal Polarization

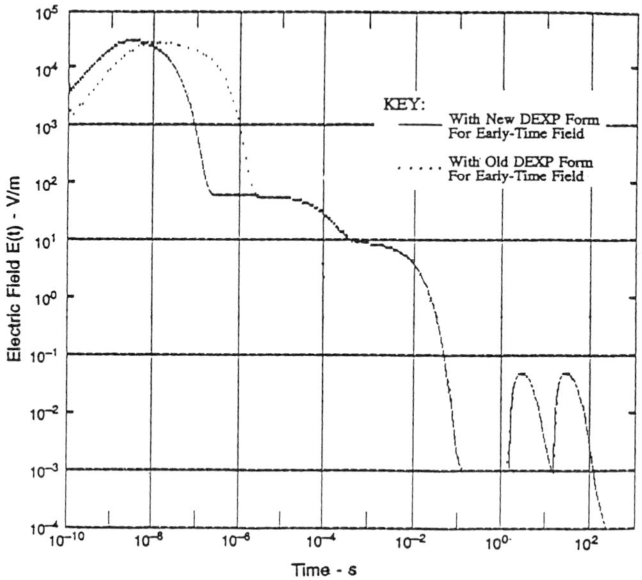

Figure 13A.14 Composite HEMP Electric Field Intensity, Vertical Polarization

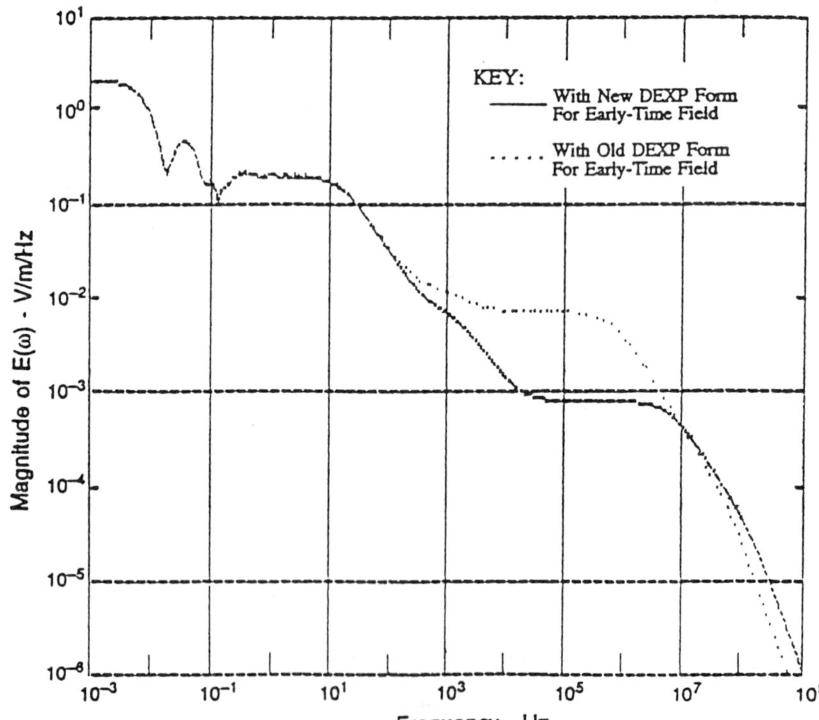

Figure 13A.15 Frequency-Domain Composite HEMP Electric Field Intensity, Vertical Polarization

13A.3 EMP COUPLING ANALYSIS

13A3.1 Introduction

A major objective of the EMPP study is to identify the necessary and sufficient measures for protecting WIS communications system against the effects of the HEMP fields defined in Section 13A.2. Such protection measures will include: shields to attenuate the incident fields so that the transients induced directly on the cabinets, racks, backplane wires, and other conductors that form integral parts of the various subsystems for WIS are sufficiently small to permit continued operation of the equipment; the installation of protective hardware (e.g., filters, surge arrestors, and other devices) to reduce the magnitude of conducted transients that otherwise might reach the WIS equipment and the imposition of effective grounding, earthing, and bonding practices to enhance the inherent efficacy of the recommended shielding and measures requires a knowledge of the excepted transients. The discussions of this section permit the determination of those transients for a variety of relevant conducting elements to be found in WIS installations.

The discussions begin with a brief description of the basic methods to be used in predicting the magnitude and time dependence of the transients expected on the various conducting elements of a WIS installation. That material is followed by discussions attempting to identify worst-case transient levels for the more important WIS conductors. In this latter effort, the intent was to derive a set of transients that would be:

a) Sufficiently complete to assure the subsequent identification of all EMPP measures needed for WIS and

b) Typical of all WIS installations so that the technical feasibility and estimated costs for the hardening measures could be established. That goal has been met; but it must be remembered that the transient levels presented herein are only typical of the worst case conditions that will actually exist at a given WIS site. Thus, proven techniques for predicting HEMP-induced transients should be applied for each site to be hardened in order to support a meaningful acceptance, verification, and hardness-maintenance program for that site. In many cases, it will be found that the design-criteria transients will be identical to those presented here; but one would be remiss in failing to verify the applicability of the levels given here when hardening a specific installation.

13A.3.2 Overview of Coupling Analysis Approach

13A.3.2.1 Early-Time and Intermediate-Time HEMP Fields

The geometry and inherent electrical/magnetic characteristics of most conducting elements of a WIS facility will be found to be too complicated to permit exact calculation of the expected transients induced by the incident HEMP fields. And even if available analytical tools were able to provide such predictions, the uncertainties in the incident fields are probably too large to justify such an effort. Thus, approximate techniques are sufficient.

For most cables, the techniques given in [3] provide the required and sufficient accuracy for predicting expected transient levels induced by the early-time and intermediate-time HEMP fields, at least for the purposes of deriving an initial design for the specific EMPPS at a given WIS site. Those techniques have been used to drive the transient levels and have been presented in this section.

One of the primary techniques employed involves the use of transmission line equations to describe the coupling and response of simple conductors to the incident HEMP fields, taking into account the effects of reflections and refractions of the incident fields by a ground of finite conductivity.

The first step in this approach is to characterize the resultant electric field at a height above the ground surface or at a depth below that surface. The analysis begins with a specification of two mutually orthogonal polarization directions for the electric field (the primary driver of induced surges for unattenuated HEMP fields). Both directions are perpendicular to the propagation vector. The vertical-polarization component of the electric field is that which lies in the plane of incidence (the plane containing the propogation vector and the local normal to the ground surface). The horizontal-polarization component is normal to the plane of incidence and, therefore, lies in the local horizontal plane.

After specification of the resultant fields, these fields may be used to describe a distributed source that drives a transmission line. The transients induced on vertical, horizontal, buried, and aerial cables can all be predicted using this approach. Admittedly, the signal propagation constants and the characteristic impedance expressions for these various cases differ one from the other, but the basic methodology is the same. A critical issue in the approach is the use of simplifying approximate expressions for the various impedances and propagation factors. Care must be used in selecting these expressions; for example, one must retain terms that provide losses in the lines or the induced transients with time. The results summarized below were all derived for lossy lines.

Another critical issue in the use of the methodology is the selection of an appropriate set of parameters that describe the electrical behavior of the ground; in particular, it is important that predictions reflect the range of conductivity values that might be actually present. To base worst-case predictions on calculations made with only a single value of the ground conductivity can lead to gross underestimates or overestimates of the actual worst-case conditions. In the analysis reported herein, transients were calculated for ground conductivity values of between 0.1 and 0001 mho/m, and the most stressing case was retained.

Another important aspect of the analysis is that the predicted transients can show a marked sensitivity to the direction of incidence of the HEMP fields. Therefore, the study included consideration of incidence directions from near-grazing to normal, relative to the local earth's surface.

In view of the above, it is felt that the results reported herein represent a valid identification of transient levels that are sufficiently close to the "true" worst-case conditions to support the design of a technically sound EMPP system.

The time dependence of the conducted transients derived in this study could, in most cases, be approximated by a superposition of damped sinusoids where each term is of the form

$$f(t) = A e^{\alpha t} \sin(\omega_0 t + \phi)$$

where the angular ringing frequency ω_0 is typically of the order of $c\pi\gamma/L$ with L being the length of a cable segment and the parameter γ is approximately unity for above-ground cables and is less than one for buried cables) corresponding to the slower propagation velocity for signals on buried cables). Consistent with this general rule, the ringing frequency for semi-infinite lines is zero. In addition, for a finite-length cable terminated at both ends in the characteristic impedance of the cable segment, no ringing is observed since there are no reflections of traveling waves at the cable terminations.

13A3.2.2 Late-Time HEMP Fields

A simpler approach was utilized to predict transients induced by the late-time HEMP (MHD) fields. Consistent with the discussions of Section 13A.2, the effects of the late-time field have been modeled in terms of a slowly time-varying gradient in the local ground potential, where the time dependence of the gradient is given by the expression for E(t) in Section 13A.2.4. The following illustrates the approach used to determine the effects of this ground potential gradient.

A conductor grounded at two points is considered and each grounding point will have associated with it a ground strap or equivalent. Assuming L to be horizontal separation between these ground straps at the points where they make concact with the earth, (it is noted that L is not necessarily the distance between the grounding points on the grounded conductor) then, the voltage difference applied to the conductor between the adjacent ground points is E(t) L. If the resistance between the ground points on the grounded cable is R and the resistance to ground for each grounding point is R_g, then the current I induced by the MHD fields on the conductor between the two grounding points is expressed as,

$$I = \frac{E(t)L}{(R + 2R_g)}$$

If, as is true in many cases, the length of the grounded conductor between adjacent grounding points is also equal to L and there is no load on the grounded conductor between these points, then the current I can be expressed as;

$$(R_0 + 2R_g/L)$$

where R_0 is the resistance per unit length of the grounded conductor. If, in addition, there is a load R_2 on the cable between its grounding points, the MHD-induced current will be given by the expression

$$I = \frac{E(t)}{[R_0 + (R_2 + 2R_g/L)]}$$

There are several special cases of interest:

a) For systems whose physical extent is no larger that a few tens of meters, as is appropriate for those confined to a single WIS site, R is usually negligible and an upper bound to the MHD-induced current is given by 0.5E(t) L/R_g for the case where, R_2=0. Since R_g will be of the order of 1 ohm and the maximum value of E(t) is 50 V/km according to Section 13A.2, the maximum MHD-induced currents will be of the order of 1 A for a cable confined to a WIS site without a load between grounding points.

b) A similar cable with a load R_2 between its grounding points is considered. The maximum MHD-induced voltage drop across the load is given by E(t) L. Even for a 50-meter distance between grounding points, this translates into a maximum MHD-induced voltage of only 2.5 V and may consititute a negligible threat even if the load is an active electronics item.

c) For long cables, the maximum MHD-induced voltage of 50L volts (where L is measured in kilometers) can be appreciable reletive to the nominal operating voltages of WIS equipment. In addition, if one has two long conductors running close to each other, only one of which is periodically grounded, the potential difference between the two adjacent conductors can become sufficiently large over a run of several kilometers to generate the possibility of breakdown the two conductors.

d) For long distances L (in km) between grounding points, an upper bound to the MHD-induced currents can be obtained from the expression 0.5E (t) L/R_g. For a ground resistance of the order of 1 ohm, this translates into a peak MHD-induced current of 25L A (i.e., a peak current of 25 A for a 1-km spacing between grounding points). However, for small-diameter conductors such as AWG 24 copper wires characteristic of twisted-pair wires that carry alarm signals in coax cables and other long-haul cables, the resistance of the cable (LR_0) can dominate the resistance to ground so that this upper bound is an unrealistic estimate of the transient. Thus, for cables with large separations between adjacent grounding points, a better upper bound to the MHD-induced current on a grounded conductor is given by

$$I = \frac{50L}{(2 + LR_0)}$$

where L is in kilometers and R_0 is in ohms/kilometer. For a separation of 1(10) km between adjacent grounding points, this yields a peak predicted MHD-induced current of 22.2(109.7) A for an AWG 24 conductor. (Neglecting the cable resistance would have yielded currents of 25(250) A for a 1(10)-km-long conductor).

e) As noted above, the resistance of long cables should be considered when calculating the expected peak MHD-induced currents on conductors grounded at widely separated points. However, when the separation between adjancent grounds is so large that the conductor's resistance between grounding points is much larger than the resistance to ground, one may even safely ignore the ground resistance in the calculation of these transients and set the induced current equal to E(t)/R_0. This expression is also appropriate when the distance between widely-spaced grounding points is unknown or beyond the control of WIS (such as for the incoming long-haul power cables). This expression would yield predicted peak currents of 196 A and 0.59 A for the AWG 00 and AWG 24 conductors treated above.

13A.3.3 Specific Coupling Analysis Results

13A.3.3.1 Incoming Power Lines

a) The Step-Down Transformer

The power lines that lead to the primary side of the step-down transformer that is usually located at the perimeter of a WIS facility can have very large transients induced on them. The worst-case short-circuit currents calculated in this study were for the early-time OLD DEXP waveform (see Section 13A.2). For an aerial cable strung 10 meters above the ground, this waveform was found to induce a 10-kA peak current with a time-to-peak of 90 ns and a full width at half maximum (FWHM) of 400 ns. The open-circuit current for the NEW DEXP (see Section 13A.2) early-time waveform was found to be about 5 kA at 16 ns, with a FWHM of about 60 ns. The corresponding open-circuit voltage was 2.1 MV.

As noted, the above early-time transients were calculated for power lines 10 meters above the ground. For a 20-meter height, maximum transients were found to be about 50% larger and occurred at times about 30% later (e.g., a maximum peak current of about 15 kA was predicted to occur at 120 ns when the OLD DEXP early-time waveform illuminated a 20-meter high power line). However, the transients exhibited about the same FWHM values as for the 10-meter case. For a 5-meter height, the maximum transients were about half those for the 10-meter case, but they exhibited the same time-to-peak and FWHM values as noted in the previous Section.

Thus, worst-case early-time power-time transients at the step-down transformer are associated with the maximum expected height of these lines above the ground. In WIS the maximum height for these lines is assumed to be 10 meters. Thus, the transients given in the first pargraph of this section can be viewed as worst-case for the early-time HEMP fields.

For the intermediate-time waveform, the corresponding worst-case peak short-circuit current was calculated to be 320 A at 34 microseconds. A maximum open-circiut voltage of 170 kV was calculated for this waveform.

The early-time transients given here are somewhat higher than those of [3] due to two factors. First, the maximum transients are generated by the vertical-polarization components of the incident field and this component can be 50% stronger in Turkey than in the U.S. Second, the analysis in this study used a better estimate of the characteristic impedance of the power cable than was cited in [3].

If the power cable is buried for a long distance before reaching the step-down transformer, the expected peak currents are reduced somewhat. For example, the OLD DEXP early-time waveform induces a maximum short-circuit current of about 5 kA instead of the 10 kA predicted for the aerial power line. Proportional reductions hold for a buried cable, the time-to-peak and the FWHM are different: for example, the current does not reach its peak for the OLD DEXP early-time waveform until 200 ns for a buried cable and the FWHM is lengthened to 600ns. These values are those calculated for a cable 1 meter deep in soil with a conductivity of 0.0001 mho/m. Insignificant differences were found for burial depths of between a few cm and 5 meters in this low-conductivity medium.

The effect of the late-time HEMP field is to produce a spatially and temporally varying ground potential, with a maximum gradient of 50 V/km, according to the discussions of Section 13A.2. The effect of this field is experienced only by conductors that are grounded; but, for these conductors, the effect is effectively independent of whether the cable is buried or aerial. Using the approach discussed in Section 13A.3.2.2, it is noted that the appropriate expression for the calculation of an upper bound to the expected LHD-induced current coming into the step-down transformer on grounded power-line conductors is given by $E(t)\,R_0$ where $E(t)$ is threat MHD field and R_0 is the resistance per unit length of the grounded conductor. This expression is used since the distance between grounding points on the incoming conductor is variable and not under the control of WIS personnel (for an AWG 4/0 copper wire; a reasonable choice to give worst-case minimum resistance and, therefore, maximum current of some 300 A on such a conductor between ground points). An AWG 1/0 cable, with its higher resistance, would experience a corresponding peak current of 150 A.

b) The WIS Building

Fortunately, the entire transient coming into the step-down transformer does not make it all the way to the WIS building. Three effects mitigate against such an occurrence. First, the transformer exhibits a frequency-dependent insertion loss. Second, the transmission characteristics of the cable between the step-down transformer and the entry to the WIS building tend to spread out the pulse that does get through the transformer. And third, lighting surge arresters on the secondary side of the transformer can reduce the surges penetrating past the transformer to the building (typically by about a factor of two). A typical case was analyzed in which the power cable from a pole-mounted transformer was led underground to a depth of one meter, then horizortally 50 meters to the building, the upward to a height of 3 meters above ground level, and then 10 meters over to its connection point to the internal AC control and distribution equipment. Even though a peak of some lo kA on the primary side of the step-down transformer for an aerial incoming 10-meter-high power line, the peak current seen at the building end of the power line from the secondary side was only about 4 kA. In fact, this was also the maximum surge level when the transformer was assumed to interface with a buried primary power line. The peak short-circuit current at the builing end of the cable run was observed to occur at about 200 ns, with a FWHM of about 500 ns. The associated open-circuit voltage for this case was about 1.5 MV.

The above values were calculated for the OLD DEXP early-time HEMP waveform. The transients for the NEW DEXP early-time waveform were approximately half as large.

The surges generated by the intermediate-time waveform were less than 100 A, even for the aerial primary power line.

Any grounding on the power line from the step-down transformer that runs over to the WIS building is nominally applied both at the transformer end and at the building end. Neglecting any perturbation that the extensive building grounding loop and other metallic objects might have on the MHD-induced ground potential gradient, this means that there will be late-time HEMP currents induced directly on the grounded conductors of this cable given by

$$I = \frac{E(t)}{(R_0 + 2R_g/L)}$$

where R_0 is the resistance per unit length of the grounded conductor, R_g is the resistance to ground, and L is the horizontal length of the run between the step-down transformer and the building grounding point. For an assumed ground resistance of the order of 1 ohm, a length of the order of 100 m or less, and a conductor no smaller than AWG 10 (all reasonable assumptions), a realistic upper bound to the expected MHD-induced current is about 2.5 A on any grounded conductor of the incoming power cable due to the presence of grounding points at its two ends. This is deemed a negligible transient.

Thus, under most circumstances, one can expect negligible transients reaching the WIS building on incoming power lines. However, there is the possibility that damage to the step-down transformer could permit the large MHD-induced currents on the incoming long-haul power lines to penatrate to the WIS installation. Therefore, it is recommended that the WIS EMPPS be designed to deal with long late-time current surges of a few hundred amperes coming in on the power lines to the facility.

In spite of the observation that the transients reaching the WIS building are generally smaller than those induced on the long-haul power cables, the induced surges are still appreciable. These surges can have two effects of concern.

First, they can cause the incoming power line to radiate significant amounts of energy to cables not directly connected to them. The magnitude of such effects has been experimentally measured in past EMP tests for those predicted for the early-time and intermediate-time HEMP fields [3]. These are the transients of primary concern, since the MHD-induced currents are of such low frequency that little radiative or inductive coupling is expected. As noted below in Section 14.3.8, the coupling loss between the incoming power cable and other conductors that approach within 3 meters of the incoming power cable was observed to be no worse than about 45 dB; for closest approach distances of between 3 and 20 meters, about 65 dB; and for cables that never get closer than 20 meters from each other, the coupling is negligible. Thus, for an incoming peak transient on the power cable of 4 kA, one can expect maximum coupling currents of about 22 A on other cables that approach within 3 meters and maximum peak currents of about 2 A on those whose nearest approach is between 3 and 20 meters.

Second, portions of the power system fed directly by the incoming power lines can experience significant transients that penetrate through the usual protection devices. From a series of EMP test on some of the applications, the largest such transients were seen on the UPS rectifier leads. For these leads, a loss coupling factor of about 41 dB was observed if no operating power arresters were in place, about 35 dB if they were operating. This coupling loss translates into current of the order of 35 A on the rectifier leads if lightning arresters are in place, about 70 A if they are not.

13A.3.3.2 Microwave Tower and Associated Conductors

The complex structure of a microwave tower with its associated waveguides and other conductors (such as an AC lead for the warning light at the top of the tower) makes it a difficult item to analyze. However, the experimental results reported in [3] indicate that the fat-dipole approach does a reasonably accurate job for the early-time HEMP threat. The reader is referred to the discussion of [3] for further details on the calculational technique.

The application of the fat-dipole approach to the WIS EMPP study leads to the following conclusions:

- Because of the larger vertical-polarization component of the incident HEMP field at early and intermediate times in a country such as Turkey than in the U.S., the peak currents that can be generated on microwave structures are about 50% larger than shown for similar structures in [3].
- As one expects, one predicts larger induced currents on tall towers than on short ones; but these predictions for a given height can vary by a factor of three depending upon the assumed fatness ratio.
- For absolute worst-case assumptions of a 100-meter tower and a fatness ratio of 6, the maximum peak-to-peak current induced by the OLD DEXP early-time waveform can be as high as 30 kA. [The fatness ratio is defined as $2\ln(2h/a_e)$, where h is the tower height and a_e is the effective radius of the tower.] For tall towers, however, a fatness ration of 10 is much more typical, yielding a maximum peak-to-peak current of 20 kA as a more realistic design criterion for a 100-m tower (taller than normally expected). According to the curves of [3], a peak current of 12 kA is associated with this peak-to-peak current value.
- The conservative approach of [3] would predict similar peak currents for the NEWDEXP early-time waveform.
- Extending the analysis to the intermediate-time regime, the intermediate-time waveform is expected to induce maximum currents of about 100 A on a 100-m tower structure. Negligible currents will be generated by the late-time fields on the primarily vertical tower.

About half of the current induced on the tower can be expected to flow to ground at the base of the tower if the ground straps are well bonded. The other half will flow toward the WIS building on the external raceway that supports the waveguides and the incidental conductors such as the lead to the tower warning light. For raceways of 5 to 20 meters in length, one can expect the OLD DEXP early-time waveform to induce an additional current of about 2 kA on a typical waveguide. However, since the waveguides, AC conduits, and raceway structure are all tied together, no more than about two or three times this current will be generated on the entire horizontal structure. Thus, a maximum peak current of 12 kA is predicted to reach the vertical tower and another 6 kA from the horizontal raceway. Bonding all conducting members at the entry to a metal plate tied to ground will keep most of this transient from penetrating to the interior of the building.

Inside the building, an additional 2 kA can be induced on each waveguide an AC line or conduit, for an assumed maximum interior horizontal run of 5 to 20 meters. This signal should be much larger than what penetrates from the outside, if the penetrating conductors are treated correctly at the entry ground plate.

The above transient level is for the OLD DEXP early-time waveform. The NEW DEXP early-time pulse gives peak currents of about half those from the OLD DEXP waveform for the horizontal runs, but approximately the same for the vertical tower. Thus, the muximum peak current induced by the NEW DEXP waveform will be approaximately 9 kA at the building entry plate: 6 kA residual peak current from the 100-m tower and 3 kA from the horizontal run from the tower to the building. As stated above, most of this current can be diverted to ground at the entry plate.

Internal to the WIS building, the NEW DEXP waveform will induce an additional current of about 6 kA on each penetrating conductor.

The intermediate-time fields will generate peak currents of about 100 A each on the tower structure and on individual horizontal conductors.

The late-time (MHD) field will be responsible for generating a small potential difference on grounded conductors between their grounding points on the tower and the grounded entry plate to the WIS building. For an assumed maximum horizontal separation of 30 m between these grounding points, a maximum induced potential difference of 1.5 V will result. Assuming a reasonable ground resistance of the order of 1 ohm, this potential difference will generate a peak current of less than 1 A on the grounded conductors between the tower and the ground entry plate to the building and the entry panel to the shielded room within the WIS facility leads to a similar prediction for the MHD transients induced on the wave guides and other grounded conductors that come from the tower and terminate at or penetrate the shielded room. Good bonding of these penetrating conductors to the grounded entry panel at the shielded 1oom will keep these transients from penetrating to the interior of the shielded enclosure. No MHD transients will be generated on these conductors inside the shielded enclosure if the grounding system within the enclosure utilizes a single earthing point (the one associated with the ground at the room's entry panel).

13A.3.3.3 Coaxial Cables

The coaxial cables to be installed in WIS may be of several configurations. These are characterized by various numbers of coax tubes in the cable, various thicknesses of the aluminum sheath surrounding the coax tubes, by the presence or absence of a steel tape armor outside the aluminum sheath, and by various numbers of copper wires inside the sheath (to carry alarm signals and power). The transients induced on the sheath, the coax tubes, and the interstitial induced in the cable are all affected by these design variations.

However, there were steps taken to reduce the number of cases treated.

a) First, it was noted that the steel tape armoring is used only where the cable is laid directly in the ground. However, for some distance before entering any building or manhole where active electronics are located, the cable is laid in conduit and the afficacy of the aluminum sheath in reducing the HEMP-induced transients on the internal conductors (the coax tubes and interstitial wires), the thinness of the steel tape, and the leaky nature that characterizes most tape-wound shields it was decided to ignore the presence of the steel in calculating the expected transients.

b) Second, since no experimental data were furnished to give the transfer impedance for the various power-and signal carrying conductors of the cable, it was necessary to use a simple analytic approximation for this quantity. The form used in the study was for a cylindrical conductor centered inside an outer cylindrical conductor. Thus, the results were insensitive to the number of coax tubes and wires actually carried in the cable. A comparison of measured data and predictions from a similar approximation for a 20-tube coax cable lent credibility to this approach [3].

It is generally agreed that the predictions below are sufficient to size the EMPPS elements and identify the protective measures needed for survival of the WIS coax transmission system.

Given the above simplifying actions, a cable is considered having an aluminum sheath with an outside diameter of 52 and a wall thicknees of either 1 or 2 mm. Inside this sheath, it is assumed there was a (centered) copper coax tube 10.7 mm in diameter with a center conductor 2.64 mm in diameter. The transients induced on the aluminum and the coax conductors were tube with a copper wire were replaced along the longitudinal axis of the aluminum sheath and the transients on this internal conductor were calculated.

The OLD DEXP early-waveform was found to produce a maximum peak short-circuit current on the aluminum sheath of about 5 kA with a corresponding open-circuit voltage of almost 1 MV. The transients reached their peak values in 170 ns and exhibited FWHM values of between 0.7 and 1 microsecond. For the NEW DEXP early-time pulse, the peak current was 1.7 kA, the maximum voltage was 0.35 MV, the time-to-peak was 30 ns, and the FWHM was only 100 ns. The intermediate-time HEMP waveform induced maximum short-circuit current of 140 A and peak open-circuit voltages of 5 kV. These maxima were reached at times of several microseconds and characterictic FWHM values of 50 to 200 microseconds were noted. The above transients were independent of the aluminum sheath thickness.

These sheath currents were used to drive the internal wires and the coax conductors. These predictions were quite dependent upon the assumed thickness of the aluminum conductor. Therefore, both sets of results are presented below.

First, a 1-mm thickness for the aluminum sheath was considered. Equal induced transients for the intersitial wires (nominally twisted-wire pairs) and the outer cylindrical conductors of the coax tubes were found. This was due to the use of a single function to describe the transfer impedance between the sheath and these conductors: that for a conductor centered within the aluminum sheath. The peak induced voltages on these interior conductors were predicted to be 210 V/km for the OLD DEXP, leading to a predicted induced potential difference of 320 V for the 1.5-km run between regenerators in the digital version of this transmission system or 960 V for the 4.5-km spacing of the analog version. The corresponding voltage level for the NEW DEXP early-time field was about 25 V/km. The peaks were predicted to occur at 8 microseconds for both these early-time waveforms; and a FWHM of 22 microseconds was observed. When the system was driven by the intermediate-time fields, the peak voltage induced on the twisted-pair wires and outer coax tube was calculated to be 27 V/km at a time of 66 microseconds, with a FWHM of 220 microseconds.

If a 2-mm aluminum thickness for the sheath be assumed, the early-time voltage levels on the internal twisted-pair wires and the outer cylindrical coax tube conductors decrease by a factor of four (i.e., to 55 V/km for the early-time OLD DEXP waveform); and the time-to-peak increases to 30 microseconds, with an associated longer FWHM of 90 microseconds. For the thicker sheath, the intermediate-time field produces an induced voltage of one-half that for the 1-mm sheath (i.e., 13 V/km) and the time-to-peak increases to 95 microseconds, but the FWHM remains at 220 microseconds.

For the inner, solid-core conductor of a coax tube, calculations show that it will experience induced voltage levels of approximately 40% of those induced on the outer conductor of the tube upon exposure to the early-time fields (i.e., 90 V/km for the OLD DEXP waveform and an assumed 1-mm thick aluminum sheath) and about onehalf the corresponding levels for the intermediate-time field (i.e., about 13 V/km for an assumed 1-mm thick aluminum sheath). The time behavior of the transients in the inner coax conductor is the same as for the outer coax conductor.

The late-time MHD fields can produce a maximum 50-V/km gradient in the ground potential along the route of a coax cable. Each grounded conductor of the coax can be expected to respond essentially like an element in a DC circuit driven by the slowly varying MHD field between adjacent ground points. The approximate 0.1-ohm/km resistance of a 2-mm thick aluminum sheath would therefore dictate peak MHD-induced currents for this conductor of about 2.5 A if it were grounded every 100 m with an assumed 1-ohm resistance to ground, or 25 A if such groundings were spaced every kilometer apart. For the higher-resistance interstitial wires, somewhat lower peak surges would be predicted: e.g., about 0.6 A for an AWG 24 wire, almost independently of the separation between grounding points.

13A.3.3.4 Twisted-Pair Communication Lines

Some interfaces between WIS and local telephone networks, as well as incoming traffic to WIS switches, may be provided via twisted-pair wires. Such cables are usually made up of a large number of individual pairs tied together in a bundle. External to the site, they may be aerial or buried. For the purposes of this analysis, it is assumed that the incoming bundles are made up of 100 pairs, characteristic of cables coming into a switching office.

Whether buried or aerial, the total current induced by the early-time and intermediate-time HEMP fields on a bundle of 100 pairs will be approximately those predicted for the incoming power lines previously discussed in Section 3.1. However, the peak current on an individual pair is significantly less. In fact, for pairs driven by the intermediate-time field, the current carried by a single cable will be just 1/N of that induced on the entire bundle, where N is the number of pairs in the bundle. For the early-time fields, however, there is a significant amount of shielding of one pair by the others so that some pairs will carry more, and some less, than 1/N of the total bulk cable current. Experience has shown that early-time transients can be as large as 8/N times the total cable-bundle current, for N of the order of 100. Thus, the following predictions are for the peak currents on individual twisted pairs in a 100-pair cable:

- Aerial Cables, OLD DEXP early-time waveform-peak short-circuit current of 800 A at 90 ns, with a FWHM of 400 ns for a 10-meter height or a peak current of 440 A for a 5-meter height (with similar time-to-peak and FWHM characteristics as for the 10-meter case).

- Buried Cables, OLD DEXP early-time waveform-peak short-circuit current of 400 A at 200 ns with a FWHM of 600 ns.
- Aerial Cables, NEW DEXP early-time waveform-peak short-circuit current of 400 A at 18 ns, with a FWHM of 60 ns for a 10-meter height or a peak current of 240 A for a 5-meter height, with the peak occuring slightly earlier (at 15 ns) than for the OLD DEXP case.
- Buried Cables, NEW DEXP early-time waveform-peak short-circuit current of 140 A at 25 ns with a FWHM of 100 ns.
- Aerial Cables, intermediate-time waveform-peak short-circuit current of 3.2 A at about 35 microseconds, wih a FWHM of 60 microseconds, rather independent of cable height.
- Buried Cables, intermediate-time waveform-peak short-circuit current of 1.4 A at about 95 microseconds.

For the late-time MHD fields, the introduction of the slowly-varying gradient in the ground potential will induce maximum peak currents of about 0.6 A in individual grounded AWG 24 wires, typical of those used in twisted-pair cables.

13A.3.3.5 Interior Unshielded Power Cables

Even with the EMPPS use of a highly shielded area to contain sensitive WIS telecominications equipment and certain portions of the power distribution and conditioning system, some power system equipment may be located outside this area; e.g., the back-up power generators and batteries of the UPS. Heavy-duty cables will be used to run between the wall of the shielded area and these UPS subsystems; and significant surges can be generated on these cables if they are not put into ducts.

Consider a primarily horizontal run for the cables at 3 m above the building floor. Predictions indicate that a peak short-circuit of about 1.6 kA would be induced in an AWG 8 conductor by the OLD DEXP early-time fields, with a corresponding open-circuit voltage of 0.4 MV. These transients are well described at late times as damped sinusoidal signals with a ringing angular frequency of c/L where L is the length of the cable. The signals with a ringing angular frequency at early times, but in some cases it was observed that the peak current was reached not at the first maximum of the transient but at a much later time-although the initial maximum was in all cases only slightly smaller than the later overall peak value. This phenomenan was only noted over a ground (floor) with a low conductivity. The apparent "delay" in the onset of the damping (caused by the cable has sufficient energy even after the onset of oscillations in a short (3-to 20-meter-long) cable that the cable continues to pick up energy from the field at a rate greater than the energy dissipation due to the losses in the cable and its terminating loads. This explanation is consistent with the greated pulse width of the resultant driving field over a low-conductivity ground, relative to that over a high-conductivity ground.

As noted, even when the overall peak value of the transient was reached after the first maximum, the induced perturbation was always close to or at its overall peak value at the time of the first maximum. The time-to-peak was specified as the time to the first maximum. This was typically about 20 ns for cables 3 m above the building floor for the OLD DEXP early-time HEMP waveform.

For the NEW DEXP early-time waveform, a maximum short-circuit current of 0.6 kA is predicted for this same conductor, together with a peak open-circuit voltage of 0.3 MV.

The time to the first maximum for this driving field is somewhat shorter (about 15 ns), consistent with the faster rise and decay of the incident field.

For exposure to the intermediate-time waveform, a maximum current of 80 A at 90 microseconds is predicted, with a corresponding maximum voltage of 400 V at about one microsecond for 3-to 20-meter-long cables strung 3 meters above the building floor (a typical height for internal cables).

The above levels are those predicted for a single AWG 8 conductor, of which five will be used in a typical power cable between the diesel generators and the entry to the shielded area. Similar time dependences for the transients will be experienced on an AWG 1/0 conductor (typical for a lead between the back-up batteries of the UPS and the shielded room), but a 25% increase in the peak short-circuit current is anticipated. This latter peak current (e.g., 2 kA for the OLD DEXP early-time waveform) is also approximately what might be expected on the five-conductor by the others in this multi-conductor cable leads us to predict peak currents on a single conductor as large as the 1.6-kA level previously cited for exposure to the OLD DEXP early-time field.

The above transients would be reduced to negligible levels if one were to install the cables in conduit that was well bonded and grounded at the wall of the shielded area and at the other terminating ends (e.g., at the batteries and the diesel generators). Peak currents on the conduits would be comparable to those on a single AWG 1/0 conductor: about 2 kA for the OLD DEXP early-time waveform.

The unshielded interior power cables will be subjected to the effects of the late-time MHD field only if they are grounded at multiple points and if the ground straps for adjacent ground points are earthed at different places. If this occurs, then a potential will be applied between these grounding points as per the discussions of Section 13A.3.2.2. However, as noted in that discussion, maximum potential differences between grounding points will be only of the order of a volt and maximum currents of the order of one ampere will be generated on the conductors discussed here.

It should be pointed out that the calculations of early-and intermediate-time transients on these and other cables interior to the WIS building were made without taking account of any shielding that might be provided by the building structure. This is, admittedly, a conservative approach but not overly so for WIS EMPPS installations in existing buildings that are often characterized by numerous and large apertures.

13A.3.3.6 Facility Support-Services Cables and Miscellaneous Penetrations

a) AC Lighting and Similar Leads

The cables feeding the perimeter and building-interior lights, those feeding wall outlets, and similar cables will experience surges like those of the cables that connect the diesel back-up generators to the rest of the UPS. Those transients were characterized by peak short-circuit currents of about 2 kA upon exposure to the OLD DEXP early-time HEMP field, about 0.6 kA for exposure to the NEW DEXP early-time waveform, and about 100 A for the intermediate-time HEMP pulse. The associated open-circuit voltages are approximately 0.5 MV, 0.3 MV and 0.4 kV, respectively.

The above levels are those expected if these cables are left unshielded or un-ducted. If these cables are placed in ducts with high-quality bonding at the terminating ends, the transients induced on the cables will be negligible, although currents approaching 2 kA will be induced on the conduits.

The transients generated by the late time MHD effect will, as per the discussions of Section 3.2.2 exist only between grounding points that utilize ground straps earthed at different positions. Maximum potential differences induced by the late-time MHD fields are expected to be of the order of a volt, with associated maximum currents of the order of one ampere for the conductors discussed here.

b) Plumbing, Sewer/Septic. and Fuel Lines

For an assumed maximum buried horizontal run of 30 meters, plumbing, sewer, and fuel lines terminated in good grounds at both ends can actually exhibit early-time peak currents of several kA in soil with a low conductivity of 0.0001 mho/m. However, the peak currents decrease rapidly with an increase in soil conductivity; e.g., peak currents of only about 100 A are expected for a soil conductivity of 0.1 mho/m. The above levels are for an assumed 1-meter depth of burial and for an incident OLD DEXP early-time HEMP waveform. Much smaller peak currents are predicted for exposure to the other waveforms; e.g., only 140 A from the intermediate waveform for a soil conductivity of 0.0001 mho/m and 3 A for an assumed value of .1-mho/m. However, the penetrating currents on these pipes can be made negligible by replacing the nominal steel pipes with nonconducting material such as PVC. The replacement of the pipes does not even have to be total: as long as the last few meters of the pipes prior to entering the building are nonconducting, no electrical transient will penetrate. This procedure is usually straightforward and removes a source of potential concern because of the possibility of arcs in the fuel supply system and because of radiation of energy from these leads into nearby cables that lead to sensitive elements of the WIS installation.

Negligible transients are expected from the late-time MHD fields for these conductors.

13A.3.3.7 Fiber Optic Cables

At the outset of the study, the fiber cables contemplated for use in WIS were identified as having several metallic members: an outer bundle of steel wires to provide armor for the cable; and interstitial conductors to carry power and alarm/signal information. Although this cable design was subsequently modified, it is of interest to note the surges predicted for its various conducting members.

First, consider such a cable buried at a 1.2-m depth. The surges generated on such conductars by the incident early-time and intermediate-time HEMP fields would be approximately equal to those induced on the aluminum sheath of the coax cable previously discussed (see Section 13A.3.3.3). This approximation should be quite good for the current induced on the steel wires. It is probably an overestimate for the surges induced on the internal conductors of such a fiber cable design since it ignores the shielding provided by the steel armor; but the shielding effectiveness of a steel-wire bundle could be quite low. Referring back to the discussion for the coax cable, it is seen that the worst-case currents expected for the metallic members of this type of cable core: 4.8 kA peak short-circuit current for exposure to the OLD DEXP early-time waveform; 1.7 kA for the NEW DEXP waveform; and 140 A for the intermediate-time HEMP filed. These peak transient values apply to cables buried anywhere from a few centimeters to 5 m below ground.

If a fiber cable of the above design is strung aerially, the same conservative approach would assign peak transients to the steel-wire bundle and the internal conductors of : 10 (5.5) kA for exposure to the OLD DEXP waveform for cables mounted at a 10(5)-meter height; 5(3) kA for the NEW DEXP early-time waveform for the same two heights above the ground; and 140 A from exposure to the intermediate-time field (regardless of the height of the cables above the ground).

Upon exposure to the late-time MHD component of the HEMP environment, the metallic members of the above-described fiber cable would experience the variation in ground potential defined in Section 13A.2 and, to a good approximation, would respond between ground points as if a quasi-static voltage equal to E(t) L were being applied by a power source, where L is the separation between the ground points. The diameter of the steel wires that would make up the cable armor was not specified; but were they to be 1.0 mm in diameter, the MHD effect would generate a peak current of about 0.4 A on a single strand, according to the discussions of Section 13A.3.2.2. Were such wire to be used in an annular bundle 1 cm in diameter, the resulting set of 32 wires would then carry a peak current of 13 A, and this current would persist for many seconds. Similar considerations for copper interstitial conductors led to predicted peak currents per grounded conductor of 0.6 A for AWG 24 wire (.51 mm in diameter), 1.5 A for AWG 20 wire (0.8 mm in diameter), or 3.8 A for AWG 16 wire (1.3 mm in diameter).

Partly in recognition of the hardening problems that such a fiber cable design would present and partly because of other factors of considerations, it has been decided to utilize a fiber cable design that contains no interstitial metallic members and no steel strength member. One of the factors that permitted this change in cable design was a decision to power each set of regenerators locally instead of feeding power along the cable. Under these circumstances, the only metallic element now being considered for the cable is steel armor. To a good approximation, the HEMP transients on this armor will be the same as on the aluminum sheath of the coax cable analyzed in Section 13A.3.3.3 and summarized above.

13A.3.3.8 Cables in Shielded Areas

a) Direct Coupling

Inherent in the treatment given above for cables and other conductors exposed to the unattenuated early-and intermediate-time HEMP fields is the assertion that the induced surges are due primarily to the incident electric field. However, for cables exposed to heavily attenuated fields (as for cables located inside a shielded steel-lined room), the surges induced directly by the residual magnetic fields often dominate those induced directly by the residual electric fields. This is because of the small attenuation provided by typical shielding materials to low-frequency magnetic fields.

Under these circumstances, it is necessary to utilize a different approach to estimate the magnitude of transients induced directly by the attenuated incident wave. The method outlined below may not be very accurate but it is generally sufficient to determine if the shielding effectiveness of a proposed shield is sufficient to disturbances within a shielded area: apertures (such as vents and doors) and currents on penatrating cables.

The first step is to specify a few characteristics of the shielded room, the shield material, and the indicent field. For our example, consider the following parameters as providing the required input:

- Incident (Unattenuated) Magnetic Field: We let H(t) = E(t)/377 where E(t) is the OLD DEXP early-time waveform previously used, with E(t) in V/m and H(t) in A/m.
- Shielded Room Dimensions: A room with a floor measuring 6 meters by 6 meters and with a 3-meter-high ceiling was considered.

- Shield Material Characterictics: The use of a steel shield with $\sigma = 1.0\text{x}10^7$ mho/m, a thickness d of 3 mm, and a relative permeability of 200 at low frequencies was considered.

The next step is to model the shield as an equivalent spherical shell, a pair of parallel plates, or a cylindrical shell. A spherical shell is most appropriate for this case. The radius of an equivalent-volume sphere is approximately 3 m. From this information, the shielding parameter C, for a spherical shell is given by $C = r/3\ d$, yielding $C = 1.67$.

For $C > 1$ (as is the case for most shields of interest), the maximum voltage induced by the attenuated magnetic field in a loop of unit area is given by $18\ A/\mu_r\ \mu_0\sigma^2\ r\ d^3$ where all units are in the mks system. In this expression, A is the low-frequency, constant spectral amplitude of the incident magnetic field: 3.7x10-5 A/m/Hz for the OLD DEXP early-time field. Thus, the voltage induced directly by the attenuated magnetic field on a loop 1 square meter in area is $3.3\text{x}10^{-7}$ V. Since the largest single loop that can be formed in the room has an area of some 36 square meters, this approximate treatment gives an estimated direct-coupling voltage of only about 10 microvolts. This is sufficiently small that attention to the apertures and the penetration currents on incoming cables is the critical issue in assuring acceptably low transients on cables in the shielded area due to the early- and intermediate-time HEMP fields.

As previously noted in Section 14.3.2, the effects of the late time MHD fields on cables inside the shielded room will be negligible if the grounding system interior to the room is earthed at a single point, nominally associated with the ground strap or bus at the entry plate to the shielded room. Thus, we anticipate no direct MHD coupling to cables within the shielded area.

b) Secondary Coupling

The effects of residual currents on penetrating conductors can, in turn, be estimated from results obtained in past tests of telecommunications facilities. From those tests, it was found that:

- For a cable that panatrates the shield wall carrying a current whose frequency content is characteristic of transients induced by the unattenuated (or only slightly attenuated) incident early-time or intermediate-time HEMP fields, one can expect coupling losses of about 45 dB for cables that approach within 3 meters.
- For cables that approach within 3 to 20 meters, the coupling loss is about 65 dB.
- For cables farther away than 20 meters, the coupling is negligible.

Thus, a cable carrying a current of 1 A into the shielded area can be expected to induce secondary currents of about 6 mA in cables that approach within 3 meters. Current of about 0.6 mA would be expected on cables whose closest approach is between 3 and 20 meters; and negligible currents can be expected on more remote cables. The validity of this general rule-of-thumb should, of course, be verified during the acceptance testing for any WIS installation. But it apprears to be useful for an initial evaluation of the many problems that might arise from currents on conductors that penetrate the shield.

13A.4 EMP HARDENING DESIGN

The Electromagnetic Pulse Protection System (EMPPS) to be designed for WIS is aimed at providing a protection of the WIS facilities and the transmission media from the detrimental effects and the transmission media from the detrimental effects of HEMP. The whole system must be hardened against the threat described in Section 13A.2, so as to avoid any possible upset or damage in the system.

All of the microwave, fiber optic and coaxial terminations are to be included inthe hardening design for the WIS transmission facilities (repeater or terminating stations) that are already existing and to-be-built-as well as the switching subsystem the elements of which are the nodal and access switches.

The EMP hardening design is addressed below in two hardening discussions. The first is the early-time and intermediate-time threat, and the other is the late-time effects of magnetohydrodynamic (MHD) EMP. The results of the generic hardening design concepts for a typical telecommunication facility are also included.

13A.4.1 Early-Time and Intermediate-Time Hemp Protection Methods

The design approaches for early-and intermediate-time HEMP focus on fast risetime, short duration EMP coupled threat defined in Section 13A.2. The elements of the design is described below for each of the terminations in a facility.

13A.4.1.1 Shielded Enclosure Design

The EMPPS primary shield design attenuates the external radiated field environment and provides a benign internal environment. Most electronic systems will survive when exposed to radiated electric field strenghts of 5 to 8 V/m. The EMPPS primary shielded enclosure reduces the radiated field strengths below the radiated susceptibility threshold levels of the equipment. The maximum radiated external field strength is 54 kV/m. The primary shielded enclosure will provide 100 dB attenuation as a design goal and 80 dB attenuation as a minimum. The resulting internal field strength is 0.5 to 5 V/m.

The primary shielded enclosure is designed as an internal shield volume within the existing or new facility. It will totally encapsulate the susceptible transmission/switching equipment. The internal design should be selected such that it:

- Minimizes exposure to ambient environment,
- Does not rely on ground rods and shield skirt for floor,
- Does not affect the exterior appearance of structure,
- Minimizes physical damage due to sabotage,
- Eliminates elaborate corrosion control measures,
- Provides easy access for inspection,
- Does not require to match contours of facility interior,
- Provides high integrity 80 dB to 100 dB shielded volume,
- Prevents thermal differential expansion between plates.

The primary shielded enclosure is constructed of solid cold rolled steel (CRS) panels. Solid CRS panels are selected based on the following:

- Characteristic high reflection and absorption coefficients,
- Greater that 80 dB attenuation with 3 mm thick plates,
- Easily weldable,
- Low-cost material,
- Protected against corrosion with zinc chromate finish,
- Reduced life cycle cost,
- High relative permeability,
- Single layer construction,
- Not relying on attenuation provided by facility wall construction techniques.

The secondary shield design concerns the entry vaults. This assembly contains the terminal protection devices and is designed for 60 dB plane-wave shielding effectiveness. This assembly prevents re-radiation of electromagnetic fields within the working volume.

13A.4.1.2 Personnel Door Design

The EMPPS is designed with two personnel doors that measure 1.0 mx2.0 m. One door can be opened with the other door closed and not compromise the shield integrity of the EMPPS. The doors are a "knife and pocket" design and incorporate two rows of beryllium copper fingerstock gasketing material around the periphery of the doors. The doors have a three-point latching mechanism. The door provides 80 dB to 100 dB attenuation. The door has a steel frame which is easily weldable into the primary enclosure volume.

13A.4.1.3 Vestibule Design

The personnel entrance tunnel or vestibule is constructed to take advantage of its inherent shielding characterictic as a waveguide-below-cutoff. The vestibule is an extension of the primary shielded volume and is constructed from the same materials and methods as the primary shielded barrier. Additional attenuation (30 dB 1.0 MHz) may be obtainable at minimum cost by constructing the vestibule 4.0 m long. The cutoff frequency is approximately 30 MHz.

13A.4.1.4 Equipment Entrance

The shielded volume is designed with a 2 mx3.5 m entrance way for equipment. A CRS panel is welded in the opening after equipment installation to maintain the shield integrity of the system.

13A.4.1.5 Grounding System Design

The EMPPS earth electrode subsystem (EES) design is compatible with the fault protection and signal reference ground subsystems. The EMPPS EES design is consistent with the guidelines of MIL-STD-188-124. The detailed ground system design is based on the following:

- Soil conductivity,
- Depth of water table,
- Frost line,
- Geological effects,
- Communication facility configuration.

The preliminary EES design consists of uniformly spaced ground rods around the new and existing transmission and switching facilities. A loop is formed by 1/0 AWG cable around the facility. Ground rods are attached to the loop every 6.0m. The ground rods are 3.0 m in length, and at least 1.9 cm in diameter. The ground rods are welded or brazed to the loop.

If the antenna structure is less than 6.0 m away, one ground loop encircles the entire facility. If the structure is greater than 6.0 m from the primary shield barrier, there is a separate ground loop around the base of the antenna tower. The antenna ground loop is attached to the facility ground loop with two separate conductors. There is one connection between the primary shielded enclosure and the EES. The fence enclosing the shielded barrier and antenna tower is connected to the EES.

13A4.1.6 Coaxial Transmission Cable Interface Protection

Coaxial transmission configurations consist of the major electronic equipment shown in Fig 13A.16. The coaxial cable entry vault contains the penetration for the coaxial bundle and provides protection devices for the coaxial tubes and twisted pairs. The overall coaxial bundle acts as a monopole antenna in the HEMP environment and couples high intensity transients onto its cable sheath. The coupled transients on the individual coaxial cable center conductors and twisted pairs are characterized by the open-circuit voltage (V_{oc}) and short-circuit current (I_{sc}).

The conducted threat on the coaxial center conductor is suppressed using a low-capacitance primary and secondary spark gap device with a breakdown voltage 20 to 50 percent higher than the normal operating voltage. The coaxial cable must be filtered as it exits the entry vault.

The conducted transients on the twisted pairs are suppressed using a primary and secondary spark gap device and filters. It is assumed the coaxial repeater stations are powered locally. The coaxial bundles are site dependent and may contain 4 to 12 coaxial tubes and 12 to 20 twisted pair.

13A4.1.7 Microwave Radio Waveguide Interface Protection

The microwave transmission configuration consists of the major electronic equipment shown in Fig. 13A.17. and Fig. 13A.18. It is assumed that there are no active electronics located at the antenna tower. The waveguide EMP protection design is the same for both existing and new facilities.

Due to skin effects, the coupled HEMP transient is predominantly present on the outer surface of the waveguide. The skin depth is a measure of the depth of penetration of a wave or current of a given frequency into a conducting surface.

Low-impedance bonding to the EES diverts the transient current to earth. The conducted transient on the waveguide is characterized by the V_{oo} and L_{sc}.

The waveguide is bonded to the earth electrode subsystem in at least four places:

- Near the antenna horn,
- At the verticle-to-horizontal transition point,
- At the WIS building waveguide entry point,
- At the primary shielded enclosure waveguide entry plate.

The waveguide is peripherally welded to the entry plate at the facility interface. If direct bonding is not practical or causes signal degradation, an indirect bonding technique may be implemented. The entry plate should be at least 1.3 cm (0.5 inch) thick to help disperse the large current densities associated with the coupled threat.

13A4.1.8 Fiber-Optic Cable Interface Protection

Fiber-optic transmission configuration consists of the major electronic equipment shown in Fig. 13A.19. The EMPPS optical fiber is immune to the coupling effects of HEMP. The EMPPS design considers the steel wire armor and any intentional conductors as primary receptors of electromagnetic energy and provides protection at the primary shielded barrier interface.

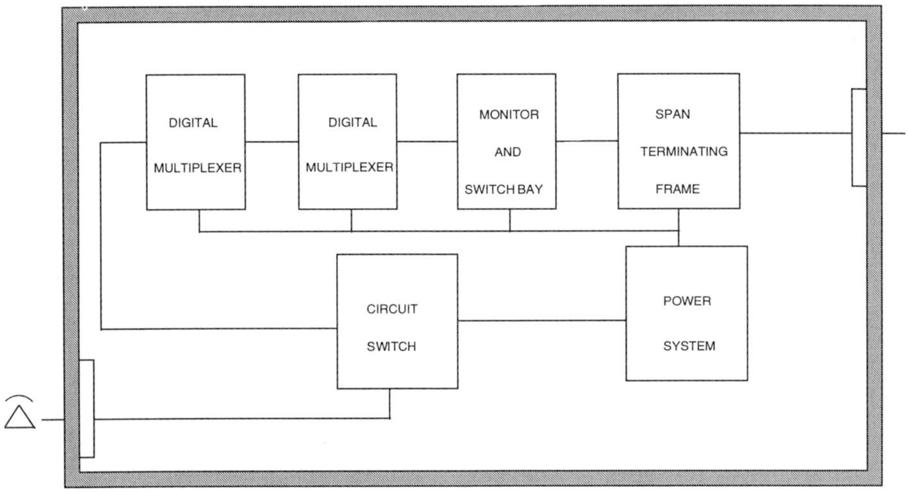

Figure 13A.16 Digital Coaxial Configuration

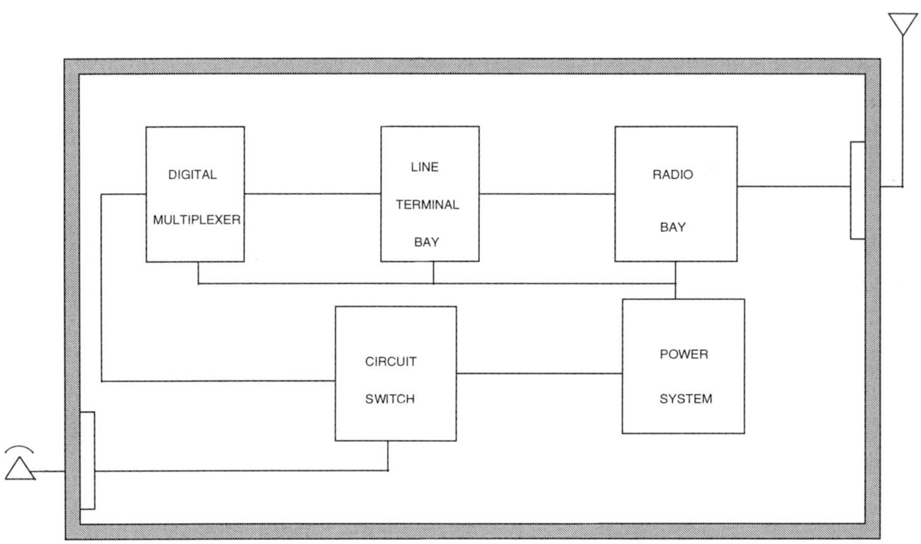

Figure 13A.17 Microwave Radio Configuration

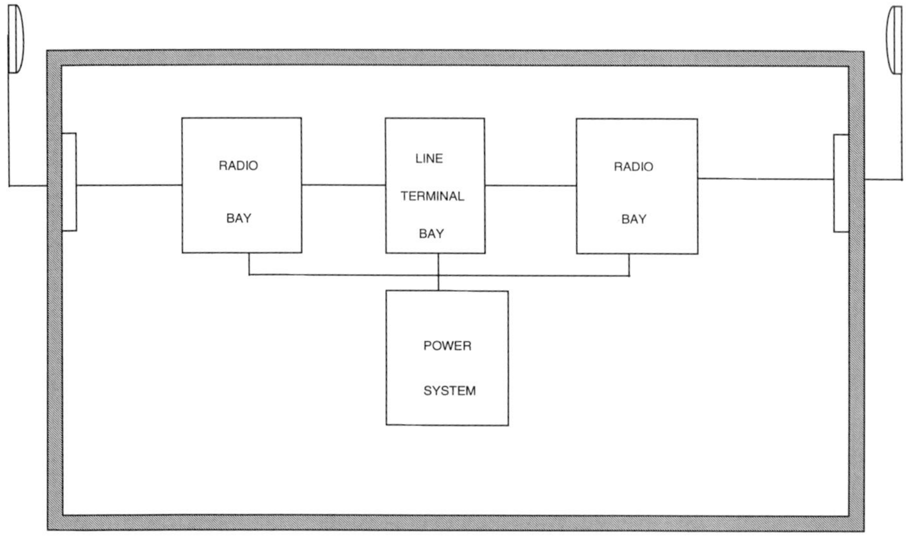

Figure 13A.18 Microwave Radio Repeater Station

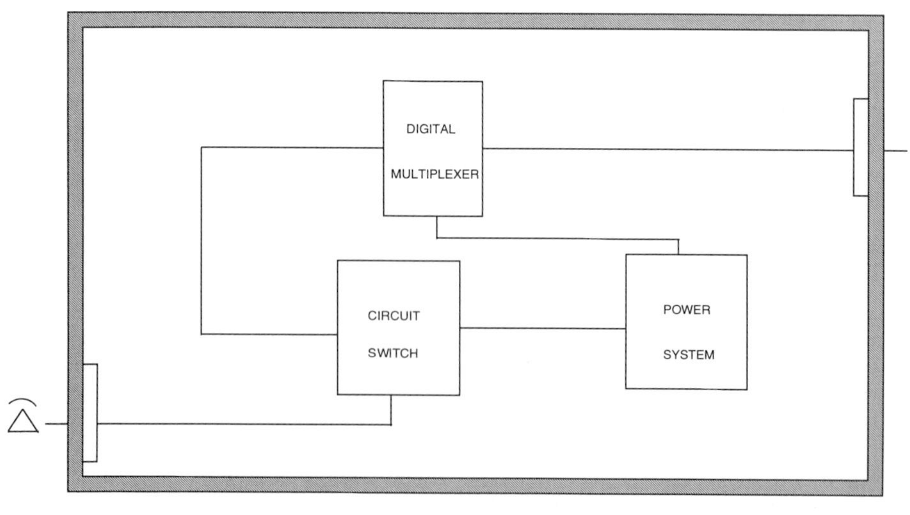

Figure 13A.19 Fiber Optic Configuration

The fiber-optic entry vault contains the penetration for the fiber-optic bundle and provides a waveguide-below-cutoff aperture for the fiber-optic lightguide. The entry vault also provides protection devices for the twisted pairs and a low impedance bond for termination of the structural members. It is assumed that the fiber-optic repeaters are powered locally, and therefore, repeater stations are less vulnerable to the effects of powerline conducted transients.

The fiber-optic bundle enters the entry vault through a circular metallic tube (waveguide-below-cutoff) designed to provide 80 dB to 100 dB radiated field attenuation. The tube has a diameter of 2.54 cm and a length of 9.3 cm. The fibers exit the entry vault through circular metallic tubes 2.54 cm in diameter and 5.6 cm in length which provide 60 dB radiated field attenuation.

The conducted threat on the twisted pair is characterized by the V_{oc} and I_{sc}. The conducted threat is suppressed by using primary and secondary spark gap devices and filter. The same type of devices described for the twisted pair in the coaxial bundle may be installed in the fiber-optic entry vault.

13A.4.1.9 Twisted Pair Interface Protection

The EMPPS design considers the commercial landlines as primary receptors of electromagnetic energy and provides protection at the primary shielded barrier interface.

The conducted threat on the twisted pair is characterized by the V_{sc}. The conducted threat is suppressed by using primary and secondary spark gap devices and filters. The same type of devices described for the twisted pair in the coaxial bundle may be used for the landline twisted pairs.

13A.4.1.10 Powerline Interface Protection

The EMPPS design considers the commercial powerlines and auxiliary backup powerlines as primary receptors of electromagnetic energy and provides protection at the primary shielded barrier interface. The powerlines for both existing and new facilities consist of three phases and a neutral.

The conducted threat on the powerlines is characterized by the V_{oc} and I_{sc}. The conducted threat is suppressed by using primary and secondary spark gap devices, metal oxide varistors (MOV's) and filters. The primary arresters are spark gap devices and the secondary arresters are MOV's. Powerlines filters follow the secondary arresters.

13A.4.1.11 Honeycomb Filter Design

The ventilation openings are treated with honeycomb filters. Detailed filter selection depends on the following:

- Maximum ventilation opening dimension,
- Pressure drop through cells,
- Attenuation requirement,
- Frequency of electromagnetic energies,
- Corrosion resistivity.

It is assumed that the ventilation opening is approximately 1mx2m. Honeycomb filters are selected over perforated sheets and wire mesh due to the high shielding qualities of honeycomb filters. The filters typically provide 80-100 dB attenuation up to 300 MHz. The cutoff frequency is 30 GHz. The individual cell size has a depth of 2.54 cm and width of 0.5 cm.

13A.4.1.12 Fortuitous Interface Protection

Fortuitous conductors are those elements of the facility not intentionally designed to conduct current. These include water piper, sewage pipes and fuel lines. Fortuitous conductors are part of a typical WIS facility, but do not penetrate the EMPPS. It is assumed that these types of conductor will be kept external of the EMPPS. These lines are treated by peripherally bonding or by replacing conductive pipe section with non-conductive members.

The coupled HEMP transient due to skin effects is predominantly present on the outer surface of these conductors. Low impedence bonding to the EES will divert the transient current to earth. The conducted transients on water pipes, sewage lines, and fuel lines are characterized by the V_{oc} and I_{sc}.

All fortuitous conductors are peripherally welded to an entry plate at the facility interface. If direct bonding is not practical or causes signal degradation, an indirect bonding technique is implemented. The entry plate is at least 1.3 cm (0.5 inch) thick to help disperse the large current densities associated with the coupled threat. The entry panel can be mounted to the primary shielded barrier by welding.

13A.4.1.13 Corrosion

The EMPPS is designed for minimum deterioration due to corrosion. The steel liner for the primary shielded enclosure should be painted with a zinc chromate finish for corrosion control. Materials to be bonded together are selected based on compatibility in the electromagnetive series.

Coatings and finishes are selected and applied based on the following characteristics:

- Dew point,
- Surface and air temperature,
- Relative humidity,
- Type of material,
- Surface configuration,
- Cost effectiveness.

13A.4.2 MHD Protection Methods

Because of the low-frequency nature of the late-time MHD portion of the HEMP environment, the required protection measures are often different from those for the early-time and intermediate-time pulses. Thus, the protection measures for the MHD threat will be discussed separately. As noted in Section 13A.3, the late-time MHD environment manifests itself by introducing a slowly time-varying spatial dependence into the ground potential, where the gradient is characterized by a peak value of 50 V/km. The following paragraphs identify the EMPP measures intended to minimize the magnitude of the transients generated by this ground potential gradient and/or to alleviate its effect on WIS equipment.

It is considered that WIS equipment located interior to the primary shielded volume (the steel-lined room housing sensitive WIS switching and transmission electronics). To prevent the introduction of MHD-induced transients into this volume and to preclude the generation of such transients interior to the volume itself, two actions are recommended.

First, all grounded conductors that penetrate into the shielded volume should be grounded at their entry into the room and the grounds from all entry panels should be earthed at the same point. Second, all grounds within the shielded room should be led to the same entry panel(s) so that all grounds within the volume share the same earthing point as the penetrating conductors upon their entry into the room.

For WIS system elements outside the primary shielded volume but still restricted to the confines of a single WIS installations (such as the AC lighting and the back-up power generations), it may not be practicable to utilize a common earthing point for all grounds. However, for these systems, the maximum expected horizontal extent of their active elements and connection cables is no greater than about 50 m. Thus, a maximum MHD-induced potential difference of about 2.5 V will be applied between any two adjacent grounding points for such systems. Many of these systems will be insensitive to such a transient, even if applied for the extended duration of the MHD threat. For example, if no active electronics load is located between two adjacent grounding points on a conductor, the MHD-induced currents of no more than a few amperes will be harmlessly conducted to ground. On the other hand, if a load containing sensitive solid-state devices is present between distinct grounding points on its various input/output connecting cables, one cannot a priori rule out the possibility of malfunction or even damage upon the continued application of a potential as small as a volt unless formal testing and/or analysis has demonstrated the equipment will survive such loading.

However, it is tacitly assumed that WIS will use commercial, off-the-shelf equipment to the largest extent possible; and the susceptibility of such equipment may well not have been determined for such a nonstandard transient. Therefore, it is recommended that; outside the primary shielded volume, for any equipment containing solid-state circuitry or components, the grounds for the equipment and the closest grounding points on grounded conductors connected to the equipment should all use a common earthing point. This action need not be taken if the vendor can prove the equipment in question to withstand a long-term 2.5V potential difference between the equipment ground (s) and the closest grounding points on the grounded input/output lines.

The above guidelines cover all MHD effects of concern to WIS except for those dealing with systems that interface with, or contain, long conductors that extend far beyond a single WIS site. There are four systems of this type of potential concern for WIS; incoming power, incoming twistedpair communications lines, coax transmission system, and the fiber-optics transmission system. Each of these is briefly discussed below.

13A.4.2.1 Power Cables

For the incoming commercial power, the cable run between the step-down transformer and the WIS installation is typically less than 100 meters in length so that the maximum difference between the ground potential at the ends of this power line will be less than 5 V. Such a level presents no operational problems to WIS. However, the long cables coming into the primary of the step-down transformer can have large MHD-induced voltages imposed on the 50-Hz power signal and could introduce large currents into the lines coming off the secondary of the transformer if, for example, arcing were to occur in the transformer. Such arcing will not occur if the lightning arresters onthe primary side of the transformer are working properly; thus normally, no significant MHD-induced surges will penetrate to a WIS installation. But, even if the surge arresters on the primary side of the transformer were to fail (e.g., under the stresses of multiple early-time HEMP surges), the circuit breaker present in the power disconnect cabinet at the entry to the facility is sufficient to prevent large MHD-induced surges from penetrating beyond this point because of the slow rise of those transients.

13A.4.2.2 Twisted-Pair Communication Lines

The twisted-pair communication lines are considered associated with, for example, local telephone lines coming into a WIS switch. The usual lightning and early-time HEMP protection is sufficient to protect against the MHD-induced surges predicted for these lines.

13A.4.2.3 Coaxial Transmission Lines

Were each of the repeater/regenerator stations powered by a local isolated power source, the impact of this effect could be easily controlled by the presence of surge arresters placed periodically along the cable between each of the conductors and the local ground plus observance of the usual grounding practices (e.g., of clamping the aluminum sheath to ground at the entry to each of the repeater stations). However, the provision of power for each of the repeaters or regenerators between the main stations by feeding power on the coax tubes presents a problem. Analysis has shown that this arrangement requires the addition of a protection circuit at each of the repeaters/regenerators: the MHDinduced currents can damage the active electronics even after activation of the system's automatic power-feed circuit. The volume requirement for the protection circuitry maybe comparable to that of the regenerator/repeater itself. The only other active protection measures needed to assure survival of the system to MHD effects would be the use of surge arresters (such as carbon blocks or gas tubes) at each repeater/regenerator location between the local ground and: (a) the twisted pairs within the coax cable and (b) the outer conducting cylinder of each coax tube. These arresters are recommended to avoid the danger of arcing between the various coaxial cable conductors in the presence of large MHD-induced voltages. This action supplements the inplace protection already afforded by the existing surge arresters between the center and outer conductors of each coax tube and by the practice of periodically grounding the aluminum sheath of the cable.

13A.4.2.4 Fibre Optic Cables

Lastly, the long-haul fiber-optics transmission system is considered. To power all fiber regenerators locally simplifies the MHD protection requirements. The only measures needed for MHD protection of the fiber system involve the installation of surge arresters (preferably gas tubes), conservatively at each splice of the fiber, between the local ground and (a) the outer steel wires of the cable armor and (b) alarm-carrying or other conductors (if any) in the cable. These latter measures are required to assure that arcing will not occur internally to the cable in the presence of the large-low-frequency MHD transients. Such arcing could damage the signal-carrying glass fiber.

The protection circuitry mentioned above for the coax transmission system is intended, not only to protect the active electronics of the repeaters/regenerators, but also to provide an automatic re-start capability of the system after the decay of the MHD-induced transients. The protective circuits do not, however, support operation through the late-time MHD disturbance: if actuated by the MHD surge, the protection circuit will de-activate the repeater/regenerator until it is safe to re-start the system. If this temporary loss of transmission capability is deemed unacceptable, it will be necessary to install an upgraded protection system that can provide power to the system during the duration of the large MHD transients. It should be noted, however, that the utility of such a system for a coaxial transmission system may be limited since the EMP-induced currents on the coax tubes can introduce much-larger-than-nominal bit-error rates into the transimitted signals.

13A.5 IMPLEMENTATION, VERIFICATION AND TESTING

13A.5.1 General

The EMP Protection System of WIS will be implemented as an integral part of the WIS implementation plan. Hardening of the facilities should therefore be realized in parallel with the construction of WIS buildings. Sites with typical functional requirements and design will be built as prototypes. These prototype sites will then be subjected to detailed verification process just after construction. Material assurance, welding, construction quality, conformity of hardware will be tested and verified in full detail. These activities are called as "Proof of Concept" (POC) or Verification of EMPPS. The results obtained in this phase will be used to improve and modify;

- Product assurance and construction plans,
- Testing program and acceptance tests,
- Hardness maintenance and surveillance program.

If the results extracted from the measurements and observations on prototypes are not found satisfactory then the engineering design of EMPPS should be rewieved. However, this improved and new design should be verified again. The results of these works will be the source for the establishment of basic principle for the serial implementation, acceptance tests hardness maintenance and surveillance programs etc. of the WIS EMPPS.

13A.5.2 Verification (Proof of Concept)

The proof of concept phase shall verify the EMP hardening performance by testing 100% of the hardening prototype sites based on the detailed design. Verification testing requires a thorough test schedule of all activities. The most important phase of the verification process is the measurement of shielding effectiveness apart from the construction features, product assurance, welding quality, etc.

The shielding effectiveness measurements of the EMPPS are to be performed at the following locations for each prototype facility;

- Near the midsection of the wall and ceiling,
- Near the upper concern of the shielded enclosure,
- Near the midsection of each door with the door opened and closed,
- For multiple doors, one door opened one closed, both opened and both closed,
- Near the midsection of each aperture, such as air vents, air conditioning ducts.

Three measurements are made over the frequency range from 10 kHz to 100 MHz for continuous wave tests. Frequency steps are no further apart than a factor of two in frequency. Measurements will be taken at 15 kHz and 1 MHz for E-field, 15 kHz H-field and 100 MHz for plane wave fields. The cumulative EMP shielding effectiveness (SE_{EMP}) is greater than the magnetic field shielding effectiveness (SE_H) and is less than the electric field shielding effectiveness (SE_E). Below approximately 10 MHz, the EMP shielding effectiveness is the average of the SE_E and SE_H. Above 10 MHz, the SE_{EMP} closely approximates the absolute plane wave measurements.

The POC development schedule consists of the following tasks:

13A.5.2.1 System Definition and Integration

System definition shall produce detailed development and procurement specifications for the sites. Detailed hardware trade studies shall optimize the final design.

A detailed WIS EMPP integration plan will be developed for system integration and design verification and an integrated test plan is to be produced.

13A.5.2.2 Design Review

Several design reviews will be held during the POC. At these reviews, design features and compliance will be discussed, and trade study results will be reviewed.

13A.5.2.3 Construction Plan

A construction plan will be prepared to include the engineering work plan, material handling and product specification.

13A.5.2.4 Product Assurance

Product Assurance shall include a quality program and configuration management. The quality program shall involve both quality control and quality assurance functions including:

- Receiving/shipping inspection,
- Construction inspection,
- Quality Control (QC) Plan,
- Test equipment calibration,
- Materials and components standards,
- Construction control,
- Customer interfacing,
- Purchase control.

13A.5.2.5 System Test and Evaluation

System Test and Evaluation shall verify EMPPS performance. Preliminary EMP, human engineering and environmental testing, if required, shall screen product performance prior to actual site installation. This activity shall include:

- Scheduling,
- Test planning,
- Test procedures,
- Test implementation,
- Reporting and briefings.

13A.5.3 Acceptance Tests

The EMPPS acceptance process follows the requirements and guidelines of the Authority. The EMPPS is verified by one or more of the following methods: similarity, inspection, test results or available vendor information. Analyses are correlated with test results for high confidence of performance.

The following chronology of EMPP and acceptance events are considered for the EMPPS.

- Visual inspection,
- Shield integrity test,
- Preliminary Acceptance Inspection,
- Complete Facility EMPP Verification Test,
- Joint Formal Acceptance Inspection.

Inspection Forms/Checklists are generated to document the inspection results to ensure traceability and to generate a historical database on the EMPPS design. Inspection Forms/Checklists and Reports are generated for the following EMPPS hardening components:

- Shields,
- Entry plate and EMP vault,
- Doors and hatches,
- Air vents,
- Electrical power lines,
- Control and monitoring lines,
- Communication lines,
- Grounding system,
- Earthing system,
- Utilies and piping.

Inspection Forms/Checklists Reports include such items as:

- Facility name,
- Inspection date,
- Name and organization of inspectors and associates,
- Description of facility and protective features,
- Overall evaluation/test predictions.

Acceptance tests include visual inspection, analysis and testing. Visual inspections and normal quality control activities during construction and installation of the EMPPS are not adequate by themselves to verify all aspects of the EMPPS. Analysis is considered prior to testing to determine potential coupling paths and transient threat levels. The analytical results are correlated with actual test results to validate the test activities. Ultimate acceptance activities require testing to give a high confidence of compliance to the specified threat.

Visual inspection is performed during and after construction. Welds will be visually inspected. A Complete Shielding Effectiveness Test is performed for all EMP hardened facilities. Shielding effectiveness tests will be performed for the walls, ceiling, doors, apertures, air vents, and other fortuitous conductors as indicated in POC according to the guidelines of MIL-STD-285.

An EMP Acceptance Test Requirement and Test Report are generated for tracking the verification activities. The test requirements contain such information as:

- Basic test objective,
- Technical requirements,
- Required measurements,
- Continuous wave test frequency requirements,
- Test equipment requirements,
- Test services,
- Description of test report contents.

The Test Report contains such information as:

- Facility identification,
- Brief description of test objectives,
- Brief discussion of test results,
- General description of test procedures,
- Description of test arrangement,
- Detailed test results,
- Discussion of shield deficiencies,
- Discussion of test difficulties.

The basic EMP test equipment set includes:

- Spectrum analyzer,
- Battery pack,
- Signal generator (0.1 Hz-13 MHz),
- Signal generator (100 kHz-990 MHz),
- Power amplifier,
- Loop antenna,
- Rod antenna,
- Dipole antenna,
- Coaxial cable assembly,
- EMP pulse generator,
- DC resistance meter,
- Antenna tripod,
- Multimeter,
- Ground resistance meter,
- RF leak detector.

The following matrix provides acceptance test methods for the EMPPS:

	INSPECTION	ANALYSIS	TEST	SIMILARITY VENDOR DATA
Shielding Effectiveness				
. primary enclosure	X	X	X	
. secondary enclosure	X	X		
Welded Seams				
. primary enclosure	X			
. secondary enclosure	X			
Bonds				
. waveguide	X		X	
. pipes	X		X	
. grounds	X		X	
. entry plate	X		X	
Apertures				
. honeycomb filter	X		X	X
. personnel door	X		X	X
Transient Suppression Devices	X		X	X
Filtering Devices	X		X	
Ground Resistance	X	X	X	
Corrosion Processes	X			

13A.5.4 Hardness Maintenance/Hardness Surveillance (HM/HS)

A Hardness Maintenance and Hardness Surveillance program is developed to ensure that the EMPPS performance is maintained throughout the life of the system. The HM/HS program considers degradation factors, maintenance cycles and replacement costs in order to provide a balanced design. The purpose of the HM/HS program is to describe the maintenance and surveallance maintained to meet the minimum acceptable requirements throughout the life of the system. The HM/HS program establishes minimum requirement goals based on the threats specified in section 13A.2. The HM/HS program shall identify the surveillance methods, surveillance schedule, test plans/proceduser for the EMPPS. The major elements of te HM/HS program consist of:

- Planning and data management,
- Hardness maintenance activiteas,
- Hardness surveillance activities,
- Hardness configuration management,
- Hardness training.

The hardening configuration management ensures that system changes do not cause degradation of EMPPS hardness and these activities include:

- Specification development on EMP hardening measures,
- Identification of Hardness Critical Items (HCI) and Hardness Critical Processes (HCP),
- Labeling HCI and HCP on system documentation,
- Discussion of engineering changes and effects on the system.

Hardness critical items include such EMPPS hardwares as shields, filters, surge arresters, and gaskets. Hardness critical processes include such EMPPS construction methods as welding, gasket installation, and door alignment.

Hardness training courses should be conducted for all EMPPS personnel covering EMP design practices and EMP criticality. Specific training courses are to be conducted for operations, maintenance, construction and HM/HS personnel.

The Hardness Maintenance program shall describe the maintenance plans/procedures that shall be employed to correct and maintain the EMPPS hardness. Hardness Maintenance consist of two tasks; those activities which consist of scheduled preventive maintenance, replacement and/or repair, and those replacement and/or repair activities which are required to correct deficiencies identified during Hardness Surveillance. Ideally, HM activities are not available, a data base is developed during system operation. The development of the data is accomplished early in the life cycle of the system by performing HM activities at short intervals, based on engineering judgement, and past experience on similar applications. As the system matures, certain areas will degrade more rapidly than others. The scheduled HM/HS activities shall be adjusted to concentrate more on those areas known to degrade more rapidly.

The HM program describe two major activities; preventive HM and corrective HM. Preventive HM activities include:

- Identification of hardware requiring preventive HM,
- Identification of preventive HM actions and schedule,
- Development of failure reporting procedures and corrective action procedures,
- Documentation of HM actions and schedule in maintenance manuals and logs,
- Development of recertification procedures of hardness after repairs are completed.

Corrective HM activities include:

- Development of failure reporting procedures and corrective action system,
- Development of constraints to be placed on repair activities,
- Implementation of recertification procedures of hardness after repairs are completed.

System HM procedures include such maintenance activities as delineated below:

- Inspection/cleaning of gaskets,
- Replacement of gasket material,
- Inspection/cleaning/refurbishment of mating metal surfaces,
- Inspection of honeycomb filters,
- Torquing of bolts, nuts for low impedance bonding,
- Torquing of bulkhead feedthrough connections,
- Inspection/adjustment of doors,
- Inspection/replacement of terminal protection devices,
- Inspection/replacement of filters.

The Hardness Surveillance program describes the surveillance plans/procedures that shall be employed to identify degradation of the EMPPS elements and the effectiveness of the HM program. The HS program consists of identifying hardware failures and providing information for assessing the effectiveness of the HM program. The HS program establishes test and inspection plans/procedures for detection of hardness failures or degradation. This includes a HS activity schedule. Test and inspection reports include the surveillance results and recommended changes.

The HS program considers similar verification methods and practices as described in the acceptance verification plans/procedure discussed. System HS include such activities as:

- Detection of faults not normally detected during system operation,
- Shielding effectiveness test,
- Pulse test of non-linear devices,
- Filter transfer function test.

The HM/HS program identifies all hardening features of the EMPPS such as:

a) Priary and seconday shielded barriers:
   - Welded or bolted seams.
b) Enclosure doors or access panels:
   - Gaskets, hinges, handles, latches, mating surfaces.
c) Conductive penetration entry panels and entry vaults:
   - Panels, access covers, gaskets, welded seams.
d) Grounding cables or bonding straps:
   - Terminations, corrosion.
e) Conductive penetrations (power, communication, control lines):
   - Shielded conduit, connectors, terminal protection devices.

13A.5.5 EMP Testing And Evaluation

EMP is of nuclear burst origin. A validation test in an actual nuclear environment is neither practical nor necessary, and perhaps not even desirable. Tests in simulated EMP environments, therefore, constitute the only practical means of evaluating the system vulnerability of hardness.

13A.5.5.1 EMP Simulator

The EMP simulator refers to a test tool or a system designed to produce a known electromagnetic field which can be used to illuminate a system. A simulator can be driven either by a pulse or a CW generator. It may be designed to simulate the plane wave of EMP originating from a distance, or a high altitude burst, or it may be designed for close-in ground burst, having a non plane wave of peculiar wave impedence. Either "the bounded wave simulator" or "the radiating simulator" are two basic EMP generating simulators. The bounded-wave simulator generally produces higher quality, more uniform field and such systems are more expensive to built. The radiating EMP simulators are more versatile and can create more realistic propagating electromagnetic fields. but such system are not desirable sources for connecting nearby electronic systems.

13A.5.5.2 EMP Hardness Testing Types

In general, several types of tests are necessary in certifying EMP hardness. The basic and important ones among such types are:

- Electromagnetic coupling tests,
- Component and subsystem characterization tests,
- Electronic susceptibility tests,
- Quality assurance tests,
- Hardness assurance and validation tests.

For each type, one needs to formulate test objectives, plans and procedures, and means for data analysis and interpretation. Often, different test facilities are involved to conduct different types of tests.

The principal objective of electromagnetic coupling tests, for example, is to verify analytical predictions of coupling of fields, voltages or currents due to EMP at specified points due to penetration through shields, apertures, etc., and due to other coupling means such as antennas, cables, and power and utility lines. Similarly, the objective of the component characterization tests is to determine component parameters, and to assess the suitability of such componenets in a system which must perform its intended mission and be "hard" to the specified EMP threat. The objective of the electronic susceptibility tests is to experimentally verify the upset and burn-out characteristics of semiconducting devices and other electronic component. The objective of the quality assurance tests, as they relate to EMP, is similar to those involving the verification of uniform quality of units in production hardwares, expecting that the hardwares involved in this case are EMP hardness modification kits and EMP protection devices. Finally, the objective of the hardness validation tests is to verify by tests or experiments that the intended system mission will be achieved, notwithstanding the anticipated EMP threats. Once the test objectives are defined, actual tests to be performed in each case be designed accordingly. The types and means of data collection, data interpretation and analysis leading to test objectives should be an integral part of the test design.

As the test objectives are somewhat unusual for the EMP threat, so are test facilities, and sensors and instrumentations used in such facilities. EMP test facilities are unusual because often such facilities have to simulate EMP which is characterized by a very large electromagnetic field with an unusually fast rise-time. Characteristic requirements for EMP facilities, sensors and instrumentations, and various existing facilities meeting these requirements, are discussed in the following section.

13A.5.5.3 Instrumentations and Sensors

Because of the unusual characteristics of EMP or simulated EMP, instrumentations and sensors suitable for EMP tests need to be designed and constructed accordingly. For example, since EMP is a short-term transient phenomenon, test equipment must be of the wideband type that can cover a frequeny range of about 100 KHz to 100 MHz.

Although the characteristics of instrumentations and sensors for EMP tests are often unusual, most EMP measurements involve traditional electrical qualities such as the voltage or current or field intensity. Typical sensors used for EMP measurements, therefore, are:

- Current probes,
- Voltage probes,
- Field-strength measuring devices.

The fields are first measured outside the building and then the points inside. The building's shielding is the difference, in decibels between the interior and exterior field strengths.

13A.5.5.4 Types of Testing Facilities

Two types of testing facilities are necessary which are Stationary Testing Facilities and Mobile Testing Facilities.

Stationary Testing Facilities have all the basic equipments and qualified personnel for EMP measurements, but these labs do the following tasks also:

- Calibration of equipment,
- Evaluation of collected data at sites,
- ID documentation for each EMPP site,
- Personnel training programs.

Mobile Testing Facilities are used primarily on production and prototype models consists of remotely controlled and fully computerized equipment set. The following tasks are performed in mobile testing area.

- Verification tests,
- Acceptance tests,
- Hardness maintenance and assurance test,
- Routine tests.

13A.5.5.5 EMP Tests and Safety Recommendations

EMP simulators and test sets pose electric shock hazards wherever they can supply pulse energies exceeding about one joule. Engineers and technicians of EMP Testing Lab should use all available means to prevent pulse shocks where energy received by the body could approach or exceed one joule. The equipments can be operated safely if appropriate precautions are taken. There are evidences that such shocks can affect heart beat. Thus, engineering responsibilities starting from selection of equipments and in the use of the equipments are very important.

13A.6 REFERENCES

[13A.1] "Technical Guidance for Implementation and Verification of EMPP," MCM-DTU/PKY-48-85, August 1985.

[13A.2] "Proceeding of the 1987 Shape EMPP Technical Symposium," 6550/SHCPE/253-87, October 1987.

[13A.3] R. SHERMAN et al., "EMP Engineering and Design Principles," AT&T Bell Laboratories, 1975.

[13A.4] A. RADASKY and J. KARZAS, "Early-time HEMP-Sample Calculations and a Method for Developing a System Design Waveform," 1987 Shape EMP Technical Symposium, July 1987.

[13A.5] "Protection of Civil Telecommunication Networks against EMP Effects," Annex to AC/121-N/446, March 1988.

[13A.6] A. RADASKY and J. KARZAS, " High Altitude EMP-The Intermediate Time Regime Phenomanology and Applications," SHAPE EMP Technical Symposium, July 1987.

INDEX

A

B

C

E

F

G

M

N

S